D0805636

SEQUENCE - EVOLUTION - FUNCTION

Computational Approaches in Comparative Genomics

SEQUENCE - EVOLUTION - FUNCTION

Computational Approaches in Comparative Genomics

by

Eugene V. Koonin

and

Michael Y. Galperin

National Center for Biotechnology Information
National Library of Medicine
National Institutes of Health

KLUWER ACADEMIC PUBLISHERS
Boston / Dordrecht / London

Distributors for North, Central and South America:
Kluwer Academic Publishers
101 Philip Drive
Assinippi Park
Norwell, Massachusetts 02061 USA
Telephone (781) 871-6600
Fax (781) 681-9045
E-Mail < kluwer@wkap.com>

Distributors for all other countries:
Kluwer Academic Publishers Group
Post Office Box 322
3300 AH Dordrecht, THE NETHERLANDS
Telephone 31 786 576 000
Fax 31 786 576 254
E-Mail < services@wkap.nl>

Electronic Services < http://www.wkap.nl>

Library of Congress Cataloging-in-Publication Data

Koonin, Eugene V.
Sequence – evolution – function : Computational approaches in comparative genomics/
By Eugene V. Koonin and Michael Y. Halperin
p. cm.
Includes bibliographical references and index.
ISBN 1-40207-274-0 (alk. paper)
1. Genomes. 2. Nucleotide sequence—Data processing. 3. Evolutionary genetics. I. Galperin, Michael Y. II. Title.

QH447 .K665 2002
572.8'6—dc21 2002034045

Printed on acid-free paper.
Printed in the United States of America

The Publisher offers discounts on this book for course use and bulk purchases. For further information, send email to <melissa.ramondetta@wkap.com>.

CONTENTS

PREFACE

The use of genome sequences to solve biological problems has been afforded its own label; for better or worse, it's called "functional genomics."
David J. Galas. Making Sense of the Sequence. Science, 2001, vol. 291, p. 1257

When the completion of the draft of the human genome sequence was announced on June 26, 2000, all the parties involved agreed that the major task of identifying the functions of all human genes was still many years ahead. In fact, even the much simpler task of mapping all the genes in the final version of the human genome sequence that should become available within the next few years remains a major problem. Identification of all protein-coding genes in the genome sequence and determination of the cellular functions of the proteins encoded in these genes can be accomplished only by combining powerful computational tools with a variety of experimental approaches from the arsenals of biochemistry, molecular biology, genetics and cell biology. Linking sequence to function and both to the evolutionary history of life is the fundamental task of new biology.

This book is devoted to the principles, methods and some achievements of computational comparative genomics, which has shaped as a separate discipline only in the last 5-7 years. Its beginnings have been modest, with only the genome sequences of viruses and organelles determined in the 1980's. These sequences were important for their respective disciplines and as a test ground for computational methods of genome analysis, but they were not particularly helpful for understanding how does an autonomous cell work. By 1992, the first chromosomes of baker's yeast and large chunks of bacterial genomes started to emerge, and researchers began pondering the question: What's in the genome? The breakthrough came in 1995 with the complete sequencing of the first genome of a cellular life form, the bacterium *Haemophilus influenzae*. The second bacterial genome, *Mycoplasma genitalium*, followed within months. The next year, the first complete genomes of an archaeon (*Methanococcus jannaschii*) and a eukaryote (yeast *Saccharomyces cerevisiae*) became available. Many more microbial genomes followed, and in 1999, the first genome of a multicellular eukaryote, the nematode *Caenorhabiditis elegans*, has been sequenced. The year 2000 brought us the complete genomes of the fruit fly *Drosophila melanogaster* and the thale-cress *Arabidopsis thaliana*, and two independent drafts of the human genome followed suit in 2001. Thus, we entered the 21st century already having at hand this 3.2 billion-letter text that has been referred to as the Book of Life, as well as a number of accompanying books on other life forms. The challenge is now to read and interpret them.

To extract biological information from enormous strings of As, Cs, Ts, and Gs, functional genomics depends on computational analysis of the sequence data. It is unrealistic to expect that every single gene or even a majority of the genes found in the sequenced genomes would ever be studied experimentally. However, using the relatively cheap and fast computational approaches, it is usually possible to reliably predict the protein-coding regions in the DNA sequence with reasonable (albeit varying) confidence and to get at least some insight into the possible functions of the encoded proteins. Such an analysis proves valuable for many branches of biology, in large part, because it assists in classification and prioritization of the targets for future experimental research.

Computations on genomes are inexpensive and fast compared to large-scale experimentation, but it would be a mistake to equate this with 'easy'. The history of annotation and comparative analysis of the first sequenced genomes convincingly (and sometimes painfully) shows that the quality and utility of the final product critically depend on the employed methods and the depth of interpretation of the results obtained by computer methods. Unfortunately, errors produced in the course of computer analysis are propagated just as easily as real discoveries, which makes development of reliable protocols and crystallization of the accumulating experience of genome analysis in easily accessible forms particularly important.

While functional annotation of genomes may be the most obvious, and in a sense, the most important purpose of computational genomics, it is not just a supporting service for experimental functional genomics, but a discipline in itself, with its own fundamental goals. The main such goal is ***understanding genome evolution***. Ultimately, understanding here means being able to reconstruct the most likely sequence of evolutionary events that produced these genomes. Attaining this goal will require many more genomes, development of new algorithms and years of careful analysis. Nevertheless, even in its infancy, comparative genomics has brought genuine revelations about evolution. We believe that the principal news that could not be easily foreseen in the pre-genomic era is the extreme diversity of the gene composition in different evolutionary lineages. This strongly suggests that, at least among prokaryotes, horizontal gene transfer and lineage-specific gene loss were major, formative evolutionary forces, rather than rare and relatively inconsequential events as assumed previously. Accordingly, the straight-forward image of evolution as the growth of the tree of life is replaced by one of a 'grove', in which vertical, tree-type growth does occur, but multiple horizontal connections are equally prominent – an incomparably more complex, but also more interesting picture of life than ever suspected before.

This book describes the computational approaches that proved to be useful in analyzing complete genomes. It is intended for a broad range of biologists, including experimental biologists and graduate and advanced undergraduate students, whose work builds upon the results of genome analysis and comprises the foundation of functional genomics. However, we attempted to make the text interesting also for practitioners of genomics itself, particularly those computational biologists whose main occupation is developing algorithms and programs for genome analysis and who could benefit from an accessible discussion of some biological implications of these methods. Most of the approaches discussed in this book have been developed during comparative analysis of the first set of completely sequenced bacterial and archaeal genomes, which are simpler and more amenable to straightforward computational dissection than the much larger eukaryotic genomes. We show, however, that the main principles remain the same for comparative genomics in general.

The book starts with a brief overview of the history of genomics. We list the completed and ongoing genome sequencing projects and show how little is actually known even about simple genomes. We then discuss the conceptual basis of comparative genomics, emphasizing the evolutionary principles of protein function assignments. The book then proceeds to discuss the databases that store and organize genomic data, with their unique advantages and pitfalls. Familiarity with these databases is useful for any biologist, but for those interested in functional or evolutionary genomics, it is essential.

The central part of the book discusses, in some depth, the principles and methods of genome analysis and annotation, including identification of genes in genomic DNA sequence and using sequence comparisons for functional annotation of predicted proteins. We introduce the most common sequence similarity search methods and discuss the ways to automate the searches and increase search sensitivity, while minimizing the error rate. The common sources of errors in functional annotation of genomes are discussed and some simple rules of thumb are provided that may help avoid them. We further focus on the approaches to functional prediction that rely on the genome context, such as examination of phyletic patterns, gene (domain) fusions and conserved gene strings (operons). The discussion is illustrated by examples from comparative genomics of prokaryotes.

The remaining parts of the book consider fundamental and practical applications of comparative genomics. In particular, in Chapter 6, we discuss the impact of comparative genomics on our current understanding of several fundamental problems of evolutionary biology and some major events of life's history.

The book is non-technical with respect to the computer methods for genome analysis; we discuss these methods from the user's viewpoint, without addressing mathematical and algorithmic details. Prior practical familiarity with the basic methods for sequence analysis is a major advantage, but a reader without such experience should be able to use the book as an introduction to these methods. Knowledge of molecular biology and genetics at the level of basic undergraduate courses is required for understanding the material; similar knowledge of microbiology is a plus. The book is accompanied by a problem set, designed to be solved by using tools available through the web. Hopefully, this will allow the reader to develop a better feeling for the practical use of the methods discussed in the text. Chapters 1 through 5 are, definitely, at the introductory level, although we attempted to include some non-trivial examples and discussion of open issues. There is considerable cross-talk between Chapters 3 and 4, which might be perceived as a degree of redundancy. We felt, however, that it was appropriate to discuss some key notions in protein analysis twice, first from a purely practical and then from a more fundamental standpoint. Chapters 6, 7 and 8 are somewhat more involved and, we hope, might be of certain interest even to experts. However, we tried to ensure that a non-expert reader would be in a position to understand the material of these chapters after reading the book from the beginning.

Probably the main purpose of any Preface is a disclaimer and apologies. So what is ***not*** in this book? First of all, we could not even think of covering the entire field of comparative genomics: this field is young, but has already branched widely, and we cannot claim even knowing of all important research directions, let alone being experts in them. We cite many publications, but, again, we could not even think of citing all the relevant ones: this would take the entire space of the book and the task still would not have been accomplished. We sincerely apologize to all those colleagues whose important work is not cited because of space considerations or, unfortunately, because of our ignorance and negligence. Most of the case studies discussed in this book are drawn from our own work. This is certainly not to imply that we believe it to be in any sense superior to the work of others, but simply because this is what we know best. However, unfortunately, there may be cases where, for the above reason, we cite and discuss our own work instead of more decisive and interesting work of other researchers, and to them our heartfelt apologies.

The parts of this book that deal with sequence and structure analysis algorithms might irk some of our colleagues involved in the development of these methods by superficiality and lack of rigor. We owe a great debt to these researchers and extend our regrets and apologies. A more technical point: most of the research discussed in this book is done with protein sequences and structures. Partly, this is because we believe that the main knowledge so far accumulated by comparative genomics has been attained through this type of analysis. The other reason, however, is that this is where our main experience is, and we apologize to the readers for not covering numerous important studies on non-coding regions of the genomes. Finally, a terminological point related to the last issue: throughout the book, we rather freely substitute proteins for the genes that encode them by talking about duplications, mutations and other evolutions of proteins. This is just for the sake of brevity; we assure the reader that we are aware of the fact that proteins actually do not undergo any of these events, only the respective genes do.

Despite of all these shortcomings and, undoubtedly, others that we are unaware of, we hope that this book will help the reader to understand the principles and approaches of comparative genomics and the potential and limitations of computational and experimental approaches to genome analysis. This should go some distance to building a bridge across the "digital divide" between biologists and computer scientists, hopefully, allowing biologists of various directions and persuasions to better grasp the peculiarities of the emerging field of Genome Biology and to learn how to benefit from the enormous amount of sequence and structural data available in the public databases.

This book has become possible thanks to our close collaboration with numerous colleagues from the NCBI and other institutions. It is, unfortunately, impossible to mention everyone, but we must gratefully acknowledge many hours of illuminating discussions over the years of interactions with L. Aravind, Peer Bork, Valerian Dolja, Mikhail Gelfand, Alexander Gorbalenya, Alexey Kondrashov, David Lipman, Arcady Mushegian, Pavel Pevzner, Igor Rogozin, and Yuri Wolf. We greatly appreciate all the work that Roman Tatusov and Darren Natale put in the COG database, which permeates this book. We thank the following colleagues for critical reading of individual chapters and helpful criticisms: Chapter 3 - Peter Cooper, Aviva Jacobs, David Wheeler and Jodie Yin, Chapters 4, 5, 6 and 8 - Igor Rogozin, Chapter 6 - Fyodor Kondrashov, and Chapter 8 - Yuri Wolf. Yuri Wolf kindly provided Figures 8.3, 8.4 and 8.5, and the entire sections 8.2 and 8.3 are largely the result of collaboration and intense discussions with Yuri Wolf and Georgy Karev. We thank L. Aravind, Trevor Fennon, Kira Makarova, Boris Mirkin and Yuri Wolf for the kind permission to cite some of our unpublished joint work. Several figures in this book come from the NCBI Entrez Genomes web site. We appreciate the work of the team that supports this site. We are grateful to our editor Joanne Tracy for her constant prodding and encouragement, not to mention editorial support, without which this book would have never come to life. Last but not least, we thank our families for their enormous patience and understanding.

The opinions expressed in this book reflect personal views of the authors and have no relation to the official positions (if any) on the issues involved held by the National Library of Medicine, National Institutes of Health, or the US Department of Health and Human Services.

Eugene Koonin
Michael Galperin

INTRODUCTION: PERSONAL INTERLUDES

In the early spring of 1980, one of the authors of this book (EVK) was an excited listener to a seminar presented in Moscow State University, the authors' alma mater, by a well-known virologist, a scientist of rare creativity, and later a good friend, Anatoly Altstein. The subject of the seminar was a model of the origin of the genetic coding and translation he had developed [23]. The model was a beauty, but it was hard to imagine how one would go about verifying it (just in case the reader is curious, an attempt to test this idea using actual molecular models of nucleotides and amino acids has been subsequently published [24]). Toward the end of the seminar, upon answering the question about validation of the model for the umpteenth time, Anatoly dropped, rather casually, something to the effect of "Not to worry, we will soon have many gene sequences to compare and will be able to reconstruct the earliest stages of evolution". This was the first occasion for the author, at the time a struggling graduate student in virology with a persistent but futile interest in evolution, to grasp the idea of comparative genomics. Even so, he remained skeptical as to the potential of sequence comparison to solve the mystery of the Origin of Life. Little did he know that his entire career in science would be dedicated to this very objective: understanding evolution through comparisons of genes, genomes, and proteins. Now, 22 years after that memorable seminar, we are still far from understanding the origin of coding. Perhaps this is the "singularity" of biological evolution that cannot be reached through reconstruction based on the comparative method. However, comparative genomics has already revealed many fascinating aspects of all subsequent stages of evolution and is bringing us breathtakingly close to the Beginning.

At about the same time and in the same building, the other author (MYG), then a graduate student in membrane biochemistry, learned about the similarity of membrane organization and energy-transducing processes in bacteria, mitochondria and chloroplasts. He became interested in the apparent paradox: if survival of any living cells depends on the integrity of its cytoplasmic membrane that maintains the electric charge and transmembrane gradients of Na^+ and K^+ ions, such a system could not have evolved in steps – a membrane that is freely permeable (or entirely impermeable) is not functional. The interest in the origin and evolution of membrane energy- coupling mechanisms has finally brought this author to – where else? – comparative genomics of microorganisms. Although the principal questions in membrane evolution still remain unresolved, we are gradually approaching a better understanding of the Beginning of Life in that respect, too.

CHAPTER 1. GENOMICS: FROM PHAGE TO HUMAN

1.1. The Humble Beginnings ...

The first genome, that of RNA bacteriophage MS2, was sequenced in 1976, in a truly heroic feat of direct determination of an RNA sequence [225]. This was followed by the genome of bacteriophage ϕX174, the first triumph of the new, rapid sequencing methods developed in the laboratories of Walter Gilbert and Fred Sanger [553,743]. These are some of the smallest known genomes with only four and ten genes, respectively. Then, in 1982, the last paper published by Sanger before he retired, announced the first relatively large genome to be sequenced, that of bacteriophage λ, probably the most famous model system of classic molecular biology [742]. Phage λ has 48,502 bases of genomic DNA and ~70 known and predicted protein-coding genes and 23 RNA-coding genes. At 70 characters per line and 43 lines per page, this sequence alone would take over 16 pages of this book. However, the listing of the λ protein-coding genes (Table 1.1) fits into just two pages and definitely conveys more information. These days, it may be hard to imagine all the excitement felt by molecular biologists 20 years ago when the λ genome was finally finished. Nevertheless, even in this era of high-throughput methods, it could be instructive to look back and address several questions: (i) is λ genome a good model of the subsequently sequenced prokaryotic and eukaryotic genomes?, (ii) how accurate was the sequence itself and the original gene assignment?, and (iii) how much more have we learned about functions of λ genes in the past 20 years?

The answer to the first question is, definitely: yes, λ genome has many features common to the genomes of cellular life forms, particularly prokaryotes. Most of the genome consists of protein-coding genes. Adjacent genes are often transcribed in the same direction and encode proteins that have similar functions and/or interact with each other (e.g. cell lysis proteins, tail components). Adjacent genes either slightly overlap or are separated by intergenic regions of varying length, typically much shorter than the genes themselves.

To answer the second question, both the sequence and gene assignments turned out to be essentially correct. The latter may not be surprising since the λ genome was annotated by researchers who had studied the phage for years, on the basis of the entire body of knowledge amassed by that time. In contemporary genome sequencing projects, such detailed analysis by highly qualified biologists with intimate knowledge of the biology of the given organism is more an exception, rather than the norm, partly because biological information on many of the sequenced organisms is simply too scarce.

Table 1.1. Protein-coding genes of bacteriophage λ

Chromosomal location, bases	DNA strand	Length, aa	Gene name	Gene product
191..736	+	181	nu1	DNA packaging protein
711..2636	+	641	A	DNA packaging protein
2633..2839	+	68	W	Head-tail joining protein
2836..4437	+	533	B	Capsid component
4418..5737	+	439	C	Capsid component
5132..5737	+	201	nu3	Capsid assembly
5747..6079	+	110	D	Head-DNA stabilization
6135..7160	+	341	E	Capsid component
7202..7600	+	132	Fi	DNA packaging
7612..7965	+	117	Fii	Head-tail joining
7977..8555	+	192	Z	Tail component
8552..8947	+	131	U	Tail component
8955..9695	+	256	V	Tail component
9711..10133	+	140	G	Tail component
10115..10549	+	144	T	Tail component
10542..13103	+	853	H	Tail component
13100..13429	+	109	M	Tail component
13429..14127	+	232	L	Tail component
14276..14875	+	199	K	Tail component
14773..15444	+	223	I	Tail component
15505..18903	+	1132	L	Tail:host specificity
18965..19585	+	206	lom	Outer host membrane
19650..20855	+	401	orf401	Tail fiber protein
20147..20767	-	206	orf206	Hypothetical protein*
21029..21973	+	314	orf314	Tail fiber
21973..22557	+	194	orf194	Fiber assembly protein
22686..23918	-	410	ea47	
24509..25399	-	296	ea31	
25396..26973	-	525	ea59	
27812..28882	-	356	int	Integration protein
28860..29078	-	72	xis	Excisionase
29118..29285	-	55	-	Hypothetical protein*
29374..29655	-	93	ea8.5	
29847..30395	-	182	ea22	
30839..31024	-	61	orf61	Hypothetical protein*
31005..31196	-	63	orf63	Hypothetical protein*

Table 1.1 – continued

31169..31351	-	60	orf60a	Hypothetical protein*
31348..32028	-	226	exo	Exonuclease
32025..32810	-	261	bet	Recombination protein
32816..33232	-	138	gam	Host-nuclease inhibitor protein
33187..33330	-	47	kil	Host-killing
33299..33463	-	54	cIII	Antitermination
33536..33904	-	122	ssb	Single-stranded DNA binding protein
34087..34287	-	66	ral	Restriction alleviation
34271..34357	-	28	orf28	Hypothetical protein*
34482..35036	+	184	imm21	Superinfection exclusion protein B
35037..35438	-	133	N	Early gene regulator
35825..36259	-	144	rexB	Exclusion
36275..37114	-	279	rexA	Exclusion
37227..37940	-	237	cI	Repressor
38041..38241	+	67	cro	Antirepressor
38360..38653	+	97	cII	Antitermination
38686..39585	+	299	O	DNA replication
39582..40283	+	233	P	DNA replication
40280..40570	+	96	ren	Ren exclusion protein
40644..41084	+	146	Nin146	
41081..41953	+	290	Nin290	
41950..42123	+	57	Nin57	
42090..42272	+	60	Nin60	
42269..42439	+	56	Nin56	
42429..43043	+	204	Nin204	
43040..43246	+	68	Nin68	
43224..43889	+	221	Nin221	
43886..44509	+	207	Q	Late gene regulator
44621..44815	+	64	orf64	Hypothetical protein*
45186..45509	+	107	S	Cell lysis protein
45493..45969	+	158	R	Cell lysis protein
45966..46427	+	153	Rz	Cell lysis protein
46459..46752	-	97	bor	Bor protein precursor
47042..47575	-	177	-	Putative envelope protein
47738..47944	+	68	-	Hypothetical protein*

Based on the data from the NCBI Entrez Genomes web site, http://www.ncbi.nlm.nih.gov/cgi-bin/Entrez/altik?gi=10119&db=Genome

A comparison of Table 1.1 with the original paper by Sanger et al. [742] shows that there is actually not much to add to the gene annotations. The use of recently developed sophisticated gene prediction programs, such as Glimmer (see ♦4.1), coupled with the analysis of the regions that are conserved between lambda and related bacteriophages, led to the conclusion that certain intergenic regions might contain additional protein-coding genes (marked by asterisks in the Table 1.1). Unfortunately, most of these genes remain uncharacterized and it is not even known whether or not they are ever expressed. It is worth noting that exactly the same doubts exist about the possible functions and/or expression of a large number of so-called "hypothetical" genes, identified in the genomes of cellular life forms by essentially the same two principal approaches (see ♦4.1).

When reading the Sanger paper now, 20 years after it appeared, one is struck by the absence of any analysis of protein sequences in this detailed, thorough work. Although the authors have done careful computational analysis of open reading frames, particularly the likely translation starts and codon usage, the very word "homolog" is not used in the article and there is no mention of any search of protein sequence databases, something that these days is, by default, an integral part of any genomic study. Not that protein sequence databases did not exist at the time: the first one, the Protein Identification Resource, was launched by Margaret Dayhoff, one of the great pioneers of computational biology, in 1965, long before genomics has even become conceivable [172,173]. However, reliable and rapid methods for searching this database still have not been developed and, more generally, database search was not a part of the culture in molecular biology at the time. And for a good reason, too. Had Sanger and his coworkers performed a PIR search, even using the methods available in 2002, they would not have found anything of interest because the sequences available at that time were few and far apart, and there were no homologs of phage λ proteins among them. Clearly, the time was not ripe for comparative genomics and, in a sense, for genomics itself because, as we will see throughout this book, the comparative approach is truly central to the genomic enterprise.

Revisiting phage λ genome after 20 years, we see a completely different "genomescape". Using the PSI-BLAST program (see ♦4.3), the search of the complete non-redundant protein sequence database maintained at the NCBI (National Center for Biotechnology Information, a division of the National Institutes of Health in Bethesda, Maryland, USA) for homologs of the 73 proteins listed as gene products of phage λ takes about an hour on a moderate power computer. Another hour was spent running selected proteins through the conserved domain search using the CDD option of the NCBI's BLAST server (see ♦4.4). Of course, we could have scoured the literature for descriptions of computational analyses of λ proteins instead. However, extracting the relevant information from databases, such as PubMed (♦3.7.), is far from trivial because, in most cases, the papers including this information dealt with more general issues and would not have λ, let alone a particular gene, mentioned in the title or abstract. Running the searches anew was much faster and easier. Besides, sequence databases are growing daily, which may substantially affect the results of searches and might even lead to new discoveries. Perusing the results, we should note that, with a few exceptions, there are now homologs readily detectable for the phage proteins. In the majority of cases, these are proteins from other related phages (sometimes integrated as prophages into the bacterial chromosome). However, 12 λ proteins show conservation in bacteria, archaea, and eukaryotes (Table 1.2). For several of these proteins whose functions have not been studied experimentally, non-trivial functional predictions become possible.

It is remarkable that some of the more interesting computational predictions remain without experimental test. Admittedly, the visibility of molecular biology of bacteriophages as a research field has not increased since the 1970's and the funds have pretty much tapered off. Good examples are the Ea59 and K genes that are predicted to encode an ATPase and a metal-dependent protease, respectively. Both are clear and readily testable predictions that have been described in print, even if briefly [296,679]. However, to our knowledge, no experimental tests of these predictions have been reported so far. Interestingly, an observation has been made during these searches that actually seems to have a novel aspect to it. The Ea31 protein was shown to contain a metal-dependent nuclease domain [50]. The stop codon of the Ea31 gene overlaps the start codon of Ea59, leading to the intriguing hypothesis that the two proteins interact and form an ATP-dependent nuclease complex. We discuss sequence analysis of Ea31 in greater detail in Chapter 4, to illustrate the process of discovery in database searches. Furthermore, this is a little example of context analysis, an increasingly important direction in genome annotation, which is covered in Chapter 5. This situation is not uncommon: computational analysis of genomes keeps yielding interesting functional predictions even years after the publication of the sequence; what is most often lacking is systematic experimental testing of these predictions.

We will come back to this dramatic rift between computational and experimental analysis of most, if not all, genomes with more numbers, but first let us step back and have a quick look into the history of genomics, which is short, but dynamic (Table 1.3). By definition, genomics requires genome sequences, and to engage in comparative genomics, one needs at least two genomes to compare. In a close analogy to the history of molecular genetics, which owes most of its early progress to bacteriophages used as model systems, comparative genomics was first practiced with the genomes of viruses. These are several orders of magnitude smaller than even the

Table 1.2 Non-trivial evolutionary connections and functional predictions for bacteriophage λ proteins

Gene product	Evolutionary conservation	Structure, domain architecture[a]	Predicted function, reference
A (TerL)	Bacteriophages, herpesviruses	A modified P-loop ATPase domain, distantly related to a vast class of helicases	ATPase subunit of the terminase, involved in DNA packaging in phage head
C	Bacteria and archaea	ClpP protease domain	Minor capsid protein, cleaves the scaffold protein during maturation
K	Bacteria, archaea and eukaryotes	Consists of an N-terminal JAB/MPN domain (predicted metalloprotease) and a C-terminal CHAP domain (Cys,His-dependent gamma-DL-glutamate-specific amidohydrolase)	Tail subunit, participates in tail assembly (based on the presence of the predicted metalloprotease JAB/MPN domain [679]) and peptidoglycan lysis (based on the presence of the peptidoglycan amidohydrolase CHAP domain [948])
Ea31	Scattered distribution in bacteria and archaea	Endo VII-colicin domain	Predicted nuclease of the McrA (HNH) family [50]
Ea59	Bacteria, archaea and eukaryotes	P-loop ATPase domain of the ABC class	Predicted ATPase [296]
Exo (RedX)	Bacteria, archaea, eukaryotes, viruses	λ exonuclease domain, distantly related to a broad variety of nucleases	A nuclease involved in phage recombination and late rolling-circle replication

Table 1.2 – continued

CI	Bacteria, archaea	N-terminal helix-turn-helix DNA-binding domain fused to a C-terminal serine protease domain of the LexA/UmuD family	Transcription repressor of genes required for lytic development
Cro	Bacteria, archaea	Helix-turn-helix DNA-binding domain	Transcription repressor of early genes
O	Bacteria, archaea	Helix-turn-helix DNA-binding domain	DNA-binding protein involved in the initiation of replication
Ren	Bacteria, archaea	Helix-turn-helix DNA-binding domain	Protein involved in exclusion of replication of heterologous genomes in λ-infected bacteria
Nin290	Bacteria, archaea, eukaryotes	PP-loop ATPase domain	Predicted ATP pyrophosphatase, role in phage replication unknown [102]
Nin221	Bacteria, archaea, eukaryotes	Calcineurin-like serine/threonine protein phosphatase domain	Protein phosphatase, role in phage replication unknown [450]

[a] Detailed descriptions of these and other domains are available in the Pfam, SMART and CDD protein domain databases (see ♦3.2) and in SCOP and CATH protein structure databases (see ♦3.3).

tiniest bacterial genomes and, in case a virus grows well, sequencing of viral genomes became a relatively straightforward enterprise already in the early 1980's. By 1983, six years after the beginning of the sequencing era, a considerable number of complete genomes of diverse small viruses of plants, animals and bacteria (bacteriophages) had been amassed and the time was ripe for the birth of comparative genomics.

Pinpointing the exact beginning of comparative genomics may be difficult. In a sense, one may say that it was born as soon as there were two genomes to compare, i.e. in 1977 when the genome of phage ϕX174 was sequenced and could be compared with the already available sequence of the RNA phage MS2. However, this was a vacuous start because the two phages had virtually nothing in common (*a propos*, this has not changed in 20 years: for all we know, these phage families are truly unrelated). It seems that comparative genomics had a real head start with two astonishing discoveries that caught most, if not all, virologists utterly by surprise. First, it has been shown that RNA-containing retroviruses (causative agents of certain leucoses in animals and humans and, as shown later, of AIDS) shared a conserved replicative enzyme, the reverse transcriptase, with two groups of DNA viruses, the hepadnaviruses (including the medically important hepatitis B virus) and caulimoviruses, infecting plants [847]. Second, it turned out that small RNA viruses infecting animals (picornaviruses, such as polio and foot-and-mouth disease) and those infecting plants (cowpea mosaic virus) shared not only significant sequence similarity that allowed the identification of homologous (orthologous) genes, but also, in part, the order of these genes in their genomes [7,56,335]. Subsequent systematic studies have revealed a complex network of homologous relationships within the vast classes of positive-strand RNA viruses and negative-strand RNA viruses. Although still disputed, the concept emerged that each of these classes was monophyletic, that is, probably evolved from a common ancestral virus [460]. These studies combined two elements that were crucial in defining the identity of the emerging discipline of comparative and evolutionary genomics.

Table 1.3. A brief timeline of genomics

Year	Event	Ref.
1962	The first theory of molecular evolution; the Molecular Clock concept (Linus Pauling and Emile Zukerkandl)	[946]
1965	Atlas of Protein Sequences, the first protein database (Margaret Dayhoff and coworkers)	[173]
1970	Needleman-Wunsch algorithm for global protein sequence alignment	[606]
1977	New DNA sequencing methods (Fred Sanger, Walter Gilbert and coworkers); bacteriophage ϕX174 sequence	[553, 743]

Table 1.3 – continued

1977	First software for sequence analysis (Roger Staden)	[797]
1977	Phylogenetic taxonomy; archaea discovered; the notion of the three primary kingdoms of life introduced (Carl Woese and coworkers)	[905]
1981	Smith-Waterman algorithm for local protein sequence alignment	[784]
1981	Human mitochondrial genome sequenced	[28]
1981	The concept of a sequence motif (Russell Doolittle)	[185]
1982	GenBank Release 3 made public	
1982	Phage λ genome sequenced (Fred Sanger and coworkers)	[742]
1983	The first practical sequence database searching algorithm (John Wilbur and David Lipman)	[892]
1985	FASTP/FASTN: fast sequence similarity searching (William Pearson and David Lipman)	[521]
1986	Introduction of Markov models for DNA analysis (Mark Borodovsky and coworkers)	[107]
1987	First profile search algorithm (Michael Gribskov, Andrew McLachlan, David Eisenberg)	[315]
1988	National Center for Biotechnology Information (NCBI) created at NIH/NLM	
1988	EMBnet network for database distribution created	
1990	BLAST: fast sequence similarity searching with rigorous statistics (Stephen Altschul, David Lipman and coworkers)	[20]
1991	EST: expressed sequence tag sequencing (Craig Venter and coworkers)	[4]
1994	Hidden Markov Models of multiple alignments (David Haussler and coworkers; Pierre Baldi and coworkers)	[71,72, 473]
1994	SCOP classification of protein structures (Alexei Murzin, Cyrus Chothia and coworkers)	[590]
1995	First bacterial genomes completely sequenced	[232,242]
1996	First archaeal genome completely sequenced	[130]
1996	First eukaryotic genome (yeast) completely sequenced	[290]
1997	Introduction of gapped BLAST and PSI-BLAST	[22]
1997	COGs: Evolutionary classification of proteins from complete genomes	[828]
1998	Worm genome, the first multicellular genome, (nearly) completely sequenced	[840]
1999	Fly genome (nearly) completely sequenced	[3]
2001	Human genome (nearly) completely sequenced	[488,870]

Firstly, the objects of analysis were complete genomes, however small, rather than individual genes and, accordingly, the notions of conservation of gene order and gene shuffling became important. Secondly, the discoveries made through these genome comparisons were completely unexpected; there was no experimental data that would prepare researchers for the startling unity of superficially unrelated viruses.

In retrospect, it is somewhat ironic that comparative genomics had to start with virus genomes (due to the experimental contingency) because viral proteins tend to evolve extremely fast and detection of conservation between distant viruses may be a non-trivial task even with advanced methods of computational sequence analysis, let alone with those available in the early 1980's. This was a challenge and perhaps a blessing in disguise. The difficulty of detecting sequence conservation among viral proteins prompted those who ventured into this area to employ approaches that later proved invaluable in comparative genomics and computational biology in general: (i) compare protein sequences, rather than nucleotide sequences directly, whenever distant relationships are involved and sensitivity is an issue, (ii) rely on multiple, rather than pairwise, comparisons, (ii) search for conserved patterns or motifs in multiple sequences, and, above all, (iii) actually look at sequences (and structures whenever these are available) and think about the potential relationships in an effort to synthesize all relevant shreds of information. This practice has been dubbed, more or less pejoratively, "sequence gazing" [341]. Sure enough, sequence and structure comparisons are prone to error and, worse, to fantasy, and these dangers had been particularly grave in the early days, before the statistical foundations of computational biology have been worked out and the rules of thumb have been established through accumulated practices. There is no doubt, however, that success stories of computational prediction of gene functions have been of much greater import and have, to a large extent, determined the very feasibility of the further progress of genomics.

The first comparative-genomic study at a larger scale, investigating the relationships between genomes that contained >100 genes each, came in 1986 [558]. The newly sequenced genome of varicella-zoster virus was carefully compared to the previously sequenced Epstein-Barr virus genome (the original Epstein-Barr genome paper [68] resembled the λ work in that no homologs were reported for any of the viral proteins because, indeed, none were to be easily identified among the sequences then available). This work, though little noticed outside virology, already included the principal elements of the comparative-genomic approach, if not the actual methods.

1.2. ...and the Astonishing Progress of Genome Sequencing

Comparative genomics of cellular life forms is in a way a "by-product" of the Human Genome Project. Probably the greatest insight of the leaders of the early stages of this project was the realization that, in isolation, the human genome would be a costly but uninterpretable string of three billion or so of A's, T's, G's and C's. Only through systematic comparisons to other genomes, may we hope to make sense of the text of this "Book of Life". As far as genomics is concerned, Theodosius Dobzhansky's famous dictum "Nothing in biology makes sense except in the light of evolution" is not some kind of evolutionist propaganda, but an entirely literal and more or less routine description of the situation. And so, in the last decade of the second millennium, the genome sequences started pouring in. Yeast chromosome III, the first respectable chunk of contiguous genome sequence [629] that became available in 1992 (quite modest, by today's standards, just ~320,000 base pairs), generated major excitement epitomized in the title of a *Nature* note describing a re-analysis of the ORFs from this chromosome: "What's in the genome?" [105]. From the analysis of this sequence and other large genome segments that started to appear in the next months, at least two notions were derived that became critical for the subsequent evolution of comparative genomics: (i) there were many more genes in the genome than anyone suspected previously on the basis of genetic or biochemical experiments, and (ii) methods of computational analysis matter – careful analysis employing multiple complementary approaches yields incomparably more information on gene functions and evolutionary relationships than any single automatic procedure.

The appearance, in August of 1995, of the complete genome sequence of the parasitic bacterium *Haemophilus influenzae* [232], ushered in the era of "real" genomics, the study of complete genomes of cellular organisms. The acceleration of genome sequencing required for this to happen was greatly facilitated by the whole-genome shotgun approach pioneered by Craig Venter, Hamilton Smith and Leroy Hood [871]. Systematic comparative approaches were tried immediately, even before the second genome came, by using the largely finished genome of *Escherichia coli* [829]. Since that point, complete genomes of bacteria and archaea have been arriving at a steady rate, which seems to be accelerating in the 3rd millennium (Figure 1.1). Starting with the second genome sequencing paper [242], reports on new genomes inevitably became comparative-gènomic studies because, as we have already mentioned, that is the only way to even start understanding "what's in the genome".

By June 1, 2002, genomes of 73 species of unicellular organisms (55 bacterial species, 16 archaea, and two eukaryotes) were completely sequenced and available in public databases. In the three parts of Table 1.4, the completely sequenced bacterial, archaeal, and eukaryotic genomes are listed in the order of decreasing size. The largest prokaryotic genomes (*Streptomyces coelicolor* among bacteria, *Methanosarcina acetivorans* among the archaea) have been sequenced only recently, which promises many interesting discoveries yet to come.

By the time of this writing (August, 2002), the first genomes of multicellular eukaryotes, the nematode worm *Caenorhabditis elegans*, the fruit fly *Drosophila melanogaster*, the thale cress *Arabidopsis thaliana*, the pufferfish *Fugu rubripes*, and *Homo sapiens* have been nearly completed (let us note that the very concept of a complete genome sequence for these organisms differs from that for prokaryotes and unicellular eukaryotes). At least a 100 more prokaryotic genomes and many eukaryotic genomes, including those of mouse and rat, were at different stages of completion. Beyond doubt, many more finished or nearly finished genome sequences exist in proprietary databases maintained by biotech companies, but since these cannot be freely analyzed, they do not count inasmuch as comparative genomics is discussed.

Any list of completed genomes rapidly becomes outdated and so will Table 1.4 even as this book appears in print. Periodically updated listings of both finished and unfinished publicly funded genome sequencing projects are available at the web sites maintained at the Institute for Genomic Research (TIGR, http://www.tigr.org/tdb/mdb/mdb.html) and at the NCBI (http://www.ncbi.nlm.nih.gov/PMGifs/Genomes/micr.html). The Chicago-based Integrated Genomics Inc. maintains Genomes OnLine Database (http://wit.integratedgenomics.com/GOLD), which lists most public as well as some private projects. In addition, web sites of the genome sequencing centers list the projects run or planned in those particular institutions (see Appendix 2).

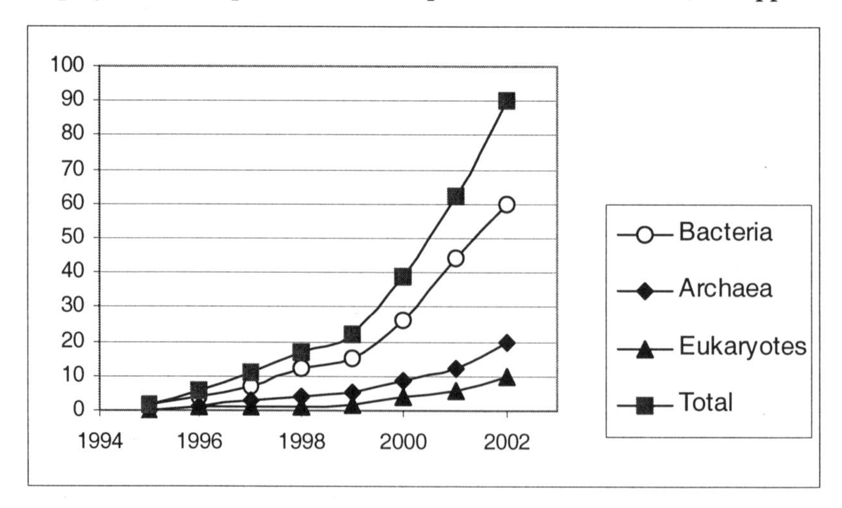

Figure 1.1. Growth of the number of completely sequenced genomes. The data are from Table 1.4. The 2002 figure is extrapolated from the 5-month results.

Table 1.4. Completely sequenced genomes (as of June 1, 2002)

Species[a]	Genome size, kb	Total no. of proteins	Year finished	Sequencing center[b]	Ref.
Bacteria					
Streptomyces coelicolor	8,668	7,567	2002	Sanger Centre	[85]
Mesorhizobium loti	7,036	6,752	2000	Kazusa Institute	[415]
Sinorhizobium meliloti	6,692	6,205	2001	EU Consortium	[137]
Nostoc sp. *(Anabaena)*	6,414	5,366	2001	Kazusa Institute	[416]
Pseudomonas aeruginosa	6,264	5,565	2000	Pathogenesis Co.	[807]
Agrobacterium tumefaciens	5,674	5,419	2001	U. Washington, Cereon Inc.	[293, 917]
Xanthomonas citri	5,176	4,312	2002	U. Sao Paulo	[167]
Xanthomonas campestris	5,076	4,181	2002	U. Sao Paulo	[167]
Salmonella typhimurium	4,857	4,451	2001	Sidney Kimmel Cancer Center	[557]
Salmonella typhi	4,809	4,600	2001	Sanger Centre	[654]
Yersinia pestis	4,654	4,008	2001	Sanger Centre	[656]
Escherichia coli	4,639	4,289	1997	U. Wisconsin	[94]
Mycobacterium tuberculosis	4,412	3,918	1998	Sanger Centre	[152]
Bacillus subtilis	4,215	4,100	1997	Institute Pasteur	[477]
Bacillus halodurans	4,202	4,066	2000	JAMST Center	[820]
Vibrio cholerae	4,033	3,827	2000	TIGR	[337]
Caulobacter crescentus	4,017	3,737	2001	TIGR	[618]
Clostridium acetobutylicum	3,941	3,672	2001	Genome Therapeutics	[622]
Ralstonia solanacearum	3,716	3,442	2002	Genoscope	[731]
Synechocystis sp.	3,573	3,169	1996	Kazusa Institute	[417]
Corynebacterium glutamicum	3,309	3,040	2002	U. Bielefeld	[831]
Mycobacterium leprae	3,268	2,720	2001	Sanger Centre	[153]
Clostridium perfringens	3,031	2,660	2002	U. Tsukuba	[767]
Listeria innocua	3,011	2,981	2001	Institute Pasteur	[286]
Listeria monocytogenes	2,945	2,855	2001	Institute Pasteur	[286]
Staphylococcus aureus	2,814	2,594	2001	Juntendo U.	[481]
Thermoanaerobacter tengcongensis	2,689	2,588	2002	Beijing Genomics Inst.	[73]
Xylella fastidiosa	2,679	2,766	2000	San Paulo State	[772]
Deinococcus radiodurans	2,649	2,580	1999	TIGR	[891]
Lactococcus lactis	2,365	2,266	2001	INRA	[97]
Pasteurella multocida	2,257	2,014	2001	U. Minnesota	[554]
Neisseria meningitidus	2,184	2,121	2000	TIGR, Sanger	[653, 837]
Fusobacterium nucleatum	2,174	2,067	2002	Integr.Genomics	[418]
Streptococcus pneumoniae	2,160	2,094	2001	TIGR	[357, 836]

Table 1.4 – continued

Brucella melitensis	2,117	2,059	2002	U. Scranton	[179]
Thermotoga maritima	1,861	1,846	1999	TIGR	[610]
Streptococcus pyogenes	1,852	1,697	2001	U. Oklahoma	[223]
Haemophilus influenzae	1,830	1,709	1995	TIGR	[232]
Campylobacter jejuni	1,641	1,654	2000	Sanger Centre	[655]
Helicobacter pylori	1,668	1,566	1997	TIGR	[848]
Aquifex aeolicus	1,551	1,522	1998	Diversa Corp.	[175]
Rickettsia conorii	1,269	1,274	2000	U. Marseille	[626]
Chlamydia pneumoniae	1,230	1,052	1999	UC Berkeley	[412]
Treponema pallidum	1,138	1,031	1998	TIGR	[243]
Rickettsia prowazekii	1,111	834	1998	Uppsala U.	[30]
Chlamydia muridarum	1,069	909	2000	TIGR	[694]
Chlamydia trachomatis	1,042	894	1998	UC Berkeley	[805]
Borellia burgdorferi	911	850	1997	TIGR	[241]
Mycoplasma pneumoniae	816	677	1996	U. Heidelberg	[347]
Ureaplasma urealyticum	752	611	2000	U. Alabama	[287]
Mycoplasma pulmonis	964	782	2001	U. Bordeaux	[144]
Buchnera sp. APS	641	564	2000	U. Tokyo	[766]
Mycoplasma genitalium	580	467	1995	TIGR	[242]
Archaea					
Methanosarcina acetivorans	5,751	4,540	2002	Whitehead Inst.	[254]
Methanosarcina mazei	4,096	3,371	2002	U. Göttingen	[181]
Sulfolobus solfataricus	2,992	2,997	2001	EU/Canada	[764]
Sulfolobus tokodaii	2,695	2,826	2001	NITE	[426]
Halobacterium sp NRC-1	2,380	2,446	2000	Inst. Syst. Biol.	[616]
Pyrobaculum aerophilum	2,222	2,605	2002	UCLA	[231]
Archaeoglobus fulgidus	2,178	2,420	1997	TIGR	[444]
Pyrococcus furiosus	1,908	2,065	2001	U. Maryland	[704]
Methanobacerium thermoautotrophicum	1,751	1,869	1997	Genome Therapeutics	[781]
Pyrococcus abyssi	1,765	1,765	2000	Genoscope	[599]
Pyrococcus horikoshii	1,739	~1,750	1998	NITE	[428]
Methanopyrus kandleri	1,695	1,691	2002	Fidelity Sistems	[779]
Aeropyrum pernix	1,670	~1,720	1999	NITE	[427]
Methanococcus jannaschii	1,665	1,715	1996	TIGR	[130]
Thermoplasma volcanuim	1,585	1,499	2000	NIBHT	[430]
Thermoplasma acidophilum	1,565	1,478	2000	MPI Biochem.	[719]

[a] Further in the book, these names are used mostly in the abbreviated form. Shading indicates obligate parasites.

[b] For the complete names of the sequencing centers, see the NCBI Entrez Genomes web site http://www.ncbi.nlm.nih.gov/entrez/query.fcgi?db=Genome, Appendix 2, or the original references.

Table 1.4 – continued

Eukaryotes					
Homo sapiens	~3,100,000	~40,000	~2002	Human Genome Project, Celera	[488, 870]
Mus musculus	~3,100,000	~40,000	~2002	Mouse Genome Project, Celera	-
Oryza sativa	~420,000	32,277	2002	Syngenta Corp.	[289]
Anopheles gambiae	~278,000	-	2002	Celera, Sanger	-
Drosophila melanogaster	~137,300	~13,500	2000	Celera, UC Berkeley	[3]
Arabidopsis thaliana	~115,400	25,498	2000	Arabidopsis Genome Project	[35]
Caenorhabditis elegans	~96,900	~19,000	1999	Sanger Centre, Washington U.	[840]
Saccharomyces cerevisiae	~11,600	~6,000	1996	European Consortium	[290]
Schizosaccharomyces pombe	~12,600	4,824	2002	Sanger Centre	[918]
Encephalitozoon cuniculi	~2,500	1,997	2001	Genoscope	[425]

The relative ease of 6- to 8-fold coverage sequencing as compared to finishing and genome annotation resulted in the availability of a number of incomplete genomes, which are not going to be finalized any time soon (see, for example, the web site of the Department of Energy Joint Genome Institute, http://www.jgi.doe.gov/JGI_microbial/html/index.html). These sequences are a treasure trove for someone who knows what to look for. Most of the data is available for searching through the NCBI BLAST page at http://www.ncbi.nlm.nih.gov/cgi-bin/Entrez/genom_table_cgi or through the web sites of the respective sequencing centers. A partial list of the major genome sequencing centers is available in Appendix 2. Of course, as new genome sequencing centers appear on the map, this listing is going to become obsolete, too. For updated listings of such centers one could look at the web sites of NCBI (http://www.ncbi.nlm.nih.gov/PMGifs/Genomes/links.html) or the National Human Genome Research Institute (http://www.genome.gov).

In addition to the whole-genome sequencing projects, there are many large-scale expressed sequence tags (EST) sequencing projects, aimed at collecting partial mRNA sequence data from eukaryotic organisms that have not yet made it to the list of priority targets for complete sequencing.

1.3. Basic Questions of Comparative Genomics

In the subsequent chapters of this book, we address many specific problems in comparative and evolutionary genomics. Right now, however, it makes sense to address some basic questions, the answers to which, as we believe, define the status of this research area.

How good is our current collection of genome sequences? Or, more precisely, how representative is it of the actual diversity of life forms? To address this issue, one has to superimpose the sequenced genomes over the taxonomy tree and see how densely populated the main branches are. When this is done with the prokaryotic part of the taxonomy, the result seems to be rather encouraging: the main bacterial and archaeal lineages are already represented by either a complete genome sequence or a genome project that is nearing completion (Table 1.5). However, this needs to be taken with a grain of salt because our knowledge of prokaryotic diversity is itself quite incomplete. Environmental molecular evolutionary studies indicate that the great majority of bacterial and archaeal species is uncultivable with the current methods [644]. Recent techniques aimed at growing these organisms [411] might eventually result in a real revolution in microbial genomics, but it will take years to unfold. Most of those species whose rRNA sequences are produced by environmental cloning fall within known bacterial and archaeal lineages suggesting that we have already sampled most of the prokaryotic diversity. However, this argument is somewhat circular because we have no idea how many prokaryotes might be not only uncultivable but also unclonable, even with the most non-specific set of PCR primers that have been tried. A case in point is the recent report of a new archaeal phylum, the Nanoarchaea [362]. With these caveats, it is fair to say that, to the best of our knowledge, the diversity of prokaryotes is reasonably well covered by genome sequences and, hence, the stage is set for prokaryotic evolutionary genomics.

The situation with eukaryotes is different in that we seem to have a better grasp of the true eukaryotic diversity and realize that the available set of genome sequences is by no means representative (Table 1.6). While certain groups (ascomycetes, nematodes, insects, mammals) are tackled by multiple genome projects, most of the early branching eukaryotic lineages are not represented among the sequenced genomes and neither are most of the animal and plant phyla, including such important groups as sponges, coelencerates, and segmented (annelid) worms. Certainly, this is no reason to postpone detailed comparative-genomic analysis, but this insufficiency of genomic data needs to be taken into account when conclusions are made on eukaryotic evolution.

Table 1.5. Coverage of the main prokaryotic phyla by genome projects

Major prokaryotic phyla	Genome sequencing	
	Completed	In progress
Archaea		
Crenarchaeota	4	3
Euryarchaeota	12	2
Bacteria		
Aquificales	1	-
CFB/Chlorobium group	-	4
Chlamydiales/Verrucomicrobia group	3	-
Chrysiogenetes	-	-
Cyanobacteria	2	2
Deferribacteres	-	-
Dehalococcoides group	-	1
Dictyoglomus group	-	-
Fibrobacter/Acidobacteria group	-	1
Bacillus/Clostridium group (low G+C gram-positive)	13	20
Actinobacteria (high G+C gram-positive)	3	8
Fusobacteria	1	-
Green non-sulfur bacteria	-	1
Planctomycetales	-	1
Proteobacteria	23	44
Alpha subdivision	7	9
Beta subdivision	2	8
Gamma subdivision	12	20
Delta subdivision	-	5
Epsilon subdivision	2	2
Spirochaetales	2	2
Thermodesulfobacteria	-	-
Thermomicrobia	-	-
Thermotogales	1	-
Thermus/Deinococcus group	1	1

[a] The taxonomy is from the NCBI Taxonomy database (see ♦3.6). The data on the finished and ongoing genome sequencing projects are from the Entrez Genomes database (http://www.ncbi.nlm.nih.gov/PMGifs/Genomes/micr.html) and the Genomes OnLine Database (http://www.genomesonline.org).

Table 1.6. Status of the eukaryotic genome projects

Major eukaryotic phyla	Representatives with ongoing sequencing projects
Acanthamoebidae	-
Acantharea	-
Alveolata	
Apicomplexa	***Babesia bovis***, Cryptosporidium parvum, Eimeria tenella, **Plasmodium falciparum**, ***P. berghei***, ***P. chabaudi***, P. vivax, P. yoelii, ***Theileria annulata***, ***Toxoplasma gondii***
Ciliophora	Paramecium tetraurelia, Tetrahymena sp.
Dinophyceae	-
Haplosporida	-
Apusomonadidae	-
Cercozoa	Chlorarachnion reptans
Core jakobids	Reclinomonas americana
Cryptophyta	**Guillardia theta** (nucleomorph genome)
Diplomonadida	***Giardia intestinalis***
Entamoebidae	***Entamoeba histolytica***
Euglenozoa	***Leishmania major***, ***Trypanosoma brucei***
Glaucocystophyceae	-
Granuloreticulosea	-
Haptophyceae	-
Heterolobosea	-
Lobosea	-
Malawimonadidae	-
Microsporidia	**Encephalitozoon cuniculi,** Spraguea lophii
Mycetozoa	***Dictyostelium discoideum***
Oxymonadida	-
Parabasalidea	-
Paramyxea	-
Pelobiontida	-
Plasmodiophorida	-
Polycystinea	-
Retortamonadidae	-
Rhodophyta	Porphyra yezoensis
Stramenopiles	Thalassiosira pseudonana
Viridiplantae	
Chlorophyta	Chlamydonas reinhardtii
Streptophyta	Alfalfa, barley, bean, coffee, corn, cotton, pine, poplar, potato, rice, sorghum, soybean, sugar cane, tomato, wheat

Table 1.6 – continued

Fungi/Metazoa group	
Aconchulinia	-
Choanoflagellida	-
Fungi	
Ascomycota	**Saccharomyces cerevisiae, Schizosaccharomyces pombe,** ***Aspergillus nidulans, A. fumigatus***, A. niger, ***Candida albicans***, Coccidioides immitis, Debaryomyces hansenii, Fusarium proliferatum, ***Neurospora crassa***, Pneumocystis carinii, Trichoderma reesei
Basidiomycota	Cryptococcus neoformans, Phanerochaete chrysosporium, Ustilago maydis
Chytridiomycota	-
Zygomycota	-
Metazoa	
Porifera (sponges)	-
Cnidaria	-
Ctenophora	-
Platyhelminthes	Schistosoma mansoni, S. japonicum
Nematoda	**Caenorhabditis elegans**, Ascaris suum, Brugia malayi, C. briggsae, Haemonchus contortus
Annelida	-
Mollusca	-
Arthropoda	**Drosophila melanogaster, Anopheles gambiae**, Aedes aegypti, A. albopictus, Amblyomma americanum, Glossina morsitans
Chordata	
Urochordata	Ciona intestinalis (sea squirt), C. savignyi
Actinopterygii	**Takifugu rubripes (fugu)**, ***Danio rerio (zebrafish)***, Oreochromis niloticus (tilapia)
Amphibia	Ambystoma mexicanum (axolotl), Xenopus tropicalis (frog), X. laevis
Crocodylidae	-
Aves (birds)	Chicken, turkey
Mammals	**Human**, **mouse**, rat, cat, chimpanzee, cow, dog

[a] The taxonomy is from the NCBI Taxonomy database (see ♦3.6). Organisms with finished or almost finished projects are shown in bold; advanced-stage projects are shown in bold and italic; not all sequencing projects for each phylogenetic lineage are listed. Absence of sequencing projects for any representative of a phylogenetic lineage, according to the Entrez Genomes and Genomes OnLine databases, is indicated by a dash.

The next question that we have to address is: ***why does comparative genomics work to give us information on gene functions and evolution?*** The general answer is provided by the neutral theory of molecular evolution [440]. Neutral evolution is fast, as convincingly demonstrated, for example, by the rapid deterioration of pseudogene sequences. Therefore, whenever we detect sequence conservation among proteins or nucleic acids from species separated by a long span of evolution (and this, in practical terms, involves any comparison between two species because these are typically separated by millions of years, time more than sufficient for a pseudogene to change beyond recognition), we can be sure that this conservation is due to the pressure of purifying selection driven by functional constraints. To put it in even simpler terms, ***what is conserved in a sequence is functionally important***. Furthermore, and less trivially, the conserved amino acids and nucleotides almost always perform the same or similar functions, at least in structural and biochemical terms, in homologous protein, RNA or DNA molecules.

These general concepts of molecular evolution indicate that comparative genomics is likely to be informative in principle, but they tell us nothing about the evolutionary distances at which it is expected to work. The theory would not have been violated in any way if only homologs from closely related species showed significant sequence similarity. However, it had been known already in the pre-genomic era that certain proteins are highly conserved even between vertebrates and bacteria, and the very first genome comparisons revealed deep evolutionary conservation for the majority of proteins. When state of the art methods for sequence comparison are applied, homologs from more than one distantly related species are detectable for 70-80% of the proteins encoded in any prokaryotic genome [827]. At present this fraction seems to be somewhat lower for some of the eukaryotes, but only because the taxonomic density of genome sequencing so far has been insufficient. Indeed, in the genomes of humans and mice, species that diverged from their common ancestor 80-100 million years ago, nearly all genes are conserved. These crucial facts show that ***genome comparisons are likely to reveal important information on the functions and evolutionary relationships of the great majority of genes in any genome.***

We have already stated that genomics would not make any sense at all without the possibility of informative genome comparison. ***Why is this so***? In principle, one could imagine that a combination of theoretical methods for deciphering a protein's three-dimensional structure from the sequence and experimental studies would allow functional identification without recourse to evolutionary analysis. However, neither of these approaches is up to the task. Some recent progress notwithstanding, there is no hope that, in the foreseeable future, *ab initio* methods become capable of correctly predicting structure of proteins on genome scale (or on any significant scale except, possibly, for some small proteins with simple folds) let alone their functions.

As for genome-wide experimental characterization of protein functions, far-reaching studies have been conducted, such as elucidation of the phenotype of all gene knockout mutants, massive study of subcellular localization and identification of protein-protein interaction in bulk for yeast *S. cerevisiae* [714,876]. However, actual determination of the biochemical activity and more so of the biological function of a protein remains a unique task and, even for model organisms, such as yeast or *E. coli*, this goal is not in sight for all gene products.

Indeed, for the great majority of organisms whose genomes have been sequenced, only a few genes have been studied experimentally (Figure 1.2), and there is no hope for substantial progress in the near future

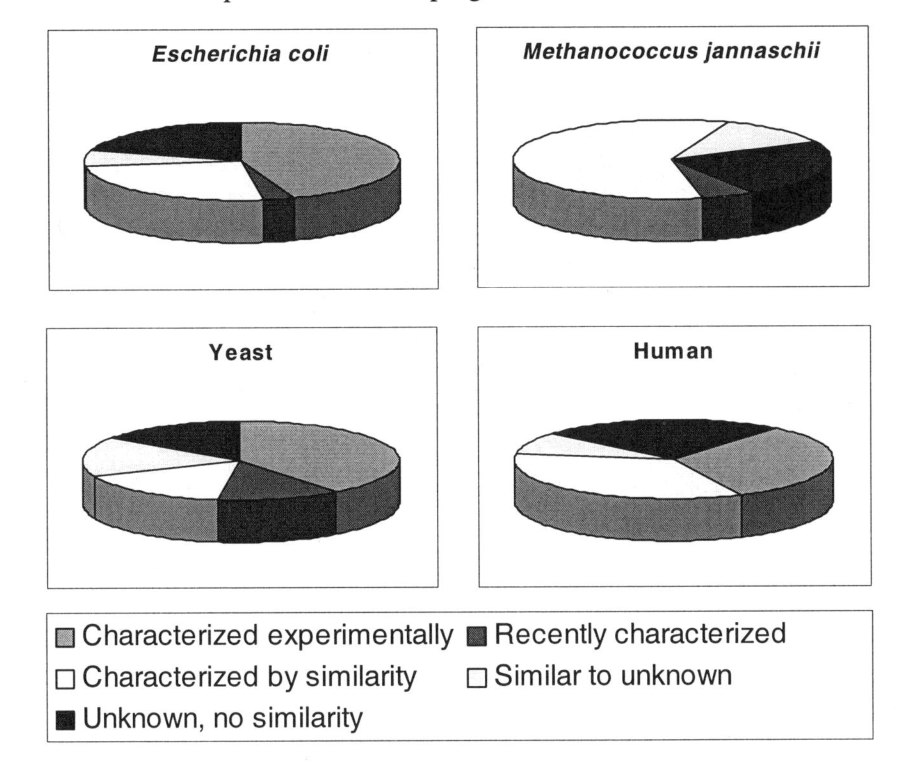

Figure 1.2. The current state of annotation of some genomes. The data were derived from the original genome sequencing papers [94,130,290,488]. The information on experimentally characterized genes of *E. coli* is from the GeneProtEC and *E. coli* Proteome databases, the corresponding data for yeast and human are from the MIPS and OMIM databases, respectively (♦3.5). The numbers of genes characterized by similarity only and similar to unknown genes are from the COG database (♦3.4); these numbers might be a slight underestimate because each COG is required to include representative of three sufficiently distant species, and those few proteins that have a homolog in only one other species are lost in this analysis.

Even for *E. coli*, the workhorse of molecular genetics for the last 50 years, less than half of the genes have been experimentally characterized. Prior to the completion of the genome of the archaeon *M. jannaschii*, only four proteins have been characterized in that organism: two flagellins, RadA recombinase, and the adenylate kinase (in Figure 1.2, this sector is just not visible).

The availability of the genome sequence spawned efforts to characterize other genes in these organisms, but so far these studies made only a limited contribution. The level of characterization of eukaryotic genomes is not much higher, although post-genomic efforts are improving the understanding of the yeast and nematode proteomes (see ♦3.5.2).

Under these circumstances, the theory of molecular evolution and, in particular, the simple connection between evolutionary conservation and function outlined above remain the crucial theoretical underpinning and the main methodology of functional genomics. The comparative approach allows researchers to predict protein functions by transferring information from functionally characterized proteins of model organisms to their uncharacterized homologs and to delineate the functionally critical parts of protein (and RNA) molecules, such as catalytic or binding sites. Naturally, the quality of these inferences depends on the sensitivity and robustness of computational methods employed by comparative genomics. These caveats notwithstanding, we will argue that comprehensive comparative analysis of genomic sequences and the proteins they encode is an absolute pre-requisite to further advances in our understanding of cell biology. Actually, we tend to believe that comparative genomics is up to something grander, namely prioritization of targets for systematic experimental studies. This approach has been partially realized in structural genomics, and we see no reason why it cannot be profitably applied in functional genomics as well. We will be quite satisfied if this book makes just a small step in this direction.

1.4. Further Reading

1. Doolittle RF. 1986. *Of Urfs and Orfs: A primer on how to analyze derived amino acid sequences*. University Science Books, San Diego.

2. Cairns J, Stent GS, Watson JD. 1992. *Phage and the Origins of Molecular Biology*. Cold Spring Harbor Laboratory Press, Cold Spring Harbor, NY.

3. Mount DW. 2000. *Bioinformatics: Sequence and genome analysis*. Cold Spring Harbor Laboratory Press, Cold Spring Harbor, NY. Chapter 1.

4. Koonin EV, Dolja VV. 1993. Evolution and taxonomy of positive-strand RNA viruses: implications of comparative analysis of amino acid sequences. *Critical Reviews in Biochemistry and Molecular Biology* 28: 375-430.

CHAPTER 2.
EVOLUTIONARY CONCEPT IN GENETICS AND GENOMICS

2.1. Similarity, homology, divergence and convergence

2.1.1. The critical definitions

In times past, gathering information on a potential partner in marriage or business routinely started with the simplest question "What family does he or she come from?" Affiliation with a certain family immediately provided a starting point for further inquiries, a general idea of what might be expected from a certain individual. Of course, families are never uniform, and classic literature from Homer to Shakespeare to Tolstoy provides ample illustrations that any expectation based solely on family history should be taken with a grain of salt. Nevertheless, absent other clues to the character of the subject in question, an educated guess could be made based on the family structure and the individual's position within that structure.

Essentially the same approach is used in predicting potential functions for a newly sequenced gene and its protein product. Since it is technically impossible to experimentally test activity of the product of every single open reading frame in every organism, understanding their cellular roles routinely relies on family history.

So how can one decide what family a given protein belongs to? Sequence analysis aims at finding important sequence similarities that would allow one to infer homology. The latter term is extensively used in scientific literature, often without a clear understanding of its meaning, which is simply common origin. Since mid-19th century, zoologists and botanists have learned to make a distinction between homologous organs (e.g. bat's wing and human's hand) and similar (analogous) organs (e.g. bat's wing and butterfly's wing). Homologous organs are not necessarily similar (at least the similarity may not be obvious); similar organs are not necessarily homologous. For some reason, this simple concept tends to get extremely muddled when applied to protein and DNA sequences [695]. Phrases like "sequence (structural) homo-logy", "high homology", "significant homology" or even "35% homology" are as common, even in top scientific journals, as they are absurd, considering the above definition. "Sequence homology" is particularly pervasive, having found its way even into the NLM's Medical Subject Heading (MeSH) system. It has been assigned as a keyword to more than 80,000 papers in MEDLINE, including, to the embarrassment of the authors, most of their own. In all of the above cases, the term "homology" is used basically as a glorified substitute for "sequence (or structural) similarity".

All this misuse of "homology", in principle, could be dismissed as an inconsequential semantic problem. One could even suggest that, after all, since it so happened that in molecular biology literature "homology" has been often used to designate quantifiable similarity between sequences (or, less often, structures), the term should be redefined, legitimizing this usage. We believe, however, that the notion of homology is of major fundamental and practical importance and, on this occasion, semantics matters. In our opinion, misuse of the term 'homology' has the potential of washing out the meaning of the very concept of common evolutionary descent [695].

A conclusion that two (or more) genes or proteins are homologous is a conjecture, not an experimental fact. We would be able to know for a fact that genes are homologous only if we could directly explore their common ancestor and all intermediate forms. Since there is no fossil record of these extinct forms, a decision on homology between genes has to be made on the basis of the similarity between them, the only observable variable that can be expressed numerically and correlated with probability. The higher the similarity between two sequences, the lower the probability that they have originated independently of each other and became similar merely by chance (see ♦4.2). Indeed, if we take two sequences of 100 amino acid residues each that have, say, 80% identical residues, we can calculate the probability of this occurring by chance, find that it is so low that such an event is extremely unlikely to have happened in the last 5 billion years, and conclude that the sequences in question must be homologous (share a common ancestry). Even for proteins that share a much lesser degree of identity, alignment of counterparts from all walks of life is often straightforward and there seems to be no reasonable doubt of homology. For example, although sequences of the ribosomal protein L36 from different species (Figure 2.1) exhibit considerable diversity and only a single amino acid residue is conserved in all the sequences, they align unequivocally and are indisputable homologs.

A real problem arises only when the similarity between two given sequences is much lower, so it is not immediately clear how to properly align them and how to calculate their degree of similarity. Even when one comes up with a figure – say, two protein sequences have 10% identical residues and additional 8% similar amino acid residues (a total of 18% similarity) – does this imply homology or not? The only reasonable answer is: it depends. This and lower levels of similarity might be indicative of homology provided that one or more of the following applies: (i) the similarity extends over a long stretch of sequence and is statistically significant by criteria known to be reliable (such as those applied in the BLAST algorithm and its derivatives); (ii) although the sequence similarity is low, the same pattern of identical and similar amino acid residues is seen in multiple sequences; (iii) the pattern of sequence similarity reflects the similarity between experimentally determined structures of the respective proteins or at least corresponds to the known key elements of one such structure.

In the rest of this chapter and in the subsequent chapters as well, we will have multiple opportunities to examine each type of evidence. Right here and now, however, it is pertinent to ponder the question: why is sequence and structural similarity considered to be evidence of homology (common origin) in the first place? Once we are confident that a particular similarity is not spurious, but rather, according to the above criteria, represents certain biological reality, is common ancestry the only explanation? The answer is: no, a logically consistent alternative does exist and involves convergence from unrelated sequences.

```
Aquifex aeolicus          MKVRSSVKK---RCAKCKIIRRKGRVMVICE-IPSHKQKTG
Bacillus subtilis         MKVRPSVKP---ICEKCKVIRRKGKVMVICE-NPKHKQKQG
Campylobacter jejuni      MKVRPSVKK---MCDKCKVVRRKGVVRIICE-NPKHKQRQG
Chlamydia trachomatis     MRVSSSIKA---PSKGDKLVRRKGRLYVINKKDPNRKQRQA
Escherichia coli          MKVRASVKK---LCRNCKIVKRDGVIRVICSAEPKHKQRQG
Helicobacter pylori       MKVRPSVKK---MCDNCKIIKRRGVIRVICA-TPKHKQRQG
Lactococcus lactis        MKVRPSVKP---ICEYCKVIRRNGRVMVICPANPKHKQRQG
Mycobacterium leprae      MKVNPSVKP---MCDKCRVIRRHRRVMVICV-DPRHKQRQG
Mycoplasma genitalium     MKVRASVKP---ICKDCKIIKRHRILRVICK-TKKHKQRQG
Rickettsia prowazekii     MKVVSSLKSLKKRDKDCQIVKRRGKIFVINKKNKRFRAKQG
Synechocystis sp.         MKVRASVKK---MCDKCRVIRRRGRVMVICSANPKHKQRQG
Treponema pallidum        MKIRTSVKV---ICDKCKLIKRFGIIRVICV-NPKHKQRQG
Thermotoga maritima       MKVQASVKK---RCEHCKIIRRKKRVYVICKVNPKHNQKQG
Vibrio cholerae           MKVRASVKK---ICRNCKVIKRNGVVRVIC-SEPKHKQRQG
Xylella fastidiosa        MKVLSSLKSAKTRHRDCKVIRRRGKIFVICKSNPRFKARQR
Yeast           ...       FKVRTSVKK---FCSDCYLVRRKGRVYIYCKSNKKHKQRQG
Rice                      MKIRASVRK---ICTKCRLIRRRGRIRVIC-SNPKHKQRQG
Fruit fly       ...       FKVKGRLKR---RCKDCYIVVRQERGYVICPTHPRHKQMSM
Mouse           ...       FKTKGVIKK---RCKDCYKVKRRGRWFILCKTNPKHKQRQM
Human           ...       FKNKTVLKK---RCKDCYLVKRRGRWYVYCKTHPRHKQRQM
```

Figure 2.1. Multiple alignment of the ribosomal protein L36 sequences. Conserved amino acid residues are shown in bold and/or shaded. The following proteins are listed: *A. aeolicus*, aq_075; *B. subtilis*, RpmJ; *C. jejuni*, Cj1591; *C. trachomatis*, CT786; *E. coli*, RpmJ; *H. pylori*, HP1297; L. *lactis*, L153863; M. *leprae*, ML1961; *M. genitalium*, MG174; *R. prowazekii*, RP456; *Synechocystis* sp., sml0006; *T. pallidum*, TP0209; *T. maritima*, TM1476; *V. cholerae*, VC2575; *X. fastidiosa*, XF2440; Yeast, YPL183w. Bacterial and yeast proteins are from COG0257, other proteins are from GenPept and have the following gi numbers: rice *O. sativa*, gi12020; fruit fly *D. melanogaster*, CG18767; mouse, gi13559402; human, gi7677060.

The functional convergence hypothesis would posit that sequence and structural similarities between proteins are observed because the shared features are strictly required for these proteins to perform their identical or similar functions. Functional convergence per se is an undeniable reality. In the broadest sense, convergence is observed, for example, between all proteins that contain disulfide bonds stabilizing their structure or between all enzymes that have the same catalytic residues (e.g. a constellation of histidines and aspartates). Even more prominent motifs associated with catalytic residues are found within different structural context and, in all likelihood, have evolved convergently [722,724]. In the case of disulfide-bonded domains, convergence can even fool sequence comparison programs, translating into statistically significant (albeit not overwhelming) sequence similarity. A rather dramatic manifestation of convergence is the recent description of a "homologous" disulfide-bonded domain in Wnt proteins and phospholipase A2 [699], which was later recognized as "mistaken identity", on the grounds of structural implausibility [77]. The classic work of Alan Wilson and colleagues comparing lysozymes from ruminants, langur monkeys and leaf-eating birds is a textbook case that reveals the nature and extent of convergence in enzymes [471,806,816]. These studies have shown beyond doubt that several amino acid residues required for functioning in the stomach have evolved independently (convergently) in different lineages of lysozymes. Importantly, however, this set of convergent positions consists of only seven amino acid residues, a small subset of the residues that comprise the lysozyme molecule.

A pan-adaptationist view of evolution would hold that functional convergence is the sole (or at least the principal) factor responsible for similarity between proteins. Formally disproving this paradigm might not be possible, but there seem to be at least two compelling arguments against it. The first one stems from the notion of a continuous gradient of similarity between proteins. The convergence explanation is implausible for closely related sequences, such as those of the same proteins (or, more precisely, orthologs; see below) from different mammalian species, which are usually 70-80% identical. For such sequences, the convergence hypothesis is equivalent to the statement that most, if not all, amino acid residues in a protein are fixed through positive selection. This runs against the neutral theory of molecular evolution, which has shown that, given the known parameters of animal populations, positive selection could not be responsible for the majority of amino acid substitutions, which are therefore effectively neutral [440]. Convergence could only be a realistic possibility for deep relationships between proteins, which involve limited similarities; indeed, the neutral theory does not preclude positive selection acting, say, on 10% of the positions in a protein. Then, the observed spectrum of similarities between proteins would have two distinct explanations: (i) divergence from common ancestors for tight families with high levels of sequence similarity and (ii) convergence from independent ancestors for larger groups of related proteins (superfamilies), in which only limited similarity is observed. While not theoretically impossible, such an opposition of two vastly different modes of evolution, with a mysterious bottleneck separating the two phases, appears extremely unlikely. This view of evolution is clearly inferior to the alternative whereby all significant similarities observed within a class of proteins are interpreted within a single theoretical framework of divergence from an ultimate common ancestor.

The second, probably most convincing argument against convergence as the principal explanation for the observed similarities between proteins has to do with the nature of structural constraints associated with a particular function. A fundamental observation is that a single function, such as catalysis of a specific enzymatic reaction, is often performed by two or more proteins that have unrelated structures [187,271]. In ♦2.2.5 we discuss this phenomenon in some detail and present several specific examples. These observations indicate that the same function does not necessarily require significantly similar structures, which means that, as a rule, there is no basis for convergent evolution of extensive sequence and structural similarity between proteins. This is not to say that unrelated enzymes that catalyze the same reaction bear no structural resemblance whatsoever. Indeed, subtle similarities in the spatial configuration of amino acid residues in the active centers are likely to exist, and these are precisely the kind of similarity that is expected to emerge due to functional convergence. These similarities, however, do not translate into structural and sequence similarity detectable by existing methods for comparison of proteins (at least in the overwhelming majority of cases). By inference, we are justified to conclude that, ***whenever statistically significant sequence or structural similarity between proteins or protein domains is observed, this is an indication of their divergent evolution from a common ancestor or, in other words, evidence of homology***. We will revisit the issue of convergence versus divergence when discussing the deepest structural connections between proteins.

Now that we have established the connection between similarity and homology, it should be emphasized that demonstration of homology is central to the interpretation of similarities between proteins. The feasibility of this conclusion, which sometimes is reached on the basis of limited similarity, is what makes sequence and structure comparison the major staples of computational biology and inspires the development of increasingly sensitive methods for such comparisons. Indeed, under the notion of homology, a sequence or structural alignment becomes a powerful tool for evolutionary and functional inferences.

Once sequences are correctly aligned, homology implies that the corresponding residues in homologous proteins are also homologous, i.e. derived from the same ancestral residue and, typically, inherit its function. If the residue in question is the same in a set of homologous sequences, we say that it is (evolutionarily) ***conserved***. Thus, homology lends legitimacy to the transfer of functional information from experimentally characterized proteins (or nucleic acids) to uncharacterized homologs, the single most common and practically important application of computational methods in molecular biology. Conversely, an alignment of non-homologous sequences is inherently meaningless and potentially misleading. Even if such an alignment attains a relatively high percentage of identity or similarity, no conclusions at all can be inferred from the (spurious, in this case) correspondence between aligned residues. This is why phrases like "significant homology"or "percent homology" are so ludicrous. Homology is a qualitative notion of common ancestry. As long as homology is established, 10% identical residues between two protein sequences could be highly meaningful and amenable to functional interpretation. In contrast, even 30% identity between two sequences that are not homologous in reality could be totally misleading.

2.1.2. Conservation of protein sequence and structure in evolution

Protein structure is conserved during evolution much better than protein sequence. There are numerous examples of proteins that show little sequence similarity, but still adopt similar structures, contain identical or related amino acid residues in their active sites and have similar catalytic mechanisms. These shared features support the notion that, despite low sequence similarity, such proteins are homologous.

Consider, for example, the structure of lysozyme, the enzyme that hydrolyzes bacterial cell walls (formal name: 1,4-beta-N-acetylmuramidase, EC 3.2.1.17). Different lysozymes are found in many organisms, from bacteriophages to mammals, and, in general, they show little sequence similarity to each other. PDB, the database of protein structures (♦3.3), includes the lysozyme from goose (PDB code 153L), which consists of 185 amino acid residues (Figure 2.2). The ***sequence*** neighbors of this protein in the protein database (♦3.1.2) are lysozymes from black swan (same length, 96% identity), ostrich (same length, 83% identity), chicken (same length, 80% identity), as well as unannotated proteins from human (44% identity), mouse (43% identity), and *B. subtilis* bacteriophage SPBc2 (25% identity in 176 aa overlap). The vertebrate proteins in this list, including the uncharacterized ones, are obvious homologs of the goose lysozyme. The phage protein is more dissimilar and, in this case, the issue of homology is worth some investigation. However, the sequence similarity between lysozymes and this phage protein is statistically significant (as can be shown, for example, using PSI-BLAST, ♦4.3.3) and their multiple alignment shows a consistent pattern of shared residues, thus establishing homology (Figure 2.2)

In contrast, the list of closest ***structural*** neighbors of goose lysozyme, according to the MMDB database (http://www.ncbi.nlm.nih.gov/Structure, ♦3.3) includes the classic chicken egg white lysozyme (e.g. PDB code 3LZT, 11% identity) and lysozymes from *E. coli* bacteriophages λ (PDB code 1AM7, 13% identity) and T4 (PDB code 149L, 11% identity). Nevertheless, a superposition of the three-dimensional structures of these three proteins clearly reveals the conserved structural core and many shared features (Figure 2.3).

A different method of structural comparison, DALI, used in the FSSP database (see ♦3.3) also identifies them as the nearest structural neighbors. Importantly, structural and sequence comparisons are a two-way street: the structural alignment shown in Figure 2.3 can be transformed into a multiple sequence alignment (Figure 2.4), in which conserved positions, including the catalytic glutamate, can be readily identified [217].

```
Goose       1 RTDCYGNVNRIDTTGASCKTAKPEGLSYCGVSASKKIAERDLQAMD
Swan        1 RTDCYGNVNRIDTTGASCKTAKPEGLSYCGVPASKTIAERDLKAMD
Ostrich     1 RTGCYGDVNRVDTTGASCKSAKPEKLNYCGVAASRKIAERDLQSMD
Chicken     1 GTGCYGSVSRIDTTGASCRTAKPEGLSYCGVRASRTIAERDLGSMN
Mouse      21 SWGCYGNIRTLDTPGASCRIGRRYGLTYCGVRASERLAEVDRPYLL
Phage    1388 DQIKSGNITQYGIVTSTTSSGGTPSSTGGSYSG-------------

Goose        RYKTIIKKVGEKLCVEPAVIAGIISRESHAGKVLKNGWGDRGNGFGLM
Swan         RYKTIIKKVGEKLCVEPAVIAGIISRESHAGKVLKNGWGDRGNGFGLM
Ostrich      RYKALIKKVGQKLCVDPAVIAGIISRESHAGKALRNGWGDNGNGFGLM
Chicken      KYKVLIKRVGEALCIEPAVIAGIISRESHAGKILKNGWGDRGNGFGLM
Mouse        RHQPTMRLVGQKYCMDPAVIAGVLSRESPGGNYVVD-LGNIGSGLGMV
Phage        KYSSYINSAASKYNVDPALIAAVIQQESGFNAKARSGVG----AMGLM

Goose        QVDKRSHKPQGTWNGEVHITQGTTILINFIKTIQKKFPSWTKDQQLKG
Swan         QVDKRSHKPQGTWNGEVHITQGTTILTDFIKRIQKKFPSWTKDQQLKG
Ostrich      QVDRRSHKPVGEWNGERHLMQGTEILISMIKAIQKKFPRWTKEQQLKG
Chicken      QVDKRYHKIEGTWNGEAHIRQGTRILIDMVKKIQRKFPRWTRDQQLKG
Mouse        KETK--FYPPTAWKSETWVSQKTQTLTSSIKEIKTRFPTWTADQHLRG
Phage        QLMPATAKSLG-VNNAYDPYQNVMGGTKYLAQQLEKFGG-----NVEK

Goose        GISAYNAGAGNVRSYARMDIGTTHDDYANDVVARAQYYKQHGY  185
Swan         GISAYNAGAGNVRSYARMDIGTTHDDYANDVVARAQYYKQHGY  185
Ostrich      GISAYNAGPGNVRSYERMDIGTTHDDYANDVVARAQYYKQHGY  185
Chicken      GISAYNAGVGNVRSYERMDIGTLHDDYSNDVVARAQYFKQHGY  185
Mouse        GLCAYSKGPNFVRSNQDLNC-----DFCNDVLARAKYFKDHGF  197
Phage        ALAAYNAGPGNVIKYGGIPPFKETQNYVKKIMA---------- 1539
```

Figure 2.2. Multiple sequence alignment of goose lysozyme and its closest homologs. Absolutely conserved amino acid residues are shown in bold, conserved hydrophobic residues are shaded.

This straightforward analysis makes us conclude that all lysozymes are homologous, which, in this case, is easy to accept given their similar, if not identical, functions. Furthermore, this analysis can be extended to a broad group of other transglycosylases, which all turn out to share a conserved catalytic domain with lysozyme and comprise a superfamily of homologous proteins [594,863]

Does structural similarity always reflect homology? For reasons discussed in the previous section, structural similarity that spans at least one complete domain most likely does. It is this type of similarity that is sought by structure comparison methods, such as VAST and DALI (♦3.3). Thus, the general rule of structure-homology correspondence seems to be straightforward: ***protein domains that have the same fold according to structure classification systems, such as SCOP or CATH, are homologs***.

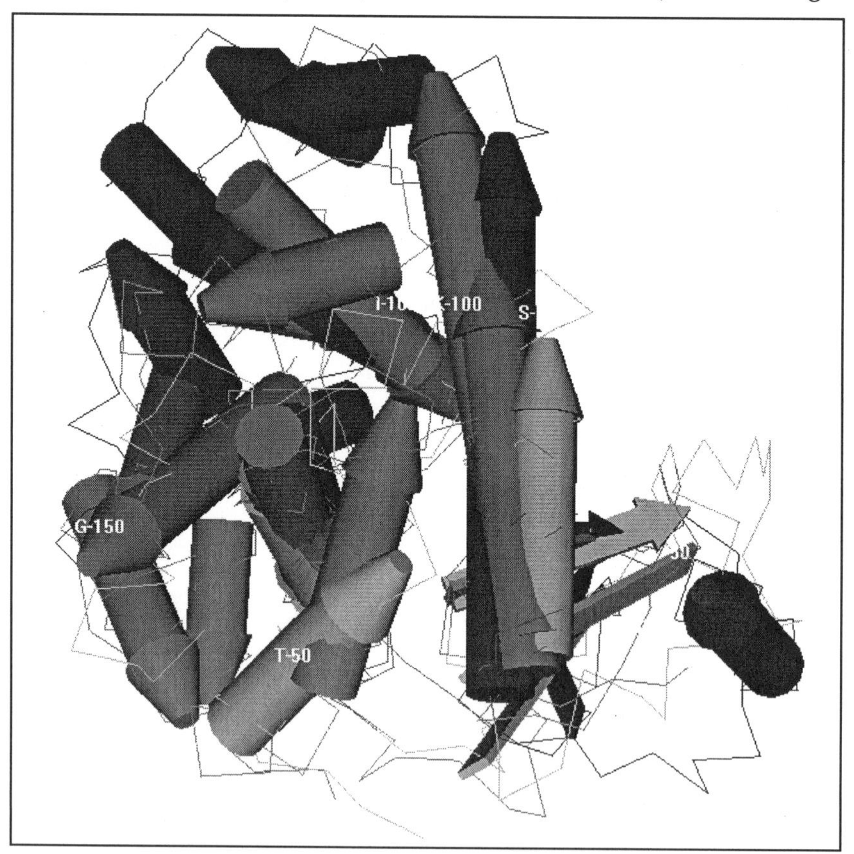

Figure 2.3. Structural alignment of goose lysozyme (PDB code 153L), chicken egg white lysozyme (3LZT) and lysozymes from *E. coli* bacteriophages λ (1AM7) and T4 (1L92). Structures of the four different types of lysozyme were aligned using VAST (http://www.ncbi.nlm.nih.gov/Structure/VAST/vast.shtml) and displayed using Cn3D (http://www.ncbi.nlm.nih.gov/Structure/CN3D/cn3d.shtml).

In principle, however, it is difficult to rule out that some common folds are so advantageous thermodynamically that they have evolved several times independently (convergently). This possibility has been considered, for example, for the triose phosphate isomerase (TIM) barrel fold, given its high stability and symmetrical, quasi-periodical organization [157].

How far does the notion of divergent evolution go? The over-reaching idea that all proteins evolved from a single primordial protein does not seem plausible. Indeed, there is no reason to believe that proteins of different structural classes, e.g. all-α (consisting exclusively of α-helices) and all-β (consisting exclusively of β-strands), have a common origin. However, certain topological changes in protein folds seem to occur during evolution [317], and the possibility of primordial common ancestry might become realistic if different folds within the same structural class are considered.

```
153L   .GEKLC.VE.PAVIAGIISRESHAG..KVLK....NGWGD...R.........
3LZT   gLDNYRgYS.LGNWVCAAKFESNFN.........tQATNR...N.........
1AM7   .mvEIN.NQrKAFLDMLAWSEGTDngrQKTRnhgyDVIVGgelftdysdhprkl
1L92   .........MNIFEMLRIDEG..........lrlKIYKdteG.........

153L   ........GNGFGLMQVDKRSH...............KP........QG..TWN
3LZT   .....tdgsTDYGILQINSRWWcndgrtpgsrnlcniPC........SAllSSD
1AM7   vtlnpklkSTGAGRYQLLSRWW...............DayrkqlglkDF..SP.
1L92   ........YYTIG.IGHLLT.........kspslnaakseldkaigrntngvIT

153L  .GEVHITQGTTILINF.IKTIQK...KFPS.WTKD..QQLKGGISAYNAGAGNVR
3LZT   ITASVNCAKKIVSDG.N......................GMNAWV.......
1AM7  ..KSQDAVALQQIKERgALPM...........idR..GDIRQAIDRCSN....iw
1L92  .KDEAEKLFNQDVDAA.VRGILRnakLKPVyDSLDavRRAAIINMVFQMGETGVA

153L  .SYARMDIGT....................THDDYANDVV....ARAQYYKQHGY
3LZT   .............................awRNRCK...gTDVQAWIRGCr
1AM7  .aslpGAGY..................gqfEHKA.DSLI....AKFKEAGgtvr
1L92  .gftnslrmlqqkrwdeaavnlaksrwynqTPNRAkrvittfrtgtwDAYK....
```

Figure 2.4. Structure-based sequence alignment of goose lysozyme (153L), chicken egg white lysozyme (3LZT) and lysozymes from *E. coli* bacteriophages λ (1AM7) and T4 (1L92). Multiple alignment, generated by the DALI program [354], was extracted from the FSSP database (http://www.ebi.ac.uk/dali/fssp/fssp.html). The residues that are structurally equivalent with ones in 153L are shown in upper case, those that are not - in lower case. The active-site Glu residue is highlighted.

Interestingly, credible relationships between certain proteins that, according to SCOP, have different folds are detectable even through PSI-BLAST searches. For example, statistically significant similarities between NAD-dependent oxidoreductases and S-adenosylmethionine-dependent methyltransferases are regularly detected in iterative database searches and the alignments produced are usually consistent with structural superpositions (N.V. Grishin and EVK, unpublished). Consequently, there is little doubt that these proteins, which formally have distinct folds, do share a common ancestry. At least in principle, such comparisons could be extended to all the numerous proteins whose structural core consists of parallel β-sheets leading to the more or less radical proposal that they all have evolved from the same primordial "Rossmann-type" domain, which possibly possessed nucleotide-binding properties [37]. The notion of divergence can be similarly extended to unite other types of structurally similar domains (e.g. different all-α-helical folds) into broad monophyletic classes. We find such generalizations attractive and credible, but caution is due and further elaboration of the methods for structure comparison, perhaps combined with theoretical analysis of evolutionary models, is required before more certainty is achieved on these potential distant evolutionary relationships. We will return to the discussion of the possible nature of primordial proteins when considering the early stages of biological evolution from a comparative-genomic perspective (see ♦6.4).

Coming back to earth, it is important to note that approximately the same level of sequence similarity that is seen between distantly related proteins whose homology is established via a combination of iterative sequence searches and structural comparisons (roughly, 8-15% identity with gaps) can be expected to exist between two randomly chosen protein sequences. We already listed above some criteria that allow one to distinguish between true evidence of homology and spurious similarities. More generally, it cannot be over-emphasized that, when this level of similarity between proteins is involved, there is no substitute (at least as of this writing) for a careful analysis of each particular relationship. Such an analysis usually pays off, allowing one to avoid false 'fundamental discoveries' and sometimes opening up new avenues of investigation.

2.1.3. Homologs: orthologs and paralogs

As discussed above, one of the main objectives of DNA and protein sequence analysis is to identify homologous sequences and to employ sequence and structure conservation to predict common biochemical activities and biological functions of proteins and non-coding sequences. The second major goal of sequence analysis is evolutionary reconstruction per se. To address each of these goals, it is critical to distinguish between two principal types of homologous relationships, which differ in their evolutionary history and functional implications. The two categories of homologs are ***orthologs***, defined as evolutionary counterparts derived from a single ancestral gene in the last common ancestor of the given two species, and ***paralogs***, which are homologous genes evolved through duplication within the same (perhaps ancestral) genome. These definitions were first introduced by Walter Fitch in 1970 [228,229] and remained virtually unknown to molecular biologists until the advent of genomics, at which time it has become clear that the distinction between the two types of homologs was crucial for understanding evolutionary relationships between genomes and gene functions. In evolutionary terms, robust identification of orthologs is essential because otherwise any evolutionary scenarios, for example, attempts to reconstruct the gene repertoire and gene order in ancestral genomes (see discussion below), are bound to be meaningless. With respect to functional analysis, orthologs typically retain the same, ancestral function, which makes transfer of functional information within a set of orthologs generally reliable. The evolutionary basis of such conservation of function among orthologs appears fairly obvious. Indeed, consider a gene (or, rather, its product) in an ancestral species that was responsible for carrying out some essential biological function. As long as the progeny of this ancestor carries a single copy of the gene in question and does not evolve or acquire an unrelated gene capable of providing the same function, it has to rely on the original gene to continue carrying out that function. This puts orthologs under strict evolutionary constraints and makes them perform the same function as long as this function remains essential for survival or at least confers a substantial selective advantage to its bearers.

In contrast, paralogs tend to evolve new functions and study of paralogous families may provide means for understanding adaptation. As first detailed by Susumu Ohno in his classic 1970 book *Evolution by Gene Duplication* [627], once paralogs emerge as a result of a gene duplication, the pressure of purifying selection decreases for either one (in Ohno's original model) or, under new, more elaborate models [448,534,877] both paralogs, which eventually enables evolution of new functions. In each sequenced genome, a substantial fraction (from 25 to 80% [374,408,484,506]) of genes belong to families of paralogs, each of which reflects functional diversification via duplications that occurred at different stages of evolution. Classic examples include animal olfactory receptors or nuclear hormone receptors, vast families in which an astonishing repertoire of specificities evolved as the result of multiple duplications.

The interplay of speciation events, leading to the divergence of orthologs, and duplications, giving rise to paralogous families, results in complex evolutionary scenarios, which may be hard to resolve (Figure 2.5). When duplication precedes speciation, each of the paralogs gives rise to a distinct line of orthologous descent. Conversely, when duplication occurs after a particular speciation event in one lineage or in both lineages independently (this can be referred to as a lineage-specific duplication or ***lineage-specific expansion*** of a paralogous family), a situation ensues whereby a one-to-one orthologous relationship cannot be delineated in principle (Figure 2.5). Instead, all one can say is that the family AB in lineage 1 is orthologous to family A'B'C' in lineage 2 or, in other words, that A and B are ***co-orthologs*** (a new term recently introduced to more accurately describe such relationships [700]) of A', B' and C' (Figure 2.5). Clearly, in such a case, the functional correspondence between the two orthologous families of paralogs is less straightforward than it is between regular, one-to-one orthologs. The relationships between homologs could become particularly tricky if some genes in certain lineages have been lost during evolution (a phenomenon referred to as ***lineage-specific gene loss***, see ♦2.2.3). In such cases, genes that, at face value, appear to be orthologous, may actually be paralogs, whereas the genuine orthologs might have been lost. Once again, functional inferences made on the basis of this type of homologous relationships require particular caution.

Reliable identification of orthologs is only possible when complete sets of genes from two or more genomes are compared. Indeed, if one of the compared genomes is incomplete, a possibility always remains that the true ortholog of the given gene is "hiding" in the unsequenced part. Even with complete genomes, identification of orthologous gene sets is not a simple task because of the complex evolutionary scenarios, which involve multiple duplications, speciations and, most importantly, lineage-specific gene loss events. In principle, complete phylogenetic analysis of all groups of homologous genes is required to decipher true orthologous relationships. This is an extremely labor-intensive task; moreover, it is well known that not all phylogenetic trees provide the required resolution. "Shortcut" approaches have been developed to circumvent the need for comprehensive phylogenetic analyses, and some of these are discussed in subsequent chapters.

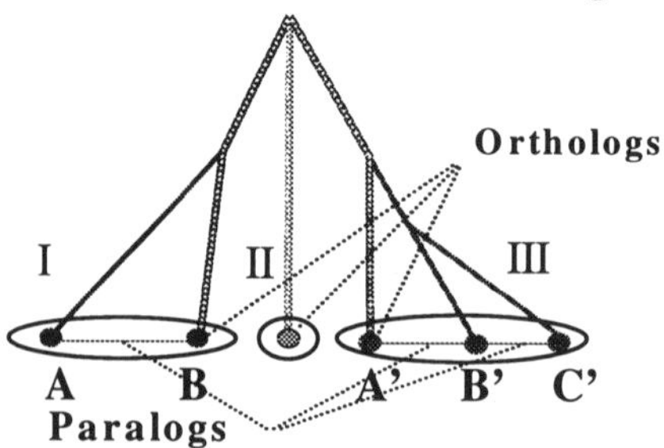

Figure 2.5. Orthologous and paralogous genes in three lineages descending from a common ancestor. Gene sets I, II, and III should be considered co-orthologous.

2.2. Patterns and Mechanisms in Genome Evolution

Although still a young discipline, comparative genomics has matured enough to allow delineating the most common and important types of events that occur during genome evolution. These include different forms of genome rearrangement, gene duplication and, more specifically, lineage-specific expansion of gene families, lineage-specific gene loss, horizontal gene transfer, and non-orthologous gene displacement.

2.2.1. Evolution of gene order

Comparison of the first completely sequenced genomes promptly showed that gene order is much less conserved than protein sequences. Genomes of closely related bacteria *Mycoplasma genitalium* and *M. pneumoniae*, for example, consist of six large segments with similar organization of genes, but the segments themselves are shifted relative to each other and partially scrambled in the two genomes [348]. Much greater differences were found between *Haemophilus influenzae* and *E. coli*, or even between *E. coli* K-12 and its pathogenic relative *E. coli* O157:H7 [669,829]. The gradient of gene order conservation is illustrated in Figure 2.6 (see color plates). In the chlamydial genomes, a genome-scale alignment is readily traceable along the main diagonal, although gaps in the alignment and two major inversions are equally obvious (Figure 2.6A). In contrast, the comparison of *E. coli* and *P. aeruginosa* looks completely disordered on the genome scale (Figure 2.6B).

In fact, any such comparison between more or less distantly related prokaryotic genomes, e.g. bacteria or archaea from different genera, would look disordered at a scale where only conservation of about a dozen genes in a row is noticeable. On a smaller scale, however, there is important conservation of gene order within operons, the units of prokaryotic gene coregulation. Extensive genome comparisons showed that, in each genome, 5% to 25% of the genes belong to conserved (predicted) operons, i.e. strings of genes that are shared with at least one relatively distant genome [916]. As should be expected, this fraction gradually increases as new genomes are sequenced. A few operons that are conserved in distantly related prokaryotes consist of genes for ribosomal proteins and some other components of the translation machinery. Other conserved operons include those encoding subunits of the H^+-ATPase and ABC-type transporter complexes [169,385,461,595].

2.2.2. Lineage-specific gene loss

A quick look at the genome sizes of the organisms with completely sequenced genomes (Table 1.4) shows that many pairs of closely related organisms have vastly different numbers of genes. Thus, *E. coli* K-12 has seven times more genes than the aphid symbiont *Buchnera* sp., which is located right next to *E. coli* in the 16S rRNA-based phylogenetic tree. Two more representatives of gamma-proteobacteria, *H. influenzae* and *P. multocida*, have 2.5 times fewer genes than *E. coli*. Substantial differences in the gene number can be found even within the same genus. The gene set of *Mycoplasma pneumoniae*, for example, includes all the 480 genes of *M. genitalium*, as well as 197 additional genes. *Mycobacterium leprae* is closely related to *M. tuberculosis*, but has at least 1,200 fewer genes [153].

The same phenomenon is seen throughout eukaryotes. Baker's yeast *S. cerevisiae*, for example, has about 6,000 genes, which is at least 2,000 genes less than in its relatives, multicellular ascomycetes such as *Aspergillus*. Furthermore, a eukaryotic intracellular parasite, microsporidian *Encephalitozoon cuniculi*, which has been identified as a derived fungus in several consistent phylogenetic studies, has only ~2,000 genes [425], which points to a truly dramatic scale of gene loss. About 300 genes were apparently lost by *S. cerevisiae* after its radiation from the common ancestor with fission yeast *S. pombe,* although the latter has even fewer genes than *S. cerevisiae* [55]. All these observations show that certain phylogenetic lineages experienced a significant gene loss, often linked to the adaptations to the parasitic lifestyle (*H. influenzae*, *P. multocida*, *M pneumoniae*, *M. genitalium*, *M. leprae*), or intracellular symbiosis (*Buchnera* sp.), or just adaptation to a constant (narrow) range of environmental conditions. Indeed, parasites might not need a complicated web of metabolic pathways for the biosynthesis of amino acids, nucleotides and cofactors as long as they can fetch those nutrients from their host.

In the same vein, the well-known absence of the biosynthetic pathways for 12 amino acids in humans and other vertebrates was probably made possible by the abundance of these amino acids in the food consumed by their common ancestor at the time of their divergence.

An analysis of gene loss in bacterial parasites showed that, in many cases, it led to the elimination of entire pathways, such as amino acid, nucleotide, and cofactor biosynthetic pathways (Chapter 7). For example, a number of parasitic bacteria lack pyrimidine biosynthesis genes that are present in their free-living relatives (Figure 2.7). This has, of course, a simple evolutionary explanation: if the necessary nutrient is available in the medium, the genes responsible for its synthesis become redundant and can be eliminated. Moreover, once at least one of these genes is lost, expression of the others would lead to the accumulation of metabolic intermediates that

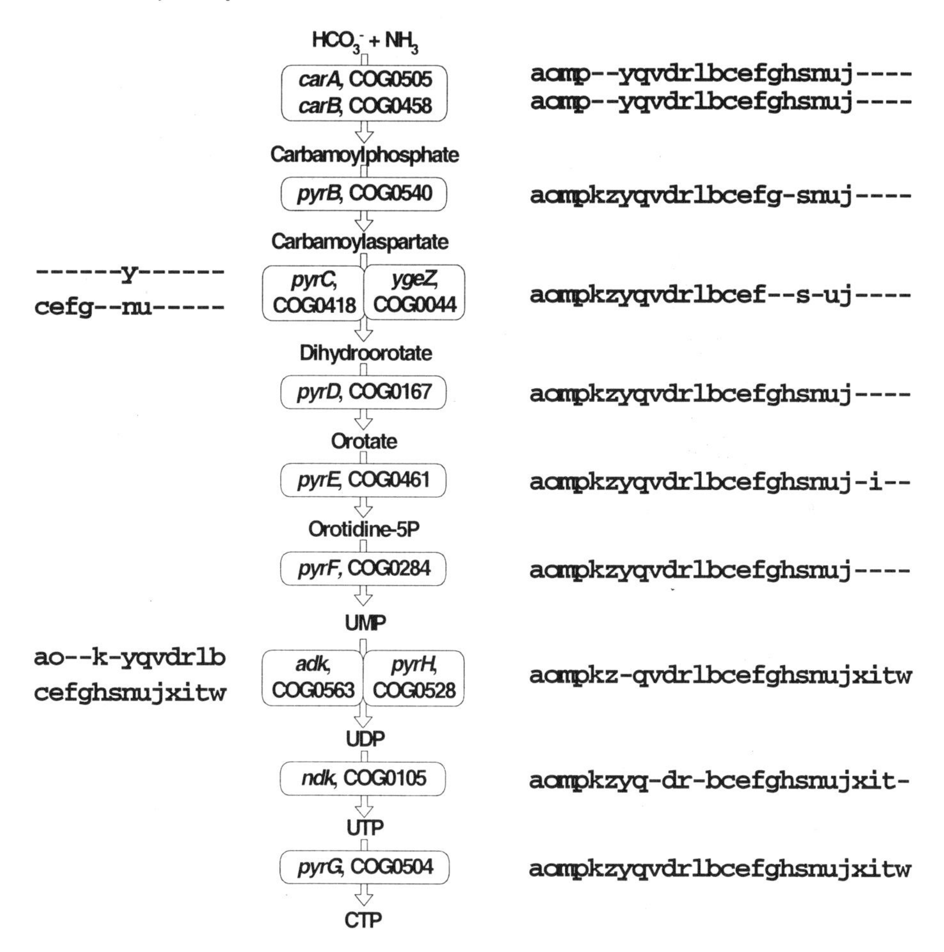

Figure 2.7. Pyrimidine biosynthesis genes in organisms with completely sequenced genomes. Each rectangle signifies an enzyme of the pyrimidine biosynthesis pathway, indicated by its gene name and COG number. Alternative enzymes catalyzing the same reaction are shown side-by-side. Each COG is accompanied by the list of organisms represented in it (the phyletic pattern, see ♦2.2.6). The species abbreviations and order are as follows: a, *Archaeoglobus fulgidus*; o. *Halobacterium* sp., m, methanogens (*M. jannaschii* and *M. thermoautotrophicum*); p, thermoplasmas (*T. acidophilum* and *T. volcanii*); k, pyrococci (*P. horikoshii* and *P. abyssi*); z, *Aeropyrum pernix*; y, yeast (*Saccharomyces cerevisiae*); q, *Aquifex aeolicus*; v, *Thermotoga maritima*; d, *Deinococcus radiodurans*; r, mycobacteria (*M. tuberculosis* and *M. leprae*); l, streptococci (*Lactococcus lactis* and *Streptococcus pyogenes*); b, bacilli (*B. subtilis* and *B. halodurans*); c, *Synechocystis* sp.; e, *Escherichia coli*; f, *Pseudomonas aeruginosa*; g, *Vibrio cholerae*; h, *Haemophilus influenzae*; s, *Xylella fastidiosa*; n, *Neisseria meningitidis*, u, *Helicobacter pylori* and *Campylobacter jejuni*; j, *Mesorhizobium loti* and *Caulobacter crescentus*, x, *Rickettsia prowazekii*; i, chlamydiae (*C. trachomatis* and *C. pneumoniae*); t, spirochetes (*Borrelia burgdorferi* and *Treponema pallidum*), and w, mycoplasmas (*M. genitalium*, *M. pneumoniae*, and *Ureaplasma urealyticum*).

can be harmful for the cell. This would result in an evolutionary pressure toward coordinated loss of all the genes in a pathway [270]. A similar trend toward coelimination of functionally connected groups of proteins, such as the signalosome and the spliceosome components, has been detected in yeast [55].

In a remarkable exception to the principle of coordinated gene loss, there are cases when only a certain (typically, upstream) part of the pathway is eliminated. Figure 2.7 shows that the complete pyrimidine biosynthesis pathway is missing in *M. genitalium* and *M. pneumoniae*, whereas *H. influenzae* lacks genes for the first three reactions of this pathway, but has the complete set of genes for all the enzymes that catalyze the conversion of dihydroorotate into CTP. Thus, while *H. influenzae* is evidently incapable of *de novo* pyrimidine biosynthesis, it has preserved certain metabolic plasticity to accommodate whatever pyrimidine it can get from its host. The same trend is seen in the even smaller genomes of *B. burgdorferi* and *C. trachomatis,* which have lost most of the pyrimidine biosynthesis genes, but still contain genes coding for the downstream steps of this pathway.

2.2.3. Lineage-specific expansion of gene families

We have already mentioned the evolutionary importance of gene duplication leading to the emergence of paralogs, which may assume new functions, sometimes substantially different from those of the ancestral gene. Genome comparisons suggest that lineage-specific expansion of paralogous gene families, which in some cases account for a sizable fraction of a genome, is one of the major mechanisms of adaptation [408,506]. Analysis of lineage-specific gene expansions can provide useful clues to the evolution of each particular lineage. Table 2.1 shows that, indeed, in pathogens *M. tuberculosis* and *H. pylori,* the most conspicuous expansions are those of genes encoding factors involved in interactions with and survival within the host organisms. In contrast, in free-living autotrophs *Synechocystis* sp. and *A. fulgidus,* the largest expansion involves signal transduction proteins, sensor histidine kinases and related ATPases.

Table 2.1. Lineage-specific expansions of paralogous families in prokaryotic genomes[a]

Species	No. copies[a]	Protein family	Likely function
M. tuberculosis	90	PPE	Surface antigen, interacts with host cells
M. tuberculosis	67	PE	Surface antigen, interacts with host cells

Table 2.1 – continued

H. pylori	34	HOP	Surface antigen, interacts with host cells
Synechocystis sp.	30	His kinase	Sensing of environmental stimuli
M. pneumoniae	25	-	Unknown
M. tuberculosis	24	MCE1	Entry and survival inside the macrophages
A. fulgidus	24	His kinase-type ATPase	Sensing of environmental stimuli
Synechocystis sp.	22	GGDEF domain	Signal transduction

[a]The data are from ref. [679].

In eukaryotes, lineage-specific expansion of certain protein families is even more evident than in prokaryotes. A comparison of the genome counts of signaling domains in the nematode *C. elegans* against the corresponding numbers in the yeast *S. cerevisiae* and some free-living bacteria and archaea (Table 2.2) shows that certain domains are dramatically expanded in *C. elegans*, even when the greater number of genes in the worm is taken into account (see also the counts of ankyrin repeats in *C. elegans* in ♦3.2.2):

Table 2.2. Expansion of signaling domains in *C. elegans*[a]

Species	Proteins	Ser/Thr/ Tyr kinase	Ser/Thr/Tyr phosphatase	BRCT	SH3	VWA	WD40
C. elegans	19,100	435	112	26	58	65	127
S. cerevisiae	6,500	116	14	10	24	3	110
E. coli	4,289	3	1	1	1	4	0
B. subtilis	4,100	4	0	1	6	5	0
M. tuberculosis	3,918	13	1	1	0	4	4
Synechocystis	3,169	12	0	1	3	4	2
A. fulgidus	2,420	4	0	0	0	2	0
M. thermoauto-trophicum	1,869	4	0	0	0	2	0
M. jannaschii	1,715	4	2	0	0	3	0
A. aeolicus	1,522	2	0	1	0	1	0

[a] The data are from ref. [679]. Domain abbreviations are as in the SMART database (see ♦3.3): BRCT, BRCA1 C-terminal domain; SH3, Src homology 3 domain; VWA, von Willebrand factor type A domain; WD40, Trp,Asp-repeat domain.

2.2.4. Horizontal (lateral) gene transfer

Horizontal (lateral) gene transfer, as opposed to the standard (vertical) transfer from ancestors to progeny, refers to acquisition of genes from organisms that belong to other species, genera, or even higher taxa. Some mechanisms of lateral gene transfer between different strains of the same species, or between closely related species, are well established and include conjugation, acquisition of plasmids, and viral (phage) infection [134]. These events are common and do not stir much controversy. After all, it was the experiment on pneumococcal transformation by heterologous DNA by Avery, MacLeod and McCarthy that proved the role of DNA in heredity. However, in the pre-genomic era, the long-range lateral gene transfer across taxa has been considered to be extremely rare and more or less unimportant in the general scheme of evolution [782]. The only instance where the fact and impact of horizontal gene transfer have been clearly recognized was the apparent massive flow of genes from the genomes of endosymbiotic organelles, mitochondria in all eukaryotes and particularly chloroplasts in plants, to the eukaryotic nuclear genome [311,312].

As soon as first comparisons of multiple, complete genome sequences representing diverse taxa had been performed, it became apparent that lateral gene transfer was too common to be dismissed as inconsequential [194]. First, horizontal gene flow between closely related species turned out to be much more pervasive than ever suspected before. Lawrence and Ochman estimate, for example, that as much as 25% of the *E. coli* genome consists of recently acquired "foreign" genes [497,625]. The actual rate of influx and loss of new genes is even faster: it appears that, in the ~100 million years since the split between *Escherichia* and *Salmonella* lineages, *E. coli* has picked up and lost as much DNA as it has now [496,497].

In addition, genome comparisons helped to uncover numerous cases of (predicted) horizontal gene transfer between organisms belonging to distinct phylogenetic lineages. Archaeal genomes presented a particularly striking picture with some genes having close homologs only among eukaryotes and others being much more similar to their bacterial homologs than to those from eukaryotes, if eukaryotic homologs were detectable at all [466]. With some exceptions, the "bacterial" and "eukaryotic" proteins in archaea were divided along functional lines, with those involved in information processing (translation, transcription and replication) showing the eukaryotic affinity, and metabolic enzymes, structural components and a variety of uncharacterized proteins appearing "bacterial" [466,540]. Because the informational components generally appear to be less prone to horizontal gene transfer [703] and in accord with the "standard model" of early evolution whereby eukaryotes share a common ancestor with archaea [906], these observations could be explained by massive gene exchange between archaea and bacteria [466]. This hypothesis was further supported by the results of genome analysis of two hyperthermophilic bacteria, *A. aeolicus* and *T. maritima*. Each of these genomes contained a significantly greater proportion of "archaeal" genes than any of the other bacterial genomes, in an obvious correlation between the similarity in the life styles of evolutionarily very distant organisms (bacterial and archaeal hyperthermophiles) and the apparent rate of horizontal gene exchange between them [52,610]. Further analyses led to the discovery of genes of clear bacterial origin in the hyperthermophilic archaeon *P. furiosus*, which proved lateral gene transfer from bacteria to archaea [184].

We believe that the demonstration of the evolutionary prominence of lateral gene transfer can be considered the single greatest change in perspective in biology brought about by comparative genomics. A new round of controversy has been sparked by the discovery of genes of possible bacterial origin in the human genome [488]. In Chapter 6, we revisit this issue and discuss implications of large-scale lateral gene transfer for the "tree of life".

2.2.5 Non-orthologous gene displacement and the minimal gene set concept

Proteins responsible for the same function in different organisms typically show significant sequence and structural conservation and can be inferred to be orthologs. However, there are exceptions to this rule. Examples of apparently unrelated enzymes with the same specificity were noted as early as 1943 when Warburg and Christian described two distinct forms of fructose-1,6-bisphosphate aldolase in yeast and rabbit muscle, respectively. These two enzymes, referred to as class I and class II aldolases, were later shown to be associated with different phylogenetic lineages and have different catalytic mechanisms and little structural similarity [95,549]. Unrelated enzymes that catalyze the same reaction have been referred to as analogous, as opposed to homologous, enzymes [228,271].

Comparative analysis of complete genomes shows that cases like this are common. Strikingly, only about 65 orthologous protein sets are universally represented in all sequenced genomes. While, in large part, this is due to lineage-specific gene loss, this number is much lower than the number of essential functions, indicating that other such functions are performed by unrelated (or at least non-orthologous) proteins in different life forms. This major evolutionary phenomenon, which came to light already in the first comparisons of sequenced genomes, was dubbed ***non-orthologous gene displacement*** [465]. The full range of mechanisms leading to non-orthologous gene displacement is not known. However, in cases when essential functions are involved, the main sequence of events appears to be clear. Since an organism cannot survive without a protein that performs an essential function, transient functional redundancy, when an organism has both forms of the respective protein, appears to be a pre-requisite of non-orthologous gene displacement [464]. Such redundancy might evolve via horizontal gene transfer or via recruitment of a protein whose original function was different from the given one (recruitment is likely to occur after gene duplication). The redundancy phase is followed by lineage-specific gene loss, resulting in non-orthologous gene displacement (Figure 2.8). In case of non-essential functions, the redundancy phase might be bypassed, with non-orthologous gene displacement evolving directly via horizontal gene transfer or recruitment.

Enzyme recruitment is a common evolutionary phenomenon leading to non-orthologous gene displacement. Typically, one of the two non-orthologous enzymes with the same catalytic activity belongs to a diverse family of enzymes and could have evolved by shifting the substrate specificity of a related but distinct enzyme [271]. A good example is the two unrelated forms of gluconate kinase. Gluconate kinases from *E. coli*, yeast and *S. pombe* form a narrow conserved group. In contrast, the gluconate kinase of *B. subtilis* belongs to the so-called FGGY family of carbohydrate kinases, which also includes glycerol kinase (GlpK), D-xylulose kinase (XylB), L-fuculose kinase, and L-xylulose kinase (LyxK). The scenario of enzyme recruitment in this case seems straightforward: a duplication of the *glpK* or *xylB* gene in the *Bacillus* lineage produced a new paralog, which accumulated several mutations resulting in a shift of substrate specificity from glycerol (or xylulose) to gluconate.

Enzyme recruitment seems to be particularly common in organisms that have adapted to novel ecological niches by developing unusual, idiosyncratic metabolic pathways. For example, most of the enzymes that are

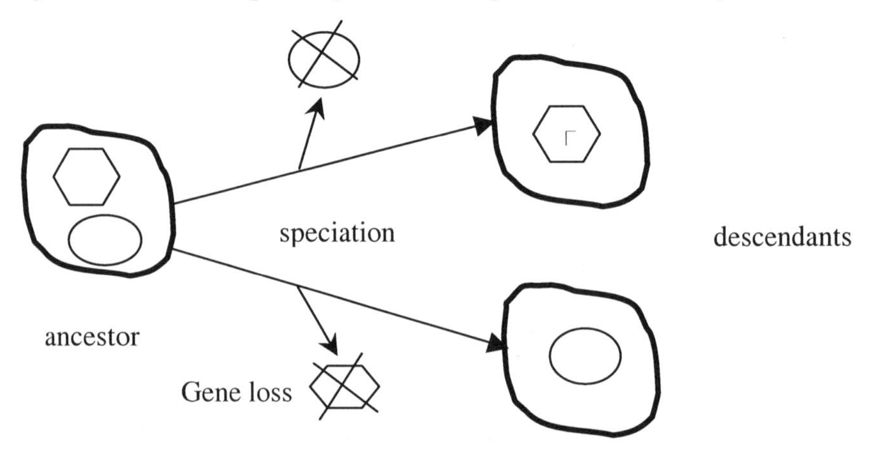

Figure 2.8. A scenario for the evolution of non-orthologous gene displacement via an ancestral redundancy stage and lineage-specific gene loss.

responsible for the biosynthesis of polyketide antibiotics in actinomycetes appear to be recent recruits from the enzymes of fatty acid biosynthesis. Similarly, enzymes that hydrolyze man-made halogenated hydrocarbons have close relatives among regular metabolic enzymes and, in all likelihood, have been recruited from this source. Perhaps the most remarkable example is the evolution of apyrase (ATP-diphosphohydrolases, EC 3.6.1.5), the enzyme secreted by blood-sucking insects into the blood of human or other mammalian victims in order to prevent or slow down blood clotting [862]. Because ADP in the blood can serve as a trigger of blood clotting, any enzyme capable of hydrolyzing it would give the hematophagous insect a substantial evolutionary advantage. As a result of this evolutionary pressure toward increasing salivary apyrase activity, insect apyrases are found in at least three different forms, which are homologous, respectively, to ATPases, 5'-nucleotidases, and inositoltriphosphate phosphatases [271,862].

It is worth noting that enzyme recruitment can be legitimately described as independent, convergent evolution of the same enzymatic activity. In Chapter 7, we look at the comparative genomic of central metabolic pathways and encounter numerous cases of non-orthologous gene displacement and, specifically, enzyme recruitment.

The idea of non-orthologous gene displacement was originally developed in conjunction with the concept of a ***minimal gene set*** for a living cell [596]. This was construed as the minimal set of genes that are essential for the functioning of a modern-type cell even under the most favorable environmental conditions, including abundance of nutrients and absence of competition. An attempt to explicitly derive a version of such a minimal gene set was undertaken by comparing the first two sequenced bacterial genomes, those of the parasites *H. influenzae* and *M. genitalium*. The straightforward logic of this reconstruction was that these two bacteria, which belong to distant phylogenetic lineages, have been independently losing genes during their adaptation to the parasitic lifestyle, and whichever common genes remain in both genomes were likely to belong to the minimal set of essential genes. It was noticed, however, that for certain essential functions (e.g. glycyl-tRNA synthetase), there was no orthologous pair of genes in the two bacteria, hence non-orthologous gene displacement had to be invoked.

The original version of the minimal gene set included 256 genes, with 16 inferred non-orthologous gene displacement cases (the magic of these numbers must not be lost on the reader: 16 is 2^2 to the power of 2; $256=16^2$ and, accordingly 256 is 2^2 to the power of 2 to the power of 2. Thus, 256 is the only number that can be represented as such a succession of powers of 2 and, at the same time, can be a reasonable approximation of a minimal gene set: 16 is obviously too few and 256^2=65,536 is, in all likelihood, much greater than the number of genes in the human genome).

Table 2.3. Examples of non-orthologous gene displacement between *M. genitalium* and *H. influenzae*

Function	*H. influenzae* gene[a], phyletic distribution	*M. genitalium* gene[a], phyletic distribution	Comments, references
Holliday junction DNA resolvase	HI0314 (*ruvC*) Most bacteria, except gram-positive	MG291.1 (*yrrK*) Most bacteria	DNA resolvase of gram-positive bacteria has not been characterized; this function for *B. subtilis* YrrK and *M. genitalium* MG291.1 remains a prediction [50].
Glycyl-tRNA synthetase	HI0924 (*glyS*), HI0927 (*glyQ*), Most bacteria	MG251 (*GRS1*) Archaea, eukaryotes, mycoplasmas, spirochetes, mycobacteria	Archaeal/eukaryotic-type (one-subunit) form of glycyl-tRNA synthetase enzyme is distantly related to the □-subunit of the bacterial form of the enzyme [556]
Asn-tRNA and Gln-tRNA synthetases	HI1354 (*glnS*) Gamma-proteobacteria, eukaryotes	MG098 (*gatC*), MG099 (*gatA*), MG100 (*gatB*), Most bacteria, archaea	Many bacteria and archaea lack dedicated Asn-tRNA and Gln-tRNA synthetases and form these aminoacyl -tRNAs through transamidation of Asp-$tRNA^{Asn}$ and Glu-$tRNA^{Gln}$ by GatABC [376]
Ribonuclease HII	HI1059 (*rnhB*) Most bacteria	MG199 (*ysgB*) Gram-positive bacteria, chlamydiae	The two forms of this enzyme are distantly related
Phosphoglycerate mutase	HI0757 (*gpmA*) Vertebrates, yeast, most bacteria	MG430 (*gpmI*) Plants, invertebrates, some bacteria	The cofactor-dependent (GpmA) and cofactor-independent (GpmI) forms of this enzyme are unrelated (see ♦2.2.6, [261,393])

a - *H. influenzae* gene names are from their *E. coli* orthologs, *M. genitalium* gene names are from their *B. subtilis* orthologs

A subsequent large-scale experimental study has shown that most of the genes included in this theoretical minimal gene set were, indeed, essential in *M. genitalium*, although a few, surprisingly, were not [364]. However, sequencing of additional genomes and the corresponding genome comparisons have clearly shown that this early reconstruction vastly underestimated the extent of non-orthologous gene displacement [452,591,674]. Indeed, as indicated above, only about 65 genes seem to be truly ubiquitous in cellular life forms, comprising perhaps 25% of the minimal set of essential functions. Therefore, it probably makes more sense to consider not so much a minimal gene set but rather a minimal set of functional requirements for cell survival. Comparative genomics shows that, for some of these requirements, a unique solution has evolved, but for the majority, evolution has come up with two or more unrelated or distantly related solutions. As discussed in ♦6.4, non-orthologous gene displacement is prominent even in the DNA replication machinery, the central functional system of all cells.

2.2.6. Phyletic patterns (profiles)

As a result of numerous lineage-specific gene losses, horizontal gene transfers and non-orthologous gene displacements, most protein families show a "patchy" distribution among the sequenced genomes. The data from the database of Clusters of Orthologous Groups of proteins (COGs, ♦3.4) show that the majority of COGs are represented in only three or four phylogenetic lineages; universal or nearly universal COGs are much less common.

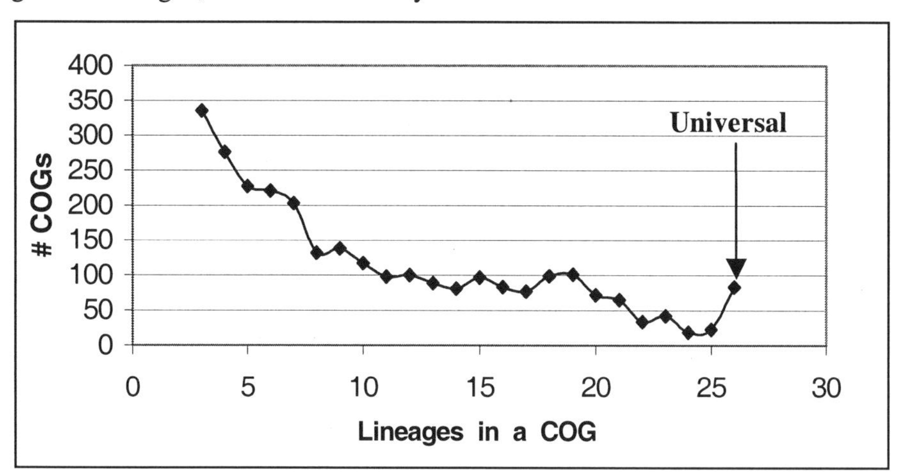

Figure 2.8. Distribution of different phylogenetic lineages in the COG database. The plot shows the number of protein families (COGs) in a release of the COG database (♦3.4), which included proteins from the given number of phylogenetic lineages out of the total of 26 lineages [827].

This distribution can be conveniently presented in the form of phyletic patterns (profiles), which show the presence or absence of a COG in each analyzed species. This approach, initially introduced as a feature of the COGs [828] and subsequently adapted, with various modifications, by several research groups [547,665,689], provides a convenient way to compare genomes and investigate the evolutionary history of individual cellular functions. For example, a quick examination of the phyletic patterns of the two distinct forms of phosphoglycerate mutase (the cofactor-dependent form GpmA and the cofactor-independent form GpmI [393]) immediately shows several interesting trends (the species symbols are the same as in Figure 2.7):

```
------y---rl--e--hsn-j-it- GpmA COG0588
-o----------bcefg---u----w GpmI COG0696
```

Firstly, the two forms have largely complementary phyletic patterns, a clear sign of non-orthologous gene displacement. Only *E. coli* encodes both forms of the enzyme (and hence shows apparent functional redundancy), whereas other organisms encode either one or the other. Secondly, several organisms do not encode either of the two forms of this enzyme. Assuming that glycolysis is an essential metabolic pathway, glycolytic enzymes should be encoded in every genome (we are aware of one exception, *Rickettsia*, which does not encode any glycolytic enzymes; see ♦7.1.1). Therefore, one might suggest that there should be an additional, third form of phosphoglycerate mutase, which is encoded in archaeal genomes and also in *T. maritima*, *A. aeolicus*, and *D. radiodurans*. Indeed, sequence analysis of those genomes shows that they all encode an uncharacterized enzyme, distantly related to alkaline phosphatase and cofactor-independent phosphoglycerate mutase. Based on the conservation of active site residues, this archaeal enzyme has been predicted to have a phosphoglycerate mutase activity [258,261]; this prediction has now been experimentally confirmed in two independent studies [308,866]. Remarkably, the phyletic pattern of the respective COG complements the union of the patterns for the two forms of phosphoglycerate mutase, which ensures the presence of at least one type of phosphoglycerate mutase in every species, except for *Rickettsia*:

```
  ------y---rl--e--hsn-j-it- GpmA COG0588
  -o----------bcefg---u----w GpmI COG0696
  a-mpkz-qvd----------------    -  COG3635
+ ----------------------------------------
  aompkzyqvdrlbcefghsnuj-itw
```

Figure 2.9. Phyletic patterns of the three forms of phosphoglycerate mutase. The species symbols are as in Figure 2.7.

This summation also shows that there is no necessity in yet another form of phosphoglycerate mutase, which has been designated GpmB in *E. coli* (see ♦3.2.1.3), but has never been experimentally demonstrated to have this activity:

```
---p--yqvdrlbcefghs--j---- GpmB COG0406
```

Indeed, recent data show that this protein does not have a phosphoglycerate mutase activity, at least in *B. subtilis*. Instead, it appears to function as a non-specific sugar phosphatase [702]. This example shows the impressive power of the comparative-genomic approach for prediction of gene functions. This methodology is discussed in greater detail later in this book (♦5.2).

2.3. Conclusions and outlook

In this chapter, we discussed some general principles of molecular evolution that are central to the comparative-genomic approaches and major evolutionary phenomena that became apparent as the result of genome comparison. The above discussion is obviously quite sketchy. However, this should be sufficient for understanding the principles underlying methods of computational genomics and the organization of various databases, which we discuss in the next two chapters. In Chapters 5 through 8, we return to problems of genome evolution at a new level and analyze some of the concepts outlined in greater depth.

2.4. Further Reading

1. Darwin C. 1859. *The Origin of Species*. Murray, London.

2. Kimura M. 1983. *The Neutral Theory of Molecular Evolution*. Cambridge University Press, Cambridge, UK.

3. Ohno S. 1970. *Evolution by Gene Duplication*. Springer, New York.

4. Graur D, Li W-H. 2000. *Fundamentals of Molecular Evolution*. Sinauer Associates, Sunderland, MA.

5. Doolittle WF. 2000. Uprooting the tree of life. *Scientific American* 282: 90-95.

6. Koonin EV, Aravind L, Kondrashov AS. 2000. The impact of comparative genomics on our understanding of evolution. *Cell* 101: 573-576.

CHAPTER 3. INFORMATION SOURCES FOR GENOMICS

The ultimate goal of genome analysis is understanding the biology of each particular organism in both functional and evolutionary terms, which requires combining disparate data from a variety of sources. Reliable information resources, compiling data on sequenced genomes and linking it to the wealth of associated functional data, are indispensable for comparative genomics. The amount of genome-related information stored in public databases and freely available to anyone with an Internet access is enormous. It has been our experience, however, that many researchers who should benefit the most from this information are not comfortable navigating these databases, let alone assessing the reliability of the data. This chapter is an attempt to bring the genomic databases closer to their principal users, molecular biologists and biochemists.

3.1. General Purpose Sequence Databases

To a computer scientist, developing a biological database might seem like a daunting task. Most fields are hard to define, and there always will be a need to create new ones. Assigning an object to a particular field is almost never final, and there are numerous exceptions to almost any rule. There is a lot of connectivity between different objects, and this, too, is subject to change. Small wonder that the problem of the optimal structure for a biological database is one of the most hotly debated topics at bioinformatics conferences and in such journals as *Bioinformatics* or *Journal of Computational Biology*. We chose not address those questions (in which none of us is a true expert) here and, instead, refer the reader to several recently published books (see Further Reading at the end of this chapter). This chapter is intended for a naïve user (a biologist, not a computer scientist) and is limited to the discussion of the relative (dis)advantages of each of the available databases for certain common tasks.

3.1.1. Nucleotide sequence databases

What makes public nucleotide sequence databases so important for modern biology? To ensure the availability of the sequence data to the general public, none of the principal scientific journals would publish a paper describing a nucleotide or protein sequence unless this sequence has been deposited in one of the three major international nucleotide sequence databases: GenBank at the NCBI (Bethesda, Maryland, USA); the European Molecular Biology Laboratory (EMBL) Nucleotide Sequence Database at the European Bioinformatics Institute (EBI) in Hinxton, near Cambridge, UK; and the DNA Database of Japan (DDBJ) at the National Institute of Genetics in Mishima, Japan. These databases form an International Nucleotide Sequence Database Collaboration and exchange updates on a daily basis, so that the DNA sequence information kept in each database is essentially the same and is arranged using common principles (see http://www.ncbi.nlm.nih.gov/collab). Although data representation in GenBank, EMBL, and DDBJ might differ slightly, each nucleotide sequence has the same accession number in all three databases. The information stored in these databases is available to the public by anonymous ftp and through the World Wide Web. This means that one can connect to the web site of any of the three databases, GenBank (http://www.ncbi.nlm.nih.gov/Entrez), EMBL (http://www.ebi.ac.uk), or DDBJ (http://www.ddbj.nig.ac.jp), and get the same nucleotide sequence using the same accession number. Thus, a sequence with a given GenBank accession number could have been originally submitted to EMBL or DDBJ, and vice versa. In everyday practice, people often refer to the public nucleotide database simply as "GenBank" when they actually mean the combination of all three public databases.

Although the nucleotide sequence data in GenBank, EMBL, and DDBJ are the same, these three databases differ in the additional services that they offer. NCBI, for example, maintains several other databases in addition to GenBank, such as the Taxonomy database (♦3.7) and PubMed (♦3.8). Accordingly, each nucleotide entry at the NCBI web site is hyperlinked to the corresponding journal article in PubMed (if available) and to the taxonomic entry for the source organism.

3.1.2. Protein sequence databases

For most of the 20th century, biologists usually had at least some idea of what they were studying, and new sequences were coming from well-defined projects that investigated a particular protein or a group of proteins. As a result, the first protein sequence database, Atlas of Protein Sequence and Structure, created by Margaret Dayhoff in the early 1960's [172,173], contained very few uncharacterized proteins and was used mainly to document and investigate sequence diversity between homologous proteins (e.g., globins or cytochromes) from diverse organisms. This trend continued for a few years even after the introduction of rapid DNA sequencing methods. However, with the rapid increase in gene sequencing rate in the early 1980-ties, more and more new protein sequences were derived from translation of anonymous pieces of DNA (or mRNA), first as a collateral benefit of sequencing the gene of interest and later through genome projects. This quantitative growth of sequence information was accompanied by a qualitative change that brought about several major problems. Although these problems could be considered just the issues of database quality control, they touch upon fundamental scientific questions.

The first problem is getting the correct protein set, i.e. correctly predicting the protein-coding regions in DNA sequences, for which there is no experimental evidence. Gene prediction historically had been one of the most important and complex aspects of computational biology (see ♦4.1), and getting the correct set of, say, human proteins still remains a daunting task. The other related and equally challenging problem is separating the wheat from the chaff, i.e. deduced protein sequences that are most likely to be correct from frameshifted fragments, sequences of pseudogenes, proteins with erroneously assigned activities, and other entries that are suspicious in one way or another, and then properly annotating them. This is an area in which the authors of this book have amassed considerable (and often painful) experience, and we try to share it with the reader in this and the next two chapters. Finally, a paramount higher-level problem is introducing some sort of database hierarchy, i.e. classifying the proteins into families and superfamilies and perhaps higher taxa according to their evolutionary relationships and organizing information in the database according to this classification. From the database angle, all these issues are aspects of database ***curation***, i.e. adding information to the entries through expert analysis. Different protein databases, including those most relevant to genomics, have adopted substantially different approaches to these problems. Usually, there is a certain trade-off between coverage and curation in a database: small, specialized databases typically deal with a single protein (super)family and are likely to be thoroughly curated. A good list of such databases is available at the ExPASy web site at http://www.expasy.org/alinks.html. Other databases strive to cover as much protein diversity as possible but offer only basic curation, if any. In this section, we briefly discuss the general-purpose databases. The more specialized ones are addressed further in this chapter.

Entrez Proteins

Currently, the principal sources of protein sequence data are translations of nucleotide sequences deposited in the GenBank\EMBL\DDBJ database. All three international databases provide these translations, but their protein sets differ in both form and content. The NCBI protein database (Entrez Proteins, http://www.ncbi.nlm.nih.gov/entrez) offers the simplest and most complete set of deduced proteins. Each protein sequence is assigned a unique gene identification (gi) number; if the sequence is changed (e.g. expanded or merged with another sequence), the new sequence is assigned a new number. Obviously, this makes the database excessively large and redundant. The size and redundancy of Entrez Proteins is further increased by incorporating the protein sequences from the PIR and SWISS-PROT databases (see below). While this ensures completeness of the database, for most practical purposes (such as database searches), NCBI maintains a non-redundant (NR) protein database, in which ***identical*** sequences from the same source organism and all their fragments are merged into a single entry. Thus, the NR database includes all sequence variants, however minor. Unfortunately, it also includes variants arising due to sequencing errors or because different databases may differently treat certain sequence features (e.g., keep or remove the initiator methionine). The completeness of Entrez Proteins makes it the ultimate resource for almost any protein sequence. Once the desired protein sequence is found, it is always useful to follow the link to "Related Sequences", which might show the same sequence from other databases, such as SWISS-PROT or PIR. Almost every Entrez Proteins entry is hyperlinked to the corresponding nucleotide sequence in GenBank (these links are absent in the records derived from PIR and SWISS-PROT). Each Entrez Proteins entry also has a link to the NCBI Taxonomy database (see ♦3.7), which allows one to examine the taxonomic position of the source organism. Many Entrez Proteins entries have links to PubMed (see ♦3.8); if the three-dimensional structure of the protein is known, there is a link to MMDB, the database of protein structures (see ♦3.4). For proteins associated with human diseases, a special database, OMIM (see ♦3.5), provides plenty of references and even some clinical information.

There are several important things one needs to know to make sequence retrieval from Entrez Proteins effective. First, each entry is assigned a unique gene index (gi) identifier, which never changes. If the same sequence is imported from a different source (SWISS-PROT, PIR, genome sequence translation), each time it receives a new gi. When the nucleotide sequence is updated, the protein sequence would also get a new gi. This makes gi the most stable identifier for a given version of a given sequence. Second, SWISS-PROT and PIR entries, imported into Entrez Proteins, are reformatted and may not be identical to the entries in those databases. Third, a search in Entrez Proteins can be best performed by specifying the search fields, such as author name [AUTH], EC number [ECNO], gene name [GENE], and organism [ORGN] and connecting them with Boolean operators AND, OR, and NOT (see ♦3.8). Two convenient search options allow one to specify the sequence length [SLEN] of the desired protein and its molecular weight [MOLWT] in Daltons (both lower and upper limits must be entered here as 6-digit numbers with leading zeros, e.g. 018500:018800[MOLWT]).

A critical aspect of the general-purpose databases, such as GenBank\EMBL\DDBJ, is that they are archival databases, which only serve as repositories of the submitted data. Curation in these databases is limited to the verification that each entry has the correct syntax and conforms to certain basic requirements, such as being free from vector contamination and actually encoding the predicted protein sequence. The responsibility for the correctness of the sequence and its annotation rests with the submitter and, accordingly, any updates or corrections must come through the submitter (third party annotation is not permitted). These ground rules are essential for preserving the integrity of the record, but they also have substantial effect on the reliability and utility of the data, which is important for all users to keep in mind (see discussion further in this chapter).

SWISS-PROT

In contrast to the Entrez Proteins, which is comprised of submitter-supplied translations of sequenced genes, the other two most commonly used protein databases, SWISS-PROT and PIR, were created and are curated by human experts. SWISS-PROT (http://www.expasy.org/sprot, mirrored on several web sites, including http://us.expasy.org/sprot), was started by Amos Bairoch at the University of Geneva (hence the 'Swiss' in SWISS-PROT) and is currently maintained by the Bairoch group in Geneva in collaboration with the EBI. SWISS-PROT strives to perform careful sequence analysis of each database entry [69]. New sequences are included into the database only after curation by expert biologists. In cases of discrepancies between several database entries for the same protein, a combined sequence is included in the database, and the variants are listed in the annotation. SWISS-PROT annotations include descriptions of the function of a protein, its domain structure, post-translational modifications, variants, reactions catalyzed by this protein, and similarities to other sequences.

The enzyme entries are cross-referenced with the ENZYME database (http://www.expasy.org/enzyme, see ♦3.6.3), the official database of the Enzyme Nomenclature Commission. As indicated above, the downside of such thorough curation is a relatively poor coverage of the protein diversity: the latest (June 6, 2002) release of SWISS-PROT contained only 110,419 entries. Wherever possible, SWISS-PROT entries are hyperlinked to various external databases, including literature citations from PubMed (see ♦3.7), nucleotide sequences from EMBL, GenBank, and DDBJ, protein motif and domain information from InterPro, Pfam, PROSITE, ProDom, and BLOCKS (see ♦3.3), and three-dimensional structures from Protein Data Bank (♦3.4). In addition to the accession number (e.g., P24182), each SWISS-PROT entry is assigned a 10-letter name, which consists of a four-letter gene or protein name and a five-letter species abbreviation (e.g. ACCC_ECOLI). These names are very convenient and are routinely used as protein identifiers in scientific literature and in this book, although, unlike accession numbers, they may change once new information becomes available. When no SWISS-PROT name is assigned to a protein, we use the Entrez Proteins gi numbers. Others, especially European researchers, routinely use identifiers from TrEMBL, a supplement to SWISS-PROT.

TrEMBL

To accommodate the growing influx of protein sequences without compromising the quality of SWISS-PROT, the protein translations of the EMBL nucleotide sequences that have not been properly curated by human annotators are put into a supplemental database, TrEMBL (Translated EMBL, http://www.expasy.org/sprot). This database serves as a kind of purgatory (or a halfway house) for SWISS-PROT [33]. Each TrEMBL entry is assigned a SWISS-PROT-type accession number that would stay with it when the sequence is finally manually checked and accepted into SWISS-PROT. To simplify curation, TrEMBL entries are even formatted in the SWISS-PROT style. However, one should be alert to the fact that TrEMBL entries are generated automatically, so their quality is not guaranteed and their annotations should not be considered as solid as those of authentic SWISS-PROT entries. In contrast to Entrez Proteins, which is updated daily, TrEMBL is produced in quarterly releases and may miss some of the latest data. On the other hand, TrEMBL is less redundant than Entrez Proteins (although it may also contain more than one entry for the same sequence). The TrEMBL release of May 31, 2002, contained 622,751 entries.

PIR

The PIR (Protein Information Resource, http://pir.georgetown.edu) database is an outgrowth of the Protein Sequence Database, originally created by Margaret Dayhoff [173] and is currently maintained at the Georgetown University in collaboration with Munich Information Center for Protein Sequences (MIPS, http://mips.gsf.de/proj/protseqdb) in Munich, Germany, and the Japanese International Protein Information Database [76]. While technically also a curated database, PIR is far less rigorous than SWISS-PROT in maintaining the quality of its annotations (our personal favorite is the annotation of the *D. radiodurans* protein DRA0097 as "probable head morphogenesis protein", see below). The advantage of PIR, however, is in its hierarchical organization. The June, 2002, release of PIR contained 283,236 entries that were classified into ~100,000 protein families and ~30,000 superfamilies. Unfortunately, as one can see from these numbers, the definitions of protein family and superfamily employed in PIR are far more narrow than those used in most of the other protein databases, particularly motif-based and structure-based ones (see ♦3.3 and ♦3.4). Thus, PIR superfamilies are often composed of very similar proteins, which may be treated by other databases as members of the same family. As a result, more distant relations between proteins (the least trivial and therefore the most interesting ones) are often not represented in PIR at all. Recently, PIR has intensified its protein classification efforts with the creation of iProClass (http://pir.georgetown.edu/iproclass, [922]), a protein classification database.

PRF

A small number of the protein database entries (< 3,000 in Entrez Proteins) come from the Protein Research Foundation (http://www.prf.or.jp/en).

3.1.3. Reliability of database entries

A critical question that emerges with any use of a database is the reliability of the information. The problem stands differently with archival and expert-curated sequence databases. Both types of databases reflect the fundamental limitation of today's genomics: only a small minority of genes in any sequenced genome or in the entire database have been characterized in direct experiments, whereas the great majority are annotated by transfer of information from the few characterized sequences on the basis of sequence similarity. With expert-curated databases, there is good reason to believe that, on most occasions, this information transfer is done responsibly and conservatively. However, one has to keep in mind that, because of the large number of sequences involved, the potential of sequence and structural analysis is rarely exploited to the fullest in these databases. The archival databases, which, because of their completeness, are searched and used for sequence retrieval most often, present an additional layer of problems caused by almost inevitable inconsistency of the approaches used by thousands of submitters for gene identification and protein annotation. Because only the submitting author can change the entry, erroneous and/or confusing annotations can linger in the databases for years. Therefore, it would be prudent to exercise certain caution before drawing any far-reaching conclusions from the sequence annotation alone, particularly when that assignment is not supported by published research. To better recognize questionable database entries, it is important to understand the common sources of unreliable and even patently wrong annotations in sequence databases; these sources are briefly discussed and exemplified below (see also ♦3.3.4). In ♦4.4.4, we discuss how to avoid making such mistakes in genome analysis and annotation.

3.1.3.1. Non-critical transfer of annotation

As we already had a chance to mention more than once, the reality of today's biological research is that only a small minority of protein sequences deposited in the databases have experimentally proven biological activity. Most of the time, the functions of the proteins encoded in the sequenced DNA are deduced on the basis of their similarity to previously characterized proteins. Hence there is a great potential for propagation of errors.

Curated databases are generally much more reliable and trustworthy. However, one should always keep in mind that all databases are compiled by humans, and not one is perfect. SWISS-PROT entries can usually be trusted, but even they can be misleading (see below). GenBank and PIR entries, particularly those coming from complete genome sequencing projects, are especially error-prone. In some cases, the result can be quite amazing. Consider, for example, the Entrez Proteins annotation of the protein DRA0097 from *D. radiodurans* (AAF12241, gi|6460535) as "head morphogenesis protein, putative". PIR curators just changed that annotation into "probable head morphogenesis protein" (PIR entry C75604). Ordinarily, as a bacterium, *D. radiodurans* would not be expected to form a head, so where could this annotation come from? It turns out that DRA0097 is closely related to the product of gene 7 of the *B. subtilis* bacteriophage SPP1, which is indeed required for the formation of the bacteriophage head, although its exact function is unknown. Non-critical transfer of the annotation of the best database hit, coupled with the truncation of the first word, produced a result that could be considered funny, if it was not virtually irreversible. The next example shows that, because of the constant data flow between GenBank and related databases, it is almost impossible to completely remove a wrong entry.

3.1.3.2. Sequence annotations from unpublished research

Let us look at another *D. radiodurans* protein, DR2227, which is annotated as phosphonopyruvate decarboxylase. Its closest relatives, also annotated as phosphonopyruvate decarboxylases, come from the complete genomes of *A. fulgidus*, *A. aeolicus*, *M. jannaschii*, *M. thermoautotrophicum*, *T. maritima*, etc., meaning that they all have been annotated on the basis of sequence similarity. Therefore these annotations should not be considered reliable: all of them may be correct, but all of them could be wrong as well. The only non-genome-project protein homologous to DR2227 is a protein from *Streptomyces hygroscopicus*, which is currently annotated as OrfZZ in Entrez Proteins (BAA93685, gi|7416071) and as BCPC_STRHY (Q54271) in SWISS-PROT. The story here is a useful illustration of the inherent conflict between the logic of scientific investigation and the tendency of the databases to provide a snapshot of the available data. In 1994, Haruo Seto and colleagues at the University of Tokyo investigated a cluster of genes responsible for the biosynthesis of bialaphos, an antibiotic produced by *S. hygroscopicus*, and sequenced a piece of DNA (GenBank accession no. D37809.1, gi|520856) that appeared to participate in this process [501]. The authors provisionally annotated one of the sequenced genes (gi|520857) as probable phosphonopyruvate decarboxylase, promptly noting that there was no experimental evidence for that annotation. In the next several years, Seto and coworkers demonstrated that phosphonopyruvate decarboxylase was a thiamin pyrophosphate-dependent enzyme, sequenced it [603,604], and in 1999 replaced the original entry with a new, corrected version (GenBank accession no. D37809.2, gi|5545270). To emphasize that the original sequence was ***not*** a phosphonopyruvate decarboxylase and that its function was unknown, that sequence was renamed OrfZZ (BAA93685).

However, in the course of those five years, the original annotation of OrfZZ as phosphonopyruvate decarboxylase, although never experimentally substantiated, made its way into several databases, including PIR and SWISS-PROT, and was used to annotate the homologs of OrfZZ encoded in the genomes of *A. fulgidus*, *A. aeolicus*, *M. jannaschii*, *T. maritima* and other prokaryotes. Even though the original incorrect annotation has been purged from the database, the ghosts it spawned still remain there and confuse scores of new annotators. As a result, several new proteins submitted to GenBank long after purging of the misannotated "phosphonopyruvate decarboxylase" were still annotated the same way, based on their similarity to misannotated proteins from *A. fulgidus*, *A. aeolicus*, *M. jannaschii*, and *T. maritima*. A detailed analysis of OrfZZ and related proteins showed that they are members of the alkaline phosphatase superfamily (♦3.3) and could function as phosphomutases, e.g. phosphoglycerate mutases [258,261]. Recently, this prediction has been experimentally confirmed [308,866] (see ♦2.2.6).

In the latter case, correcting the wrong annotation has been relatively straightforward, as there had been no experimental evidence whatsoever that the sequenced protein (OrfZZ) actually had the phosphonopyruvate decarboxylase activity. Such cases are caused by the strict requirement that no manuscript is accepted for publication unless the new sequences described in that manuscript are submitted to GenBank. When a manuscript describing a new sequence without sufficient experimental verification gets (justifiably) rejected at the stage of peer review, the sequence with its preliminary annotation would still linger in the database. Eventually, newly sequenced homologs of this sequence may get annotated as "protein related to" whatever was in that preliminary annotation. As the example above shows, such cases can become quite pervasive.

While the benefits of data exchange between the databases are obvious, it makes errors difficult to weed out. For example, even though the originally incorrect assignment of the protein P28176 (gi|401236) as thymidylate synthase was corrected in SWISS-PROT and the symbol TYSY_MYCTU was re-assigned to the (correct) SWISS-PROT entry O33306 (gi|2624286), it remained in Entrez Proteins until June of 2001. In a similar case, although annotation of the *M. tuberculosis* protein Rv3018c (SWISS-PROT entry P31500, gi|399410) as dihydrofolate reductase has been recognized as erroneous in the *M. tuberculosis* genome annotation (gi|2791615) and this ORF has been included in TrEMBL under the new identifier O53265, this wrong annotation was lingering in both SWISS-PROT and in PIR (see Table 3.1) until June 2002.

The simple lesson from these cases is that one can trust an annotation of a protein in the database without further analysis only when this protein (or its close homolog) has been experimentally characterized and there is a trustworthy publication that supports the functional assignment.

3.1.3.3. Sequences with misinterpreted function

Unfortunately, a considerable number of erroneously annotated database entries seem to be "supported" by at least some experimental data. The most common scenario is apparently as follows. When a researcher clones a new gene, he or she usually looks for an open reading frame (ORF) that would complement an existing (known) mutation or produce an increase in the desired enzymatic activity. These effects, of course, can be caused by suppression of the mutation, provision of a missing cofactor or a transcriptional regulator, as well as a number of other mechanisms. For example, an ORF that complemented the *hemG* mutation in *E. coli* (HEMG_ECOLI, P27863) was initially correctly referred to as a "gene involved in the protoporphyrinogen oxidase activity" [746], but was later assumed to code for protoporphyrinogen oxidase itself [619], even though it represented a small flavodoxin-like protein, which usually comprises only one of the several subunits of the dehydrogenase complex (Table 3.1). The apparent published experimental confirmation ("Cloning and identification of the *hemG* gene encoding protoporphyrinogen oxidase of *Escherichia coli* K-12", [619]) makes such cases very difficult to recognize. The simplest way to identify them we could think of is based on the observation that such cases usually result in the database having two or more completely unrelated sequences assumed to perform the same function. While non-orthologous gene displacement resulting in such situations is common in nature, particularly among prokaryotes (see ♦2.1.5), each such case should be viewed with certain suspicion. Table 3.1 lists several cases where the available experimental evidence does not seem sufficiently convincing to justify the current annotation of the protein. It seems likely that the functions of most of these proteins have been predicted erroneously.

Another group of misleading database entries includes cases where annotation of a protein, while technically correct, does not contain any useful biological information and should not be used for assigning functions to its homologs. Thus, *M. jannaschii* protein MJ1618, originally annotated as polyketide synthase CurC, is indeed homologous to one of the ORFs in an operon that encodes a polyketide synthase, which is responsible for the

Table 3.1. Examples of questionable functional assignments in curated protein databases

Questionable protein[a]				Enzyme with proven activity, but an unrelated sequence	Comment	Ref.
SWISS-PROT entry, name	PIR entry	Assigned enzyme activity	Closest characterized homolog			
P27863 HEMG_ECOLI	JC2513	Protoporphyrinogen oxidase	FLAV_CLOAB	HEMG_BACSU	A flavodoxin component	[619, 746]
P09127 HEMX_ECOLI	S02185	Uroporphyrin-III C-methyltransferase	-	CYSG_ECOLI	No similarity to methyltransferases	[745]
P48012 YLEU_DEBOC	S55845	Isopropylmalate dehydrogenase	BUD3_YEAST	LEU3_ECOLI	Corrected in SWISS-PROT, still wrong in PIR	[379]
P31500 DYR_MYCTU	S21834	Dihydrofolate reductase	-	DYRA_ECOLI	Uncharacterized PCR product	-
P27242 RFAK_ECOLI	C42981	N-acetylglucosamine transferase	AAC45593	RFAK_SALTY	Mis-annotated based on operon alignment	[443]
P54578 TGT_HUMAN	S68430	Queuine tRNA-ribosyltransferase	UBPF_YEAST	TGT_ECOLI	Most likely a ubiquitine C-terminal hydrolase	[182]
P46417 LGUL_SOYBN	S47177	Lactoylglutathione lyase	GTXA_TOBAC	LGUL_HUMAN	Most likely a glutathione S-transferase	-
P21204 PHEB_BACSU	D32804	Chorismate mutase	-	CHMU_BACSU	Discussed by authors, PIR used to have a warning note	[849]
P13337 VG29_BPT4	GBBPT4	Folylpolyglutamate synthase	-	FOLC_ECOLI	Correct annotation added, the old (wrong one) remains, too	[382]

[a] The latest release of SWISS-PROT in June 2002, when this book was already written, has corrected some of these errors.

biosynthesis of an antibiotic, curamycin, in *Streptomyces curacoi* [86]. However, such an annotation is clearly flawed because (i) *M. jannaschii* evidently does not produce this antibiotic and (ii) polyketide synthase is a complex of several enzymes with different biochemical activities. In contrast, a detailed analysis of MJ1618 shows that it has statistically significant sequence similarity to several enzymes of the cupin superfamily [200], including phosphomannose isomerases and, with reasonable confidence, can be annotated as a probable phosphohexomutase. Even annotating MJ1618 simply as a member of the cupin superfamily would make more sense than the "polyketide synthase" assignment.

The cases of "lost meaning" are especially common when the function of an experimentally characterized protein is complex and requires several words to explain, which does not fit into the preconfigured annotation fields. When such an entry is used to annotate an entire family of homologous proteins, confusion is almost inevitable. For example, in 1993, Michael Yarmolinsky and colleagues characterized two genes involved in the maintenance of bacteriophage P1 in the bacterial cell in the prophage form and gave them nice tongue-in-cheek names. The gene encoding the killer protein, which is responsible for cell death when the prophage is lost, was named "death on curing" (*doc*), whereas the gene encoding its antagonist was named "prevent host death" (*phd*) [502]. These puns did not go unnoticed, and homologs of these proteins in other organisms are now annotated either as Doc and Phd proteins, which is not very helpful for those unfamiliar with the original paper, or just as "analogues" (gi|1359617, gi|1359618, gi|1359619, gi|1359620), which is even less useful.

Although the problem of misleading database entries is quite serious, one cannot help enjoying these and other examples, which defy the rather common notion that sequence annotation is a tedious business. The following two items are remarkable for their attempts to properly reflect the uncertainty of the annotation:
gi|1968785, cDNA 5' end similar to similar to arrest-defective protein isolog (*Homo sapiens*), and gi|6522905, very hypothetical protein (*Schizosaccharomyces pombe*).

And here are some stimulating entries from the current version of Entrez Proteins:
gi|8953396|CAB96669 - separation anxiety protein-like (*Arabidopsis*)
gi|1935023|CAA73071 - automembrane protein H (*Yersinia enterocolitica*)
gi|2584763|CAA74752 - brute force protein (*H. pylori*)
gi|1591460|AAB98740 - hemerythrin sipunculid (*M. jannaschii*)
gi|1590909|AAB98132 - centromere/microtubule-binding protein cbp5 (*M. jannaschii*)
gi|2144179|JC4991 - detergent sensitivity rescuer (*C. glutamicum*)
gi|1788039 - Periplasmic protein related to spheroblast formation (*E. coli*)

gi|2650057|AAB90673 - DR-beta chain MHC class II (*A. fulgidus*)
gi|1334443|CAA24069 - inside intron 7 (*Saccharomyces cerevisiae*)
gi|1171589|CAA64574 - frameshift (*Plasmodium falciparum*)

Finally, an *E. coli* protein with gi|537235 has the following remarkable annotation: "Kenn Rudd identifies as gpmB [*Escherichia coli*]". Although this protein is only distantly related to the *E. coli* phosphoglycerate mutase GpmA and, according to the latest results, is a broad-specificity phosphatase ([702], see ♦2.2.6 and ♦7.1), this is probably still a better way to introduce tentative annotations than any of the examples above. If readers of this book come across other exciting examples of creative protein annotation, the authors will be happy to hear about them.

3.1.3.5. Is there a way out?

The problem of unreliable annotation is real and serious (although apparently not as drastic as sometimes predicted [89]) and is being dealt with through a number of approaches. Entrez Proteins, for example, now offers a new graphical viewer, BLink, which allows the user to list all the sequence neighbors of the given protein with BLAST similarity scores over the certain value. In addition, BLink shows the annotations of all those hits. Comparing different annotations of closely related proteins is a good way to select the correct one. Of course, as in the phosphonopyruvate decarboxylase case described above, all those annotations might be wrong. Another recent development in Entrez Proteins is the "Domains" link that shows the user the conserved domains from CDD (♦3.2) that are found in the protein in question. While GenBank annotation may still characterize a protein as "conserved hypothetical", the "Domains" view may offer new clues to its functions.

Curated databases constantly work at improving their annotations. Thus, the most recent release of SWISS-PROT has finally corrected a number of mistakes listed above. PIR descriptions now include the protein annotations from SWISS-PROT and NCBI's RefSeq, which helps identify the remaining discrepancies.

Finally, the ultimate way out may be through specialized domain and protein family databases that are discussed in the next section and in Chapters 4 and 5. By annotating protein families, rather than individual proteins, these databases are capable of taking care of the most common sources of annotation problems.

3.2. Protein Sequence Motifs and Domain Databases

The terms "protein sequence motif" and "protein domain" are widely used in biological literature for describing certain parts of proteins. The exact meaning of each of these terms is not easy to define because both are used in several, partially overlapping contexts. We would broadly define a protein sequence motif as a set of conserved amino acid residues that are important for protein function and located within a certain (short) distance from one another. These motifs can often provide clues to the functions of otherwise uncharacterized proteins. A protein domain is a structurally compact, independently folding unit that forms a stable three-dimensional structure and shows a certain level of evolutionary conservation. Typically, a conserved domain contains one or more motifs. Many proteins consist of a single protein domain, whereas others contain several domains or include additional, non-globular parts, e.g. signal peptides in membrane and secreted proteins. Some protein domains are "promiscuous" and can be found in association with a variety of other domains. Therefore, during protein sequence analysis, it is often advantageous to deal with one domain at a time. To facilitate annotation of multi-domain proteins, several popular databases contain extensive listings and descriptions of all identified protein domains. In subsequent chapters, we discuss the concepts of motif and domain and especially multidomain proteins in greater depth.

3.2.1. Motif databases

PROSITE: from patterns to profiles

The oldest and best known sequence motif database is PROSITE (http://www.expasy.org/prosite, mirrored in the US at http://us.expasy.org/prosite), maintained by Amos Bairoch and tightly integrated with SWISS-PROT [220]. For many years, PROSITE has been a collection of sequence motifs, which were represented and stored as UNIX regular expressions. For example, the famous P-loop motif, first described in 1982 by John Walker and colleagues as "Motif A" and found later in many ATP- and GTP-binding proteins (see ♦4.3.3), corresponds to a flexible loop, sandwiched between a β-strand and an α-helix and interacting with β- and γ-phosphates of ATP or GTP [880]. This motif is represented in PROSITE as

[AG]-x(4)-G-K-[ST] (PROSITE entry PS00017),

which means that the first position of the motif can be occupied by either Ala or Gly, the second, third, fourth, and fifth positions can be occupied by any amino acid residue, and the sixth and seventh positions have to be Gly and Lys, respectively, followed by either Ser or Thr.

This approach to describing sequence motifs has both advantages and disadvantages. On the plus side, a comparison of a given sequence against all the patterns in the database can be performed very fast even with limited computational resources. Virtually any user could download the whole database (less than 5 Mb) and use it on the home computer. On the other hand, regular expressions cannot fully account for the whole sequence diversity and necessarily exclude certain deviant, but closely related, sequences (see ♦4.3.3). An attempt to relax the motifs to accommodate this sequence diversity makes some motifs quite fuzzy and, as a result, almost useless. For example, possible sites for N-glycosylation of an Asn residue

N-{P}-[ST]-{P} (PS00001),

where {P} means any amino acid other than proline, or for phosphorylation of protein Ser and Thr residues

[RK](2)-x-[ST] (PS00004),
[ST]-x-[RK] (PS00005), and
[ST]-x(2)-[DE] (PS00006)

can be found in almost every protein. To improve description of such motifs, PROSITE authors have started supplementing patterns with rules and profiles (matrices).

A rule is a textual description of a complex pattern that allows one to indicate not just what amino acid residues are permitted in a particular position, but also which of these residues are most frequent (i.e. best conserved). For example, PROSITE pattern PS00008 for the N-terminal myristoylation site

G-{EDRKHPFYW}-x(2)-[STAGCN]-{P}

is supplemented by the following rule:

- The N-terminal residue must be glycine.
- In position 2, uncharged residues are allowed. Charged residues, proline and large hydrophobic residues are not allowed.
- In positions 3 and 4, most, if not all, residues are allowed.
- In position 5, small, uncharged residues are allowed (Ala, Ser, Thr, Cys, Asn and Gly); serine is favored.
- In position 6, proline is not allowed.

Here, "serine is favored" clearly indicates that not all small, uncharged residues are equal in position 5, but how strongly is it favored? To answer this question, one has to go to a more complex system of notation, such as a profile (matrix). For example, ankyrin repeats (PROSITE pattern PS50088), which are responsible for the interaction of p53 with the p53-binding protein [297], of NFkB with its inhibitor IkBα [389], and for many other important protein-protein interactions, are too diverse to be described by even a complex pattern or set of rules. Instead, it is easier to align all known ankyrin repeats and calculate the frequency of each amino acid residue at each position of the alignment. This operation would produce a matrix that would have 20 frequency numbers for the first position, 20 numbers for the second one, and so on. If the alignment contains gaps, the frequency of a gap in any given position would give us the 21st number. Also, because some sequences come from acid hydrolysis, which converts Asn into Asp and Gln into Glu, there traditionally are two more letters, B (either Asn or Asp) and Z (either Gln or Glu). In addition, X would stand for an unknown amino acid residue. As a result, one would end up with a matrix of the size 24xL, where L is the length of the motif. Actually, for the purposes of sequence comparison, rather than frequencies, it is more convenient to use their logarithms. PROSITE, like other tools (see ♦3.2), uses log-odds position-specific scoring matrices (PSSMs; see also ♦4.3 for further discussion of PSSMs and their use in sequence searches), which look as follows:

	A	B	C	D	E	F	G	H	I	K	L	M	N	P	Q	R	S	T	V	W	Y	Z
B	-10	9	-23	9	5	-18	-12	2	-20	-1	-17	-12	9	-15	2	-2	-1	-4	-19	-26	-8	2
G	-3	-4	-28	-4	-11	-26	40	-13	-33	-11	-25	-17	2	-18	-12	-12	-1	-15	-26	-22	-23	-11
R	-9	-3	-22	-5	-2	-11	-16	-3	-15	-1	-12	-7	0	-18	-1	3	-6	-7	-14	-16	-4	-2
T	0	0	-13	-7	-8	-11	-16	-14	-11	-9	-13	-11	3	-11	-8	-9	17	32	-4	-30	-10	-8
P	8	-17	-23	-16	-7	-21	-15	-20	-11	-12	-16	-12	-16	35	-11	-19	-4	-5	-13	-27	-22	-
L	-9	-29	-20	-31	-22	11	-30	-21	21	-28	41	19	-28	-28	-20	-20	-26	-9	13	-19	1	-21
H	-15	-5	-25	-6	-5	-12	-20	55	-17	-11	-11	3	1	-20	3	-5	-10	-14	-18	-24	13	-4
L	-8	-19	-20	-21	-14	1	-25	-12	5	-15	9	5	-18	-23	-12	-13	-15	-7	3	-8	5	-14
A	39	-11	-7	-20	-12	-16	-4	-19	-9	-12	-9	-9	-10	-13	-11	-19	8	0	0	-20	-16	-12
A	16	-17	1	-24	-18	-11	-17	-22	3	-17	-2	-1	-15	-20	-16	-20	0	0	11	-27	-14	-18
R	-5	-8	-20	-10	-2	-14	-16	-6	-12	1	-10	-3	-5	-17	1	3	-4	-5	-10	-20	-7	-2
N	-5	-2	-20	-6	-2	-13	-12	0	-15	-1	-14	-7	3	-17	-1	0	-1	-3	-14	-21	-4	-2
G	-3	1	-25	-1	-8	-26	29	-10	-31	-8	-26	-17	8	-17	-8	-8	1	-12	-25	-24	-22	-8
H	-12	10	-23	9	3	-19	-14	18	-22	0	-19	-11	13	-17	5	1	-1	-7	-22	-27	-3	2
L	-6	-13	-21	-14	-8	-6	-22	-12	0	-11	1	0	-12	-13	-9	-10	-8	-3	1	-22	-5	-10
E	-7	9	-25	13	16	-25	-12	-3	-23	3	-20	-15	5	-6	5	-2	1	-5	-21	-29	-16	10
V	-4	-22	-9	-28	-22	-1	-27	-20	16	-20	11	11	-19	-23	-18	-19	-12	-3	17	-24	-6	-21
V	3	-24	-11	-28	-23	-1	-24	-24	17	-20	12	8	-22	-24	-21	-20	-10	-2	23	-24	-7	-22
K	-9	1	-25	2	11	-24	-17	-2	-22	15	-19	-10	2	-13	10	14	-4	-7	-18	-25	-11	9
L	-7	-18	-20	-21	-13	4	-24	-12	4	-14	10	5	-16	-22	-12	-11	-13	-5	2	-13	5	-13
L	-10	-29	-19	-31	-21	12	-30	-20	20	-28	42	19	-28	-29	-20	-20	-27	-10	11	-18	1	-21
L	-8	-26	-18	-29	-21	3	-30	-20	22	-24	28	15	-24	-26	-18	-19	-21	-8	15	-21	-2	-21
E	-5	7	-24	8	14	-25	-11	-2	-23	6	-20	-13	5	-11	9	4	3	-4	-20	-27	-14	11
H	-3	-3	-20	-8	-4	-12	-14	4	-15	1	-13	-7	2	-18	-1	3	-3	-6	-14	-21	-3	-3
G	-2	-3	-27	-4	-13	-27	44	-14	-34	-13	-27	-18	4	-18	-13	-13	1	-14	-26	-23	-24	13
A	26	-11	-11	-18	-11	-16	-5	-18	-7	-12	-9	-8	-10	-11	-12	-18	5	-1	1	-23	-16	-12
D	-11	26	-24	30	10	-28	-8	-1	-27	0	-26	-21	20	-11	1	-4	5	-4	-24	-35	-17	5
T	-4	-24	-21	-26	-19	-4	-27	-23	16	-18	8	6	-21	-8	-18	-18	-13	-4	15	-23	-8	-20
N	-9	22	-21	17	4	-20	-8	2	-21	-2	-22	-16	26	-15	-1	-4	5	-1	-22	-33	-15	1
A	4	-12	-17	-16	-10	-10	-18	-14	1	-11	-1	0	-11	-17	-8	-12	-4	-2	3	-23	-8	-10
R	-6	-9	-20	-10	-3	-16	-18	-10	-11	3	-11	-5	-7	-13	0	6	-5	-4	-7	-23	-10	-3

Figure 3.1. Position-specific scoring matrix for ankyrin repeats. The first column represents the consensus sequence of the ankyrin repeat. From PROSITE entry PS50088 ANK_REPEAT (see also the documentation in PDOC50088).

Clearly, while the above form of presentation might be perfect for a computer, it is challenging for a human to comprehend. We would be more comfortable with something perhaps less precise, but capturing the crucial features of a motif. There are several ways to achieve this. Probably the most convenient one is a Sequence Logo [752], in which the height of each letter indicates the degree of its conservation, whereas the total height of each column represents the statistical importance of the given position (Figure 3.2)

Despite of all the shortcomings of sequence patterns, PROSITE remains a very convenient tool for rapid protein sequence analysis. The textual descriptions of the sequence motifs and protein families that are characterized by these motifs are of special value. They offer a unique perspective of the functional diversity of proteins that may be quite similar sequence-wise. A reader of this book would likely benefit from spending ample time looking through PROSITE documentation files (http://www.expasy.org/cgi-bin/prosite-list.pl). Another useful exercise is to take some well-characterized protein sequences, perhaps those most familiar to the reader, and search them for PROSITE patterns using the ScanProsite tool (http://www.expasy.org/tools/scnpsite.html). Finally, those interested in creating their own sequence patterns should take a look at the "optimal way to deduce motifs" picture at http://www.expasy.org/images/cartoon/prosite.gif. This nice cartoon clearly explains why, with the exception of PROSITE, all other motif databases do not attempt to create representative patterns for the selected motifs and keep their data simply as sets of alignments or matrices.

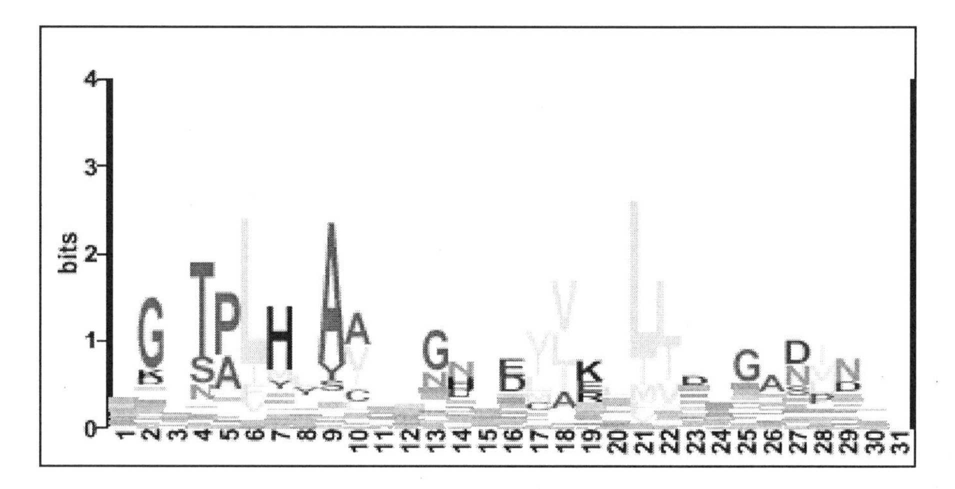

Figure 3.2. The conserved motif of the ankyrin repeat. The picture was drawn using the Sequence Logo program [752] in the web-based implementation by Steven Brenner and colleagues (http://weblogo.berkeley.edu) [949]. Amino acid residues numbering is as in yeast protein AKR1 (P39010, AKR1_YEAST).

BLOCKS

The BLOCKS database (http://www.blocks.fhcrc.org, mirrored at http://bioinformatics.weizmann.ac.il/blocks) was developed by Steven Henikoff and coworkers at the Fred Hutchinson Cancer Center in Seattle, WA, and is based on a completely different approach than PROSITE [345]. Each "block" in this database is a short ungapped multiple alignment of a conserved region in a family of proteins. These blocks were originally derived from proteins with PROSITE entries, but were later expanded using data from many different sources. A part of the BLOCKS database entry for the ATP-grasp superfamily of proteins, which includes biotin carboxylase, carbamoyl phosphate synthetase, succinyl-CoA synthetase, D-alanine-D-alanine ligase, and several other enzymes ([262], see ♦3.3) is shown below.

Obviously, this sequence block can be easily converted to a PSSM. The BLOCKS alignments were used in developing the BLOSUM series of amino acid residue substitution matrices, which are currently employed in most sequence similarity search methods (see ♦4.2.1). Recently, the database has been updated and now includes blocks derived from Pfam, ProDom, PRINTS, and Domo motif and/or domain databases (see below).

```
Block BP00180A
DE    LIGASE SYNTHETASE CARBAMOYL
ACCC_BACSU|P49787   ( 104) GPSADAISKMG   39
ACCC_METJA|Q58626   ( 104) GPNPDAIEAMG   52
CARB_BACSU|P25994   ( 276) GIEGGCNVQLA   57
CARY_BACSU|P18185   ( 954) GTFASWMEQEG   51
DUR1_YEAST|P32528   ( 735) GPSGDIIRGLG   17
PCCA_HUMAN|P05165   (  27) GSVGYDPNEKT  100
PUR2_ECOLI|P15640   (  93) GPTAGAAQLEG   37
PURK_PSEAE|P72158   (   9) GQLGRMLALAG   26
PURT_PASHA|P46927   ( 261) GIFGVELFVCG   46
PYR1_DROME|P05990   (  33) GVGGEVVFQTG   18
SUCC_METJA|Q57663   (  52) GKAGGILFASN   54
YFIQ_ECOLI|P76594   ( 125) NSLGLLAPWQG   59
ACCC_ECOLI|P24182   ( 104) GPKAETIRLMG   27
ACCC_HAEIN|P43873   ( 104) GPTADVIRLMG   23
ACCC_PSEAE|P37798   ( 104) GPTAEVIRLMG   23
CARB_ECOLI|P00968   ( 397) ALRGLEVGATG   28
CPSM_HUMAN|P31327   ( 429) GSGGLSIGQAG    8
CPSM_RAT|P07756     ( 429) GSGGLSIGQAG    8
PYR1_DICDI|P20054   ( 372) GSGGLSIGQAG    8
PYR1_HUMAN|P27708   ( 400) GSGGLSIGQAG    8
PYR1_YEAST|P07259   ( 445) GSGGLSIGQAG    8
COA1_HUMAN|Q13085   (  60) SDLGISALQDG   23
```

Figure 3.3. A part of the BLOCKS database entry BP00180 "Ligase synthetase carbamoyl" for the proteins of ATP-grasp superfamily (see Table 3.2).

The BLOCKS server allows one to search a given protein or nucleotide sequence against the blocks in the database; a nucleotide sequence will be translated in all six reading frames and each translation will be checked. The BLOCKS database also has an important feature that allows the user to submit a set of sequences, create a new block, and search this block against the database. This option can be especially useful in cases where a standard database search finds several homologous proteins with no known function.

Finally, an attractive feature of BLOCKS is that each sequence block in the database can be used for creating sequence logos, similar to the one in Figure 3.2. This option allows one to visualize the degree of sequence conservation in each block, which helps to memorize the principal conserved residues of each enzyme family covered in the database.

PRINTS

The PRINTS database (http://www.bioinf.man.ac.uk/dbbrowser/PRINTS), also referred to as "PRINTS-S: the database formerly known as PRINTS", is, like BLOCKS, a collection of conserved sequence fragments in protein sequences [61,62]. In contrast to the BLOCKS database, PRINTS would list several conserved sequence blocks for each protein, which results in much smaller families than in BLOCKS. One can compare a sequence or even a library of sequences against the whole database using BLAST (♦4.3.3), making it a useful tool for identifying distant relationships among proteins. PRINTS data are now incorporated into the EBI's InterPro database (http://www.ebi.ac.uk/interpro, see ♦3.2.3) and can be searched at the InterPro web site.

3.2.2. Domain databases

Pfam

The Pfam database [80] was jointly developed by three groups in UK, USA, and Sweden and is now available at the web sites of the Sanger Centre (http://www.sanger.ac.uk/Software/Pfam), Washington University in St. Louis (http://pfam.wustl.edu), and the Karolinska Institute in Stockholm (http://www.cgr.ki.se/Pfam), as well as on the web site of INRA in France (http://pfam.jouy.inra.fr). Pfam contains protein sequence alignments that were constructed using hidden Markov models (HMMs, see ♦4.3.3). In contrast to Entrez Proteins, SWISS-PROT and PIR, which include full-length protein sequences, Pfam is a ***protein domain*** database. This means that a typical Pfam entry is not a protein sequence as in Entrez Proteins, SWISS- PROT and PIR, or a sequence pattern as in PROSITE, but an alignment of the most conserved portions ("domains") of many related proteins from SWISS-PROT and TrEMBL databases. Although a typical Pfam alignment consists of 20-30 sequences, the entries PF00516 (glycoprotein GP120) and PF00096

(C2H2-type zinc finger) include more than 10,000 sequences each. Altogether, more than 60% of the proteins in SWISS-PROT are included in one or more Pfam alignments.

In addition to complete alignments, Pfam provides "seed alignments". These include fewer proteins which are, nevertheless, sufficiently different to reflect the diversity of the members of each given Pfam family. Besides multiple sequence alignments, each Pfam entry contains a HMM for the corresponding family, which combines a PSSM (see above) with a measure of the probability of the appearance of a given amino acid at a given position as a result of a mutation (see ♦4.3.3). The availability of precomputed HMMs for each protein family in Pfam allows a relatively fast and sensitive search of any given protein against the Pfam database.

Because proteins often include more than one conserved domain, correctly identifying domains and their boundaries is a necessary pre-requisite to a detailed sequence analysis. By storing complete domain alignments, rather than selected amino acid patterns or blocks, Pfam preserves the sequence information in its entirety. This makes it a powerful tool for domain identification. As with PROSITE, simple browsing of Pfam description files provides an informative tour of important protein families.

Another useful feature of Pfam is that it now includes a supplement, referred to as Pfam-B, for entries that are only being considered for inclusion in Pfam. Pfam-B serves as a temporary storage for those entries, which have not yet been manually curated, much like the repository TrEMBL provides for SWISS-PROT. Browsing Pfam-B entries and deciding whether they really belong to the corresponding Pfam family can be a useful training exercise, which might even result in unexpected findings. Pfam data have been incorporated into the EBI's InterPro (http://www.ebi.ac.uk/interpro) and the NCBI's CDD (http://www.ncbi.nlm.nih.gov/Structure/cdd/cdd.shtml) databases (see ♦3.2.3) and can be searched at their respective web sites. It is important to remember, however, that the search tool at InterPro (HMMer-based) is similar, but not identical to the one used at the Pfam web site, whereas CDD

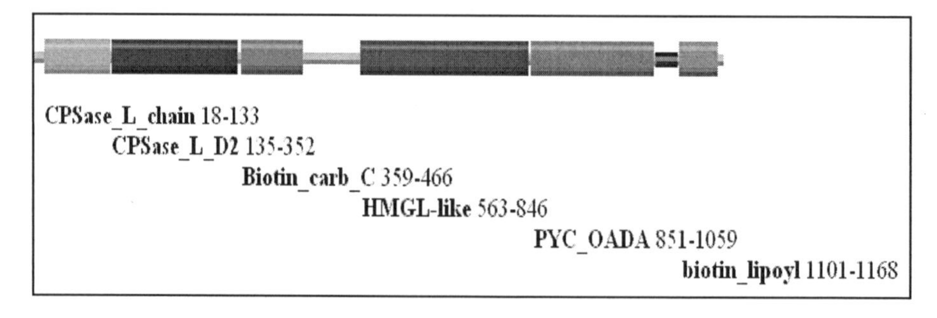

Figure 3.4. Pfam representation of the conserved domains in yeast pyruvate carboxylase PYC1_YEAST (P11154).

uses a completely different algorithm (RPS-BLAST, see ♦4.3.3). As a result, one should not be surprised by differences in the Pfam search outputs with the same query sequence at those three web sites.

SMART

Like Pfam, Simple Modular Architecture Research Tool (SMART, http://smart.embl-heidelberg.de and http://smart.ox.ac.uk), developed by Peer Bork's group at EMBL and Chris Ponting at the University of Oxford, consists of multiple domain alignments and the accompanying HMMs that are used to search the database [507,757]. Although a much smaller database than Pfam, SMART concentrates on most common domains, particularly those involved in various forms of signal transduction. SMART alignments have been curated with greatest care and attempt has been made to include even the most divergent representatives of each domain. This makes SMART a highly reliable and sensitive tool for domain identification.

SMART also includes an excellent graphical tool which, in addition to displaying all the SMART domains found in a given protein, shows predicted signal peptides, transmembrane segments and regions of low complexity identified during the SMART search (see Chapter 4 for the discussion of methods used for the identification of these features in proteins). The June 2002 release of SMART contained alignments of 639 domains. The power of extensive HMM searches, performed by the SMART team, becomes clear from the following example. The PROSITE profile for ankyrin repeats (see above) is said to correctly recognize all 134 occurrences of this repeat in the SWISS-PROT database. In contrast, SMART reports as many as 4489 statistically significant occurrences of this repeat in 1158 proteins in the non-redundant database, including 108 ankyrin repeats in *C. elegans* alone. Data from SMART have been incorporated into the EBI's InterPro database (http://www.ebi.ac.uk/interpro) and NCBI's CDD database (http://www.ncbi.nlm.nih.gov/Structure/cdd/cdd.shtml, see ♦3.2.3) and can be searched at their web sites, which, however, cannot rival the superior graphics capabilities of the original SMART web site (Figure 3.8).

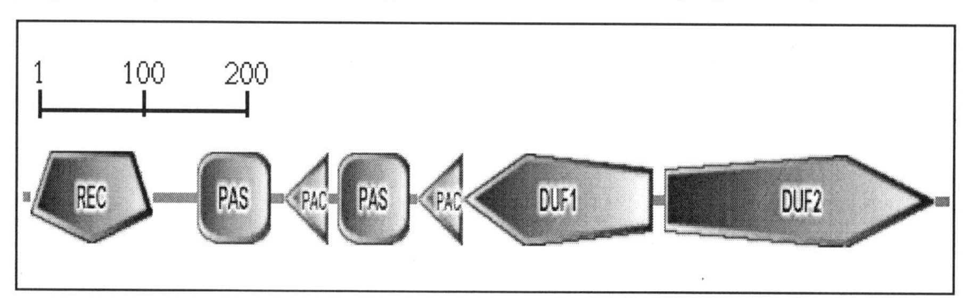

Figure 3.5. SMART representation of the domain organization of the predicted activator of nitrogen fixation NifL from *Synechocystis* sp.

ProDom

In contrast to Pfam and SMART, which are manually curated, the ProDom database (http://www.toulouse.inra.fr/prodom.html), developed by Jérôme Gouzy, Florence Corpet, and Daniel Kahn in Toulouse, France, is created largely automatically, based on the results of PSI-BLAST searches of SWISS-PROT and TrEMBL databases [158,159]. As an automatic compilation of homologous domains, ProDom relies on fairly high threshold values for domain assignments. As a result, homologous sequences may end up being assigned to different domain families (see [260] for an example). Nevertheless, thanks to its colorful images, ProDom offers an easy and convenient way to visualize domain organization of proteins. Importantly, ProDom allows one to display all the proteins that share at least one domain with the given protein. This useful option is included in SWISS-PROT, which links its entries to ProDom. ProDom is also extensively interlinked with Pfam and provides a good graphical option for viewing Pfam alignments. ProDom data have been incorporated into InterPro (http://www.ebi.ac.uk/interpro, ♦3.3.3) and are available through its unified interface.

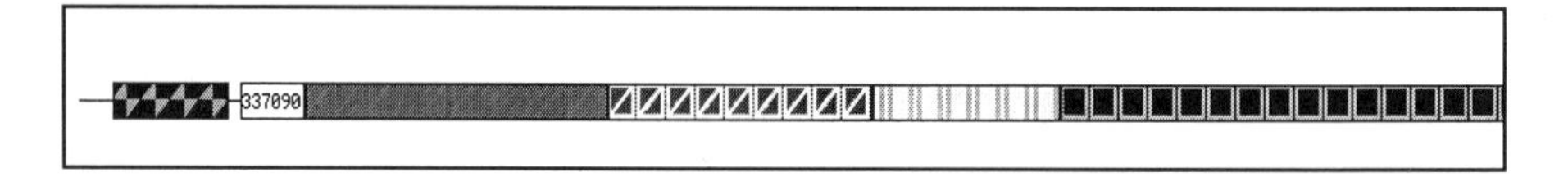

Figure 3.6. ProDom representation of the conserved domains in yeast pyruvate carboxylase PYC1_YEAST (only the first 900 aa are shown, compare to Figure 3.4).

COGs

Although the Clusters of Orthologous Groups of proteins (COG) database, maintained at the NCBI (http://www.ncbi.nlm.nih.gov/COG, [827]) is primarily a "genome-oriented" database and is described in more detail later in this chapter (♦3.5), we also mention it here because a comparison of orthologous proteins from phylogenetically distant organisms provides a powerful way to identify sequence motifs that are common to those proteins. This makes the COG database a convenient tool for motif and domain search, particularly because the annotation of many COGs is itself based primarily on their conserved motifs. Comparing a protein sequence against the proteins included in the COG database using the COGnitor program (http://www.ncbi.nlm.nih.gov/COG/xognitor.html) often allows one to identify conserved sequence motifs that are hard to recognize by other means.

3.2.3. Integrated motif and domain databases

The rapid growth of the domain-based databases, such as Pfam, SMART, ProDom, and others, made them a valuable resource for sequence similarity searches, conveniently supplementing EBI's SP-TrEMBL and the NCBI's non-redundant protein database (see also Chapters 4 and 5). In an effort to incorporate domain databases into their web sites, EBI and NCBI have created their own integrated domain databases, InterPro and Conserved Domain Database (CDD), respectively.

InterPro

Integrated Resource of Protein Families, Domains and Sites (InterPro, http://www.ebi.ac.uk/interpro) is an EBI database unifying protein sequences from SWISS-PROT with the data on functional sites and domains from PROSITE, PRINTS, ProDom, Pfam, and SMART databases [34]. InterPro entries are assigned their unique accession numbers and include functional descriptions and literature references. Each InterPro entry lists its matches in SWISS-PROT and TrEMBL. The family, domain and functional site definitions of InterPro are expected to greatly simplify the automated annotation of TrEMBL by increasing both its efficiency and reliability. Notably, InterPro was used as the principal protein annotation resource during the analysis of the draft sequence of the human genome [488].

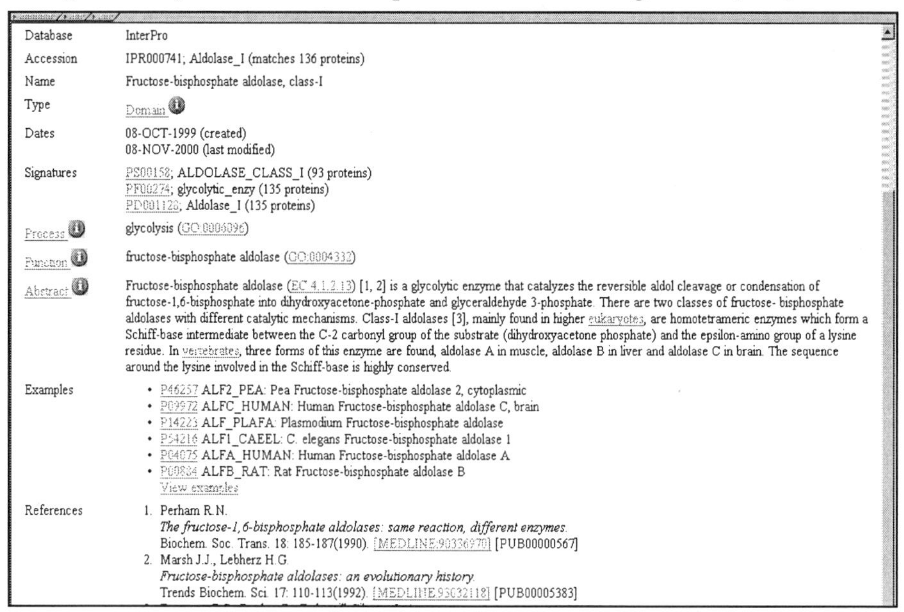

Database	InterPro
Accession	IPR000741; Aldolase_I (matches 136 proteins)
Name	Fructose-bisphosphate aldolase, class-I
Type	Domain
Dates	08-OCT-1999 (created) 08-NOV-2000 (last modified)
Signatures	PS00158; ALDOLASE_CLASS_I (93 proteins) PF00274; glycolytic_enzy (135 proteins) PD001128; Aldolase_I (135 proteins)
Process	glycolysis (GO:0006096)
Function	fructose-bisphosphate aldolase (GO:0004332)
Abstract	Fructose-bisphosphate aldolase (EC 4.1.2.13) [1, 2] is a glycolytic enzyme that catalyzes the reversible aldol cleavage or condensation of fructose-1,6-bisphosphate into dihydroxyacetone-phosphate and glyceraldehyde 3-phosphate. There are two classes of fructose- bisphosphate aldolases with different catalytic mechanisms. Class-I aldolases [3], mainly found in higher eukaryotes, are homotetrameric enzymes which form a Schiff-base intermediate between the C-2 carbonyl group of the substrate (dihydroxyacetone phosphate) and the epsilon-amino group of a lysine residue. In vertebrates, three forms of this enzyme are found, aldolase A in muscle, aldolase B in liver and aldolase C in brain. The sequence around the lysine involved in the Schiff-base is highly conserved.
Examples	• P46257 ALF2_PEA: Pea Fructose-bisphosphate aldolase 2, cytoplasmic • P09972 ALFC_HUMAN: Human Fructose-bisphosphate aldolase C, brain • P14223 ALF_PLAFA: Plasmodium Fructose-bisphosphate aldolase • P54216 ALF1_CAEEL: C. elegans Fructose-bisphosphate aldolase 1 • P04075 ALFA_HUMAN: Human Fructose-bisphosphate aldolase A • P00884 ALFB_RAT: Rat Fructose-bisphosphate aldolase B View examples
References	1. Perham R.N. *The fructose-1,6-bisphosphate aldolases: same reaction, different enzymes.* Biochem. Soc. Trans. 18: 185-187(1990). [MEDLINE:90336970] [PUB00000567] 2. Marsh J.J., Lebherz H.G. *Fructose-bisphosphate aldolases: an evolutionary history.* Trends Biochem. Sci. 17: 110-113(1992). [MEDLINE:93032118] [PUB00005383]

Figure 3.7. The InterPro entry for class I fructose-1,6-bisphosphate aldolase. Links to the PROSITE entry PS00158, Pfam entry PF00274, and ProDom entry PD001128 at the top of the page and to the PROSITE document PDOC00143 and to the Blocks entry IPB000741at the bottom of the page are underlined.

CDD

The NCBI's Conserved Domain Database and Search Service (CDD, http://www.ncbi.nlm.nih.gov/Structure/cdd/cdd.shtml, [545]) is a collection of multiple alignments of protein domains from Pfam and SMART databases, supplemented with some alignments created by NCBI's own researchers. The alignments downloaded from Pfam and SMART are trimmed to leave only those positions that are represented in at least 50% of all aligned sequences, which determines the length of the consensus and the size of the corresponding PSSM. A compilation of these PSSMs can be used as a database for rapid sequence similarity search (CD-Search, Figure 3.8) using reverse position-specific BLAST (RPS-BLAST, see ♦4.3.3).

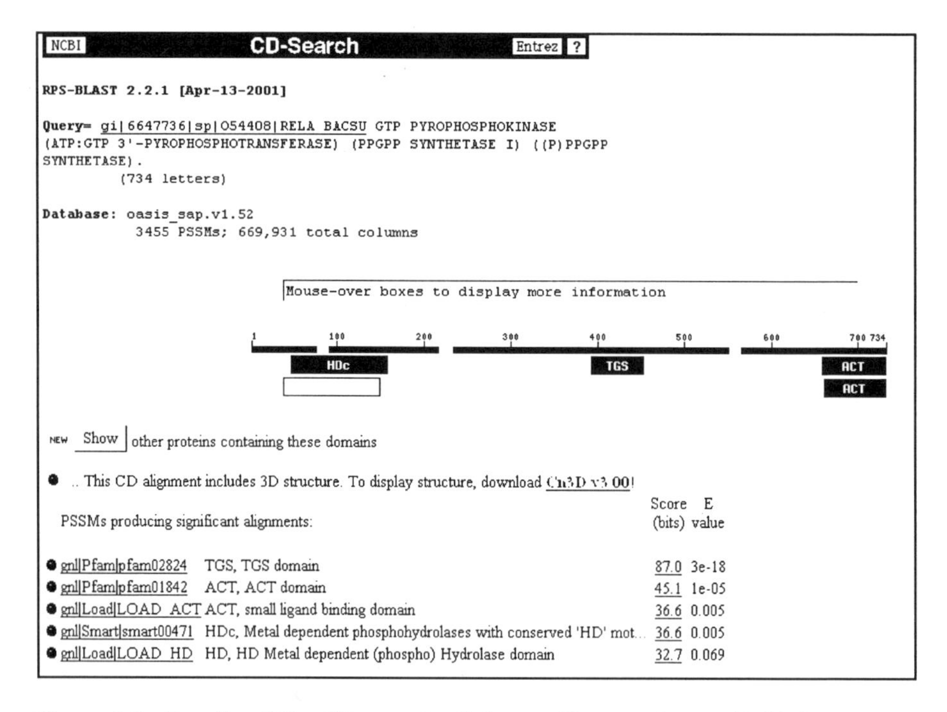

Figure 3.8. Results of the CD-search of the pppGpp synthetase RelA from *Bacillus subtilis* (O54408, RELA_BACSU). The boxes indicate RPS-BLAST hits to the N-terminal metal-dependent phosphohydrolase HD superfamily [40] from SMART (top) and from LOAD (bottom), to the centrally located TGS domain [909] from Pfam, and to the C-terminal ACT domain [43] from Pfam (top) and from LOAD (bottom). Links to the corresponding domain alignments are underlined.

3.3. Protein Structure Databases

Three-dimensional (3D) protein structures are much harder to determine than primary sequences, but they are, at least in some respects, more informative. Knowledge of atomic coordinates leads to elucidation of the active site architecture, packing of secondary structural elements, patterns of surface exposure of side-chains and relative positions of individual domains. Structural information is available only for a limited number of proteins, comprising ca. 600 distinct protein folds.

PDB

Protein Data Bank (PDB) is a public repository of 3D structures of proteins and nucleic acids. Until recently, PDB was housed by Brookhaven National Laboratory; it is now maintained by Research Collaboratory for Structural Bioinformatics (RCSB), which unites groups at the San Diego Supercomputer Center (http://www.rcsb.org/pdb), Rutgers University (http://rutgers.rcsb.org/pdb/), and the National Institute of Standards and Technology (http://nist.rcsb.org/pdb/). PDB is mirrored around the world, including fully supported mirror sites in UK, Singapore, Japan, and Brazil.

Just as every nucleotide sequence has to be deposited in GenBank prior to publication, atomic coordinates of all proteins and nucleic acids whose structure has been solved have to be deposited in PDB. However, processing of the submissions in PDB differs from that in GenBank in several important aspects. First, nucleotide (and protein) sequences submitted to GenBank are released to the public immediately after the publication of the paper that describes these sequences, if not earlier. The structures submitted to the PDB may remain "on hold" for up to a year after the publication. This delay has been instituted to allow successful processing of patent applications spawned by the determination of 3D structures of important drugs and drug targets. This policy is under review, as many researchers argue for release upon publication, which is the standard in sequence databases. In any case, the list of structures awaiting release is available at the PDB web site (http://www.rcsb.org/pdb/status.html). Those willing to test their skills in predicting the 3D structures can download protein sequences whose 3D structures have been determined and submitted to PDB, but are still kept on hold, and subsequently compare the predictions with released structures.

Just as GenBank automatically checks newly deposited nucleotide sequence to ensure that it indeed encodes the protein it is claimed to encode, has a correct taxonomic assignment and contains all the required fields, the PDB submission process (ADIT) includes a number of tests that automatically validate certain parameters, such as bond distances, torsion angles, names of heteroatoms, etc. This validation procedure helps to ensure the quality of the newly submitted structures. Finally, unlike Entrez Proteins,

PDB does not index structures by the degree of their similarity. This task is performed by other databases, such as MMDB, FSSP, SCOP, or CATH, each of which relies on its own approach to protein structure comparison.

MMDB

The Molecular Modeling Database (MMDB), maintained by the NCBI Structure group (http://www.ncbi.nlm.nih.gov/Structure), is tightly linked to Entrez Proteins and offers the same convenient links to similar protein sequences, Taxonomy and PubMed databases. In addition, for each given entry, it allows the user to access the list of structural neighbors, calculated using the VAST algorithm, developed by Steven Bryant and colleagues at the NCBI [283,535]. VAST (Vector Alignment Search Tool) searches for topologically similar fragments (α-helices, β-strands) in proteins, which is useful in comparing distantly related proteins that have no detectable sequence similarity. The output of a VAST search can be ranked by percent identity between the aligned sequences, the length of aligned region, the RMSD (root mean square distance) between the superimposed elements, or VAST scores and probability values. VAST allows the user several ways to view the structural alignment of the selected proteins and generate structure-based sequence alignments, which can be especially useful for the identification and analysis of distantly related proteins (see ♦3.1).

FSSP

The Fold classification based on Structure-Structure alignment of Proteins database (FSSP, http://www.ebi.ac.uk/dali/fssp), created by Liisa Holm and Chris Sander at the EBI, is produced by all-against-all structural comparisons of proteins with known three-dimensional structures using the DALI program [353,354]. DALI aligns protein structures based on the minimal RMSD of the carbon atoms in the main polypeptide chain (Cα atoms). For convenience, proteins with closely related structures are clustered together, and only structures representing substantially different proteins are compared and listed in the database. FSSP allows one to search for structural neighbors of each representative structure or to see the list of all indexed structures, rendered in a hierarchical format. Like MMDB, FSSP is convenient for generating structure-based sequence alignments of distantly related proteins. However, because these two databases use radically different approaches to the structural comparisons of proteins, they can sometimes complement each other. In cases of low structural similarity, it may be useful to compare the lists of neighbors of the given structure generated by both algorithms and examine the common hits and discrepancies between the two.

SCOP

In contrast to MMDB and FSSP, which both use automated procedures to generate their lists of structural neighbors, the Structural Classification Of Proteins (SCOP) is a manually curated database of protein structures, developed at the MRC Laboratory of Molecular Biology in Cambridge, England [522,590]. SCOP (http://scop.mrc-lmb.cam.ac.uk/scop, mirrored at http://scop.berkeley.edu) is a fully hierarchical database that classifies all protein structures into families of related proteins, structural superfamilies, folds, and structural classes (see also ♦8.1). All known structures are divided into eight classes, namely, all alpha proteins, all beta proteins, alpha and beta proteins with beta-alpha-beta units (α/β), alpha and beta proteins ($\alpha+\beta$) with segregated alpha and beta regions, multi-domain proteins (alpha and beta), membrane and cell surface proteins, small proteins, and coiled-coil proteins. Each class contains folds, which are further divided into superfamilies.

Manual curation of the protein taxonomy in SCOP supplements automatic structural comparisons with case-by-case analysis that take into account results of sequence comparisons, conserved sequence motifs, functional data and other information. As a result, SCOP assignments are often used as the ultimate authority on the structural similarity and evolutionary relatedness of proteins. Indeed, many SCOP superfamilies include proteins that, in addition to structural similarity, share other common features, such as similar substrate-binding sites, common enzymatic mechanisms, and so on (Table 3.2, see [468,846]). Typically, all proteins assigned by SCOP to the same fold show enough structural similarity to be considered homologous, although this may be questioned for some common folds (see ♦2.1.2).

To simplify assignment of new protein structures to folds and superfamilies, SCOP now offers a possibility to compare a protein sequence against the database, which allows one to determine its nearest relative with known 3D structure. In cases of sufficient sequence similarity, such comparison may yield important structural information.

Table 3.2. Common features of the proteins from several structural superfamilies

Superfamily	PDB codes	Member enzymes	Common properties of all proteins of the superfamily	Ref.
Acid phosphatase	1vnc 1eoi 1qi9	Phosphatidic acid phosphatase, diacylglycerol pyrophosphate phosphatase, bromoperoxidase, chloroperoxidase, glucose-6-phosphatase	Common structural core, common $Kx_6RPx_{12\text{-}54}PSGHx_{31\text{-}54}SRx_5Hx_3D$ sequence motif, common catalytic mechanism	[381, 566, 612, 810]
Alkaline phosphatase	1alk 1fsu 1ejj	Alkaline phosphatase, phosphoglycerate mutase, phosphopentomutase, streptomycin-6-phosphatase, phosphonoacetate hydrolase, phosphoglycerol transferase, nucleotide pyrophosphatase, various sulfatases	Common structural core, common metal-binding residues, common catalytic mechanism that includes phosphorylation (sulfatation) of the active-site Ser/Thr/Cys residue	[98, 258, 261, 529]
Amido-hydrolase	1add 1pta 1fkx 1ubp	AMP deaminase, adenine deaminase, cytosine deaminase, hydantoinase, dihydroorotase, allantoinase, aminoacylase, chlorohydrolase, imidazolonepropionase, phosphotriesterase, urease, formylmethanofuran dehydrogenase	Common structural core, common metal-binding His and Asp residues, common catalytic mechanism	[355]
ATP-grasp	1gsh 2dln 1bnc 1jdb 1scu	Phosphoribosylamine-glycine ligase, phosphoribosylglycinamide formyltransferase, phosphoribosylaminoimidazole carboxylase, tubulin-tyrosine ligase, protein S6-glutamate ligase, malate thiokinase, ATP-citrate lyase	Common structural core, common ATP-binding residues, common catalytic mechanism that includes formation of a phosphoacyl intermediate	[59, 262, 588, 841]

Aldolase class I	1nal 1dhp 1ald 1eua 1onr 1jcj 1b4k 1gg0	*N*-acetylneuraminate lyase, dihydrodipicolinate synthase, fructose 1,6-bisphosphate aldolase, 2-keto-3-deoxy-6-phosphogluconate aldolase, dehydroquinate dehydratase, transaldolase, deoxyribose-phosphate aldolase, 5-amino-levulinic acid dehydratase, 3-deoxy-D-arabino-heptulosonate-7-phosphate synthase, 3-deoxy-D-manno-octulosonate 8-phosphate synthase	Common structural core, conserved Lys residues in the active sites, common catalytic mechanism that includes formation of Schiff-base intermediate	[17,339, 396,499]
Double-stranded beta helix (cupin)	1fi2 1pmi 1dzr 1eyb 2phl 2cau 1fxz	Oxalate oxidase, phosphomannose isomerase, homogentisate 1, 2-dioxygenase, 1-hydroxy-2-naphthoate dioxygenase, cysteine dioxygenase, 3-hydroxyanthranilate 3,4-dioxygenase, dTDP-4-dehydrorhamnose 3,5-epimerase, ectoine synthetase, transcriptional regulators AraC, CelD, and RhaS, nuclear regulatory proteins CENP-C and pirin, seed storage proteins phaseolin, canavalin, vicilin, and glutelin	Common structural core, partly conserved metal-binding residues	[200,285, 437]
Enolase	1one 2mnr 1muc 1bqg 1kcz 1fhv	Enolase, mandelate racemase, muconate lactonizing enzyme, glucarate dehydratase, o-succinylbenzoate synthase L-alanine-DL-glutamate epimerase, methylaspartate ammonia-lyase, galactonate dehydratase, starvation sensing protein RspA	Common structural core, common metal-binding residues, common catalytic mechanism that includes metal-assisted proton abstraction from the α-carbon atom, forming an enolic intermediate	[64,323, 490,511, 608,751]

CATH

The CATH database (http://www.biochem.ucl.ac.uk/bsm/cath_new), created by Janet Thornton and colleagues at the University College, London [633,661], also is a hierarchical classification of protein domain structures. CATH clusters proteins at four major levels, Class (C), Architecture (A), Topology (T) and Homologous superfamily (H). CATH classification also includes manual curation of each group, based on the secondary structure content (class), orientation of secondary structure elements (architecture), topological connections between them (topology), and, finally, sequence and structure comparisons (homologous superfamily level).

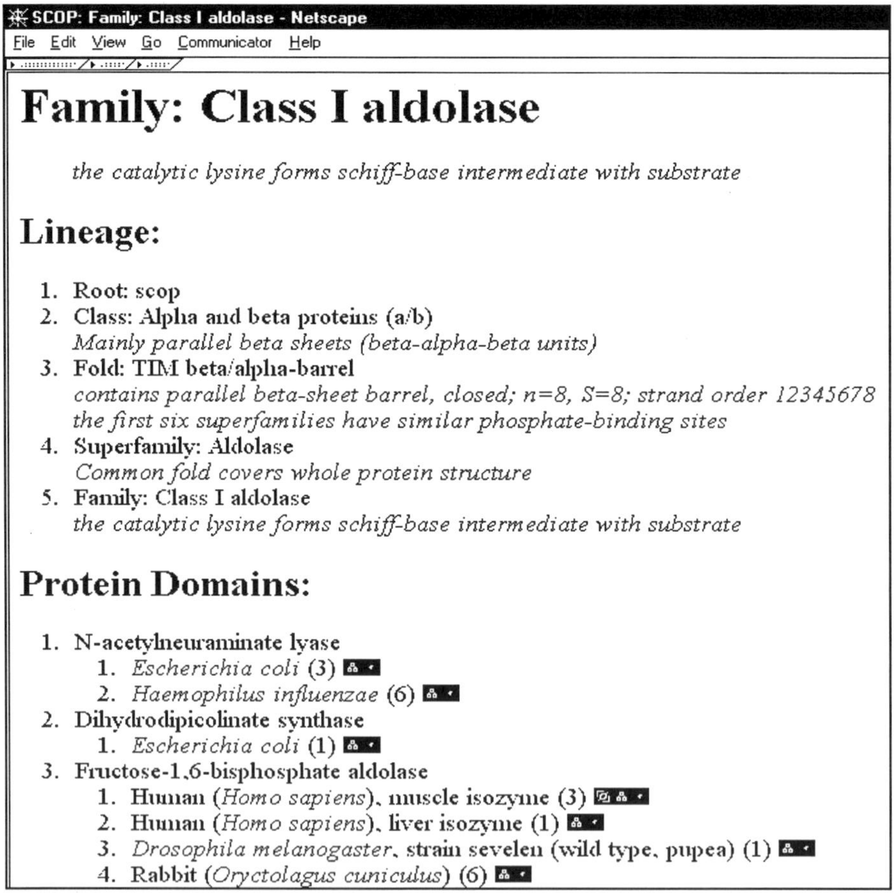

SCOP: Family: Class I aldolase - Netscape

File Edit View Go Communicator Help

Family: Class I aldolase

the catalytic lysine forms schiff-base intermediate with substrate

Lineage:

1. Root: scop
2. Class: Alpha and beta proteins (a/b)
   Mainly parallel beta sheets (beta-alpha-beta units)
3. Fold: TIM beta/alpha-barrel
   contains parallel beta-sheet barrel, closed; n=8, S=8; strand order 12345678
   the first six superfamilies have similar phosphate-binding sites
4. Superfamily: Aldolase
   Common fold covers whole protein structure
5. Family: Class I aldolase
   the catalytic lysine forms schiff-base intermediate with substrate

Protein Domains:

1. N-acetylneuraminate lyase
   1. *Escherichia coli* (3)
   2. *Haemophilus influenzae* (6)
2. Dihydrodipicolinate synthase
   1. *Escherichia coli* (1)
3. Fructose-1,6-bisphosphate aldolase
   1. Human (*Homo sapiens*), muscle isozyme (3)
   2. Human (*Homo sapiens*), liver isozyme (1)
   3. *Drosophila melanogaster*, strain sevelen (wild type, pupea) (1)
   4. Rabbit (*Oryctolagus cuniculus*) (6)

Figure 3.9. Some representatives of class I (metal-independent) aldolase family in the SCOP database. This SCOP family unites enzymes with rather low sequence conservation (according to FSSP alignments, dihydrodipicolinate synthase and fructose-1,6-bisphosphate aldolase have only 10% identical residues), which have, however, a common catalytic mechanism that includes substrate binding to a conserved Lys residue resulting in a formation of Schiff-base intermediate. This serves as supplemental evidence for inferring homology of these enzymes.

3.4. Specialized Genomics Databases

Since the World Wide Web makes genome sequences available to anyone with an Internet access, there are a variety of databases that offer more or less convenient access to essentially the same sequence data. However, there are several convenient web sites that provide useful additional information, such as phylogenetic relationships, operon organization, functional predictions, 3D structure, or metabolic reconstructions.

Entrez Genomes

Since complete genome sequences are, in fact, nothing more than extremely long nucleotide sequences, one can always retrieve them from the NCBI FTP site (ftp://ncbi.nlm.nih.gov/Entrez/Genomes) or the FTP sites of the appropriate sequencing center. The other two public databases, DDBJ and EMBL, maintain their own genome retrieval systems, referred to as EBI Genomes (http://www.ebi.ac.uk/genomes) and Genome Information Broker (http://gib.genes.nig.ac.jp). It is important to emphasize that, as with all records from archival databases, these genomes represent original submissions and are not immune to the errors mentioned earlier in this chapter. Sometimes, submitters update genome sequences and/or their annotations. This was done, for example, for *E. coli, H. influenzae*, *M. genitalium*, and *M. pneumoniae* genomes. For most genomes, however, the sequence and its annotation remain unchanged. In order to provide updated (and unified) versions of complete genomes, NCBI has recently initiated the Reference Sequences project (http://www.ncbi.nlm.nih.gov/LocusLink/refseq.html, RefSeq) that links the lists of gene products with some valuable sequence analysis information, such as predicted functions for uncharacterized gene products, frameshifted proteins, and so on (Figure 3.10).

Because NCBI also maintains a special BLAST page for searching unfinished genome sequences, contributed by various genome-sequencing centers (http://www.ncbi.nlm.nih.gov/Microb_blast/unfinishedgenome.html), many researchers are confused about the status of all these different databases. One needs to clearly understand the distinction between the three kinds of data maintained at the NCBI. Complete genome entries in GenBank are kept exactly as submitted and can be changed only by submitters themselves. Genome entries in the Genomes division of Entrez, which can be identified by their NC_xxxxxx RefSeq accession numbers, comprise GenBank entries that have been curated by the NCBI staff. These entries are supplemented by various tables that present precomputed data on the taxonomic distribution of the best hits in the database for each protein in the given genome (see Chapter 6), COG assignments of these proteins (see the next page), neighbors with known three-dimensional structures, and results of sequence comparison against the CDD (see ♦3.3).

Finally, unfinished genome data, submitted by various genome sequencing centers (Appendix 2), are available only for BLAST searches. They are protected from unauthorized access just like any GenBank submission, held until publication. The authors of this book, for example, have no more access to those data than anybody else in the world. For each BLAST hit, the user can get the corresponding DNA sequence and up to 1 kb of flanking sequence on each side. This allows the users to effectively search unfinished genome sequences for any protein (gene) of interest, at the same time preventing them from any large-scale genome analysis.

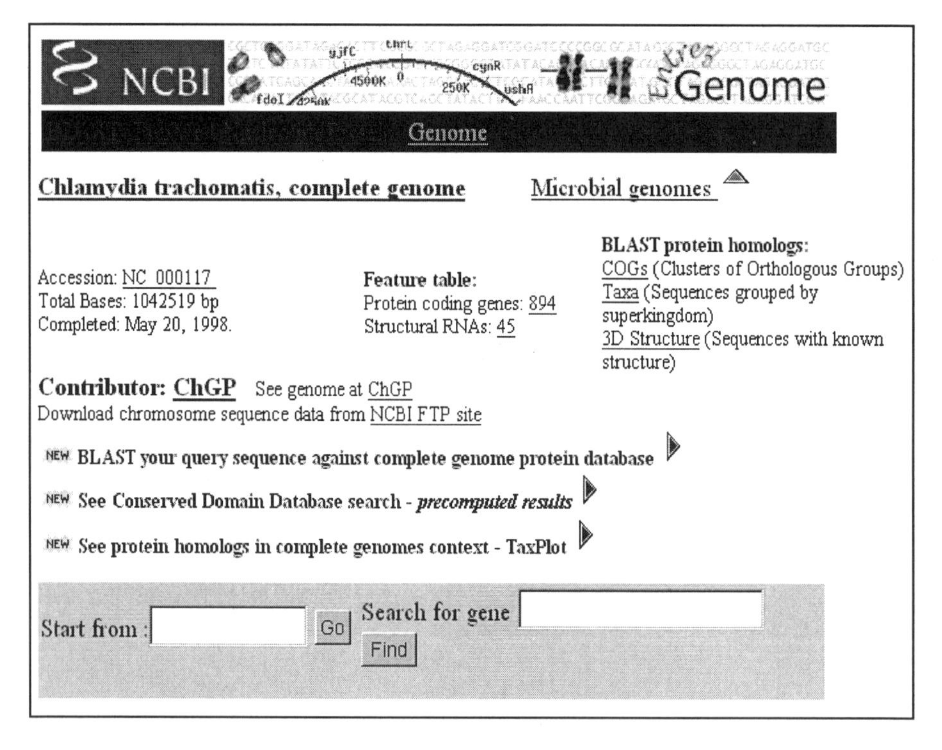

Figure 3.10. The front page of NCBI Entrez Genomes web site for *Chlamydia trachomatis* genome. The sequence data can be retrieved by following the link to the GenBank entry of the complete genome, to *Chlamydia* genomes project at UC Berkeley, or to the NCBI FTP site. One can also view the positions of RNA- and protein-coding genes or search for genes based on their position or gene name. Additional links show the results of an RPS-BLAST comparison of *C. trachomatis* proteins against the CDD; the taxonomic affinity of the best hits; close homologs with known 3D structure; and gene location and functions of *C. trachomatis* proteins represented in the COG database. Finally, one can compare any protein sequence against the set of *C. trachomatis* proteins.

COGs

The Clusters of Orthologous Groups of proteins (COG) database ([827,828], http://www.ncbi.nlm.nih.gov/COG), already mentioned in the preceding section, has been designed to facilitate comparative genomic analysis and improve functional assignments of individual proteins. The latest COG release consists of 4,620 clusters of inferred orthologs from the completely sequenced genomes of bacteria, archaea, and unicellular eukaryotes. Each COG contains sets of proteins from at least three phylogenetic lineages.

Very briefly, the COG construction procedure included the following main steps: (i) all-against-all protein sequence comparison using BLAST, (ii) detection and clustering of obvious paralogs, i.e. proteins from the same genome that are more similar to each other than to any proteins from other species, (iii) detection of triangles of mutually consistent, genome-specific best hits, taking into account the paralogous groups detected at step (ii), (iv) merging triangles with a common side to form COGs, and (v) a case-by-case analysis of each COG to eliminate potential false-positives. Since orthologs typically perform the same function, delineation of orthologous families from diverse species allows the transfer of functional annotation from better-studied organisms to less-studied ones. The COGs are classified into 18 functional groups, which include uncharacterized conserved proteins and

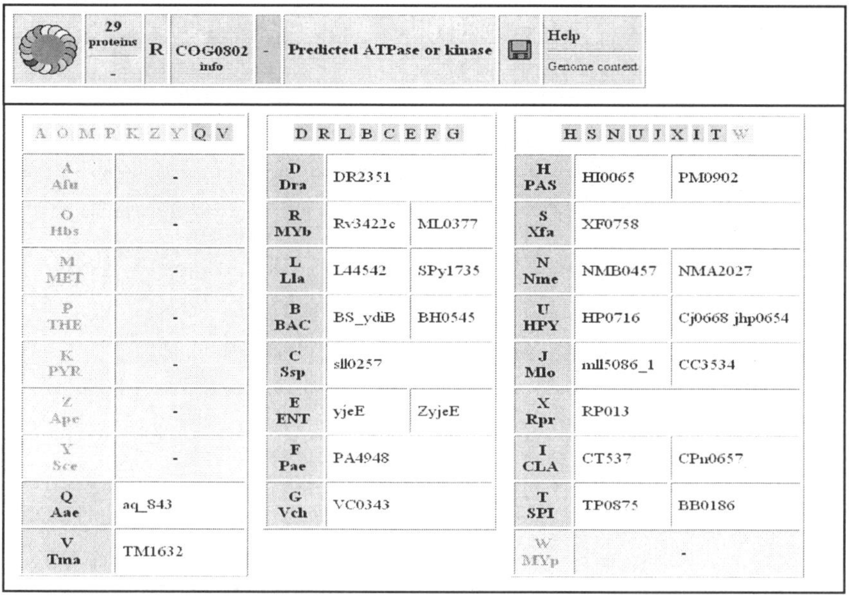

29 proteins -	R	COG0802 info	-	Predicted ATPase or kinase	Help / Genome context

A O M P K Z Y Q V	
A Afu	-
O Hbs	-
M MET	-
P THE	-
K PYR	-
Z Ape	-
Y Sce	-
Q Aae	aq_843
V Tma	TM1632

D R L B C E F G		
D Dra	DR2351	
R MYb	Rv3422c	ML0377
L Lla	L44542	SPy1735
B BAC	BS_ydiB	BH0545
C Ssp	sll0257	
E ENT	yjeE	ZyjeE
F Pae	PA4948	
G Vch	VC0343	

H S N U J X I T W		
H PAS	HI0065	PM0902
S Xfa	XF0758	
N Nme	NMB0457	NMA2027
U HPY	HP0716	Cj0668 jhp0654
J Mlo	mll5086_1	CC3534
X Rpr	RP013	
I CLA	CT537	CPn0657
T SPI	TP0875	BB0186
W MYp	-	

Figure 3.11. GOG database page for COG0802, for which only a general functional prediction was available. Note that this COG is represented in each sequenced bacterial genome, except for three mycoplasmas (bottom right cell of the table), but absent in archaea and yeast (top left corner of the table). The figure is from the September 2001 release of the COGs.

proteins for which only a general functional assignment (typically, prediction of biochemical activity but not the actual biological function) appeared appropriate (Figure 3.11).

The COG database is particularly useful for functional predictions in borderline cases where the protein sequence similarity is relatively low. Due to the diversity of proteins in COGs, sequence similarity searches against the COG database (available at http://www.ncbi.nlm.nih.gov/COG/xognitor.html or from ORF finder, http://www.ncbi.nlm.nih.gov/gorf, see ♦3.2) can sometimes suggest a possible function for a protein that otherwise has no clear database hits. This database also offers convenient tools for comparative analysis of complete genomes, particularly phyletic pattern analysis that we use widely throughout this book (see ♦4.2).

KEGG

The Kyoto Encyclopedia of Genes and Genomes (KEGG, http://www.genome.ad.jp/kegg) is a part of the GenomeNet web site, created by Minoru Kanehisa and colleagues at the Kyoto University Institute for Chemical Research for comprehensive analysis of complete genomes. KEGG aims at using genome sequences for complete reconstruction of cellular metabolism and its regulation [414]. The KEGG web site presents a comprehensive set of metabolic pathway charts, both general and specific for each of the sequenced genomes. The enzymes that have already been identified in a particular organism are color-coded, so that one can easily trace the pathways that are likely to be present or absent (Figure 3.12).
For each of the metabolic pathways that it covers, KEGG also provides the lists of orthologous genes from all sequenced genomes that code for the enzymes participating in those pathways. It is also indicated whenever these

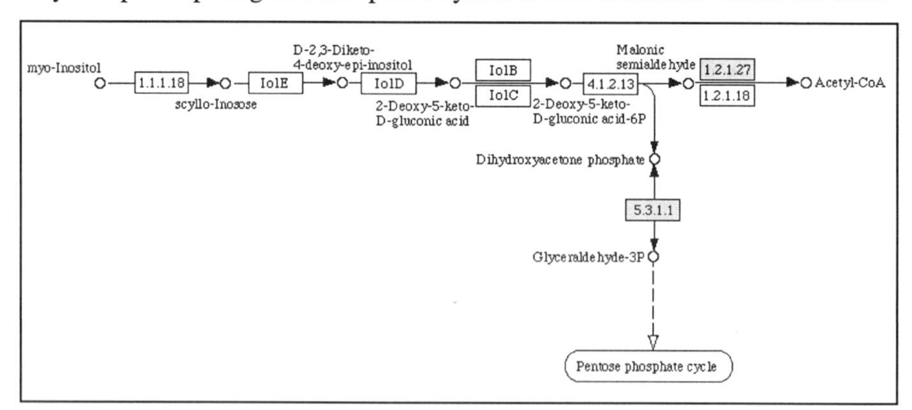

Figure 3.12. The KEGG pathway for inositol metabolism. Each rectangle signifies an enzyme, identified by the EC number, if available, or the gene name. Each circle signifies a chemical substance (an intermediate of the pathway). Selecting the "Homo sapiens" option showed that only two enzymes of the pathway (indicated by two darkened rectangles) have been identified in humans.

genes are adjacent in the genome and form likely operons. A convenient search tool allows the user to compare two complete genomes and identify all cases where conserved genes in both organisms are adjacent or located close (within five genes) to each other. The KEGG site is continuously updated and is currently the best source of data for the analysis of metabolism in various organisms.

WIT/ERGO

The WIT (What Is There) database was originally developed by Ross Overbeek and Evgeni Selkov at the Argonne National Laboratory in Argonne, IL [642]. It is currently maintained in two different variants, as the public WIT database at the Argonne web site (http://wit.mcs.anl.gov) and as ERGO (http://ergo.integratedgenomics.com/ERGO) at the Integrated Genomics' web site, which is largely closed to the public. Like KEGG, this system combines diverse tools that assist in functional annotation. WIT/ERGO is best known for its operon search tool [640,641], see ♦4.6.2. Like COGs, WIT/ERGO delineates clusters of orthologous proteins and uses these clusters to assign functions to the uncharacterized members of each cluster. In contrast to other databases discussed in this section, which perform analysis of complete genomes only, WIT/ERGO also includes proteins from many partially sequenced genomes. This allows this system to offer many more sequences of the same protein from different organisms than any other database, which facilitates detection of additional members of the respective protein families and increases the utility of operon analysis. An interesting feature of WIT/ERGO is that it allows registered users to submit their own functional annotations and comments. Eventually, this might lead to true "community annotation" projects that would offer everybody an opportunity to participate in the process.

MBGD

The Microbial Genome Database for Comparative Analysis (http://mbgd.genome.ad.jp) at the University of Tokyo is another convenient tool for comparative analysis of completely sequenced microbial genomes. Like COGs, MBGD stores precomputed results of similarity searches between all the ORFs in the complete genomes and attempts to classify them into homology clusters. In contrast to COGs, however, MBGD assigns homology relationships based solely on BLAST searches with the arbitrary cut-off P-value of 10^{-2} (see ♦4.2). MBGD contains a hierarchical list of cellular functions, classified into 16 principal functional groups, and allows one to list the genes that are responsible for a particular function in any given genome. After selecting the gene of interest, the user can search for homologs of this gene among all other sequenced microbial genomes.

PEDANT

The Protein Extraction, Description and ANalysis Tool (PEDANT, http://pedant.gsf.de), maintained at MIPS, is a useful web resource that presents results of extensive cross-genome comparisons using a variety of popular tools [245]. The available complete genomes and a number of unfinished genome sequences are analyzed using standard PEDANT queries, such as EC numbers, PROSITE patterns, Pfam domains, BLOCKS, and SCOP domains. Because these queries comprise some of the most common questions asked in genome comparisons, PEDANT can be used as a convenient entry point into the field of comparative genome analysis. For example, if you want to find out how many proteins in *H. pylori* have known (or confidently predicted) 3D structure or how many NAD^+-dependent alcohol dehydrogenases (EC 1.1.1.1) are encoded in the *C. elegans* genome, PEDANT provides an easy way to do that (Figure 3.10). Although PEDANT does not allow the users to enter their own queries, the variety of data available at this web site makes it an important tool for comparative and functional genomics.

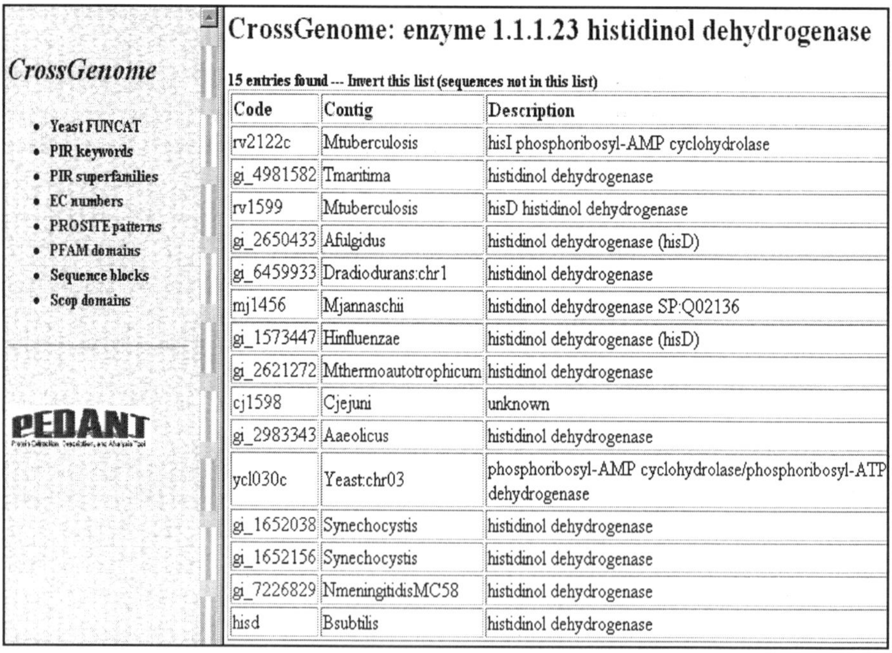

CrossGenome

- Yeast FUNCAT
- PIR keywords
- PIR superfamilies
- EC numbers
- PROSITE patterns
- PFAM domains
- Sequence blocks
- Scop domains

PEDANT

CrossGenome: enzyme 1.1.1.23 histidinol dehydrogenase

15 entries found --- Invert this list (sequences not in this list)

Code	Contig	Description
rv2122c	Mtuberculosis	hisI phosphoribosyl-AMP cyclohydrolase
gi_4981582	Tmaritima	histidinol dehydrogenase
rv1599	Mtuberculosis	hisD histidinol dehydrogenase
gi_2650433	Afulgidus	histidinol dehydrogenase (hisD)
gi_6459933	Dradiodurans:chr1	histidinol dehydrogenase
mj1456	Mjannaschii	histidinol dehydrogenase SP:Q02136
gi_1573447	Hinfluenzae	histidinol dehydrogenase (hisD)
gi_2621272	Mthermoautotrophicum	histidinol dehydrogenase
cj1598	Cjejuni	unknown
gi_2983343	Aaeolicus	histidinol dehydrogenase
ycl030c	Yeast:chr03	phosphoribosyl-AMP cyclohydrolase/phosphoribosyl-ATP dehydrogenase
gi_1652038	Synechocystis	histidinol dehydrogenase
gi_1652156	Synechocystis	histidinol dehydrogenase
gi_7226829	NmeningitidisMC58	histidinol dehydrogenase
hisd	Bsubtilis	histidinol dehydrogenase

Figure 3.13. PEDANT list of histidinol dehydrogenases (EC 1.1.1.23) encoded in completely sequenced genomes. The list on the left panel shows precomputed categories for cross-genome search in PEDANT.

TIGR Databases

The Institute for Genomic Research (http://www.tigr.org) maintains several useful databases, including the **Comprehensive Microbial Resource** (CMR, http://www.tigr.org/tigr-scripts/CMR2/CMRHomePage.spl), devoted to the analysis of bacterial and archaeal genomes [673], the TIGR Parasites Database (http://www.tigr.org/tdb/parasites) that provides links to protozoan sequencing projects under way at TIGR, and TIGR Gene Indices (http://www.tigr.org/tdb/tgi.shtml) that integrate the data from eukaryotic genome sequencing and EST projects.

The CMR combines information on all publicly available completely sequenced genomes with pre-publication data on the genomes sequenced at TIGR. It offers a variety of search and display options, including a convenient genome browser that lists, for each gene, the evidence on which the annotation is based (e.g., HMM match, BLAST match, or PROSITE match). It also allows the user to align the DNA sequences of any two microbial genomes using MUMmer [178]. The Restriction Digest Tool searches the genomic sequences for cutting sites recognized by most commonly used restriction endonucleases. The results can be displayed in a variety of formats, including a genomic map showing the cutting sites, a list of the predicted restriction fragments, DNA sequences of these fragments, and an image showing predicted positions of these fragments in an agarose gel.

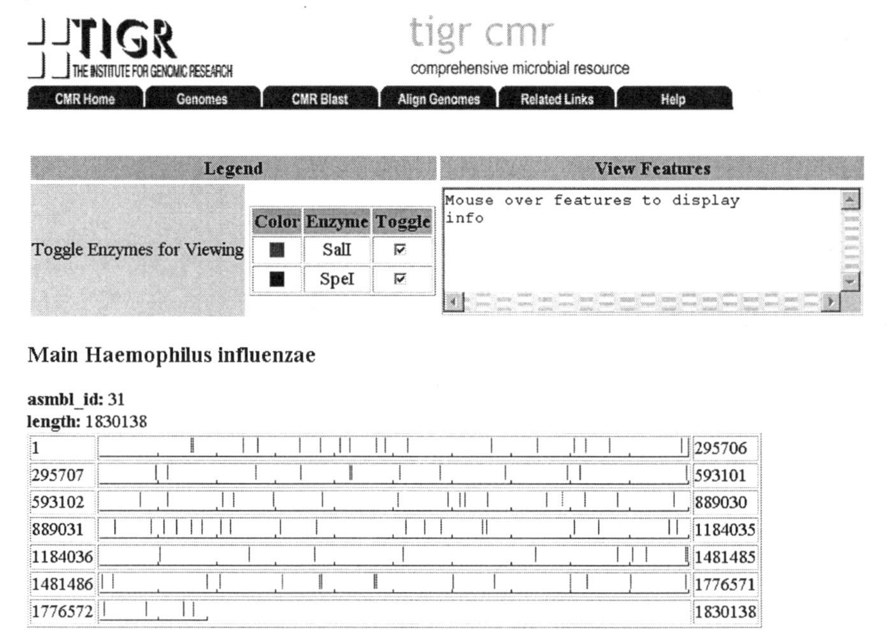

Figure 3.14. A TIGR Comprehensive Microbial Resource map of the *H. influenzae* chromosome showing the positions of SalI and SpeI restriction sites.

Other CMR options include the possibility to retrieve from various genomes genes with the same biological role (e.g. genes involved in amino acid biosynthesis), genes encoding enzymes with the same EC number, common name, and other options similar to those in PEDANT. One can also search for predicted proteins with particular properties, such as isoelectric point, molecular weight, or the number of predicted transmembrane regions.

The CMR also includes its own version of clusters of orthologs similar to COGs and TIGRFAMs, protein clusters built using the HMM searches of protein sets encoded in the complete genomes with Pfam profiles. Finally, the CMR has a well-organized list of all the transfer and ribosomal RNAs encoded in the complete microbial genomes.

TIGR Gene Indices (http://www.tigr.org/tdb/tgi.shtml) currently contain tentative consensus sequences (clustered EST sequences) from 12 protists, including *Cryptosporidium parvum*, *Dictyostelium discoideum*, *Leishmania major*, *Plasmodium falciparum*, *Toxoplasma gondii*, and *Trypanosoma brucei*, six fungi, including *Saccharomyces cerevisiae*, *Schizosaccharomyces pombe*, *Neurospora crassa*, and *Aspergillus nidulans*, 12 plants, including *Arabidopsis*, barley, maize, potato, rice, soybean, tomato, wheat, and cotton, and 13 animal species, such as *C. elegans*, *Drosophila*, human, mouse, rat, pig, and zebrafish. Groups of tentative consensus sequences from different organisms that encode homologous proteins form the **TIGR Orthologous Gene Alignment** database (TOGA, http://www.tigr.org/tdb/toga/toga.shtml). TOGA differs from COGs and other similar databases in that, instead of proteins, it clusters and aligns DNA sequences. This results in small clusters, composed of closely related, most likely indeed orthologous sequences (see ♦3.1). However, many orthologous genes end up being assigned to different TOGAs.

An important application of TOGAs is the identification of orthologs of human disease genes, i.e. genes, mutations in which cause hereditary diseases. These data are organized in a single large table http://www.tigr.org/tigr-scripts/nhgi_scripts/human_disease.pl?name=human_disease that lists genes involved in the pathogenesis of a variety of human disorders. Each human disease gene is hyperlinked to the respective OMIM entry (see ♦3.5) and is accompanied by inferred orthologs from other organisms.

3.5. Organism-specific Databases

In addition to general genomics databases, numerous databases center on a particular organism or a group of organisms. While most of these databases are useful in some respect, those devoted to model organisms, such as *E. coli, B. subtilis*, yeast, *C. elegans*, *Drosophila*, and mouse, are probably the ones most widely used for functional assignments in other, less thoroughly studied organisms. For someone involved in functional genomics, it is important to be able to quickly verify the reliability of each database entry. Thus, if one has reasons to doubt the database annotation of a particular gene or protein (see ♦3.2), it often helps to check whether a functional assignment made from studies of a particular model organism is accepted by the community of researchers studying that organism. In addition, organism-specific web sites may contain additional information that is hard to fit into the standard annotation scheme (e.g. viability of mutants, availability of clones or results of two-hybrid experiments). The following list is by no means complete or even representative; however, it covers the databases for model organisms that the authors find most useful in their own work and that are likely to similarly help other researchers.

3.5.1. Prokaryotes

Escherichia coli

The importance of *E. coli* for molecular biology is reflected in the large number of databases dedicated to this bacterium. The research groups of Fred Blattner at the University of Wisconsin-Madison (http://www.genome.wisc.edu) and Hirotada Mori at the Nara Institute of Technology (http://ecoli.aist-nara.ac.jp), which independently sequenced the *E. coli* genome, maintain useful web sites devoted to the post-genomic analysis of *E. coli* genes.

Since the Blattner group has recently completed sequencing the genome of enteropathogenic *E. coli* O157:H7 and is currently involved in genome sequencing of other enteric pathogens, such as *E. coli* K1, *Shigella flexneri*, *Salmonella typhi*, and *Yersinia pestis*, their web site is most useful as a source of data on these bacteria. It also contains a list of *E. coli* genes that have been amplified using gene-specific primer pairs and are now available to other researchers. There is also a partial list of genes shown to be essential for growth in *E. coli* (http://magpie.genome.wisc.edu/~chris/ essential.html).

The group led by Mori coordinates the Japanese **GenoBase** project (http://ecoli.aist-nara.ac.jp/docs/genobase/index.html), aimed at elucidating the functions of *E. coli* genes that currently remain uncharacterized. Their web site provides a convenient link from the genomic data to the Kohara restriction map of *E. coli* and allows one to search for the Kohara clones that cover the region of interest. The Japanese National Institute of Genetics maintains another useful database, called **Profiling of *Escherichia coli* Chromosome** (PEC, http://shigen.lab.nig.ac.jp/ecoli/pec), which contains a detailed description of each *E. coli* gene, including its location, Kohara clone that covers this gene, information on whether it is essential, results of PSI-BLAST searches of its product against the PDB, PROSITE motifs and Pfam domains present in this protein, and many other pieces of valuable information.

EcoGene (http://bmb.med.miami.edu/ecogene), a database of *E. coli* genes, created by Kenneth Rudd, currently at the University of Miami, aims at providing curated sequences of *E. coli* proteins. This is a good place to look for frameshifted and potentially mistranslated proteins. For each *E. coli* gene, EcoGene provides a short description of its function, including alternative gene names and relevant references.

A useful web site (http://web.bham.ac.uk/bcm4ght6), aptly named "**The *E. coli* index**", is maintained by Gavin Thomas at the University of Sheffield. It contains good links devoted to clinical strains of *E. coli*, but the major attraction is the list of recent functional assignments in *E. coli*. The compilation of genes that have been annotated since the completion of the genome sequence can be found in the "Completing the *E. coli* proteome" section (http://web.bham.ac.uk/bcm4ght6/genome.html), whereas the "What's new" section (http://web.bham.ac.uk/bcm4ght6/gennew.html) lists the latest experimental results.

The web site of the ***E. coli* Genetic Stock Center** at Yale University (http://cgsc.biology.yale.edu) lists all the mutant strains of *E. coli* available in its collection. It also provides gene linkage and functional information.

The GeneProtEC (http://genprotec.mbl.edu) database, created by Monica Riley at Woods Hole, the Encyclopedia of *E. coli* Genes and Metabolism (**EcoCyc**, http://www.ecocyc.org), developed by Peter Karp, and **RegulonDB** (http://www.cifn.unam.mx/Computational_Biology/regulondb), maintained by Julio Collado-Vides, are interconnected database devoted, respectively, to metabolic and regulatory pathways of *E. coli*.

Finally, the **Colibri** (http://bioweb.pasteur.fr/GenoList/Colibri) database at the Institut Pasteur is specifically designed for a molecular biologist doing experimental work on *E. coli*. It has a good web site with a convenient feature allowing the user to download the DNA sequence of a given *E. coli* gene with up to 1 kb upstream and downstream sequence. This can be useful for designing PCR primers, searching for the convenient restriction sites, delineating promoters and transcription regulator-binding sites, and many other applications. However, the Colibri web site has not been updated for a long time, because of which its functional information is likely to be not as up to date as in other *E. coli* databases.

Bacillus subtilis

B. subtilis is a popular model organism for microbiological studies. Its genome, like that of *E. coli*, is the subject of an ongoing functional annotation project. In contrast to *E. coli*, the data collection is largely centralized, with the **Subtilist** web site maintained at the Institut Pasteur (http://bioweb.pasteur.fr/GenoList/SubtiList), serving as a clearing house for all new information concerning *B. subtilis* genome. Like Colibri, Subtilist allows the user to download the DNA sequence of a given *B. subtilis* gene with flanking regions, which can be useful for experiment design.

For phenotypes of various mutants, one can use the **Micado** (a.k.a. MadBase) database (http://locus.jouy.inra.fr) at INRA, France. This site also lists 110 *B. subtilis* genes that had been previously mapped but have not been identified in the complete genome.

Mechanisms of sporulation and its regulation being some of the most actively studied properties of *B. subtilis*, there is a useful web-based index of ***B. subtilis* sporulation genes**, maintained by Simon Cutting at the Royal Holloway University of London (http://www.rhul.ac.uk/Biological-Sciences/cutting/index.html).

Synechocystis sp.

The cyanobacterium *Synechocystis* sp. PCC6803 was one of the first bacterial genomes to be sequenced [417]. The Kazusa DNA Research Institute in the Japanese Prefecture of Chiba, which carried out the *Synechocystis* genome sequencing, maintains **CyanoBase** (http://www.kazusa.or.jp/cyano), a database devoted to the post-genomic studies of cyanobacterial genes. Although most of the CyanoBase gene assignment data can be found elsewhere (in GenBank, COGs, KEGG, WIT, and many other databases), this site contains a useful list of *Synechocystis* mutants, sorted in the order of the chromosomal locations of the corresponding genes. For each mutant, the list includes whatever functional information is available and provides the address of the researcher that has constructed this mutant. This resource is expected to grow rapidly, boosted by the recent completion of the genome of the second cyanobacterium, *Anabaena* (*Nostoc*) sp. PCC7120 [416].

3.5.2. Unicellular eukaryotes

Unicellular eukaryotes are targeted by a number of ongoing genome sequencing projects (see e.g. http://www.sanger.ac.uk/Projects/Protozoa), which generate a substantial amount of sequence data. Accordingly, there exist extensive databases and numerous web sites dedicated to *Candida albicans*, *Dictyostelium discoideum*, *Entamoeba histolytica*, *Leishmania major*, *Neurospora crassa*, *Plasmodium falciparum*, *Pneumocystis carinii*, and other unicellular eukaryotes (see Appendix 2). However, only yeasts *Saccharomyces cerevisiae* and *Schizosaccharomyces pombe* have been sufficiently studied biochemically to generate a database that would be useful in annotating other genomes. For this reason, yeast databases remain the major source of data for genome annotation for most of other eukaryotes, including protozoa.

Yeast

The baker's yeast *Saccharomyces cerevisiae* was the first eukaryote whose genome had been completely sequenced [290]; it is arguably the best characterized of all eukaryotic organisms. Several databases are specifically dedicated to functional analysis of yeast genome, including three major ones, the Saccharomyces Genome Database (SGD) at Stanford University (http://genome-www.stanford.edu/Saccharomyces), the Yeast Database at MIPS (http://mips.gsf.de/proj/yeast), and Yeast Protein Database (YPD) at Proteome, Inc. (http://www.proteome.com/databases). All three resources, SGD, MIPS, and YPD, provide useful up-to-date information on the current status of the yeast genome analysis, including periodically updated lists of proteins with known or predicted functions, phenotypes of mutants (if available), protein-protein interactions, gene expression patterns, and other data. For each gene, there is a list of appropriate references that help in understanding its cellular role, even if the exact function remains unknown. Although there is a substantial overlap in the data between these three databases, it is often useful to check each of them when searching for information about a particular yeast protein. The SGD entries are interlinked with the yeast Gene Registry (http://genome-www.stanford.edu/Saccharomyces/registry.html) that keeps a complete list of all standard and non-standard names of *S. cerevisiae* genes. For the researchers who experimentally characterize yeast genes, this list includes useful links to SGD Gene Naming Guidelines and Gene Registry Form. It also has a link to Global Gene Hunter (http://genome-www.stanford.edu/cgi-bin/SGD/geneform), a simple but convenient search engine that looks for the given yeast gene in SGD, YPD, PIR, SWISS-PROT, GenBank, PubMed, and Sacch3D (yeast protein structures) databases.

The MIPS yeast database serves as a resource for new results coming from the multinational EUROFAN project [199]. YPD is a commercial site, but is free for academic users.

There are several other useful sites for yeast genome analysis. TRIPLES, TRansposon-Insertion Phenotypes, Localization, and Expression in *Saccharomyces* database (http://ygac.med.yale.edu), maintained by Michael Snyder's laboratory at Yale University, tracks the expression of transposon-induced mutants and the cellular localization of yeast proteins, tagged with the Tn3-derived minitransposon developed in the Snyder lab. This database also offers a convenient search for the phenotypes of insertion mutants, including insertions into unannotated short (< 100 codons) open reading frames.

Yeast Mitochondrial Protein Database (http://bmerc-www.bu.edu/mito) at Boston University presents a useful compilation of information regarding both proteins encoded in the mitochondrial genome and those encoded within the nuclear genome and post-translationally imported into the mitochondria.

Ron Davis' lab at Stanford University (http://genomics.stanford.edu) maintains the *Saccharomyces* Genome Deletion Project, aimed at creating and characterizing PCR-generated deletion mutants in every yeast gene. Although the complete database is currently open only to members of the consortium, the strains generated in the course of the project are available to other researchers and can be searched through the project web site. The Davis lab also maintains the *Saccharomyces* Cell Cycle Expression Database, which presents the available data on the changes in the mRNA transcript levels during the yeast cell cycle. A list of regulatory elements and transcriptional factors in yeast is kept in the Saccharomyces *cerevisiae* Promoter Database (http://cgsigma.cshl.org/jian) at Cold Spring Harbor Laboratory.

3.5.3. Multicellular eukaryotes

The Human Genome Project and related projects on complete genome sequencing of model organisms, such as nematode worm, fruit fly, pufferfish, mouse, and rat, resulted in proliferation of web sites that attempt to make use of genomic sequence data. Only a few of them, however, are concerned with sequence annotation, that is, specialize in predicting genes and evaluating their probable functions. In this section, we review only those databases that are likely to help a beginner gene hunter in finding functional assignments.

Thale cress *Arabidopsis thaliana*

Arabidopsis thaliana, the first plant whose genome has been sequenced [35], is widely used as a model organism in plant biology. **The Arabidopsis Information Resource** (TAIR, http://www.arabidopsis.org), a collaboration between the Carnegie Institution of Washington Department of Plant Biology at Stanford University and the National Center for Genome Resources, a nonprofit organization in Santa Fe, New Mexico, serves as the principal resource on the *Arabidopsis* biology [360]. The primary sources for Arabidopsis genome annotation are the **TIGR** *Arabidopsis thaliana* Database (http://www.tigr.org/tdb/e2k1/ath1), **MIPS** *Arabidopsis thaliana* database (http://mips.gsf.de/proj/thal/db/index.html), and Stanford\Penn\PGEC database of *Arabidopsis thaliana* Annotation (DAtA, http://sequence-www.stanford.edu/ara/SPP.html). Useful web sites are also maintained at the Kazusa DNA Research Institute (KAOS, http://www.kazusa.or.jp/kaos) and Cold Spring Harbor Laboratory http://nucleus.cshl.org/protarab).

Worm *Caenorhabditis elegans*

The nematode worm *Caenorhabditis elegans* has been one of the favorite models for developmental biology for many years. With the availability of the (almost) complete genome of *C. elegans*, it is now becoming a target of functional genomics efforts.

WormBase (http://www.wormbase.org or http://wormbase.sanger.ac.uk) is a unified public resource on *C. elegans* biology, jointly maintained by researchers from CalTech, Cold Spring Harbor Laboratory, Washington University, The Sanger Centre, and CNRS (France), with contributions from scientists from all over the world [804]. WormBase combines mapping and sequencing data with phenotypic information on *C. elegans*. It has a powerful search engine that allows one to search the database by allele name, gene name (predicted or confirmed), cosmid or YAC clone name, author name, or GenBank accession number. WormBase has a convenient sequence viewer that displays positions of predicted curated and uncurated genes, results of transcriptional profiling, and RNA inhibition (RNAi) experiments. WormBase also contains a Pedigree Browser showing the complete cell lineages for the male and hermaphrodite organisms and information on each cell.

WormPD (http://www.proteome.com/databases/index.html), like YPD, is a protein database maintained at Proteome Inc. [160]. It is a useful resource for annotation of *C. elegans* proteins that is being continuously updated. WormPD has a convenient search engine that allows one to search the database by keywords and/or categories (organismal role, biochemical function and cellular role, mutant phenotype, subcellular localization, molecular environment, post-translational modification, number of introns in the gene, and chromosomal location of the gene) as well as by properties of the predicted proteins (isoelectric point, molecular weight, codon adaptation index, and the number of potential transmembrane segments). By following the "WormPD Facts" link, the user can retrieve the updates made within the last week.

As the work on the *C. elegans* genome sequence continues, the web sites of the Sanger Centre (http://www.sanger.ac.uk/Projects/C_elegans) and the Washington University (http://genome.wustl.edu/gsc/Projects/C.elegans) continue to serve as valuable data sources for sequence updates.

Fruit fly *Drosophila melanogaster*

FlyBase (http://flybase.bio.indiana.edu/), produced by a consortium of researchers at Harvard University, University of Cambridge, Indiana University, UC Berkeley, and the EBI and mirrored in Japan, Taiwan, Australia, France, and Israel, is the ultimate data source on *Drosophila melanogaster* and related species. It contains a wide variety of *Drosophila*-related links, including one to the Insect Biology and Ecology site at Cornell University (http://www.nysaes.cornell.edu/ent/biocontrol/info/primer.html), which provides the introductory information on *Drosophila* and other insects. Another good site for an introduction to the *Drosophila* world is the *Drosophila* Virtual Library (http://ceolas.org/fly/).

GadFly (Berkeley Drosophila Genome Project, http://www.fruitfly.org) is another comprehensive resource that allows the user to search Drosophila genome annotations by name, chromosomal position, molecular function, or protein domain. Research results on the development and functioning of *Drosophila* nervous system are collected by **FlyBrain** database at the University of Arizona in Tucson (http://flybrain.neurobio.arizona.edu), which is mirrored at the web sites of University of Freiburg, Germany (http://flybrain.uni-freiburg.de/) and National Institute for Basic Biology in Okazaki, Japan (http://flybrain.nibb.ac.jp)

InterActive Fly (http://sdb.bio.purdue.edu/fly/aimain/1aahome.htm) is a superb collection of information on tissue and organ development in *Drosophila*, compiled by Thomas and Judith Brody [121] and hosted at the Society for Developmental Biology web site. It lists development-related genes by name (in alphabetical order), by biochemical function (e.g. transcription factors, receptors), and by developmental pathways (maternal genes or zygotically transcribed genes). For convenience, there is a separate listing of the most recent additions to the database. Arguably, the most interesting part of the database is the listing of 36 evolutionarily conserved developmental pathways, common for *Drosophila* and other organisms, such as vertebrates (http://sdb.bio.purdue.edu/fly/aimain/aadevinx.htm).

***Drosophila* microarray project** (http://quantgen.med.yale.edu) aims to define the gene expression patterns of *Drosophila* genes *in vivo* using DNA microarrays. It can be useful for predicting the function(s) of an unknown gene on the basis of co-expression with a previously characterized gene (see ♦3.8.2).

Finally, **Drosophila Community Portal** at CyberGenome Technologies (http://www.cybergenome.com/drosophila) contains a good collection of protocols for experimental work on *Drosophila*.

Human

Although a full description of the considerable bioinformatics activity spawned by the Human Genome Project is outside the scope of this book, several useful databases certainly deserve a mention.

Because, unfortunately, a substantial part of our knowledge about human genes comes from the analysis of hereditary diseases, **Online Mendelian Inheritance in Man** (OMIM™, http://www.ncbi.nlm.nih.gov/Omim), a catalog of human genes and genetic disorders, is probably the most important resource on the functions of human genes. This database is based on the book *Mendelian Inheritance in Man* by Victor McKusick and colleagues of Johns Hopkins University. The on-line version of the text and the database were developed at the NCBI. Recently, OMIM has become accessible through Entrez and now it can be queried using Entrez retrieval system (http://www.ncbi.nlm.nih.gov/entrez/query.fcgi?db=OMIM) just like other NCBI databases. OMIM is supplemented with the OMIM Morbid Map (http://www.ncbi.nlm.nih.gov/htbin-post/Omim/getmorbid), an alphabetic list of all the disease genes described in OMIM with their cytogenetic map locations. Because it is intended for use by physicians and patients who might be unfamiliar with the Entrez system, OMIM has its own extensive help file (http://www.ncbi.nlm.nih.gov/entrez/Omim/omimhelp.html), which contains detailed descriptions of the possible search strategies and databases linked to OMIM. In addition, there is a detailed list of frequently asked questions (http://www.ncbi.nlm.nih.gov/entrez/Omim/omimfaq.html).

Genes and Disease (http://www.ncbi.nlm.nih.gov/disease) section of the NCBI web site features a collection of simplified descriptions, which are similar to those in OMIM but are easier to comprehend, contain fewer references and are intended for a more general audience than OMIM. This is a good site for introductory reading on most common human genetic diseases, such as Alzheimer disease, phenylketonuria, Marfan syndrome, diastrophic dysplasia, muscular dystrophy, and many others. This site also contains useful information on the genetic roots of cancer, atherosclerosis, and obesity.

LocusLink (http://www.ncbi.nlm.nih.gov/LocusLink) is an NCBI resource that provides a simple unified query interface to curated portions of human, mouse, rat, fruit fly, and zebrafish genomes. It can be used to search the RefSeq records that contain a variety of genetic information, such as official nomenclature, sequence accession numbers, EC numbers, UniGene clusters, dbSNP links, and STS marker links, and other data. RefSeq records are created by a combination of automated data processing with subsequent manual curation by NCBI staff, which also adds links to the relevant publications in PubMed (see ♦3.8). The latest release of LocusLink included 20,582 human records, 32,014 mouse records, 4,164 records for rat, 18,879 records for *Drosophila,* and 1,194 records for zebrafish.

LocusLink entries are also interlinked with **HomoloGene** (http://www.ncbi.nlm.nih.gov/HomoloGene), a collection of homologous genes in human, mouse, rat, fruit fly, zebrafish, and cow genomes, obtained from published reports and by nucleotide sequence comparisons between ESTs from each pair of organisms. It includes over 7,000 putative orthologs in human, mouse and rat genomes. In contrast, there are only ~200 putative orthologs found in human, rodent (mouse or rat), and zebrafish.

Genomic Information for Eukaryotic Organisms database, **euGenes** (http://iubio.bio.indiana.edu/eugenes), maintained at the Center for Genomics and Bioinformatics at Indiana University, Bloomington, presents data automatically collected from the primary databases and available through a single convenient interface [284]. Information available through euGenes includes gene name, gene symbol, its chromosomal location, function, structure and sequence similarity information for the gene product. Table 3.3 shows summary data from euGenes on the status of sequencing and annotation of eukaryotic genomes. Although these numbers are preliminary and not necessarily reliable, they offer a glimpse of the future comparative genomics of eukaryotes.

Table 3.3 Gene number statistics from euGenes

Organism	Genes			
	Total records	Located on the genome	Experimentally verified	Homologous to others
Human	53,210	84%	39%	39%
Mouse	36,433	-	-	94%
Zebrafish	1,583	-	-	89%
Fruit fly	25,728	51%	65%	61%
Mosquito	12,687	100%	9%	66%
Worm	22,705	90%	22%	43%
Weed	26,819	100%	-	30%
Yeast	7,226	92%	68%	34%

3.6. Taxonomy, Protein Interactions, and Other Databases

3.6.1. Taxonomy databases

NCBI Taxonomy

To organize the sequence data in accordance with the existing phylogenetic classification of organisms, NCBI maintains its own Taxonomy database (http://www.ncbi.nlm.nih.gov/Taxonomy), which contains the names of all organisms that are represented in GenBank. The NCBI taxonomy database attempts to provide a consensus, up-to-date taxonomy tree based on a variety of sources, including published literature, web-based databases, and advice of sequence submitters and outside taxonomy experts.

The database has a hierarchical structure with six root-level taxa, Archaea, Eubacteria, Eukaryota, Viroids, Viruses, and Unclassified (the latter group - for uncultured environmental samples). For convenience, plasmids and other synthetic constructs are grouped together as “Other”. The Taxonomy database offers a convenient way to extract nucleotide or protein sequences from all organisms that belong to a particular genus, family or a higher taxon. The user simply needs to follow the tree to the desired taxon. In case the user is unsure about the exact spelling of an organism name, there is a nifty "phonetic search" option that will search for similarly sounding names, so that an unfortunate researcher that entered "Drozofila" as a search pattern would not be completely lost.

The Taxonomy database offers a useful tool that allows one to construct and display a taxonomy tree for a selected set of organisms. For the organisms that are most commonly used in molecular biology, such a tree can be obtained simply by going to the Taxonomy database home page in Entrez (http://www.ncbi.nlm.nih.gov/entrez/query.fcgi?db=Taxonomy), selecting the desired organisms and clicking on the “Display Common tree” option. After that, the user can edit the resulting tree by adding and deleting species and selecting a complete or abbreviated lineage for each of them. Of course, it has to be kept in mind that the tree obtained from this database is only a taxonomic dendrogram, rather than a true phylogenetic tree. Nevertheless, it offers a convenient view of the taxonomic relationships between the selected organisms, which often reflects the actual phylogeny. The same tool can be reached from the BLink page for any protein in Entrez Proteins (see ♦3.2.2) by selecting the “Common Tree” option.

Ribosome Database Project

For those who want to see an actual phylogenetic tree of small subunit rRNA sequences, the place to look is the Ribosomal Database Project at Michigan State University (RDP, http://rdp.cme.msu.edu, mirrored at the Japanese National Institute of Genetics, http://wdcm.nig.ac.jp/RDP). The latest release of RDP provides numerous precomputed phylogenetic trees for various groups of organisms, accompanied by a sensible tutorial. These trees range from incredibly large ones, such as the full prokaryotic tree with 7322 nodes, full eukaryotic tree with 2055 nodes, and full mitochondrial tree with 1503 nodes, to general trees for the domains *Bacteria* (197 nodes) and *Archaea* (107 nodes), to more specific trees, covering, for example, only the genus *Escherichia* (105 nodes) or genera *Treponema* and *Spirochaeta* (132 nodes). The trees can be viewed and edited using a Java-based viewer and saved as pictures or in the standard nested tree format that can be read using TreeView [646] and other programs. A phylogenetic tree for organisms with completely sequenced genomes is not yet available, although 16S rRNA sequences from most of them are included into at least some precomputed trees.

3.6.2. Signal transduction, regulation, protein-protein interaction and other useful databases

TRANSFAC

The Transcription Factor database (TRANSFAC, http://transfac.gbf.de, also available at http://www.gene-regulation.de) compiles data on eukaryotic regulatory DNA elements and protein factors interacting with them [897,898]. It is maintained by Edgar Wingender and colleagues in Braunschweig, Germany, and mirrored at several sites around the world. The database consists of six tables that cover transcription factor sites of various eukaryotes. The SITE table lists 4504 individual regulatory sites within 1078 eukaryotic genes. It also contains 3494 artificial sequences derived from mutagenesis studies, in vitro selection procedures starting from random oligonucleotide mixtures, etc., and 417 consensus binding sequences, mostly taken from [219]. The GENE table provides short descriptions of each of these 1078 genes, the FACTOR table (2785 entries) describes the proteins that bind these sites, the CLASS table lists 39 classes of transcriptional factors, and the CELL table lists the cellular sources of these proteins. Finally, the MATRIX table (309 entries) provides nucleotide frequency matrices for some of the transcription factor binding sites.

TRANSFAC also includes a hierarchical Classification of Transcription factors (http://transfac.gbf.de/TRANSFAC/cl/cl.html).

BRITE

The Biomolecular Relations in Information Transmission and Expression database (http://www.genome.ad.jp/brite_old), a part of KEGG (see ♦3.5), has long been known as a useful collection of regulatory pathways, including cell cycle control pathways for human and yeast, developmental pathways of *Drosophila*, and enzyme regulatory mechanisms from KEGG.

This database has been recently expanded (http://www.genome.ad.jp/brite) and is now intended to serve as a collection of diverse data on all possible kinds of relations between any two proteins. It includes data on generalized protein-protein interactions (e.g. from KEGG pathway diagrams), experimental data on protein-protein interactions obtained from yeast two-hybrid systems, sequence similarity relations calculated using the Smith-Waterman algorithm (see ♦3.3.2.1), expression similarity relations uncovered by microarray gene expression profiles, and cross-reference links between database entries. Because this site contains data from two large-scale studies of protein-protein interactions in yeast [384,859], it is currently most useful for the analysis of yeast protein function.

DIP

The Database of Interacting Proteins (DIP, http://dip.doe-mbi.ucla.edu) is a compilation of experimentally demonstrated protein-protein interactions. It was created by David Eisenberg and colleagues at the UCLA-DOE Laboratory of Structural Biology and Molecular Medicine to provide a tool for understanding protein function and protein-protein relationships, properties of networks of interacting proteins, and protein evolution [925,926]. The data on protein-protein interactions in DIP come primarily from yeast two-hybrid experiments, although other experimental techniques, such as co-purification, immunoprecipitation (co-immunoprecipitation), binding to affinity columns, in vitro binding assays and others, are also represented. The DIP database consists of three hyperlinked tables that list: (i) protein information, (ii) protein–protein interactions, and (iii) details of experiments. An additional table links DIP to the YPD database (see ♦3.6.2). The latest release of DIP lists 3472 interactions between 2659 proteins, reported in 1020 publications [925]. Although more than 80% of those interactions have been reported in a single experiment, they offer useful hints to the functions of otherwise uncharacterized proteins. Besides, the fraction of confirmed protein-protein interactions in DIP is steadily growing, such that, with time, the utility of this database is most likely to increase.

BIND

The Biomolecular Interaction Network Database (BIND, http://www.bind.ca) was originally developed by Chris Hogue at the Samuel Lunenfeld Research Institute at the Mount Sinai Hospital in Toronto and Francis Ouellette at the Center for Molecular Medicine and Therapeutics of the University of British Columbia in Vancouver. It was recently transferred to a new non-profit company, Blueprint Worldwide, to initiate arguably the most ambitious project in the biological database building. BIND strives to unify protein sequence data with the information on protein-protein interactions and signal transduction pathways and plans to incorporate virtually all interactions between molecules, including proteins, nucleic acids, and small molecules [66]. In addition, there are plans to include photochemical reactions and conformational changes in proteins. Although BIND is only beginning to grow and its first pathway entries might seem cumbersome, it seems to have a great potential.

BioCarta

BioCarta (http://www.biocarta.com) has assembled an impressive list of pathways (http://www.biocarta.com/genes/allPathways.asp), which are presented as appealing colorful images. The site offers a place for users' comments, discussion of the pathway itself, and submission of new pathways. If nothing else, the readers should visit this web site just to enjoy its graphics.

EPD

Eukaryotic Promoter Database (EPD, http://www.epd.isb-sib.ch, [668]) was developed by Philipp Bucher and colleagues at the Swiss Institute for Experimental Cancer Research (ISREC). EPD is a curated non-redundant collection of 1390 eukaryotic promoters with experimentally determined transcription start sites. Each entry contains a description of the initiation site, cross-references to other databases (EMBL\GenBank\DDBJ, LocusLink, Unigene, RefSeq, SWISS-PROT), and bibliographic references.

3.6.3. Biochemical databases

Biochemical Pathways Map

Anyone working on genome annotation should have a firm grasp of cell biochemistry and, ideally, should be able to quickly recall properties and functions of hundreds of different proteins. Since very few of us actually remember all the biochemical pathways, there are several useful resources that allow one to take a quick look at the biochemical pathways and figure out whether a particular annotation is plausible.

For many years, almost every molecular biology laboratory had a wall chart of biochemical pathways, created by the now retired biochemist Gerhard Mihal at Boehringer Mannheim Corp. A hyperlinked version of this chart is now available at the ExPASy web site (http://www.expasy.org/cgi-bin/search-biochem-index). For convenience, the chart is split into 120 fields, each representing a small fraction of the map. One can take a look at the whole map or examine one or two adjacent fields. The names of the enzymes and metabolites on this map can be searched as keywords. In addition, the enzyme names are hyperlinked to the ENZYME database (see below), allowing one to associate a reaction with the amino acid sequence of the enzyme.

As previously mentioned (see ♦3.5), a useful collection of metabolic pathways is available at the KEGG web site (http://www.genome.ad.jp/kegg). KEGG charts are much simpler and cover individual pathways, such as glycolysis or TCA cycle.

ENZYME

The ENZYME database (http://www.expasy.org/enzyme, mirrored at http://us.expasy.org/enzyme) has already been mentioned in the description of SWISS-PROT (♦3.2.2). ENZYME is a convenient source of information on the official nomenclature of enzymes, based on the recommendations of the Nomenclature Committee of the International Union of Biochemistry and Molecular Biology (http://www.chem.qmw.ac.uk/iubmb/enzyme). ENZYME lists all the enzymes that have been assigned Enzyme Commission (EC) numbers and describes them with respect to the EC number, recommended and alternative names, catalytic activity, cofactors (if any) and the diseases associated with the deficiency of the enzyme (if known). Enzymes with known sequences are linked to the corresponding SWISS-PROT entries. The names of substrates and products of the catalyzed reactions are linked to their structures in the Klotho (http://www.biocheminfo.org/klotho) database.

The Nomenclature Committee web site, mentioned above, also contains some useful information on enzymes, including lists of newly approved enzymes (http://www.chem.qmw.ac.uk/iubmb/enzyme/newenz.html), retracted EC numbers, reaction schemes, and references to the original publications.

KLOTHO

Klotho: Biochemical Compounds Declarative Database (http://www.biocheminfo.org/klotho), developed by Toni Kazic and colleagues, is a listing of 439 (bio)chemical compounds, shown in a variety of representations, including Fischer diagrams, smiles strings, and actual 3D structures.

BRENDA

BRENDA (http://www.brenda.uni-koeln.de) is a comprehensive enzyme information system maintained by Dietmar Schomburg and colleagues at the Institute of Biochemistry of the University of Köln. In addition to the information listed in ENZYME, each enzyme entry in BRENDA is associated with up to 20 extra parameters, such as specific activity, turnover number, K_M for various substrates, pH range, pH optimum, and pH stability, temperature range, temperature optimum, and temperature stability, inhibitors, molecular weight, and many others. Each entry is accompanied by extensive bibliography. While these data are extremely interesting from a purely enzymological standpoint, they also prove invaluable when one needs to evaluate the substrate specificity of an enzyme encoded in a newly sequenced genome or to decide whether a given gene product can catalyze a particular reaction (see ♦4.2).

LIGAND

The LIGAND database (http://www.genome.ad.jp/dbget/ligand.html) is part of the GenomeNet site, maintained by the Kanehisa laboratory at the Kyoto University. LIGAND is a site dedicated to enzymes and their substrates and tightly interlinked with KEGG (see ♦3.5). Its entries are somewhat similar to those of ENZYME, but contain a much larger library of structures of enzyme substrates, specifically drawn for this database using the ISIS/Draw program (MDL Information Systems, http://www.mdli.com). This allows the user to view and, if necessary, save those structures, which definitely helps to understand the function(s) of each particular enzyme. In addition, for each metabolic enzyme, LIGAND lists its representation in the completely sequenced genomes.

AAindex

The AAindex (http://www.genome.ad.jp/dbget/aaindex.html) is yet another database from the Kanehisa Laboratory that provides an exhaustive listing of various amino acid indices and similarity matrices [429]. Amino acids can be grouped on the basis of their physico-chemical and biochemical properties, such as the propensity to form an α-helix, a turn or a β–strand, hydrophobicity, polarity, bulkiness, and many others. AAindex currently lists 434 amino acid indices that all come handy for one particular task or the other. In addition, amino acids can be grouped based on their exchangeability in protein sequences, similar to the matrix shown on Figure 3.1. Again, these matrices can be very different depending on the evolutionary distances between proteins, on whether they are soluble or membrane-bound, globular or non-globular, and so on (see ♦3.3.1.1). AAindex currently lists 66 such amino acid substitution matrices that all can be used for evaluating sequence similarity between protein sequences in different contexts.

PMD

Protein Mutant Database (PMD, http://pmd.ddbj.nig.ac.jp), maintained at the DDBJ, is a collection of literature references that describe various mutations, naturally occurring in proteins or induced by mutagenesis. PMD allows the user to submit a protein sequence that will be compared against the sequences in the PMD using straightforward text matching. If a match is found, the mutated amino acid residues will be indicated, linked to the articles that describe the respective mutations. However, the text matching tool in PMD is not particularly powerful and cannot recognize sequences with <30% identity. Nevertheless, PMD is the only database, other than SWISS-PROT, that consistently records mutation data, which could be useful in delineating the active sites of poorly characterized enzymes.

3.7. PubMed

PubMed (http://www.ncbi.nlm.nih.gov/PubMed, or just http://pubmed.gov) is definitely the most widely used database in biology. As of the time of this writing, PubMed lists more than 11 million scientific articles, which makes the ability to find the relevant reference promptly a useful skill that requires at least some experience. Because the NCBI web site contains a PubMed overview, a vast PubMed help file, a list of frequently asked questions about PubMed, and even a detailed down-to-earth tutorial (http://www.nlm.nih.gov/bsd/pubmed_tutorial/m1001.html), we consider here only several of the less trivial aspects of PubMed searches.

The first thing to know about PubMed is that, although it contains over 11 million citations, it does not and has never been intended to cover all the biological literature. For example, PubMed has a poor coverage of plant science, environmental research, and many other areas of biology that are not immediately related to human health. Also, PubMed lists very few papers published before 1965 (some papers from 1958 through 1965 are kept in the OldMEDLINE database, which is available through the NLM Gateway, http://gateway.nlm.nih.gov/gw/Cmd).

The full list of journals, which are indexed by PubMed, is available through the Journal Browser (http://www.ncbi.nlm.nih.gov/entrez/jrbrowser.cgi), which also allows browsing selected journals issue by issue.

3.7.1. Specifying the terms in PubMed search

By default, PubMed looks for a combination of all the terms entered in a query (i.e. each term is treated as a required string). The simplest way to find a reference is to enter into the search field as many relevant terms as possible. This method works surprisingly well, especially for topics that happen to deviate from the mainstream health research. However, entering popular search terms like "AIDS" and "drug" would return more than 15,000 citations and force one to narrow down the search. PubMed has a special option, called "Single Citation Matcher", which allows one to enter various bits and pieces of citation information (the author, year, title, and journal of the publication).

Boolean operators

For more complex queries, it is best to specifically indicate the field(s) to be searched and to connect them with appropriate Boolean operators AND, OR, and NOT (PubMed requires that these be in caps). The operator AND is used in PubMed by default, so it only needs to be put in complex Boolean search patterns that contain different fields (see below). This operator requires that both terms connected by it appear in the citation; it is used to narrow down the search space. The operator OR allows either of the specified search terms to appear in the output; it is often used to expand the search space to include synonyms or otherwise similar subjects. The operator NOT is used to exclude certain terms from the search. The use of Boolean operators can dramatically improve search efficiency, especially when used in combination with terms from appropriately selected fields. Most of the fields used by PubMed are self-explanatory, but some are not. Use of the fields illustrated below is not entirely straightforward, but may be convenient under certain circumstances.

Affiliation [AFFL]

Looking for publications by an author [AUTH] with a common last name, such as Smith or Green, can be frustrating. For example, a search for publications by Janet L. Smith (Smith JL[AUTH]), a Purdue University biochemist specializing in amidotransferases, returns 930 papers authored by various John L. Smiths from all over the world. However, entering the search pattern

Smith JL[AUTH] AND Purdue[AFFL]

allows one to retrieve a collection of papers by Janet L. Smith on various topics, not just amidotransferases, at the same time avoiding the sea of irrelevant citations. Of course, this search will miss the papers for which Purdue University is not entered as an affiliation, including the recent review of amidotransferase mechanisms [936], but that is a different headache.

Journal [JOUR] and Publication Date [PDAT]

It has probably happened to everyone: just before leaving for vacation, you read a particularly interesting paper in, say, *Trends in Biochemical Sciences*, but completely forgot what it was about. What you need to do is to browse back issues of the journal and try to figure out which paper it was. Short of going to the library, one can use the Journal Browser to retrieve all the papers from that journal

"Trends Biochem Sci"[JOUR]

and further limit the output to the papers published in July and August 2000:

"2000/07"[PDAT]:"2000/08"[PDAT]

to come up with a simple search pattern:

"Trends Biochem Sci"[JOUR] AND ("2000/07"[PDAT]:"2000/08"[PDAT]), which would narrow your search down to just 21 papers. The publication date search can also be entered through the Limits function as described below.

Enzyme Classification [ECNO]

When searching for information on a specific enzyme or a group of enzymes, it often turns out to be convenient to simply use the EC number as the search parameter. For example, when searching for data on NADP-dependent alcohol dehydrogenases, the last thing one would like to do is to enter NADP, dehydrogenase, and alcohol (1194 citations). NADP, alcohol, and dehydrogenase (in that order) would return only 455 citations, because MESH system would recognize "alcohol dehydrogenase" as a single search term. In contrast, entering 1.1.1.2[ECNO] as the search term would return only 162 citations, most of which would be relevant to the topic.

Limits

Because the sheer number of publications broadly related to a particular topic in the database can be overwhelming, PubMed offers the user the opportunity to limit the search to certain values in particular fields. This alleviates the need to remember the syntax of the examples mentioned above and allows the user to construct fairly complex search patterns. This feature allows one to select articles published in a specific language and further specify the type of articles to retrieve, e.g. review papers only. Although the preset limits are geared mostly towards clinical studies, there are several options useful for biologists. For example, Limits allow one to directly enter the range of acceptable publication dates for the articles to look for. Importantly, the Entrez date parameter specifies the date when the new citation was added to the database. By using the Entrez date, one can search for papers added to PubMed during the last week, month, or any other period.

The Limits option is also convenient for the retrieval of protein and nucleotide sequences. For proteins, it allows one to search by gene location (genomic, mitochondrial, or chloroplast DNA) and the database (GenBank, EMBL, DDBJ, PDB, SWISS-PROT, PIR, or RefSeq). For nucleotide sequences, in addition, it offers the option of excluding patents, sequences of ESTs, STSs, GSS, and/or working draft sequences. This allows the user to significantly reduce the noise caused by the redundancy of GenBank protein and nucleotide databases.

3.7.2. Interpretation of the search pattern

Often enough, a PubMed search would not find the reference that should be there or would return references that seem to have nothing to do with the entered search pattern. One of the reasons for this is that PubMed does not simply scan all the abstracts for the word or phrase entered by the user. Instead, it first searches precompiled indexes of terms in four main lists. It starts by looking for a match in the Medical Subject Heading (MeSH) table. If it does not find a match, it looks in the Journals Table, then in the Phrase List and, finally, in the Author Index. As soon as PubMed finds a match in one of those four lists, the search stops. Thus, if one enters "Silver" as the search pattern, PubMed would not even look for papers authored by Simon Silver from the University of Illinois at Chicago or any other researcher with that last name. Instead, PubMed would interpret "silver" as a MeSH term and would ignore it in all other lists. After receiving the report that PubMed has found as many as 23,486 references, very few of which have Silver as an author, one could click "Details" and find out that the word "Silver" was translated by PubMed as

("silver"[MeSH Terms] OR silver[Text Word]).

As discussed above, to search for papers authored by Silver, one would need to enter the search pattern "Silver[AUTH]" by typing it or going through the "Limits" option. If the author's initial is known, one could simply enter "silver s", which would be interpreted as a name. Finally, to search for Silver's papers on Ag^{+}-resistance, one could use the pattern "silver s silver" and end up with a list of 13 references, 7 of which would be relevant.

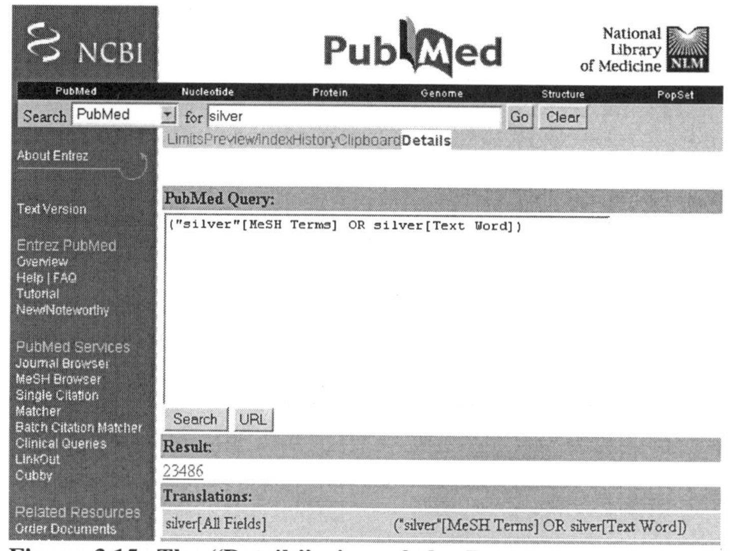

Figure 3.15. The "Details" view of the PubMed search for "silver". Note the "Translations" line at the bottom.

The option of pressing "Details" to find out how the search terms have been interpreted by PubMed offers an easy way to avoid a lot of confusion. It also allows the user to modify the search so that PubMed would look exactly for what the user wants. Consider the following real-life example: you used Triton X-100 to solubilize your protein and want to find an easy way to measure residual Triton X-100 in your sample. Simply entering "triton determination" would bring 6,663 citations, some of which are literally out of this world. Pressing "Details" shows that your search pattern has been interpreted as follows:

(("neptune"[MeSH Terms] OR triton[Text Word]) AND
("analysis"[Subheading] OR determination[Text Word]))

Where in the world did "neptune" come from and what does it have to do with triton? Looking up "neptune" as a MeSH term brings you the following comprehensive answer:

"Neptune: The eighth planet in order from the sun. It is one of the five outer planets of the solar system. Its two natural satellites are Nereid and Triton. Year introduced: 1995."

Even if this entry leaves you puzzled about the impact of the planet Neptune on contemporary medical science, it at least explains the link of triton to neptune. As a matter of fact, the PubMed search engine is trying to be as helpful as possible: should you enter triton X-100, it would not need any quotation marks to correctly interpret it as

("octoxynol"[MeSH Terms] OR triton X-100[Text Word]).

Looking up "octoxynol" as a MeSH term reveals yet another reason why there are so many irrelevant references in the output:

"Octoxynol: Nonionic surfactant mixtures varying in the number of repeating ethoxy (oxy-1,2-ethanediyl) groups. They are used as detergents, emulsifiers, wetting agents, defoaming agents, etc. Octoxynol-9, the compound with 9 repeating ethoxy groups, is a spermatocide."

This means that you might want to disallow searching for octoxynol; otherwise, you might get distracted by the contraceptive uses of Triton X-100 that have never been mentioned in biochemistry textbooks. In the long run, the easiest way is usually to simply include in the search pattern as many relevant terms as possible, hoping that their intersection retrieves the relevant publications. In this case, the pattern

triton X-100[Text Word] AND detergent removal method

returns only 39 citations, two of which are definitely relevant. Pressing the "Related articles" link brings more papers of the same kind and finally allows one to move from searching to reading.

3.7.3. NCBI Bookshelf

None of us is equally proficient in all areas of biology. However, for most cancer researchers, relative ignorance in algology or mycology can be easily forgiven. Not so for genome annotators, who encounter database hits from *Synechocystis* or *Dictyostelium* on a daily basis and need to be able to quickly decide whether those hits and their annotations are plausible and whether those annotations could be applied to human genes. The NCBI Bookshelf (http://www.ncbi.nlm.nih.gov/entrez/query.fcgi?db=Books) project aims at putting classical textbooks on the web and hyperlinking them with PubMed abstracts. The third edition of "Molecular Biology of the Cell" by Bruce Alberts, Dennis Bray, Julian Lewis, Martin Raff, Keith Roberts, and James D. Watson, has been available on-line since 2001. In the last two years, it has been joined by ten other books on such diverse topics as immunobiology, developmental biology, retroviruses and cancer medicine. While availability of all these book on-line is an important development per se, their value for PubMed users goes far beyond that. By clicking the "Books" link from the abstract view, the reader can link the terms in the respective abstract to the same terms in either of the books on the bookshelf. Then, just by clicking on the obscure term, the user can jump to the book paragraph that mentions this term and see it in the proper context. There is a possibility to directly search the bookshelf from Entrez, so that the reader can use a term that is even not in PubMed. Obviously, this tool works best for subjects that are specifically covered in the books available so far; there are still very few links to algae- or fungi-related topics. Nonetheless, this is a start of a very promising trend that should help researchers to deal with unfamiliar terminology in even most complex PubMed entries.

3.8. Conclusions and Outlook

In this chapter, we gave only a perfunctory and non-technical overview of the databases that, in our opinion, are most important for researchers working in genomics. For more detail and particularly information on technical aspects of database architecture, the reader should refer to the sources listed below and other relevant literature. Appropriate information resources are necessary for any type of research, but in genomics, the quality of the employed databases affects the science more directly than in many other areas. Throughout this chapter, we emphasized the critical distinction between archival and curated ("value-added") databases. It would be a grave mistake to think that the latter are unconditionally "better" than the former. The two types of databases perform fundamentally different and equally essential functions. Archival databases ensure the integrity of the edifice of genomics and will exist as long as the field itself. However, it is the other type of databases, the expert-curated, specialized ones that are currently in the phase of explosive growth and, in our opinion, the future of genomics critically depends on these resources. In Chapters 4 and 5, we shall see how these databases are already transforming comparative-genomic research.

3.9. Further Reading

1. Nucleic Acids Research, 1998-2002, January 1st issues

2. *Computer Methods for Macromolecular Sequence Analysis*. 1996. Doolittle R.F., ed. (Methods in Enzymology, vol. 266). Academic Press, San Diego.

3. *Analysis of Amino Acid Sequences*. 2000. Bork P., ed. (Advances in Protein Chemistry, vol. 54). Academic Press, San Diego. The following articles are relevant for this chapter:

   Apweiler R. *Protein sequence databases*, pp. 31-71

   Bateman A., Birney E. *Searching databases to find protein domain organization*, pp. 137-157

   Kanehisa M. *Pathway databases and higher order function*, pp. 381-408.

4. *Bioinformatics: Databases and Systems*. 1999. Letovsky S.L., ed. Kluwer Academic Publishers, Boston.

CHAPTER 4.
PRINCIPLES AND METHODS OF SEQUENCE ANALYSIS

We aligned sequences by eye.
Troy CS, MacHugh DE, Bailey JF,
Magee DA, Loftus RT, Cunningham P,
Chamberlain AT, Sykes BC, Bradley DG.
Genetic evidence for Near-Eastern
origins of European cattle.
Nature, 2001, vol. 410, p. 1091

This chapter is the longest in the book as it deals with both general principles and practical aspects of sequence and, to a lesser degree, structure analysis. Although these methods are not, in themselves, part of genomics, no reasonable genome analysis and annotation would be possible without understanding how these methods work and some practical experience with their use. Inappropriate use of sequence analysis procedures may result in numerous errors in genome annotation (we have already touched upon this subject in the previous chapter and further discuss it in Chapter 5). We attempted to strike a balance between generalities and specifics, aiming to give the reader a clear perspective of the computational approaches used in comparative and functional genomics, rather than discuss any one of these approaches in great detail. In particular, we refrained from any extensive discussion of the statistical basis and algorithmic aspects of sequence analysis because these can be found in several recent books on computational biology and bioinformatics (see ♦4.8) and, no less importantly, because we cannot claim advanced expertise in this area. We also tried not to duplicate the "click here"-type tutorials, which are available on many web sites. However, we deemed it important to point out some difficult and confusing issues in sequence analysis and warn the readers against the most common pitfalls. This discussion is largely based on our personal, practical experience in analysis of protein families. We hope that this discussion might clarify some aspects of sequence analysis that "you always wanted to know about but were afraid to ask". Still, we felt that it would be impossible (and unnecessary) to discuss all methods of sequence and structure analysis in one, even long, chapter and concentrated on those techniques that are central to comparative genomics. We hope that after working through this chapter interested readers will be encouraged to continue their education in methods of sequence analysis using more specialized texts, including those in the 'Further Reading' list (♦4.8).

4.1. Identification of Genes in a Genomic DNA Sequence

4.1.1. Prediction of protein-coding genes

Archaeal and bacterial genes typically comprise uninterrupted stretches of DNA between a start codon (usually ATG, but in a minority of genes, GTG, TTG, or CTG) and a stop codon (TAA, TGA, or TAG; alternative genetic codes of certain bacteria, such as mycoplasmas, have only two stop codons). Rare exceptions to this rule involve important but rare mechanisms, such as programmed frameshifts. There seem to be no strict limits on the length of the genes. Indeed, the gene *rpmJ* encoding the ribosomal protein L36 (Figure 2.1) is only 111 bp long in most bacteria, whereas the gene for *B. subtilis* polyketide synthase PksK is 13,343 bp long. In practice, mRNAs shorter than 30 codons are poorly translated, so protein-coding genes in prokaryotes are usually at least 100 bases in length. In prokaryotic genome-sequencing projects, open reading frames (ORFs) shorter than 100 bases are rarely taken into consideration, which does not seem to result in substantial under-prediction. In contrast, in multicellular eukaryotes, most genes are interrupted by introns. The mean length of an exon is ~50 codons, but some exons are much shorter; many of the introns are extremely long, resulting in genes occupying up to several megabases of genomic DNA. This makes prediction of eukaryotic genes a far more complex (and still unsolved) problem than prediction of prokaryotic genes.

4.1.1.1. Prokaryotes

For most common purposes, a prokaryotic gene can be defined simply as the longest ORF for a given region of DNA. Translation of a DNA sequence in all six reading frames is a straightforward task, which can be performed on line using, for example, the Translate tool on the ExPASy server (http://www.expasy.org/tools/dna.html) or the ORF Finder at NCBI (http://www.ncbi.nlm.nih.gov/gorf/gorf.html).

Of course, this approach is oversimplified and may result in a certain number of incorrect gene predictions, although the error rate is rather low. Firstly, DNA sequencing errors may result in incorrectly assigned or missed start and/or stop codons, because of which a gene might be truncated, overextended or missed altogether. Secondly, on rare occasions, among two overlapping ORFs (on the same or the opposite DNA strand), the shorter one might be the real gene. The existence of a long "shadow" ORF opposite a protein-coding sequence is more likely than in a random sequence because of the statistical properties of the coding regions. Indeed, consider the simple case where the first base in a codon is a purine and the third base is a pyrimidine (the RNY codon pattern). Obviously, the mirror frame in the complementary strand would follow the same pattern, resulting in a deficit of stop codons [235]. Figure 4.1 shows the ORFs of at least 100 bp located in a 10 kb fragment of the *E. coli* genome (from 3435250 to 3445250) that encodes potassium transport protein TrkA, mechanosensitive channel MscL, transcriptional regulator YhdM, RNA polymerase alpha subunit RpoA, preprotein translocase subunit SecY, and ribosomal proteins RplQ (L17), RpsD (S4), RpsK (S11), RpsM (S13), RpmJ (L36), RplO (L15), RpmD (L30), RpsE (S5), RplR (L18), RplF (L6), RpsH (S8), RpsN (S14), RplE (L5), and RplX (L24). Although the two ORFs in frame +1 (top line, on the right) are longer (207 aa and 185 aa) than the ORFs in frame -3 (bottom line, 117aa, 177 aa, 130 aa, and 101 aa), it is the latter that encode real proteins, namely the ribosomal proteins RplR, RplF, RpsH, and RpsN.

Because of these complications, it is always desirable to have some additional evidence that a particular ORF actually encodes a protein. Such evidence comes along many different lines and can be obtained using various methods, e.g. the following ones:

- the ORF in question encodes a protein that is similar to previously described ones (search the protein database for homologs of the given sequence);
- the ORF has a typical GC content, codon frequency or oligonucleotide composition (calculate the codon bias and/or other statistical features of the sequence, compare to those for known protein-coding genes from the same organism);
- the ORF is preceded by a typical ribosome-binding site (search for a Shine-Dalgarno sequence in front of the predicted coding sequence);
- the ORF is preceded by a typical promoter (if consensus promoter sequences for the given organism are known, check for the presence of a similar upstream region).

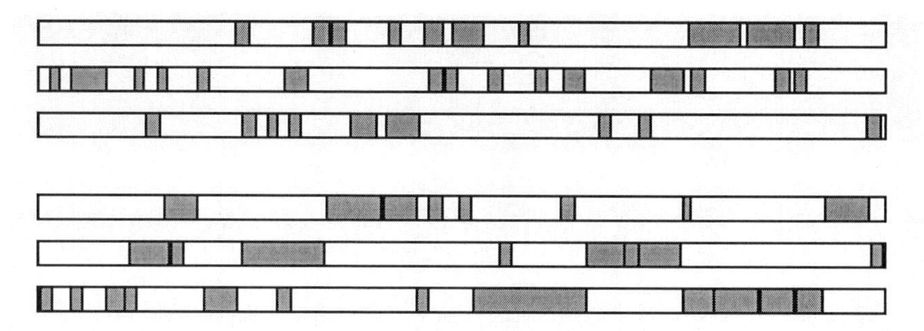

Figure 4.1. Open reading frames of ≥100 bp encoded on a 10 kb fragment of the *Escherichia coli* K12 genome from 3435250 to 3445250. The figure was generated using the program ORF finder at the NCBI web site (http://www.ncbi.nlm.nih.gov/gorf/gorf.html). The six horizontal lines represent frames 1, 2, 3, -1, -2, and -3, respectively. ORFs in each frame are shown as dark boxes.

The most reliable of these approaches is a database search for homologs. In several useful tools, DNA translation is seamlessly bound to the database searches. In the ORF finder, for example, the user can submit the translated sequence for a BLASTP or TBLASTN (see ♦4.4) search against the NCBI sequence databases. In addition, there is an opportunity to compare the translated sequence to the COG database (see ♦3.4). A largely similar Analysis and Annotation Tool (http://genome.cs.mtu.edu/aat.html) developed by Xiaoqiu Huang at Michigan Tech [361] also compares the translated protein sequences to **nr** and SWISS-PROT; in addition, it checks them against two cDNA databases, the dbEST at the NCBI and Human Gene Index at TIGR.

Other methods take advantage of the statistical properties of the coding sequences. For organisms with highly biased GC content, for example, the third position in each codon has a highly biased (very high or very low) frequency of G and C. FramePlot, a program that exploits this skew for gene recognition [380], is available at the Japanese Institute of Infectious Diseases (http://www.nih.go.jp/~jun/cgi-bin/frameplot.pl) and at the TIGR web site (http://tigrblast.tigr.org/cmr-blast/GC_Skew.cgi). The most useful and popular gene prediction programs, such as GeneMark and Glimmer (see ♦3.1.2), build Markov models of the known coding regions for the given organism and then employ them to estimate the coding potential of uncharacterized ORFs.

Inferring genes based on the coding potential and on the similarity of the encoded protein sequences to those of other proteins represents the intrinsic and extrinsic approaches to gene prediction [110], which ideally should be combined. Two programs that implement such a combination, developed specifically for analysis of prokaryotic genomes, are ORPHEUS (http://pedant.gsf.de/orpheus [249]) and CRITICA ([67], source code at http://www.math.uwaterloo.ca/~jhbadger/). Several other algorithms that incorporate both these approaches are aimed primarily at eukaryotic genomes and are discussed further in this section.

4.1.1.2. Unicellular eukaryotes

Genomes of unicellular eukaryotes are extremely diverse in size, the proportion of the genome that is occupied by protein-encoding genes and the frequency of introns. Clearly, the smaller the intergenic regions and the fewer introns are there, the easier it is to identify genes. Fortunately, genomes of at least some simple eukaryotes are quite compact and contain very few introns. Thus, in yeast *S. cerevisiae*, at least 67% of the genome is protein-coding, and only 233 genes (less than 4% of the total) appear to have introns [660]. Although these include some biologically important and extensively studied genes, e.g. those for aminopeptidase APE2, ubiquitin-protein ligase UBC8, subunit 1 of the mitochondrial cytochrome oxidase COX1, and many ribosomal proteins, introns comprise less than 1% of the yeast genome. The tiny genome of the intracellular eukaryotic parasite *Encephalitozoon cuniculi* appears to contain introns in only 12 genes and is practically prokaryote-like in terms of the "wall-to-wall" gene arrangement [425]. Malaria parasite *Plasmodium falciparum* is a more complex case, with ~43% of the genes located on chromosome 2 containing one or more introns [272]. Protists with larger genomes often have fairly high intron density. In the slime mold *Physarum polycephalum*, for example, the average gene has 3.7 introns [851]. Given that the average exon size in this organism (165±85 bp) is comparable to the length of an average intron (138±103 bp), homology-based prediction of genes becomes increasingly complicated.

Because of this genome diversity, there is no single way to efficiently predict protein-coding genes in different unicellular eukaryotes. For some of them, such as yeast, gene prediction can be done by using more or less the same approaches that are routinely employed in prokaryotic genome analysis. For those with intron-rich genomes, the gene model has to include information on the intron splice sites, which can be gained from a comparison of the genomic sequence against a set of ESTs from the same organism. This necessitates creating a comprehensive library of ESTs that have to be sequenced in a separate project. Such dual EST/genomic sequencing projects are currently under way for several unicellular eukaryotes (see Appendix 2).

4.1.1.3. Multicellular eukaryotes

In most multicellular eukaryotes, gene organization is so complex that gene identification poses a major problem. Indeed, eukaryotic genes are often separated by large intergenic regions, and the genes themselves contain numerous introns, many of them long. Figure 4.2 shows a typical distribution of exons and introns in a human gene, the X chromosome-located gene encoding iduronate 2-sulfatase (IDS_HUMAN), a lysosomal enzyme responsible for removing sulfate groups from heparan sulfate and dermatan sulfate. Mutations causing iduronate sulfatase deficiency result in the lysosomal accumulation of these glycosaminoglycans, clinically known as Hunter's syndrome or type II mucopolysaccharidosis (OMIM entry 309900) [896]. A number of clinical cases have been shown to result from aberrant alternative splicing of this gene's mRNA, which emphasizes the importance of reliable prediction of gene structure [631].

Obviously, the coding regions comprise only a minor portion of the gene. In this case, positions of the exons could be unequivocally determined by mapping the cDNA sequence (i.e. iduronate sulfatase mRNA) back to the chromosomal DNA. Because of the clinical phenotype of the mutations in the iduronate sulfatase gene, we already know the "correct" mRNA sequence and can identify various alternatively spliced variants as mutations. However, for many, perhaps the majority of the human genes, multiple alternative forms are part of the regular expression pattern [118,576], and correct gene prediction ideally should identify all of these forms, which immensely complicates the task.

Ideally, gene prediction should identify all exons and introns, including those in the 5'-untranslated region (5'-UTR) and the 3'-UTR of the mRNA, in order to precisely reconstruct the predominant mRNA species. For practical purposes, however, it is useful to assemble at least the coding exons correctly because this allows one to deduce the protein sequence.

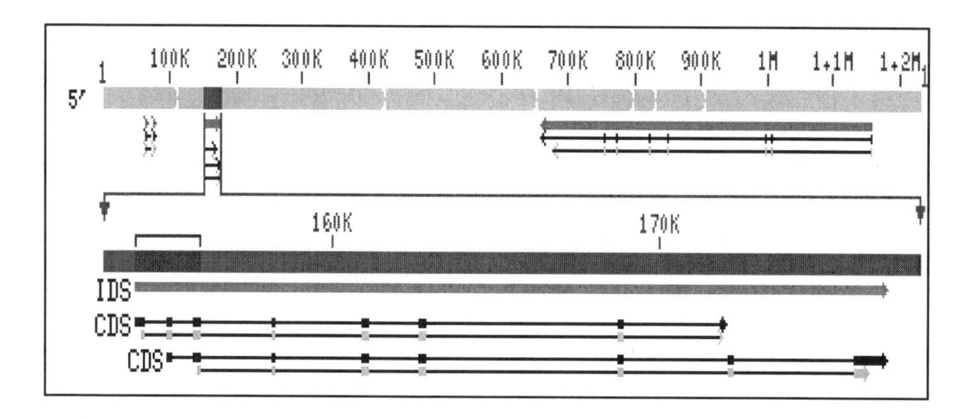

Figure 4.2. Organization of the human iduronate 2-sulfatase gene.
This gene is located in positions 152960-177995 of human X chromosome and encodes a 550-amino acid precursor protein that contains a 25-aa N-terminal signal sequence, followed by eight amino acids that are removed in the course of protein maturation. Mutations in this gene cause mucopolysaccharidosis type II, also known as Hunter's disease, which results in tissue deposits of chondroitin sulfate and heparin sulfate. The symptoms of Hunter's disease include dysostosis with dwarfism, coarse facial features, hepatosplenomegaly, cardiovascular disorders, deafness, and, in some cases, progressive mental retardation (See OMIM 309900). The top line indicates the X chromosome and shows the location of the iduronate sulfatase gene (thick line in the middle). Thin lines on the bottom indicate two alternative transcripts. Exons are shown with small rectangles. The square bracket above the iduronate sulfatase gene marks the region of the gene shown in figure 4.3.

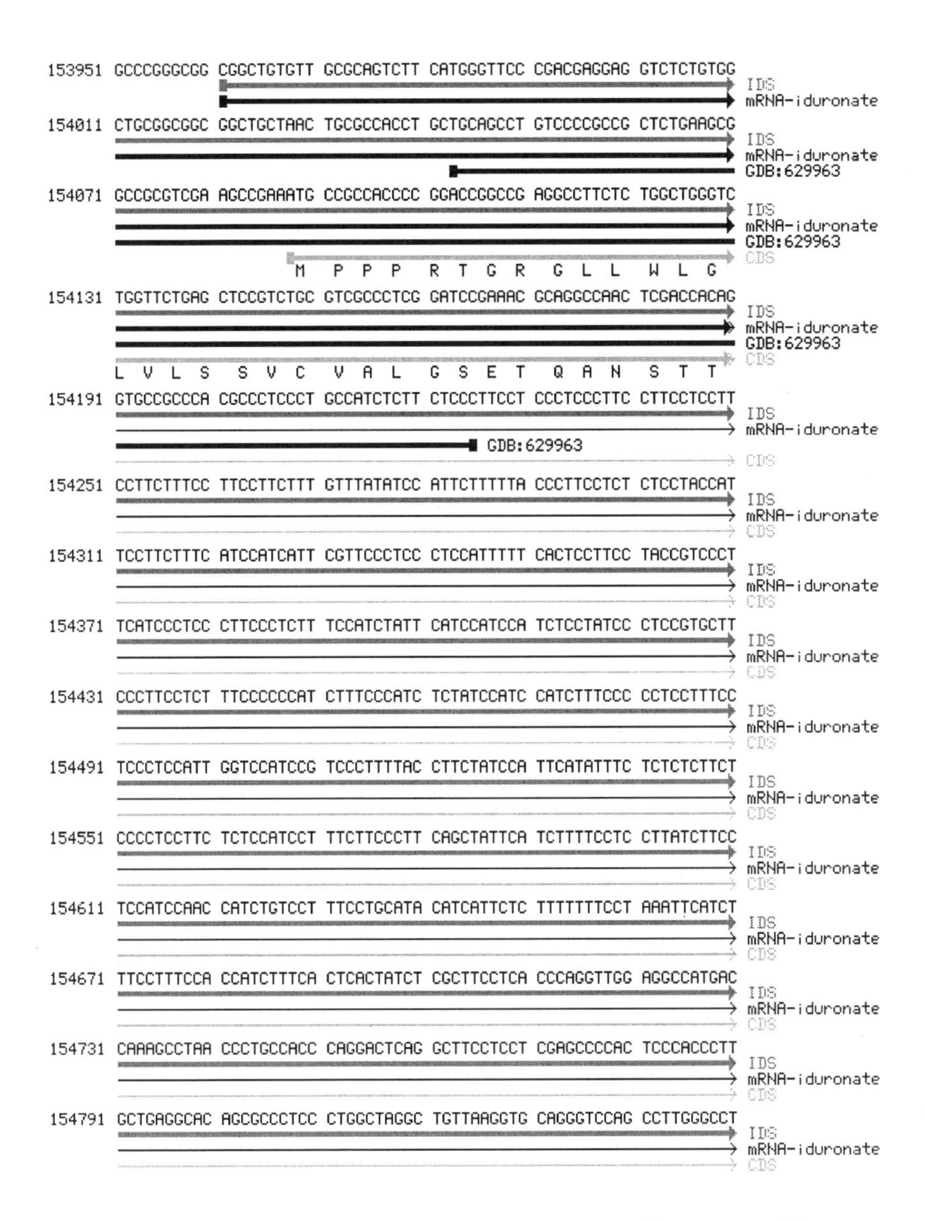

Figure 4.3. Sequence of the first two exons of human iduronate sulfatase gene. The figure shows the DNA sequence of the positions 15391-15571 of human X chromosome. The iduronate sulfatase mRNA and its coding sequence are shown as thick lines; the corresponding amino acid residues are shown underneath. "Variation" indicates the positions of mapped mutations causing type II mucopolysaccharidosis.

Correct identification of the exon boundaries relies on the recognition of the splice sites, which is facilitated by the fact that the great majority splice sites conform to consensus sequences that include two nearly invariant dinucleotides at the ends of each intron, a GT at the 5' end and an AG at the 3' end. Non-canonical splice signals are rare and come in several variants [329,582]. In the 5' splice sites, the GC dinucleotide is sometimes found instead of GT. The second class of exceptions to the splice site consensus includes so-called "AT-AC" introns that have the highly conserved /(A,G)TATCCT(C,T) sequence at their 5' sites. There are additional variants of non-canonical splice signals, which further complicate prediction of the gene structure.

The available assessments of the quality of eukaryotic gene prediction achieved by different programs show a rather gloomy picture of numerous errors in exon/intron recognition. Even the best tools correctly predict only ~40% of the genes [697]. The most serious errors come from genes with long introns, which may be predicted as intragenic sequences, resulting in erroneous gene fission, and pairs of genes with short intergenic regions, which may be predicted as introns, resulting in false gene fusion. Nevertheless, most of the popular gene prediction programs discussed in the next section show reasonable performance in predicting the coding regions in the sense that, even if a small exon is missed or overpredicted, the majority of exons are identified correctly.

Another important parameter that can affect ORF prediction is the fraction of sequencing errors in the analyzed sequence. Indeed, including frameshift corrections was found to substantially improve the overall quality of gene prediction [133]. Several algorithms were described that could detect frameshift errors based on the statistical properties of coding sequences [224]. On the other hand, error correction techniques should be used with caution because eukaryotic genomes contain numerous pseudogenes, and non-critical frameshift correction runs the risk of wrongly "rescuing" pseudogenes. The problem of discriminating between pseudogenes and frameshift errors is actually quite complex and will likely be solved only through whole-genome alignments of different species or, in certain cases, by direct experimentation, e.g., expression of the gene(s) in question.

4.1.2. Algorithms and software tools for gene identification

As discussed in the previous section, recognizing genes in the DNA sequences remains one of the most pressing problems in genome analysis. Several different approaches to gene prediction have been developed and there are several popular programs that are most commonly used for this task (see Table 4.1). Some of these tools perform gene prediction *ab initio*, relying only on the statistical parameters in the DNA sequence for gene identification. In contrast, homology-based methods rely primarily on identifying homologous sequences in other genomes and/or in public databases using BLAST or Smith-Waterman algorithms. Many of the commonly used methods combine these two approaches.

The absence of introns and relatively high gene density in most genomes of prokaryotes and some unicellular eukaryotes provides for effective use of sequence similarity searches as the first step in genome annotation. Genes identified by homology can be used as the training set for one of the statistical methods for gene recognition and the resulting statistical model can then be used for analyzing the remaining parts of the genome. In most eukaryotes, the abundance of introns and long intergenic regions makes it difficult to use homology-based methods as the first step unless, of course, one can rely on synteny between several closely related genomes (e.g., human, mouse, and rat). As a result, gene prediction for genome sequences of multicellular eukaryotes usually starts with *ab initio* methods, followed by similarity searches with the initial exon assemblies.

A detailed comparison of the algorithms and tools for gene prediction is beyond the scope of this book. We would only like to emphasize that each of these methods has its own advantages and limitations, and none of them is perfect. Therefore, it is advisable to use at least two different programs for gene prediction in a new DNA sequence, especially if it comes from a eukaryote or a poorly characterized prokaryote. A comparison of predictions generated by different programs reveals the cases where a given program performs the best and helps achieving consistent quality of gene prediction. Such a comparison can be performed, for example, using the TIGR Combiner program (http://www.tigr.org/softlab), which employs a voting scheme to combine predictions of different gene-finding programs, such as GeneMark, GlimmerM, GRAIL, GenScan, and Fgenes.

We describe only several computational tools that are most commonly used for gene prediction in large-scale genome annotation projects.

GeneMark

GeneMark (http://opal.biology.gatech.edu/genemark, mirrored at the EBI web site http://www.ebi.ac.uk/genemark) was developed by Mark Borodovsky and James McIninch in 1993 [108]. GeneMark was the first tool for finding prokaryotic genes that employed a non-homogeneous Markov model to classify DNA regions into protein-coding, non-coding, and non-coding but complementary to coding. It has been shown previously that, by multivariate codon usage analysis, the *E. coli* genes could be classified into so-called typical, highly typical and atypical gene sets, with the latter two groups apparently corresponding to highly expressed genes and horizontally transferred genes [562]. Accordingly, more than one Markov model was required to adequately describe different groups of genes in the same genome [109,110].

Like other gene prediction programs (see below), GeneMark relies on organism-specific recognition parameters to partition the DNA sequence into coding and non-coding regions and thus requires a sufficiently large training set of known genes from a given organism for best performance. The program has been repeatedly updated and modified and now exists in separate variants for gene prediction in prokaryotic, eukaryotic, and viral DNA sequences [88,528,768].

Glimmer

Gene Locator and Interpolated Markov Modeler (GLIMMER, http://www.tigr.org/softlab), developed by Steven Salzberg and colleagues at Johns Hopkins University and TIGR, is a system for finding genes in prokaryotic genomes. To identify coding regions and distinguish them from noncoding DNA, Glimmer uses interpolated Markov models, i.e. series of Markov models with the order of the model increasing at each step and the predictive power of each model separately evaluated [735]. Like GeneMark, Glimmer requires a training set, which is usually selected among known genes, genes coding for proteins with strong database hits and/or simply long ORFs. Glimmer is used as the primary gene finder tool at TIGR, where it has been applied to the annotation of numerous microbial genomes [241,243,610,891].

Recently, Salzberg and coworkers developed GlimmerM, a modified version of Glimmer specifically designed for gene recognition in small eukaryotic genomes, such as the malaria parasite *Plasmodium falciparum* [736].

Grail

Gene Recognition and Assembly Internet Link (GRAIL, http://compbio.ornl.gov), developed by Ed Uberbacher and coworkers at the Oak Ridge National Laboratory, is a tool that identifies exons, polyA sites, promoters, CpG islands, repetitive elements and frameshift errors in DNA sequences by comparing them to a database of known human and mouse sequence elements [858]. Exon and repetitive element prediction is also available for Arabidopsis and Drosophila sequences.

Grail has been recently incorporated into the Oak Ridge genome analysis pipeline (http://compbio.ornl.gov/tools/pipeline), which provides a unified web interface to a number of convenient analysis tools. For prokaryotes, it offers gene prediction using Glimmer (see above) and Generation programs, followed by BLASTP searches of predicted ORFs against SWISS-PROT and NR databases and a HMMer search against Pfam. There is also an option of BLASTN search of the submitted DNA sequence against a variety of nucleotide sequence databases.

For human and mouse sequences, the Oak Ridge pipeline offers gene prediction using GrailEXP and GenScan (see below), also followed by BLASTP searches of predicted ORFs against SWISS-PROT and NR databases and a HMMer search against Pfam. Again, the user can perform BLASTN search of the submitted DNA sequence against a variety of nucleotide sequence databases, as well as search for CpG islands, repeat fragments, tRNAs, and BAC-end pairs. As discussed above, the possibility to directly compare gene predictions made by two different programs is a valuable feature, which is available at the Oak Ridge web site.

GenScan

GenScan (http://genes.mit.edu/GENSCAN.html) was developed by Chris Burge and Samuel Karlin at Stanford University and is currently hosted in the Burge laboratory at the MIT Department of Biology. This program uses a complex probabilistic model of the gene structure that is based on actual biological information about the properties of transcriptional, translational and splicing signals. In addition, it utilizes various statistical properties of coding and noncoding regions. To account for the heterogeneity of the human genome that affects gene structure and gene density, GenScan derives different sets of gene models for genome regions with different GC content [131,132]. Its high speed and accuracy make GenScan the method of choice for the initial analysis of large (in the megabase range) stretches of eukaryotic genomic DNA. GenScan has being used as the principal tool for gene prediction in the International Human Genome Project [488].

GeneBuilder

GeneBuilder (http://www.itba.mi.cnr.it/webgene) performs *ab initio* gene prediction, using numerous parameters, such as GC content, dicodon frequencies, splicing site data, CpG islands, repetitive elements, and others. It also utilizes a unique approach that is based on evaluating relative frequencies of synonymous and nonsynonymous substitutions to identify likely coding sequences. In addition, it performs BLAST searches of predicted genes against protein and EST databases, which helps to refine the boundaries of predicted exons using the BLAST hits as guides. The program allows the user to change certain parameters, which permits interactive gene structure prediction. As a result, GeneBuilder is sometimes able to predict the gene structure with a good accuracy even when the similarity of the predicted ORF to a homologous protein sequence is low [569,708].

Table 4.1. Software tools for *ab initio* gene prediction

Program(s)	Author(s), WWW site	Program versions available for:	Refs.
GeneMark, GenMark.hmm	Mark Borodovsky, http://opal.biology.gatech.edu/GeneMark	Human, mouse, rat, chicken, *C. elegans*, *Drosophila*, rice, *Arabidopsis*, yeast, many bacteria and archaea	[88,108, 528,768]
Glimmer, GlimmerM	Steven Salzberg, http://www.tigr.org/softlab	Many bacteria and archaea, *Plasmodium*, *Aspergillus*, rice, *Arabidopsis*	[177,735, 736]
Grail, GrailEXP	Edward Uberbacher, http://compbio.ornl.gov	Human, mouse, *Drosophila*, *Arabidopsis*, *E. coli*	[858]
GenScan	Christopher Burge, http://genes.mit.edu/GENSCAN.html	Human, *Arabidopsis*, maize	[131,132]
GeneBuilder	Igor Rogozin, Luciano Milanesi http://www.itba.mi.cnr.it/webgene/	Human, mouse, rat, fugu, *Drosophila*, *C. elegans*, *Arabidopsis*, *Aspergillus*	[569,708]
Genie	David Kulp, David Haussler, http://www.cse.ucsc.edu/~dkulp/cgi-bin/genie	Human	[698]

GeneID	Roderick Guigo, http://www1.imim.es/software/geneid	Human, *Drosophila*	[657]
GeneFinder, Fgenes, Fgenesh	Victor Solovyev, http://genomic.sanger.ac.uk/gf/gf.shtml	Human, *Drosophila*, *C. elegans*, *Arabidopsis*, yeast	[730]
HMMgene	Anders Krogh, Anders Pedersen, Søren Brunak, http://www.cbs.dtu.dk/services	Human (vertebrates), *C. elegans*, *Arabidopsis*	[472]
GeneFinder, MZEF	Michael Zhang, http://www.cshl.org/genefinder	Human, mouse, *Arabidopsis*, *S. pombe*	[940]
GeneParser	Erik Snyder http://mcdb.colorado.edu/~eesnyder/ GeneParser.html	Available only for download	[789]

Splice site prediction

Programs for predicting intron splice sites, which are commonly used as subroutines in the gene prediction tools, can also be used as stand-alone programs to verify positions of splice sites or predict alternative splicing sites. Such programs (Table 4.2) can be particularly useful for predicting non-coding exons, which are commonly missed in gene prediction studies.

Recognition of the splice sites by these programs usually relies on statistical properties of exons and introns and on the consensus sequences of splicing signals. A detailed study of the performance of one such program, SpliceView, showed that, although the fraction of missed splicing signals was relatively low (~5%), the false positive rate was quite high (typically, one potential splicing signal per 150-250 bases). One should note, however, that such false-positive signals might correspond to rare alternative splice forms or cryptic splice sites (splice sites that are not active in normal genes and become activated as a result of mutations in major splicing signals) [710].

Table 4.2. Software tools for prediction of splicing sites

Program	Author, WWW site	Training sets	Refs.
HSPL	Victor Solovyev, http://genomic.sanger.ac.uk/gf/gf.shtml	Human	[792]
NNSplice	Martin Reese, http://www.fruitfly.org/seq_tools/splice.html	Human, *Drosophila*	[696]
SpliceView	Igor Rogozin, Luciano Milanesi http://www.itba.mi.cnr.it/webgene/	Human, yeast, *Drosophila*, *Arabidopsis*	[710]
NetGene2	Søren Brunak, http://www.cbs.dtu.dk/ services	Human, *C. elegans*, *Arabidopsis*	[123]
GeneSplicer	Steven Salzberg, http://www.tigr.org/softlab	Human, *Arabidopsis*	[672]
SpliceProximalCheck	Thangavel Thanaraj http://industry.ebi.ac.uk/~thanaraj/MZEF-SPC.html	Human	[838]

Combining various gene prediction tools

While the first step of gene identification in long genomic sequences utilizes *ab initio* programs that can rapidly and with reasonable accuracy predict multiple genes, the next step validates these predictions through similarity searches. Predicted genes are compared to nucleotide sequence databases, including EST databases, and protein sequences encoded by these predicted genes are compared to protein sequence databases. These data are then combined with the information about repetitive elements, CpG islands, and transcription factor binding sites, and used for further refinement of gene structure. Thus, homology information is ultimately incorporated into every gene prediction pipeline (see above). There are, however, several programs that primarily rely on similarity search for gene prediction (Table 4.3). Although differing in details, they all search for the best alignment of the given piece of DNA to the homologous nucleotide or protein sequences in the database.

Table 4.3. Software tools for homology-based gene prediction

Program	Authors, WWW site	Comment	Refs.
INFO	Michael Laub http://elcapitan.ucsd.edu/~info	Compares the given DNA sequence against GenBank	[492]
ORFGene	Rogozin and Milanesi http://www.itba.mi.cnr.it/webgene/	Compares translated DNA against SWISS-PROT	[569]
PipMaker	Webb Miller http://bio.cse.psu.edu	Aligns similar regions in two DNA sequences	[759]
AAT	Xiaoqiu Huang http://genome.cs.mtu.edu/aat.html	Searches the given DNA sequence against dbEST or HGI, or a protein against NR or SWISS-PROT	[361]
Procrustes	A.Mironov, M.Gelfand, P. Pevzner, http://www-hto.usc.edu/software/procrustes	Explores all possible exon assemblies in the given DNA sequence and finds the multi-exon structure with the best fit to a homologous protein	[276]
GeneWise	Ewan Birney, Richard Durbin http://www.sanger.ac.uk/Software/Wise2	Compares a given protein sequence to a genomic DNA sequence, allowing for introns and frameshift errors	[93]

4.2. Principles of Sequence Similarity Searches

As discussed in the previous section, initial characterization of any new DNA or protein sequence starts with a database search aimed at finding out whether or not homologs of this gene (protein) are already available, and if they are, what is known about them. Clearly, looking for exactly the same sequence is quite straightforward. One can just take the first letter of the query sequence and search for its first occurrence in the database, and then check if the second letter of the query is the same in the subject. If it is indeed the same, the program could check the third letter, then the fourth, and continue this comparison to the end of the query. If the second letter in the subject is different from the second letter in the query, the program should search for another occurrence of the first letter, and so on. This will identify all the sequences in the database that are identical to the query sequence (or include it). Of course, this approach is primitive computation-wise, and there are sophisticated algorithms for text matching that do it much more efficiently [92].

```
Query: 1 MK  → Query: 1 MKV  → Query: 1 MKVR  → Query: 1 MKVRA
Sbjct: 1 MK    Sbjct: 1 MKV    Sbjct: 1 MKVR    Sbjct: 1 MKVRA

...  →  Query: 1 MKVRASVKKLCRNCKIVKRDGVIRVICSAEPKHKQRQG
        Sbjct: 1 MKVRASVKKLCRNCKIVKRDGVIRVICSAEPKHKQRQG
```

Note that, in the example above, we looked only for sequences that ***exactly match*** the query. The algorithm would not even find a sequence that is identical to the query with the exception of the first letter. To find such sequences, the same analysis should be conducted with the fragments starting from the second letter of the original query, then from the third one, and so on.

```
Query1: 1  KVRASVKKLCRNCKIVKRDGVIRVICSAEPKHKQRQG
Query2: 1   VRASVKKLCRNCKIVKRDGVIRVICSAEPKHKQRQG
Query3: 1    RASVKKLCRNCKIVKRDGVIRVICSAEPKHKQRQG
Query4: 1     ASVKKLCRNCKIVKRDGVIRVICSAEPKHKQRQG
```

Such search quickly becomes time-consuming, and we are still dealing only with identical sequences. Finding **close relatives** would introduce additional conceptual and technical problems. Let us assume that sequences that are 99% identical are definitely homologous. What should one select as the threshold to consider sequences not to be homologous: 50% identity, 33%, or perhaps 25%? These are legitimate questions that need to be answered before one goes any further. The example of two lysozymes (see ♦2.1.2) shows that sequences with as low as 8% identity may belong to orthologous proteins and perform the same function.

As a matter of fact, when comparing nucleic acid sequences, there is very little one could do. All the four nucleotides, A, T, C, and G, are found in the database with approximately the same frequencies and have roughly the same probability of mutating one into another. As a result, DNA-DNA comparisons are largely based on straightforward text matching, which makes them fairly slow and not particularly sensitive, although a variety of heuristics have been developed to overcome this [571].

Amino acid sequence comparisons have several distinct advantages over nucleotide sequence comparisons, which, at least potentially, lead to a much greater sensitivity. Firstly, because there are 20 amino acids but only four bases, an amino acid match carries with it >4 bits of information as opposed to only two bits for a nucleotide match. Thus, statistical significance can be ascertained for much shorter sequences in protein comparisons than in nucleotide comparisons. Secondly, because of the redundancy of the genetic code, nearly one third of the bases in coding regions are under a weak (if any) selective pressure, and represent noise, which adversely affects the sensitivity of the searches. Thirdly, nucleotide sequence databases are much larger than protein databases because of the vast amounts of non-coding sequences coming out of eukaryotic genome projects and this further lowers the search sensitivity. Finally, and probably most importantly, unlike in nucleotide sequence, the likelihoods of different amino acid substitutions occurring during evolution are substantially different and taking this into account greatly improves the performance of database search methods as described below. Given all these advantages, comparisons of any coding sequences are typically carried out at the level of protein sequences; even when the goal is to produce a DNA-DNA alignment (e.g. for analysis of substitutions in silent codon positions), it is usually first done with protein sequences, which are then replaced by the corresponding coding sequences. Direct nucleotide sequence comparison is indispensable only when non-coding regions are analyzed.

4.2.1. Substitution scores and substitution matrices

The fact that each of the 20 standard protein amino acids has its own unique properties means that the likelihood of the substitution of each particular residue for another residue during evolution should be different. Generally, the more similar the physico-chemical properties of two residues the greater the chance that the substitution will not have an adverse effect on the protein's function and, accordingly, on the organism's fitness. Hence, in sequence comparisons, such a substitution should be penalized less than a replacement of amino acid residue with one that has dramatically different properties. This is, of course, an over-simplification because the effect of a substitution depends on the structural and functional environment where it occurs. For example, a cysteine to valine substitution in the catalytic site of an enzyme will certainly abolish the activity and, on many occasions, will have a drastic effect on the organism's fitness. In contrast, the same substitution within a β-strand may have little or no effect. Unfortunately, in general, we do not have *a priori* knowledge of the location of a particular residue in the protein structure, and even with such knowledge, incorporating it in a database search algorithm is an extremely complex task. Thus, a generalized measure of the likelihood of amino acid substitutions is required so that each substitution is given an appropriate score (weight) to be used in sequence comparisons. The score for a substitution between amino acids i and j always can be expressed by the following intuitively plausible formula, which shows how likely a particular substitution is given the frequencies of each the two residues in the analyzed database:

$$S_{ij} = k\ln(q_{ij}/p_i p_j) \tag{4.1}$$

where k is a coefficient, q_{ij} is the observed frequency of the given substitution and p_i, p_j are the background frequencies of the respective residues. Obviously, here the product $p_i p_j$ is the expected frequency of the substitution and, if $q_{ij}=p_i p_j$ $(S_{ij}=0)$, the substitution occurs just as often as expected. The scores used in practice are scaled such that the expected score for aligning a random pair of amino acid sequences is negative (see below).

There are two fundamentally different ways to come up with a substitution score matrix, i.e. a triangular table containing 210 numerical score values for each pair of amino acids, including identities (diagonal elements of the matrix; Figures 4.4 and 4.5). As in many other situations in computational biology, the first approach works *ab initio*, whereas the second one is empirical. One *ab initio* approach calculates the score as the number of nucleotide substitutions that are required to transform a codon for one amino acid in a pair into a codon for the other. In this case, the matrix is obviously unique (as long as alternative genetic codes are not considered) and contains only 4 values, 0,1,2 or 3. Accordingly, this is a very coarse grain matrix that is unlikely to work well. The other *ab initio* approach assigns scores on the basis of similarities and differences in physico-chemical properties of amino acids. Under this approach, the number of possible matrices is infinite, and they may have as fine a granularity as desirable, but a degree of arbitrariness is inevitable because our understanding of protein physics is insufficient to make informed decisions on what set of properties "correctly" reflects the relationships between amino acids.

Empirical approaches, which historically came first, attempt to derive the characteristic frequencies of different amino acid substitutions from actual alignments of homologous protein families. In other words, these approaches strive to determine the actual likelihood of each substitution occurring during evolution. Obviously, the outcome of such efforts critically depends on the quantity and quality of the available alignments, and even now, any alignment database is far from being complete or perfectly correct. Furthermore, simple counting of different types of substitutions will not suffice if alignments of distantly related proteins are included because, in many cases, multiple substitutions might have occurred in the same position. Ideally, one should construct the phylogenetic tree for each family, infer the ancestral sequence for each internal node and then count the substitutions exactly. This is not practicable in most cases and various shortcuts need to be taken.

Several solutions to these problems have been proposed, each resulting in a different set of substitution scores. The first substitution matrix, constructed by Dayhoff and Eck in 1968 [172], was based on an alignment of closely related proteins, so that the ancestral sequence could be deduced and all the amino acid replacements could be considered occurring just once. This model was then extrapolated to account for more distant relationships (we will not discuss here the mathematics of this extrapolation and the underlying evolutionary model [174]), which resulted in the PAM series of substitution matrices (Figure 4.4). PAM (Accepted Point Mutation) is a unit of evolutionary divergence of protein sequences, corresponding to one amino acid change per 100 residues. Thus, for example, the PAM30 matrix is supposed to apply to proteins that differ, on average, by 0.3 change per aligned residue, whereas PAM250 should reflect evolution of sequences with an average of 2.5 substitutions per position. Accordingly, the former matrix should be employed for constructing alignments of closely related sequences, whereas the latter is useful in database searches aimed at detection of distant relationships. Using an approach similar to that of Dayhoff, combined with rapid algorithms for protein sequence clustering and alignment, Jones, Taylor and Thornton produced the series of the so-called JTT matrices [403], which are essentially an update of the PAMs.

The PAM and JTT matrices, however, have obvious limitations because of the fact that they have been derived from alignments of closely related sequences and extrapolated to distantly related ones. This extrapolation may not be fully valid because the underlying evolutionary model might not be adequate and the trends that determine sequence divergence of closely related sequences might not apply to the evolution at larger distances.

In 1992, Steven and Jorja Henikoff developed a different series of substitution matrices [342] using conserved ungapped alignments of related proteins from the BLOCKS database (♦3.2.1). The use of these alignments offered three important advantages over the alignments used for constructing the PAM matrices. First, the BLOCKS collection obviously included a much larger number and, more importantly, a much greater diversity of protein families than the collection that was available to Dayhoff and coworkers in the 1970's. Second, coming from rather distantly related proteins, BLOCKS alignments better reflected the amino acid changes that occur over large phylogenetic distances and thus produced substitution scores that represented sequence divergence in distant homologs directly, rather than through extrapolation. Third, in these distantly related proteins, BLOCKS included only the most confidently aligned regions, which are likely to best represent the prevailing evolutionary trends. These substitution matrices, named the BLOSUM (= BLOcks SUbstitution Matrix) series, were tailored to particular evolutionary distances by ignoring the sequences that had more than a certain percent identity. In the BLOSUM62 matrix, for example, the substitution scores were derived from the alignments of sequences that had no more than 62% identity; the substitution scores of the BLOSUM45 matrix were calculated from the alignments that contained sequences with no more than 45% identity. Accordingly, BLOSUM matrices with high numbers, such as BLOSUM80, are best suited for comparisons of closely related sequences (it is also advisable to use BLOSUM80 for database searches with short sequences, see ♦4.4.2), whereas low-number BLOSUM matrices, such as BLOSUM45, are better for distant relationships. In addition to the general- purpose PAM, JTT and BLOSUM series, some specialized substitution matrices were developed, for example, for integral membrane proteins [404], but they never achieved comparable recognition.

Several early studies found the PAM matrices based on empirical data consistently resulted in greater search sensitivity than any of the *ab initio* matrices (see [186]). An extensive empirical comparison showed that: (i) BLOSUM matrices consistently outperformed PAMs in BLAST searches and (ii) on average, BLOSUM62 (Fig 4.5) performed best in the series [343]; this matrix is currently used as the default in most sequence database searches. It is remarkable that so far, throughout the 30 plus year history of amino acid substitution matrices, empirical matrices have consistently outperformed those based on theory, either physico-chemical or evolutionary. This is not to say, of course, that theory is powerless in this field, but to point out that we currently do not have a truly adequate theory to describe protein evolution. Clearly, the last word has not been said on amino acid substitution matrices and one can expect that eventually the BLOSUM series will be replaced by new matrices based on greater amounts of higher quality alignment data and more realistic evolutionary models. A recently reported maximum-likelihood model for substitution frequency estimation has already been claimed to describe individual protein families better than the Dayhoff and JTT models [889]. It remains to be seen how this and other new matrices perform in large-scale computational experiments on real databases. AAindex (http://www.genome.ad.jp/dbget/aaindex.html, ♦3.6.3), lists 66 different substitution matrices, both *ab initio* and empirical, and there is no doubt that this list will continue to grow [429].

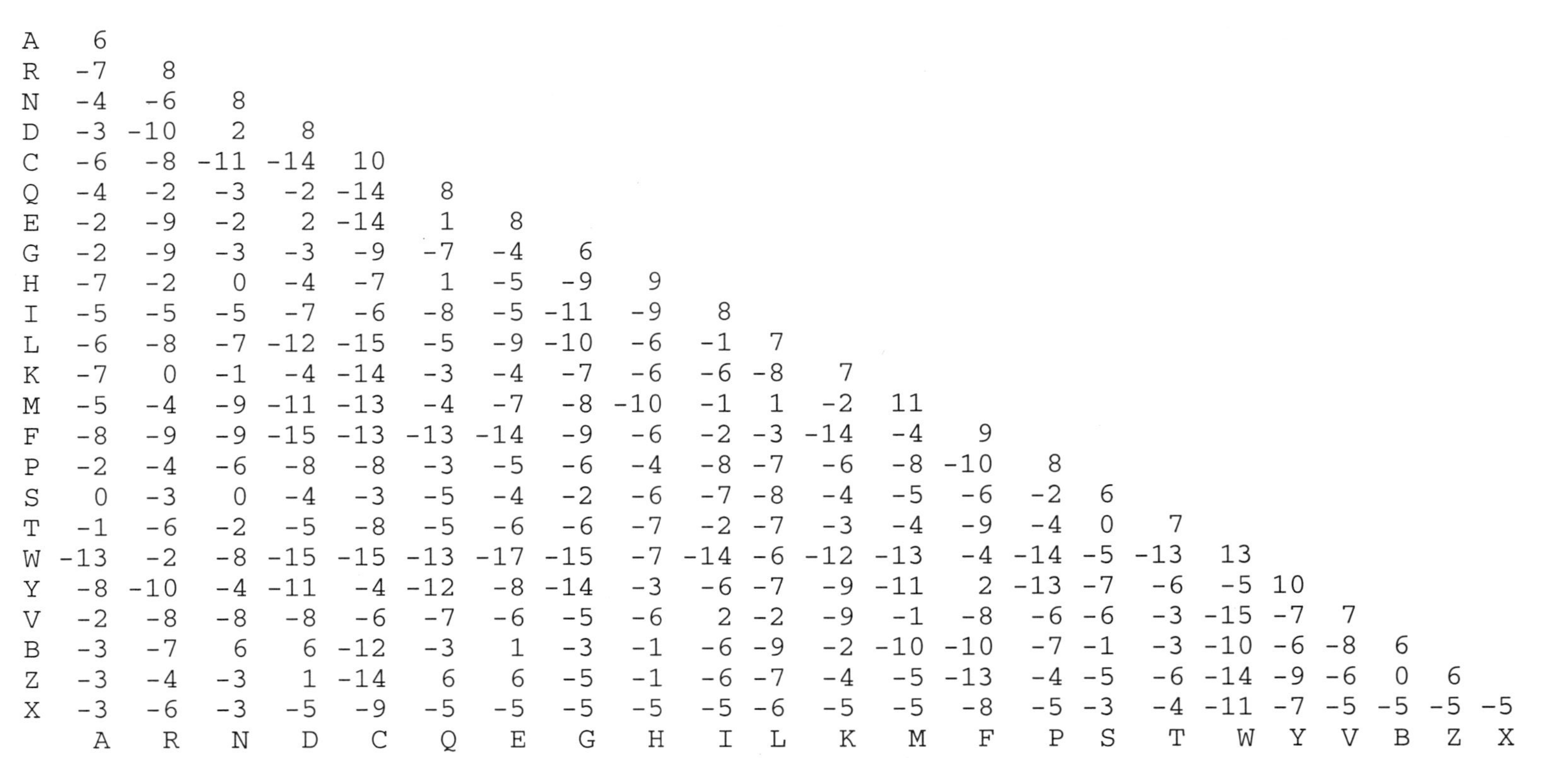

	A	R	N	D	C	Q	E	G	H	I	L	K	M	F	P	S	T	W	Y	V	B	Z	X
A	6																						
R	-7	8																					
N	-4	-6	8																				
D	-3	-10	2	8																			
C	-6	-8	-11	-14	10																		
Q	-4	-2	-3	-2	-14	8																	
E	-2	-9	-2	2	-14	1	8																
G	-2	-9	-3	-3	-9	-7	-4	6															
H	-7	-2	0	-4	-7	1	-5	-9	9														
I	-5	-5	-5	-7	-6	-8	-5	-11	-9	8													
L	-6	-8	-7	-12	-15	-5	-9	-10	-6	-1	7												
K	-7	0	-1	-4	-14	-3	-4	-7	-6	-6	-8	7											
M	-5	-4	-9	-11	-13	-4	-7	-8	-10	-1	1	-2	11										
F	-8	-9	-9	-15	-13	-13	-14	-9	-6	-2	-3	-14	-4	9									
P	-2	-4	-6	-8	-8	-3	-5	-6	-4	-8	-7	-6	-8	-10	8								
S	0	-3	0	-4	-3	-5	-4	-2	-6	-7	-8	-4	-5	-6	-2	6							
T	-1	-6	-2	-5	-8	-5	-6	-6	-7	-2	-7	-3	-4	-9	-4	0	7						
W	-13	-2	-8	-15	-15	-13	-17	-15	-7	-14	-6	-12	-13	-4	-14	-5	-13	13					
Y	-8	-10	-4	-11	-4	-12	-8	-14	-3	-6	-7	-9	-11	2	-13	-7	-6	-5	10				
V	-2	-8	-8	-8	-6	-7	-6	-5	-6	2	-2	-9	-1	-8	-6	-6	-3	-15	-7	7			
B	-3	-7	6	6	-12	-3	1	-3	-1	-6	-9	-2	-10	-10	-7	-1	-3	-10	-6	-8	6		
Z	-3	-4	-3	1	-14	6	6	-5	-1	-6	-7	-4	-5	-13	-4	-5	-6	-14	-9	-6	0	6	
X	-3	-6	-3	-5	-9	-5	-5	-5	-5	-5	-6	-5	-5	-8	-5	-3	-4	-11	-7	-5	-5	-5	-5

Figure 4.4. The PAM30 substitution matrix. The numbers indicate the substitution scores for each replacement. The greater the number the lesser the penalty for the given substitution. Note the high penalty for replacing Cys and aromatic amino acids (Phe, Tyr, and Trp) with any other residues and, accordingly, the high reward for conservation of these residues (see the diagonal elements).

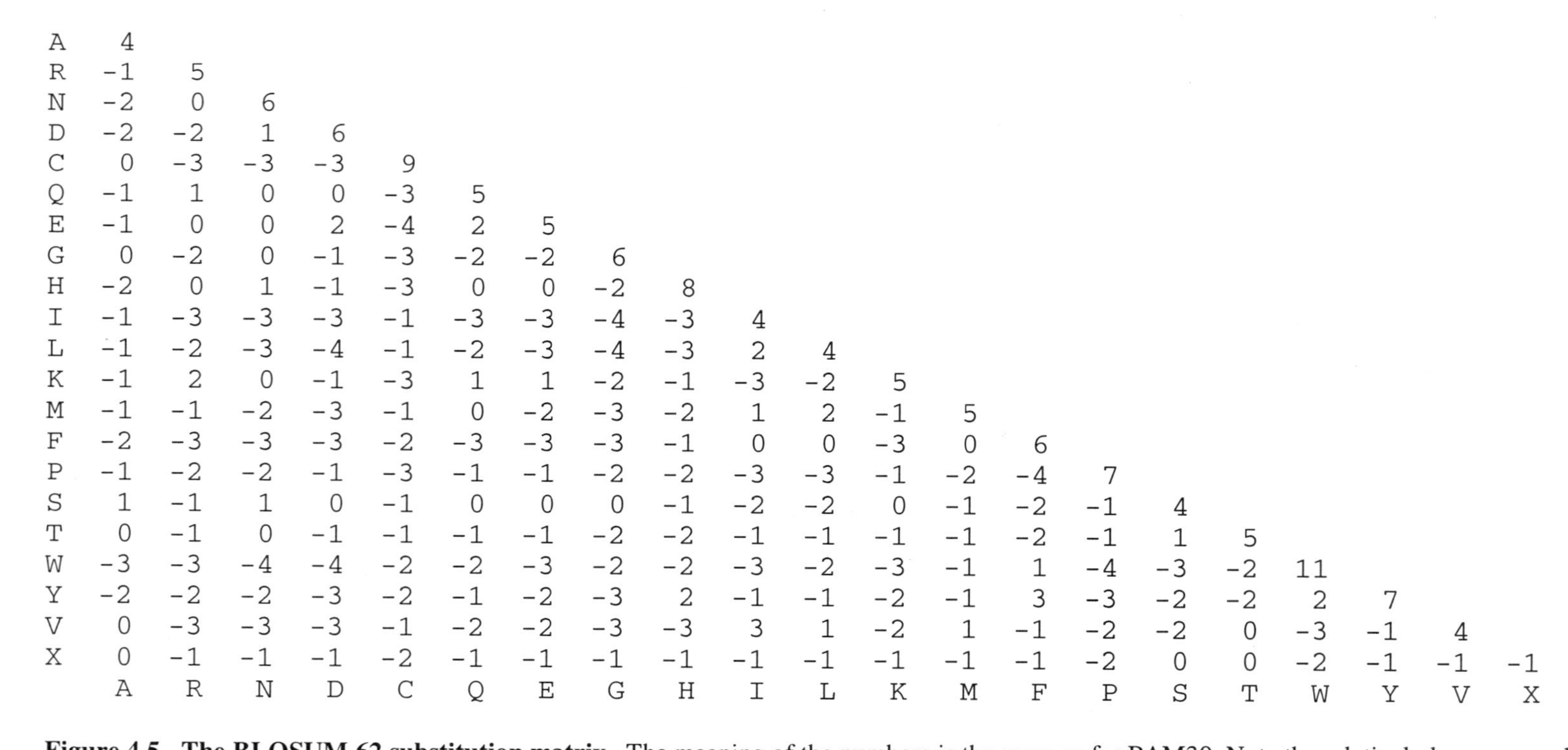

	A	R	N	D	C	Q	E	G	H	I	L	K	M	F	P	S	T	W	Y	V	X
A	4																				
R	-1	5																			
N	-2	0	6																		
D	-2	-2	1	6																	
C	0	-3	-3	-3	9																
Q	-1	1	0	0	-3	5															
E	-1	0	0	2	-4	2	5														
G	0	-2	0	-1	-3	-2	-2	6													
H	-2	0	1	-1	-3	0	0	-2	8												
I	-1	-3	-3	-3	-1	-3	-3	-4	-3	4											
L	-1	-2	-3	-4	-1	-2	-3	-4	-3	2	4										
K	-1	2	0	-1	-3	1	1	-2	-1	-3	-2	5									
M	-1	-1	-2	-3	-1	0	-2	-3	-2	1	2	-1	5								
F	-2	-3	-3	-3	-2	-3	-3	-3	-1	0	0	-3	0	6							
P	-1	-2	-2	-1	-3	-1	-1	-2	-2	-3	-3	-1	-2	-4	7						
S	1	-1	1	0	-1	0	0	0	-1	-2	-2	0	-1	-2	-1	4					
T	0	-1	0	-1	-1	-1	-1	-2	-2	-1	-1	-1	-1	-2	-1	1	5				
W	-3	-3	-4	-4	-2	-2	-3	-2	-2	-3	-2	-3	-1	1	-4	-3	-2	11			
Y	-2	-2	-2	-3	-2	-1	-2	-3	2	-1	-1	-2	-1	3	-3	-2	-2	2	7		
V	0	-3	-3	-3	-1	-2	-2	-3	-3	3	1	-2	1	-1	-2	-2	0	-3	-1	4	
X	0	-1	-1	-1	-2	-1	-1	-1	-1	-1	-1	-1	-1	-1	-2	0	0	-2	-1	-1	-1

Figure 4.5. The BLOSUM 62 substitution matrix. The meaning of the numbers is the same as for PAM30. Note the relatively lower reward for conservation of Cys, Phe, Tyr, and Trp and lower penalties for replacing these amino acids than in the PAM30 matrix. This trend is even stronger in lower series members (e.g. BLOSUM45) because drastic amino acid changes are more likely at larger evolutionary distances.

4.2.2. Statistics of protein sequence comparison

It is impossible to explain even the basic principles of statistical analysis of sequence similarities without invoking some mathematics. To introduce these concepts in the least painful way, let us consider the same protein sequence (*E. coli* RpsJ) as above

```
Query: 1 MKVRASVKKLCRNCKIVKRDGVIRVICSAEPKHKQRQG 38
```

and check how many times segments of this sequence of different lengths are found in the database (we chose fragments starting from the second position in the sequence because nearly every protein in the database starts with a methionine). Not unexpectedly, we find that the larger the fragment, the smaller the number of exact matches in the database (Table 4.4).

Table 4.4. Dependence of the number of exact database matches on the length of the query word.

Sequence	Occurrences in the database
KV	488,559
KVR	28,592
KVRA	2,077
KVRAS	124
KVRASV	23
KVRASVK	8
KVRASVKK	4
KVRASVKKL	1
KVRASVKKLC	1

Perhaps somewhat counterintuitively, a 9-mer is already unique. With the decrease in the number of database hits, the likelihood that these hits are biologically relevant, i.e. belong to homologs of the query protein, increases. Thus, 13 of the 23 occurrences of the string KVRASV and all 8 occurrences of the string KVRASVK are from RpsJ orthologs.

The number of occurrences of a given string in the database can be roughly estimated as follows. The probability of matching one amino acid residue is 1/20 (assuming equal frequencies of all 20 amino acids in the database; this not being the case, the probability is slightly greater). The probability of matching two residues in a row is then $(1/20)^2$, and the probability of matching n residues is $(1/20)^n$. Given that the protein database currently contains $N \sim 2\text{x}10^8$ letters, one should expect a string of n letters to match approximately $N\text{x}(1/20)^n$ times, which is fairly close to the numbers in Table 4.4.

Searching for perfect matches is the simplest and, in itself, obviously insufficient form of sequence database search, although, as we shall see below, it is important as one of the basic steps in currently used search algorithms. As repeatedly mentioned above, the goal of a search is finding homologs, which can have drastically different sequences such that, in distant homologs, only a small fraction of the amino acid residues are identical or even similar. Even in close homologs, a region of high similarity is usually flanked by dissimilar regions like in the following alignment of *E. coli* RpmJ with its ortholog from *Vibrio cholerae*:

```
E. coli RpmJ:  1 MKVRASVKKLCR---NCKIVKRDGVIRVICSAEPKHKQRQG
                 MKV +S+K         +C+IVKR G + VIC + P+ K  Q
Vibrio VC0879: 1 MKVLSSLKSAKNRHPDCQIVKRRGRLYVICKSNPRFKAVQR
```

In this example, the region of highest similarity is in the middle of the alignment, but including the less conserved regions on both sides improves the overall score (taking into account the special treatment of gaps, which is introduced below). Further along the alignment, the similarity almost disappears so that inclusion of additional letters into the alignment would ***not*** increase the overall score or would even decrease it. Such fragments of the alignment of two sequences whose similarity score cannot be improved by adding or trimming any letters are referred to as ***high-scoring segment pairs (HSPs).*** For this approach to work, the expectation of the score for random sequences must be negative, and the scoring matrices used in database searches are scaled accordingly (see Figs. 4.4 and 4.5).

So, instead of looking for perfect matches, sequence comparisons programs actually search for HSPs. Once a set of HSPs is found, different methods, such as Smith-Waterman, FASTA, or BLAST, deal with them in different fashions (see below). However, the principal issue that any database search method needs to address is identifying those HSPs that are unlikely to occur by chance and, by inference, are likely to belong to homologs and to be biologically relevant. This problem has been solved by Samuel Karlin and Stephen Altschul who showed that maximal HSP scores follow the extreme value distribution [421]. Accordingly, if the lengths of the query sequence (m) and the database (n) are sufficiently high, the expected number of HSPs with a score of at least S is given by the formula

$$E = Kmn2^{-\lambda S} \qquad (4.2)$$

Here, S is the so-called raw score calculated under a given scoring system, and K and λ are natural scaling parameters for the search space size and the scoring system, respectively. Normalizing the score according to the formula:

$$S' = (\lambda S - lnK)/ln2 \quad (4.3)$$

gives the bit score, which has a standard unit accepted in information theory and computer science. Then,

$$E = mn2^{-S'} \quad (4.4)$$

and, since it can be shown that the number of random HSPs with score ≥S' is described by Poisson distribution, the probability of finding at least one HSP with bit score ≥S' is

$$P = 1 - e^{-E} \quad (4.5)$$

Equation (4.5) links two commonly used measures of sequence similarity, the probability (P-value) and expectation (E-value). For example, if the score S is such that three HSPs with this score (or greater) are expected to be found by chance, the probability of finding at least one such HSP is $(1-e^{-3}) \sim 0.95$. By definition, P-values vary from 0 to 1, whereas E-values can be much greater than 1. The BLAST programs (see below) report E-values, rather than P-values, because E-values of, for example, 5 and 10 are much easier to comprehend than P-values of 0.993 and 0.99995. However, for $E < 0.01$, P-value and E-value are nearly identical.

The product mn defines the ***search space***, a critically important parameter of any database search. Equations (4.2) and (4.4) codify the intuitively obvious notion that the larger the search space the higher the expectation of finding an HSP with a score greater than any given value. There are two corollaries of this that might take some getting used to: (i) the same HSP may come out statistically significant in a small database and not significant in a large database; with the natural growth of the database, any given alignment becomes less and less significant (but by no means less important because of that) and (ii) the same HSP may be statistically significant in a small protein (used as a query) and not significant in a large protein.

Clearly, one can easily decrease the E-value and the P-value associated with the alignment of the given two sequences by lowering n in equation (4.2), i.e. by searching a smaller database. However, the resulting increase in significance is false, although such a trick can be useful for detecting initial hints of subtle relationships that should be subsequently verified using other approaches. It is the experience of the authors that the simple notion of E(P)-value is often misunderstood and interpreted as if these values applied just to a single pairwise comparison (i.e., if an E-value of 0.001 for an HSP with score S is reported, then, in a database of just a few thousand sequences, one expects to find a score >S by chance). It is critical to realize that the size of the search space is already factored in these E-values and the reported value corresponds to the database size at the time of search (thus, it is certainly necessary to indicate, in all reports of sequence analysis, which database was searched, and desirably, also on what exact date).

Speaking more philosophically (or futuristically), one could imagine that, should the genomes of all species that inhabit this planet be sequenced, it would become almost impossible to demonstrate statistical significance for any but very close homologs in standard database searches. Thus, other approaches to homology detection are required that counter the problems created by database growth by taking advantage of the simultaneously increasing sequence diversity and, as discussed below, some have already been developed.

The Karlin-Altschul statistics has been rigorously proved to apply only to sequence alignments that do not contain gaps, whereas statistical theory for the more realistic gapped alignments remains an open problem. However, extensive computer simulations have shown that these alignments also follow the extreme value distribution to a high precision; therefore, at least for all practical purposes, the same statistical formalism is applicable [19,22].

Those looking for a detailed mathematical description of the sequence comparison statistics can find it in the "Further Reading" list at the end of the chapter. A brief explanation of the statistical principles behind the BLAST program, written by Stephen Altschul, can be found online at http://www.ncbi.nlm.nih.gov/BLAST/tutorial/Altschul-1.html.

4.2.3. Protein sequence complexity. Compositional bias

The existence of a robust statistical theory of sequence comparison, in principle, should allow one to easily sort search results by statistically significance and accordingly assign a level of confidence to any homology identification. However, a major aspect of protein molecule organization substantially complicates database search interpretation and may lead to gross errors in sequence analysis. Many proteins, especially in eukaryotes, contain ***low (compositional) complexity regions***, in which the distribution of amino acid residues is non-random, i.e. deviates from the standard statistical model [919,921]. In other words, these regions typically have biased amino acid composition, e.g. are rich in glycine or proline, or in acidic, or basic amino acid residues. The ultimate form of low complexity is, of course, a homopolymer, such as a Q-linker [920]. Other low-complexity sequences have a certain amino acid periodicity, sometimes subtle, such as, for example, in coiled-coil and other non-globular proteins (e.g., collagen or keratin).

The notion of compositional complexity was encapsulated in the SEG algorithm and the corresponding program, which partitions protein sequences into segments of low and high (normal) complexity [921]. An important finding made by John Wootton is that low-complexity sequences correspond to non-globular portions of proteins [919]. In other words, a certain minimal level of complexity is required for a sequence to fold into a globular structure. Low-complexity regions in proteins, although devoid of enzymatic activity, have important biological functions, most often promoting protein-protein interactions or cellular adhesion to various surfaces and to each other.

In a detailed empirical study, a set of parameters of the SEG program was identified that allowed reasonably accurate partitioning of a protein sequence into predicted globular and non-globular parts. The ***mastermind*** protein of *Drosophila* is a component of Notch-dependent signaling pathway and plays an important role in the development of the nervous system of the fruit fly [755,785]. In spite of this critical biological function, this protein consists mostly of stretches of only three amino acid residues, Gln, Asn, and Gly and is predicted to have a predominantly non-globular structure (Figure 4.6). Recently discovered human homologs of ***mastermind*** are also involved in Notch-dependent transcriptional regulation and similarly appear to be almost entirely non-globular [442,923].

```
>gi|126721|sp|P21519|MAM_DROME NEUROGENIC PROTEIN
MASTERMIND
                                  1-13    MDAGGLPVFQSAS
qaaavaqqqqqqqqqqhlnlqlhqqhlglh   14-85
lqqqqqlqlqqqqhnaqaqqqqiqvqqqqq
                  qqqqqqqqqqqh
                                 86-88    SPY
nanlgatggiagitggngaggptnpgavpt   89-121
                           apg
                                122-195   DTMPTKRMPVVDRLRR
                                          RMENYRRRQTDCVPRY
                                          EQAFNTVCEQQNQETT
                                          VLQKRFLESKNKRAAK
                                          KTDKKLPDPS
        qqhqqqqhqqqqqhqqhqqhqq  196-217
                                218-258   AQTMLAGQLQSSVHVQ
                                          QKFLKRPAEDVDNGPD
                                          SFEPPHKLP
nnnnnsnsnnnngnananngggngsntgnnt 259-460
nnngnstnnnggsnnngsenltkfsveivq
qlefttspansqpqqistnvtvkaltntsv
ksepgvggggggggggnsgnnnnngggggg
gngnnnnnggdhhqqqqqhqhqqqqqqqgg
glgglgnngrgggpggmatgpggvagglgg
        mgmppnmmsaqqksalgnlanl
                                461-465   VECKR
```

```
epdhdfpdlgsldkdggggqfpgfpdllgd  466-755
dnsenndtfkdlinnlqdfnpsfldgfdek
plldiktedgikveppnaqdlinslnvkse
gglghgfggfglgldnpgmkmrggnpgnqg
gfpngpnggtggapnaggnggnsgnlmseh
plaaqtlkqmaeqhqhknamggmggfprpp
hgmnpqqqqqqqqqqqqqqaqqqhgqmmgq
gqpgryndygggfpndfglgpngpqqqqqq
aqqqqpqqqhlppqfhqqkgpgpgagmnvq
          qnfldikqelfyssqndfdl
                                756-758  KRL
qqqqamqqqqqqqhhqqqqpkmggvpnfnk  759-1594
qqqqqqvpqqqlqqqqqqqqqqqqqqqqys
pfsnqnpnaaanflncpprggpngnqqpgn
laqqqqqpgagpqqqqqrgnagngqqnnpn
tgpggntpnapqqqqqqqsttttlqmkqtq
qlhisqqgggaqgiqvsagqhlhlsgdmks
nvsvaaqqgvffsqqqaqqqqqqqqpggtn
gpnpqqqqqqphggnagggvgvgvgvgn
ggpnpgqqqqqpnqnmsnanvpsdgfslsq
sqsmnfnqqqqqqaaaqqqqvqpnmrqrqt
qaqaaaaaaaaaaqaqaaanasgpnvplmq
qpqvgvgvgvgvgvgvgnggvvggpgsg
gpnngamnqmggpmggmpgmqmggpmnpmq
mnpnaagptaqqmmmgsgaggpgqvpgpgq
gpnpnqakflqqqqmmraqamqqqqqhmsg
arppppeynatkaqlmqaqmmqqtvggggv
gvggvgvgvgvggvgganggrfpnsaaqaa
amrrmtqqpippsgpmmrpqhamymqqhgg
agggprtgmgvpygggrggpmggpqqqqrp
pnvqvtpdgmpmgsqqewrhmmmtqqqtqm
gfggpgpggpmrqgpggfnggnfmpngapn
gaagsgpnaggmmtgpnvpqmqltpaqmqq
qlmrqqqqqqqqqqqqhmgpgaannmqmqq
llqqqqsggggnmmasqmqmtsmhmtqtqq
qitmqqqqqfvqsttttthqqqqmmqmgpg
gggggggpgsannnnggggggaagggnsas
tiasassisqtinsvvansndfgleflfdnl
pvdsnfstqdlinsldndnfnlqdfn
                                1595-1596 MP
```

Figure 4.6. Sequence of the Drosophila *mastermind* protein: partitioning into predicted non-globular (left column) and globular (right column) regions. The SEG program was run with the parameters optimized for detection of non-globular regions: window length 45, trigger complexity 3.4, extension complexity 3.7. See [919,921] for details. Asn, Gln, and Gly residues are shown in bold. Note that, because the existence of globular domains consisting of <50 amino acids is unlikely, *mastermind* probably contains only one globular domain, between amino acid residues 122 and 195. Prediction of short segments as 'globular' appears to be a SEG artifact.

Low-complexity regions represent a major problem for database searches. Since the λ parameter of equation (4.2) is calculated for the entire database, Karlin-Altschul statistics breaks down when the composition of the query or a database sequence or both significantly deviates from the average composition of the database. The result is that low-complexity regions with similar composition (e.g. acidic or basic) often produce "statistically significant" alignments that have nothing to with homology and are completely irrelevant. The SEG program can be used to overcome this problem in a somewhat crude manner: the query sequence, the database or both can be partitioned into normal complexity and low-complexity regions and the latter are masked (i.e. amino acid symbols are replaced with the corresponding number of X's). For the purpose of a database search, such filtering is usually done using short windows so that only the segments with a strongly compositional bias are masked. Low-complexity filtering has been indispensable for making database search methods, in particular BLAST, into reliable tools [18]. Without masking low-complexity regions, false results would have been produced for a substantial fraction of proteins, especially eukaryotic ones (an early estimate held that low-complexity regions comprised ~15% of the protein sequences in the SWISS-PROT database [919]). These false results would have badly polluted any large-scale database search and the respective proteins would have been refractory to any meaningful sequence analysis. For these reasons, for several years, SEG filtering had been used as the default for BLAST searches to mask low-complexity segments in the query sequence. However, this procedure is not without its drawbacks. Not all low-complexity sequences are captured and false-positives still occur in database searches. The opposite problem also hampers database searches for some proteins: when short low-complexity sequences are parts of conserved regions, statistical significance of an alignment may be underestimated, sometimes grossly.

In a recent work of Alejandro Schäffer and colleagues, a different, less arbitrary approach for dealing with compositionally biased sequences was introduced [748]. This method, called composition-based statistics, recalculates the λ parameter and, accordingly, the E values [see equation (4.2)] for each query and each database sequence, thus correcting the inordinately low ("significant") E-values for sequences with similarly biased amino acid composition. This improves the accuracy of the reported E-values and eliminates most false-positives. Composition-based statistics is currently used as the default for the NCBI BLAST. In ♦4.4.2, we will discuss the effect of this procedure on database search outcome in greater detail and using specific examples.

4.3. Algorithms for Sequence Alignment and Similarity Search

4.3.1. The basic alignment concepts and principal algorithms

As discussed in the previous sections, similarity searches aim at identifying the homologs of the given query protein (or nucleotide) sequence among all the protein (or nucleotide) sequences in the database. Even in this general discussion, we repeatedly mentioned and, on some occasions, showed sequence alignments. An alignment of homologous protein sequences reveals their common features that are ostensibly important for the structure and function of each of these proteins; it also reveals poorly conserved regions that are less important for the common function, but might define the specificity of each of the homologs. In principle, the only way to identify homologs is by aligning the query sequence against all the sequences in the database (below we will discuss some important heuristics that allow an algorithm to skip sequences that are obviously unrelated to the query), sorting these hits based on the degree of similarity, and assessing their statistical significance that is likely to be indicative of homology. Thus, before considering algorithms and programs used to search sequence databases, we must briefly discuss alignment methods themselves.

It is important to make a distinction between a ***global*** (i.e. full-length) ***alignment*** and a ***local alignment***, which includes only parts of the analyzed sequences (subsequences). Although, in theory, a global alignment is best for describing relationships between sequences, in practice, local alignments are of more general use for two reasons. Firstly, it is common that only parts of compared proteins are homologous (e.g. they share one conserved domain, whereas other domains are unique). Secondly, on many occasions, only a portion of the sequence is conserved enough to carry a detectable signal, whereas the rest has diverged beyond recognition. Optimal global alignment of two sequences was first realized in the Needleman-Wunsch algorithm, which employs dynamic programming [606]. The notion of optimal local alignment (the best possible alignment of two subsequences from the compared sequences) and the corresponding dynamic programming algorithm were introduced by Smith and Waterman [784]. Both of these are $O(n^2)$ algorithms, i.e. the time and memory required to generate an optimal alignment are proportional to the product of the lengths of the compared sequences (for convenience, the sequences are assumed to be of equal length n in this notation). Optimal alignment algorithms for multiple sequences have the $O(n^k)$ complexity (where k is the number of compared sequences). Such algorithms for k >3 are not feasible on any existing computers, therefore all available methods for multiple sequence alignments produce only approximations and do not guarantee the optimal alignment.

It might be useful, at this point, to clarify the notion of optimal alignment. Algorithms like Needleman-Wunsch and Smith-Waterman guarantee the optimal alignment (global and local, respectively) for ***any*** two compared sequences. It is important to keep in mind, however, that this optimality is a purely formal notion, which means that, given a scoring function, the algorithm outputs the alignment with the highest possible score. This has nothing to with statistical significance of the alignment, which has to be estimated separately (e.g., using the Karlin-Altschul statistics as outlined above), let alone biological relevance of the alignment.

For better or worse, alignment algorithms treat protein or DNA as simple strings of letters without recourse to any specific properties of biological macromolecules. Therefore it might be useful to illustrate the principles of local alignments using a text free of biological context as an example. Below is the text of stanza I and IV of one of the most famous poems of all times; we shall compare them line by line, observing along the way various problems involved in sequence alignment (the alignable regions are shown in bold):

I

"Once upon a midnight dreary, while I p**ondere**d, weak and weary,
Over many a quaint and curious volume of **forg**otten **lore**,
While I nodded, nearly **napping**, sudd**enly** there **came** a t**apping**,
As of some one ge**ntly** r**apping,** r**apping at my chamber door**.
"'Tis some visitor," I muttered, "**tapping at my chamber door**-
Only this, **and nothing more**."

IV

"Presently my soul grew str**onger**; hesitating then no l**onger**,
"Sir," said I, "or Madam, truly your **forg**iveness I imp**lore**;
But the fact is I was **napping**, and so g**ently** you **came** r**apping**,
And so fai**ntly** you came t**apping,** t**apping at my chamber door**,
That I scarce was sure I heard you"- here I opened wide the door;-
Darkness there, **and nothing more**. "

It is easy to see that, in the first two lines of the two stanza, the longest common string consists of only five letters, with one mismatch:

```
...I pondered ...                              (I)
...stronger...
```

The second lines align better, with two similar blocks separated by spacers of variable lengths, which requires gaps to be introduced, in order to combine them in one alignment:

...of **forg**ott**en**--- - ---**lore** (II)
your **forg**iv-**en**ess I imp**lore**

In the third lines, there are common words of seven, four and six letters, again separated by gaps:

...**napping** sud - **den-ly** there **came** a t**apping**, (III)
...**napping** and so **gen**t**ly** you-- **came** - r**apping**

The fourth lines align very well, with a long string of near identity at the end:

As of **so**me one ge**ntly** --- ---- r**apping** r**apping at my chamber door** (IV)
An d- **so**-- --f ai**ntly** you came t**apping** t**apping at my chamber door**

In contrast, there is no reasonable alignment between the fifth lines, except for the identical word 'door'. Obviously, however, the fourth line of the second stanza may be aligned not only with the fourth (IV), but also with the fifth line of the first stanza:

... I muttered **tapping at my chamber door** (IV')
... came tapping **tapping at my chamber door**

Alignments (IV) and (IV') can thus be combined to produce a multiple alignment:

...r**apping** r**apping at my chamber door** (IV'')
...t**apping** t**apping at my chamber door**
...------- t**apping at my chamber door**

Finally, sixth lines of the two stanza could be aligned at their ends:

Only **th**is- **and nothing more** (V)
Darkness **there and nothing more**

This simple example seems to capture several important issues that emerge in sequence alignment analysis. Firstly, remembering that an optimal alignment can be obtained for any two sequences, we should ask: which alignments actually reflect homology of the respective lines? The alignments III, IV, IV' (and the derivative IV'') and V seem to be relevant beyond reasonable doubt. However, are they really correct? In particular, aligning en-ly/ently in III and ntly/ntly in IV require introducing gaps into both sequences? Is this justified? We cannot answer this simple question without a statistical theory for assessing the significance of an alignment, including a way to introduce some reasonable gap penalties.

The treatment of gaps is one of the hardest and still unsolved problems of alignment analysis. There is no theoretical basis for assigning gap penalties relative to substitution penalties (scores). Deriving these penalties empirically is a much more complicated task than deriving substitution penalties as in PAM and BLOSUM series because, unlike the alignment of residues in highly conserved blocks, the number and positions of gaps in alignments tend to be highly uncertain (see, for example alignment IV: is it correct to place gaps both before and after 'so' in the second line?). Thus, gap penalties typically are assigned on the basis of two notions that stem both from the existing understanding of protein structure and from empirical examinations of protein family alignments: (i) deletion or insertion resulting in a gap is much less likely to occur than even the most radical amino acid substitution and should be heavily penalized and (ii) once a deletion (insertion) has occurred in a given position, deletion or insertion of additional residues (gap extension) becomes much more likely. Therefore a linear function:

$$G=a+bx, \quad a>>b \tag{4.6}$$

where a is the gap opening penalty, b is the gap extension penalty and x is the length of the gap is used to deal with gaps in most alignment methods. Typically, a=10 and b=1 is a reasonable choice of gap penalties to be used in conjunction with the BLOSUM62 matrix. Using these values, the reader should be able to find out whether or not gaps should have been introduced in alignments III and IV above. In principle, objective gap penalties could be produced through analysis of distributions of gaps in structural alignments, and such a study suggested using convex functions for gap penalties [84]. However, this makes alignment algorithms much costlier computationally and the practical advantages remained uncertain, so linear gap penalties are still universally employed.

The feasibility of alignments (IV) and (IV') creates the problem of choice: which of these is the correct alignment? Alignment (IV) wins because it clearly has a longer conserved region. What is, then, the origin of line 5 in the first stanza and, accordingly, of alignment (IV')? It is not too difficult to figure out that this is a repeat, a result of duplication of line 4 (this is what we have to conclude given that line 4 is more similar to the homologous line in the second stanza). Such duplications are common in protein sequences, too, and often create major problems for alignment methods.

We concluded that lines 3, 4 and 6 in each stanza of "Raven" are homologous, i.e. evolved from common ancestors with some subsequent divergence. In this case, the conclusion is also corroborated by the fact we recognize the English words in these lines and see that they are indeed nearly the same and convey similar meanings, albeit differing in nuances. What about alignments (I) and (II)? The content here tells us that no homology is involved, even though alignment (II) looks "believable". However, it would not have been recognized as statistically significant in a search of any sizable database, such as, for example, the "Complete poems of Edgar Allan Poe" at the American Verse Project of the University of Michigan Humanities Text Initiative (http://www.hti.umich.edu/index-all.html).

Is this similarity purely coincidental, then? Obviously, it is not. This is a case of convergence, a phenomenon whose role in molecular evolution we already had a chance to discuss in Chapter 2. Of course, in this case, the source of convergence is known: Edgar Allan Poe deliberately introduced these similarities for the sake of rhyme and alliteration. Such a force, to our knowledge, does not exist in molecular evolution, but analogous functional constraints act as its less efficient substitute.

Most of the existing alignment methods utilize modifications of the Smith-Waterman algorithm. Although it is not our goal here to discuss the latest developments in sequence alignment, the reader has to keep in mind that this remains an active research field, with a variety of algorithms and tools being developed, which at least claim improvements over the traditional ones appearing at a high rate. Just one recent example is BALSA, a Bayesian local alignment algorithm that explores series of substitution matrices and gap penalty values and assesses their posterior probabilities, thus overcoming some of the shortcomings of the Smith-Waterman algorithm [885].

Pairwise alignment methods are important largely in the context of a database search. For analysis of individual protein families, ***multiple alignment*** methods are critical. We believe that anyone routinely involved in protein family analysis would agree that, so far, no one has figured out the best way to do it. As indicated above, optimal alignment of more than three sequences is not feasible in the foreseeable future; so all the available methods are approximations. The main principle underlying popular algorithms is hierarchical clustering that roughly approximates the phylogenetic tree and guides the alignment (to our knowledge, this natural idea was first introduced by Feng and Doolittle [221]). The sequences are first compared using a fast method (e.g. FASTA, see below) and clustered by similarity scores to produce a guide tree. Sequences are then aligned step-by-step in a bottom-up succession, starting from terminal clusters in the tree and proceeding to the internal nodes until the root is reached. Once two sequences are aligned, their alignment is fixed and treated essentially as a single sequence with a modification of dynamic programming. Thus, the hierarchical algorithms essentially reduce the $O(n^k)$ multiple alignment problem to a series of $O(n^2)$ problems, which makes the algorithm feasible but potentially at the price of alignment quality. The hierarchical algorithms attempt to minimize this problem by starting with most similar sequences where the likelihood of incorrect alignment is minimal, in the hope that the increased weight of correctly aligned positions precludes errors even on the subsequent steps. The most commonly used method for hierarchical multiple alignment is Clustal, which is currently used in the ClustalW or ClustalX variants [392,843] (available, e.g., at http://www.ebi.ac.uk/clustal, http://clustalw.genome.ad.jp, and http://www.bork.embl-heidelberg.de/Alignment/alignment.html).

Clustal is fast and tends to produce reasonable alignments even for protein families with limited sequence conservation provided the compared proteins do not differ in length too much. A combination of length differences and low sequence conservation tends to result in gross distortions of the alignment. The T-Coffee program is a recent modification of Clustal that incorporates heuristics partially solving these problems [623].

4.3.2. Sequence database search algorithms

4.3.2.1. Smith-Waterman

Any pairwise sequence alignment method in principle can be used for database search in a straightforward manner. All that needs to be done is to construct alignments of the query with each sequence in the database, one by one, rank the results by sequence similarity and estimate statistical significance.

The classic Smith-Waterman algorithm is a natural choice for such an application and it has been implemented in several database search programs, the most popular one being SSEARCH written by William Pearson and distributed as part of the FASTA package [663]. It is currently available on numerous servers around the world. The major problem preventing SSEARCH and other implementations of Smith-Waterman algorithm from becoming the standard choice for routine database searches is the computational cost, which is orders of magnitude greater than it is for the heuristic FASTA and BLAST methods (see below). Since extensive comparisons of the performance of these methods in detecting structurally relevant relationships between proteins failed to show a decisive advantage of SSEARCH [117], the fast heuristic methods dominate the field. Nevertheless, on a case-by-case basis, it is certainly advisable to revert to full Smith-Waterman search when other methods do not reveal a satisfactory picture of homologous relationship for a protein of interest. On a purely empirical and even personal note, the authors have not had much success with this, but undoubtedly, even rare findings may be important. A modified, much faster version of the Smith-Waterman algorithm has been implemented in the MPSRCH program, which is available at the EBI web site (http://www.ebi.ac.uk/MPsrch).

4.3.2.2 FASTA

FASTA, introduced in 1988 by William Pearson and David Lipman [664], was the first database search program that achieved search sensitivity comparable to that of Smith-Waterman, but was much faster. FASTA looks for biologically relevant global alignments by first scanning the sequence for short exact matches called "words"; word search is extremely fast. The idea is that almost any pair of homologous sequences is expected to have at least one short word in common. Under this assumption, the great majority of the sequences in the database that do not have common words with the query can be skipped without further examination with a minimal waste of computer time. The sensitivity and speed of the database search with FASTA are inversely related and depend on the "k-tuple" variable, which specifies the word size; typically, searches are run with k=3, but, if high sensitivity at the expense of speed is desired, one may switch to k=2.

Subsequently, Pearson introduced several improvements to the FASTA algorithm [662,664], which are implemented in the FASTA3 program available on the EBI server at http://www2.ebi.ac.uk/fasta3. A useful FASTA-based tool for comparing two sequences, LALIGN, is available at http://fasta.bioch.virginia.edu/fasta/lalign2.htm.

4.3.2.3. BLAST

Basic Local Alignment Search Tool (BLAST®) is the most widely used method for sequence similarity search; it is also the fastest one and the only one that relies on a complete, rigorous statistical theory [18-20,22].

Like FASTA and in contrast to the Smith-Waterman algorithm, BLAST employs the word search heuristics to quickly eliminate irrelevant sequences, which greatly reduces the search time. The program initially searches for a word of a given length W (usually 3 amino acids or 11 nucleotides, see ♦4.4.2) that scores at least T when compared to the query using a given substitution matrix. Word hits are then extended in either direction in an attempt to generate an alignment with a score exceeding the threshold of "S". The "W" and "T" parameters dictate the speed and sensitivity of the search, which can thus be varied by the user.

The original version of BLAST (known as BLAST 1.4) produced only ungapped local alignments, for which rigorous statistical theory is available. Although this program performed well for many practical purposes, it repeatedly demonstrated lower sensitivity than the Smith-Waterman algorithm and the FASTA program, at least when run with the default parameters [662]. The new generation of BLAST makes alignments with gaps, for which extensive simulations have demonstrated the same statistical properties as proved for ungapped alignments (see above).

The BLAST suite of programs is available for searching at the NCBI web site (http://www.ncbi.nlm.nih.gov/BLAST) and many other web sites around the world. It has three programs that work with nucleotide queries and two programs using protein queries (Table 4.5). The BLASTX, TBLASTN and TBLASTX programs are used when either the query or the database or both are uncharacterized sequences and the location of protein-coding regions is not known. These programs translate the nucleotide sequence of the query in all 6 possible frames and run a protein sequence comparison analogous to that in BLASTP.

Table 4.5. Use of BLAST programs for database searches

Program	Query sequence	Query type used for the database search	Database used for the search
BLASTN	DNA	DNA	DNA
BLASTP	Protein	Protein	Protein
BLASTX	DNA	Translated DNA	Protein
TBLASTN	Protein	Protein	Translated DNA
TBLASTX	DNA	Translated DNA	Translated DNA

A version of gapped BLAST, known as WU-BLAST, with a slightly different statistical model, which, in some cases, may lead to a greater search sensitivity, is supported by Warren Gish at Washington University in St. Louis (http://blast.wustl.edu/).

Recently, the BLAST suite was supplemented with BLAST2sequences, (http://www.ncbi.nlm.nih.gov/blast/bl2seq/bl2.html), a tool for comparing just two nucleotide or protein sequences [830].

Because of its speed, high selectivity and flexibility, BLAST is the first choice program in any situation when a sequence similarity search is required and, importantly, this method is used most often as the basis for genome annotation. Therefore we consider the practical aspects of BLAST use in some detail in ♦4.4. Before that, however, we need to introduce some additional concepts that are critical for protein sequence analysis.

4.3.3. Motifs, domains and profiles

4.3.3.1. Protein sequence motifs and methods for motif detection

Let us ask a very general question: what distinguishes biologically important sequence similarities from spurious ones? Looking at just one alignment of the query and its database hit showing more or less scattered identical and similar residues as in this, already familiar alignment:

```
E. coli RpmJ:    1 MKVRASVKKLCR---NCKIVKRDGVIRVICSAEPKHKQRQG
                   MKV +S+K        +C+IVKR G + VIC + P+ K  Q
Vibrio VC0879:   1 MKVLSSLKSAKNRHPDCQIVKRRGRLYVICKSNPRFKAVQR
```

it might be hard to tell one from the other. However, as soon as we align more homologous sequence, particularly from distantly related organisms, as it is done for L36 in Figure 2.1, we will have a clue as to the nature of the distinction. Note two pairs of residues that are conserved in the great majority of L36 sequences: Cx(2)Cx(12)Cx(4-5)H [here x(n) indicates n residues whose identity does not concern us]. Those familiar with protein domains might have already noticed that this conserved pattern resembles the pattern of metal-coordinating residues in the so-called Zn-fingers and Zn-ribbons, extremely widespread metal-binding domains, which mediate protein-nucleic acid and protein-protein interactions. Indeed, L36 has been shown to bind Zn^{2+} and those very cysteines and histidines are involved [112,333]. Such constellation of conserved amino acid residues associated with a particular function is called a sequence ***motif*** (see also ♦3.2). Typically, motifs are confined to short stretches of protein sequences, usually spanning 10 to 30 amino acid residues. The notion of a motif, arguably one of the most important concepts in computational biology, was first explicitly introduced by Russell Doolittle in 1981 [185]. Fittingly and, to our knowledge, quite independently, the following year, John Walker and colleagues [880] described what is probably the most prominent sequence motif in the entire protein universe, the phosphate-binding site of a vast class of ATP/GTP-utilizing enzymes, which subsequently has been named P-loop [744]. Discovery of sequence motifs characteristic of a vast variety of enzymatic and binding activities of proteins proceeded first at an increasing and then, apparently, at a steady rate [103]), and the motifs, in the form of amino acid patterns, were swiftly incorporated by Amos Bairoch in the PROSITE database (♦3.2.1)

The P-loop, which we already encountered in ♦3.2.1, is usually presented as the following pattern of amino acid residues:

[GA]x(4)GK[ST]

Note that there are two strictly conserved residues in this pattern and two positions where one of two residues is allowed. By running this pattern against the entire protein sequence database using, for example, the FPAT program available through the ExPASy server program or any other pattern-matching program (even the UNIX 'grep' command will do), one immediately realizes just how general and how useful this pattern is. Indeed, such a search retrieves sequences of thousands of experimentally characterized ATPases and GTPases and their close homologs. However, only about one half of the retrieved sequences are known or predicted NTPases of the P-loop class, whereas the rest are false-positives (EVK, unpublished). This is not surprising given the small number of residues in this pattern, which results in the probability of chance occurrence of about

$$(1/10)(1/20)(1/20)(1/10) = 2.5 \times 10^{-5}$$

(this is an approximate estimate because the actual amino acid frequencies are not taken into account, but it is close enough). With the current database size of about $3.2\text{x}10^{8}$ residues, the expected number of matches is about 8,000!

This simple calculation shows that this and many other similar patterns, although they include the most conserved amino acid residues of important motifs, are insufficiently selective to be good diagnostic tools. The specificity of a pattern can be increased by taking into account adjacent residues that tend to have conserved properties. In particular, for the P-loop pattern, it can be required that there are at least three bulky, hydrophobic residues among the five residues upstream of the first glycine (structurally, this is a hydrophobic β-strand in ATPases and GTPases). This would greatly reduce the number of false-positives in a database search, but would require a more sophisticated search method (as implemented, for example, in the GREF program of the SEALS package [878], see ♦5.1.2). Still, this does not solve the problem of motif identification. Figure 4.7 shows the alignment of a small set of selected P-loops that were chosen for their sequence diversity. Obviously, not even a single amino is conserved in all these sequences, although they all represent the same motif that has a conserved function and, in all likelihood, is monophyletic, i.e. evolved only once. Given this lack of

RecB_Ecoli	ERLIE**A**SAGT**GKT**FTIAALYLRLL
SbcC_Ecoli	LFAIT**G**PTGA**GKT**TLLDAICLALY
MutS_Ecoli	MLIIT**G**PNMG**GKS**TYMRQTALIAL
Adk_Ecoli	RIILL**G**APGA**GK**GTQAQFIMEKYG
MCM2_HUMAN	NVLLC**G**DPGT**AKS**QFLKYIEKVSS
UvsX_T4	LLILA**G**PSKSF**KS**NFGLTMVSSYM
CmpK_Mjan	VITVS**G**LAGS**GTT**TLCRNLAKHYG
Pta_Ecoli	IMLIPTGTSV**G**L**T**SVSLGVIRAME

Figure 4.7. Alignment of P-loops from diverse ATPases and GTPases. The most conserved residues are shown in bold.

strict conservation of amino acid residues in an enzymatic motif, this trend is even more pronounced in motifs associated with macromolecular interactions, in which invariant residues are exception rather than norm. Pattern search remains a useful first-approximation method for motif identification, especially because a rich pattern collection, PROSITE (♦3.2.1), can be searched using a rapid and straightforward program like SCANPROSITE (http://www.expasy.org/tools/scnpsite.html). However, by the very nature of the approach, patterns are either insufficiently selective or too specific and, accordingly, are not adequate descriptions of motifs.

The way to properly capture the information contained in sequence motifs is to represent them as amino acid ***frequency profiles***, which incorporate the frequencies of each of the 20 amino acid residues in each position of the motif. Even in the absence of invariant residues, non-randomness of a motif may be quite obvious in a profile representation (Figure 4.9).

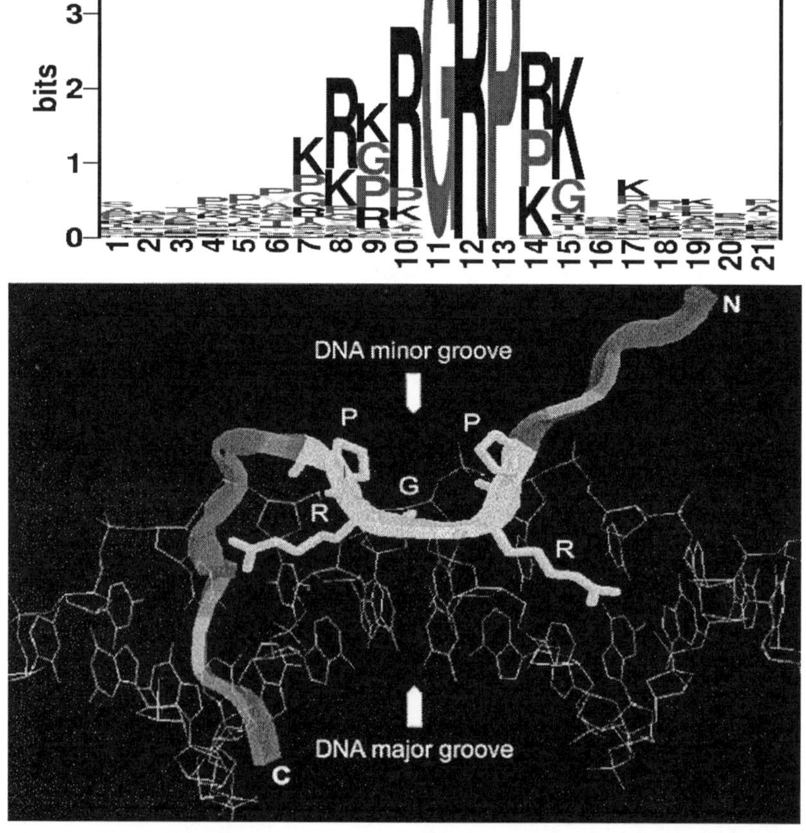

Figure 4.8. Profile representation of a conserved sequence motif and the corresponding 3D structure of the DNA-binding AT-hook domain [48]. The pictorial form of the profile was produced using the Sequence Logo method.

Utilization of frequency profiles for database searches had a profound effect on the quality and depth of sequence and structure analysis. The principles and methods that made this possible are discussed in the next section.

4.3.3.2. Protein domains, PSSMs and advanced methods for database search

Sequence motifs are extremely convenient descriptors of conserved, functionally important short portions of proteins. However, motifs are not the natural units of protein structure and evolution. Such distinct units are protein ***domains***. In structural biology, domains are defined as structurally compact, independently folding parts of protein molecules. In comparative genomics and sequence analysis in general, the central, "atomic" objects are parts of proteins that have distinct evolutionary trajectories, i.e. occur either as stand-alone proteins or as parts of variable ***domain architectures*** (we refer to the linear order of domains in protein sequences as domain or multidomain architecture), but are never split into parts. Very often, probably in the majority of cases, such units of protein evolution exactly correspond to structural domains. However, in some groups of proteins, an evolutionary unit may consist of two or more domains. For example, from a purely structural viewpoint, trypsin-like proteases have two domains. However, at least so far, separation of these domains has not been observed, and therefore, they should be treated as a single evolutionary unit. It might be desirable to propose a special name for these units of protein evolution, but, to our knowledge, this has not been done and, in comparative-genomic literature, including this book, they are commonly referred to as domains. On rare occasions, a domain consists of a single motif, as in the case of AT-hooks shown in Figure 4.8. However, much more often, domains are relatively large, comprising 100 to 300 amino acid residues and including two or more distinct motifs. Motifs are highly conserved patches in multiple alignments of domains that tend to be separated by regions of less pronounced sequence conservation and often of variable length (Figure 4.9A); in other words, motifs may be conceptualized (and visualized) as peaks on sequence conservation profiles. In the 3D structure of most domains, the distinct motifs are juxtaposed and function together, which explains their correlated conservation. Figure 4.9B illustrates the juxtaposition of motifs that center around the two catalytic residues in the alignment of the catalytic domain of caspase-related proteases from Figure 4.9A.

The notion of protein motifs has been employed directly in algorithms that construct multiple sequence alignments as a chain of motifs separated by unaligned regions. The first of such methods, Multiple Alignment Construction and Analysis Workbench (MACAW), originally used a BLAST-like

```
Csp1_Hs      KTSDSTFLVFMSHGIRE------GICGKKHSEQVPDILQ-LNAIFNMLNT--3-PSLKDKPKVIIIQACRGD-SPGVVWF
Csp2_Hs      RVTDSCIVALLSHGVE-------GAIYGVDG----KLLQ-LQEVFQLFDN--3-PSLQNKPKMFFIQACRGDETDRGVDQ
Csp3_Hs      SKRSSFVCVLLSHGEE-------GIIFGTN-----GPVD-LKKITNFFRG--3-RSLTGKPKLFIIQACRGTELDCGIET
Csp9_Hs      GALDCCVVVILSHGCQASHLQFPGAVYGTDG----CPVS-VEKIVNIFNG--3-PSLGGKPKLFFIQACGGEQKDHGFEV
Csp10_Hs     ADGDCFVFCILTHGRF-------GAVYSSDE----ALIP-IREIMSHFTA--3-PRLAEKPKLFFIQACQGEEIQPSVSI
CED3_Ce      G--DSAILVILSHGEE-------NVIIGVDD----IPIS-THEIYDLLNA--3-PRLANKPKIVFVQACRGERRDNGFPV

PC_Hs        DKGVYGLLYYAGHGYEN-----FGNSFMVPVD---APNPYRSENCLCVQN--5-QEKETGLNVFLLDMCRKRNDYDDTIP
PC_Ce        GNGVYAVFYFVGHGFEV-----NGQCYLLGVD---APADAHQPQHSMSMD--6-RHKTPDLNLLLLDVCRKFVPYDAISA
PC_Dd        QSYIEVVVYYAGHGRSD-----NGNLKLIMT----DGNPVQLSIIASTLT--2-IKNSDSLCLFIVDCCRDGENVLPFHY
Mlr2366_Ml   YNADLAVIFYAGHGMQV-----DGKNYL-------IPVDADLTSPAYLKT-11-LPADPAVGVIILDACRDNPLGRTLAA
Mlr1804_Ml   IGADMAVFYYAGHALQY-----NGQNLL-------LPVDTRISSAKEVAA-12-KNDPVGVKVFILDACRNNPVAKEKGL
Mll2372_Ml   RGADVALFFYAGHGLQV-----SGKNYL-------LPVDAALEDETSLDF-11-MSRETSIRLVFLDACRDNPLADVLAK
Mlr3463_Ml   EGAGVGLFYYAGHGLQV-----DGRNYI-------VPVDAKLDMPVKLQL-11-MEQQTKVSLVFLDACRNNPFARSLSR
Mll5190_Ml   KGADVALVYFSGHGVEI-----SGDNRL-------LPVDADASSVDQLDK-12-VAATAKVGLIVLDACRSDPFSASSGD
Mlr1170_Ml   EGADVAFIYYSGHGIEA------GGEN------YLVPVDADVSSLKDAGQ-11-LKKTVPVTIMLLDACRTNPFPADAVV
YOR197w_Sc   QPNDSLFLHYSGHGGQTED--LDGDEEDGM-DDVIYPVDFETQGPIIDDE--8-PLQQGVRLTALFDSCHSGTVLDLPYT

MC1_At       TAGDSLVFHYSGHGSRQRN--YNGDEVDGY-DETLCPLDFETQGMIVDDE--7-PLPHGVKLHSIIDACHSGTVLDLPFL
MC2_At       KPGDSLVFHFSGHGNNQMD--DNGDEVDGF-DETLLPVDHRTSGVIVDDE--7-PLPYGVKLHAIVDACHSGTVMDLPYL
MC3_At       KPGDVLVVHYSGHGTRLPA--ETGEDDDTGYDECIVPCD-MNLITDDEFR--4-KVPKEAHITIISDSCHSGGLIDEAKE
Mlr3300_Ml   QRDDFVYLHLSGHGAQQPER-AKGDETDGLDE-IFLPVDIEKWINRDAGV-15-IRNKGAFVWAVFDCCHSGTATRAVEV
MCH_Rsph     EPGGIFLMSYAGHGAQIGDFDEGDGPDRDRLDETLCLHD-AMLV-DDELY--4-AFREGVRVVAVFDSCHSGSILRASAN
MCH_Gsul     GKGDIFMLSYSGHGGQVP---DTSNDEPDGVDETWCLFD-GELI-DDELY--4-KFAAGVRVLVFSDSCHSGTVVKMAYY
```

Figure 4.9. Conserved catalytic motifs in the caspase-like superfamily of proteases. A. Multiple alignment of the catalytic motifs around the two active residues (His and Cys) of caspases (Csp, top group), paracaspases (PC, middle group), and metacaspases; see more about these proteases in ♦6.4.4. Note the conservation in the stretches preceding each of the catalytic residues and corresponding to the two main β-strands of the caspase domain (see panel B). The species abbreviations are Hs, human; Ce, *C. elegans*; Dd, Dictyostellium, Ml, *M. loti*; Sc, yeast, At, *Arabidopsis*, Rsph, *Rhodobacter sphaeroides*, Gsul, *Geobacter sulfurreducens*.

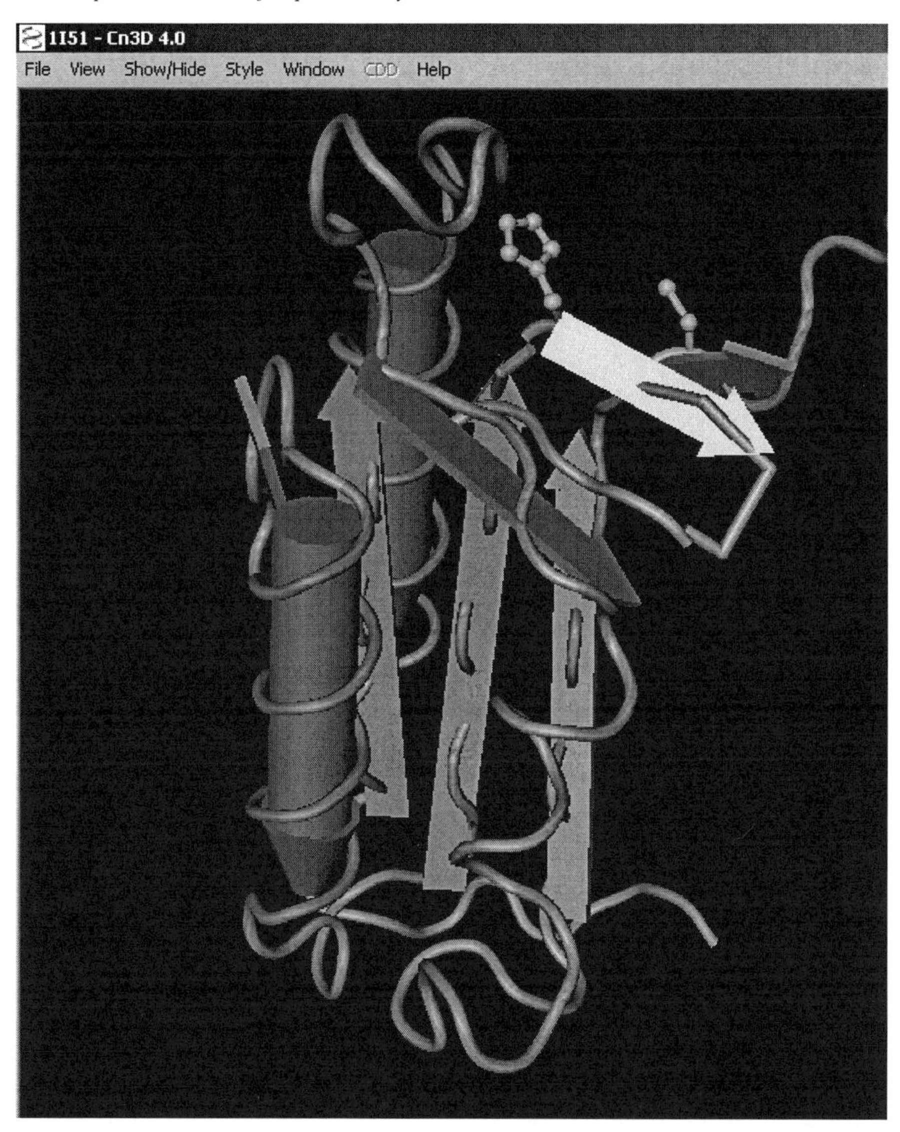

Figure 4.9. Conserved catalytic motifs in the caspase-like superfamily of proteases.

B. Structure of the human caspase-7 (PDB entry 1I51). The figure is generated using Cn3D (see http://www.ncbi.nlm.nih.gov/Structure/CN3D/cn3d.shtml) representation of the MMDB (♦3.3) entry. The conserved histidine and cysteine residues are shown in ball-and-stick representation in the upper part of the structure.

method for approximately delineating conserved sequence blocks (motifs) and then allowed the user to determine whether or not inclusion of additional alignment columns increased the significance of the block alignment [756].

MACAW is a very convenient, accurate and flexible alignment tool; however, the algorithm is $O(n^k)$ and, accordingly, becomes prohibitively computationally expensive for a large number of sequences [756]. MACAW is an interactive tool that embodies the important notion that completely automatic methods are unlikely to capture all important motif in cases of subtle sequence conservation, particularly in proteins that substantially differ in length. For many occasions, it remains the method of choice when careful alignment analysis is required, although, in the current situation of explosive growth of sequence data, the computational cost severely limits MACAW's utility. Subsequently, Charles Lawrence, Andrew Neuwald and coworkers adapted the Gibbs sampling strategy for motif detection and developed the powerful (if not necessarily user-friendly) PROBE method that allows delineation of multiple, subtle motifs in large sets of sequences (http://bayesweb.wadsworth.org/gibbs/gibbs.html [493,614,615]). Importantly, Gibbs sampler is an $O(n)$ algorithm, which allows analysis of large numbers of sequences. Gibbs sampling has been incorporated into MACAW as one of the methods for conserved block detection. In principle, this should enable MACAW to efficiently align numerous sequences. In practice, the authors find it problematic to identify relevant motifs among he numerous blocks detected by Gibbs sampler.

Arguably, the most important methodological advance based on the concepts of domains and motifs was the development of ***position-specific weight matrices (PSSMs)*** and their use in database searches as an incomparably more powerful substitute for regular matrices, such as BLOSUMs and PAMs. A PSSM is a rectangular table, which consists of n columns (n is the number of positions in the multiple alignment for which the PSSM is made) and 20 rows and contains, in each cell, the score (weight) for the given amino acid in the given position of the multiple alignment (see Figure 3.1). In the simplest case, this score can be the frequency of the amino acid in the given position. It is easy to realize, however, that, on most occasions, residue frequencies taken from any given alignment are unlikely to adequately describe the respective domain family. Firstly, we certainly never know the full range of family members and, moreover, there is no evidence that we have a representative set. Therefore, if a residue is missing in a particular alignment column, this does not justify a 0 score in a PSSM. In reality, a PSSM never includes a score of exactly 0, although scores for some residues might be extremely low and rounding sometimes may result in 0 values. Instead, a finite score is assigned to the missing residue using so-called regularizers, i.e. various mathematical techniques that strive to derive the correct distribution of amino acids for a given position on the basis of a limited sample [422,826]. It is easy to realize that the score given to a missing residue depends on two factors: the distribution actually found in the sample of available superfamily members and the size of the sample. Clearly, if a set of 1,000 diverse sequences invariably contains, for example, a serine residue in a particular position, the probability of finding any other residue in this position is extremely low. Nevertheless, threonine, as a residue that is structurally close to serine and, according to substitution matrices like BLOSUMs and PAMs, is often exchangeable with serine in proteins, certainly should receive a higher score than, say, lysine. One, of course, could argue that an invariant serine is most likely to be part of a catalytic center of an enzyme and as such is more likely to be replaced by cysteine than by threonine (such replacements in enzymes, e.g. proteases and acyltransferases, are well documented, e.g. [295,338]). This level of sophistication seems to be beyond the capabilities of current automatic methods for PSSM generation, although, in principle, a PSSM for a particular domain could be tailored manually. Another aspect of PSSM construction that requires formal treatment beyond calculating and regularizing amino acid residue scores stems from the fact that many protein families available to us are enriched with closely related sequences (this might be the result of a genuine proliferation of a particular subset of a family or could be caused by sequencing bias). Obviously, an overrepresented subfamily will sway the entire PSSMs toward detection of additional closely related sequences and hamper the performance. To overcome this problem, different weighting schemes are applied to PSSMs to downweigh closely related sequences and increase the contribution of diverse ones. Optimal PSSM construction remains an important problem in sequence analysis and even small improvements have the potential of significant enhancing the power of database search methods. Some of the recent developments that we do not have the opportunity to discuss in detail here seem to hold considerable promise [650,691]. Once a PSSM is constructed, using it in a database search is straightforward and not particularly different from using a single query sequence combined with a regular substitution matrix, e.g. BLOSUM62. The common database search methods, such as BLAST, can work equally well with a PSSM and the same statistics applies.

To our knowledge, the notion of a PSSM and its use for detecting weak sequence similarities was first introduced by Michael Gribskov, Andrew McLachlan and David Eisenberg in 1987 [315]. However, their method was initially of limited utility because it depended on a pre-constructed multiple sequence alignment and consequently could not be used with the speed and ease comparable to those of using FASTA or BLAST. An important additional step was combining the use of PSSMs with iterative search strategy. To our knowledge, this was first done by Gribskov [314]. Under this approach, after the first run of a PSSM-based similarity search against a sequence database, newly detected sequences (with the similarity to PSSM above a certain cut-off) are added to the alignment, the PSSM is rebuilt and the cycle is repeated until no new members of the family are detected. This approach was implemented in a completely automated fashion in the Motif Search Tool (MoST) program, which also included a rigorous statistical method for evaluating resulting similarities, but only worked with ungapped alignment blocks [826].

A decisive breakthrough in the evolution of PSSM-based methods for database search was the development of the ***Position-Specific Iterating (PSI)-BLAST*** program [22]. This program first performs a regular BLAST search of a protein query against a protein database. It then uses all the hits with scores greater than a certain cut-off to generate a multiple alignment and create a PSSM, which is used for the second search iteration. The search goes on until convergence or for a desired number of iterations. Obviously, the first PSI-BLAST iteration must employ a regular substitution matrix, such as BLOSUM62, to calculate HSP scores. For the subsequent iterations, the PSSM regularization procedure was designed in such a way that the contribution of the initial matrix to the position-specific scores decreases, whereas the contribution of the actual amino acid frequencies in the alignment increases with the growth of the number of retrieved sequences. PSI-BLAST also employs a simple sequence-weighting scheme [344], which is applied for PSSM construction at each iteration. Since its appearance in 1997, PSI-BLAST has become the most common method for in-depth protein sequence analysis. The method owes its success to its high speed (each iteration takes only slightly longer than a regular BLAST run), the ease of use (no additional steps are required, the search starts with a single sequence, and alignments and PSSMs are constructed automatically on the fly), and high reliability, especially when composition-based statistics is invoked. The practical aspects of using PSI-BLAST are considered at some length in ♦4.3.5.

Hidden Markov Models (HMMs) of multiple sequence alignments are a popular alternative to PSSMs [72,202,351,473]. HMMs can be trained on unaligned sequences or pre-constructed multiple alignments and, similarly to PSI-BLAST, can be iteratively run against a database in an automatic regime. A variety of HMM-based search programs are included in the HMMer2 package (http://hmmer.wustl.edu [208]; Sean Eddy's web site displays a recommendation to pronounce the name of this package "hammer" as in "a more precise mining tool than a BLAST"). HMM search is slower than PSI-BLAST, but there have been reports of greater sensitivity of HMMs (e.g. [652]). In the extensive, albeit anecdotal experience of the authors, the results of protein superfamily analysis using PSI-BLAST (with a few "tricks" discussed in ♦4.4) and HMMer2 are remarkably similar.

The availability of techniques for constructing models of protein families and using them in database searches naturally leads to a vision of the future of protein sequence analysis. The methods discussed above, such as PSI-BLAST and HMMer, start with a protein sequence and gradually build a model that allows detection of homologs with low sequence similarity to the query. Clearly, this approach can be reversed such that a sequence query is run against a pre-made collection of protein family models. In principle, if models were developed for all protein families, the problem of classifying a new protein sequence would have been essentially solved. In addition to family classification, regular database searches like BLAST also provide information on the most closely related homologs of the query, thus giving an indication of its evolutionary affinity. In itself, a search of a library of family models does not yield such information, but an extension of this approach is easily imaginable whereby a protein sequence, after being assigned to a family through PSSM and HMM search, is then fit into a phylogenetic tree. Searching the COG database may be viewed as a rough prototype of this approach (♦3.3). Such a system seems to have the potential of largely replacing current methods with an approach that is both much faster and more informative. Given the explosive growth of sequence databases, transition to searching databases of protein family models as the primary sequence analysis approach seems inevitable in a relatively near future. Only for discovering new domains will it be necessary to revert to searching the entire database, and since the protein universe is finite (see ♦8.1), these occasions are expected to become increasingly rare.

Presently, sequence analysis has not reached such an advanced stage, but searches against large, albeit far from complete, databases of domain-specific PSSMs and HMMs have already become extremely useful approaches in sequence analysis. Pfam, SMART and CDD, which were introduced in ♦3.2, are the principal tools of this type. Pfam and SMART perform searches against HMMs generated from curated alignments of a variety of proteins domains. The CDD server compares a query sequence to the PSSM collection in the CDD (see ♦3.2.3) using the ***Reversed Position-Specific (RPS)-BLAST*** program [545]. Algorithmically, RPS-BLAST is similar to BLAST, with minor modifications; Karlin-Altschul statistics applies to E-value calculation for this method. RPS-BLAST searches the library of PSSMs derived from CDD, finding single- or double-word hits and then performing ungapped extension on these candidate matches. If a sufficiently high-scoring ungapped alignment is produced, a gapped extension is done and the alignments with E-values below the cut-off are reported. Since the search space is equal to nm where n is the length of the query and m is the total length of the PSSMs in the database (which, at the time of writing, contains ~5,000 PSSMs), RPS-BLAST is ~100 times faster than regular BLAST.

Pattern-Hit-Initiated BLAST (PHI-BLAST) is a variant of BLAST that searches for homologs of the query that contain a particular sequence pattern [942]. As discussed above, pattern search often is insufficiently selective. PHI-BLAST partially rectifies this by first selecting the subset of database sequences that contain the given pattern and then searching this limited database using the regular BLAST algorithm. Although the importance of this method is not comparable to that of PSI-BLAST, it can be useful for detecting homologs with a very low overall similarity to the query that nevertheless retain a specific pattern (see ♦4.5).

Stand-alone (non-web) BLAST. The previous discussion applied to the web version of BLAST, which is indeed most convenient for analysis of small numbers of sequences, and is, typically, the only form of database search used by experimental biologists. However, the web-based approach is not suitable for large-scale searches requiring extensive post-processing, which are common in genome analysis. For these tasks, one has to use the stand-alone version of BLAST, which can be obtained from NCBI via ftp and installed locally under the Unix or Windows operation systems. Although the stand-alone BLAST programs do not offer all the conveniences available on the web, they do provide some additional and useful opportunities. In particular, stand-alone PSI-BLAST can be automatically run for the specified number of iterations or until convergence.

With the help of simple additional scripts, the results of stand-alone BLAST can be put to much use beyond the straightforward database search. Searches with thousands of queries can be run automatically, followed with various post-processing steps; some of these will figure in the next chapter on genome annotation. The BLASTCLUST program (written by Ilya Dondoshansky in collaboration with Yuri Wolf and EVK), which is also available from NCBI via ftp and works only with stand-alone BLAST, allows clustering sequences by similarity using the results of an all-against-all BLAST search within an analyzed set of sequences as the input. It identifies clusters using two criteria: (i) level of sequence similarity, which may be expressed either as percent identity or as score density (number of bits per aligned position), and (ii) the length of HSP relative to the length of the query and subject (e.g. one may require that, for the given two sequences to be clustered, the HSP(s) should cover at least 70% of each sequence). BLASTCLUST can be used, for example, to eliminate protein fragments from a database or to identify families of paralogs.

4.4. Practical Issues: How to Get the Most out of BLAST

BLAST in all its different flavors remains by far the most widely used sequence comparison program. For this reason and also given extensive personal experience of using BLAST for a variety of projects, we describe the practical aspects of using BLAST in considerable detail. Some of this information and additional hints on BLAST usage are available online at the NCBI web site as BLAST tutorial (http://www.ncbi.nlm.nih.gov/Education/BLASTinfo/information.html). A simple description of the statistical foundations of BLAST, written by Stephen Altschul, is available at the NCBI web site as BLAST course (http://www.ncbi.nlm.nih.gov/BLAST/tutorial/Altschul-1.html).

4.4.1. Setting up the BLAST search

The BLAST web site has been recently redesigned in order to simplify the selection of the appropriate BLAST program to perform the desired database search. The user only needs to select the type of the query (nucleotide or protein sequence) and the type of the database (protein or nucleotide). These selections automatically determine which of the BLAST programs is used for the given search. The discussion below deals primarily with BLASTP, but all the same considerations apply to BLASTX, TBLASTN and TBLASTX (used only in exceptional cases).

Selecting a subsequence and a database

The default is to search with the entire query sequence. On many occasions, however, it is advantageous to use as the query only a portion of a protein sequence, e.g. one domain of a multidomain protein. To this end, one can indicate the range of the amino acid position to be used as the query or simply paste the fragment of interest into the query window.

NCBI, EBI, and other centers offer the users a wide variety of nucleotide and protein databases for various searches. For example, instead of the non-redundant (**nr**) database, which is used as default, **Swissprot, Month** (subset of the nr database contaning only the sequences added to GenPept in the last 30 days), or pdb (sequences of the proteins whose structures are present in the PDB (see ♦3.4). Furthermore, a recently implemented option of the NCBI BLAST allows one to limit the BLAST search by a taxonomy tree node (see ♦3.6.1) or any other properly formatted Entrez query. This option provides for setting up an infinite variety of special-purpose searches, saves a large amount of computer and human time and leads to an increase of search sensitivity because of the reduced search space (this, of course has to be remembered when E-values are interpreted). For example, if one needs to compare a query protein only to the two-component signaling systems in Cyanobacteria (e.g., if one suspects that the protein in question is a cyanobacterial histidine kinase), the search can be easily adapted by indicating "histidine kinase" AND "Cyanobacteria [ORGN]" in the Entrez window. Of course, this needs to be done with caution because it is likely that not all histidine kinases that are actually encoded in the cyanobacterial genomes have been annotated as such. Therefore it might be a better idea to search all cyanobacterial proteins, which is still many times faster than searching the entire non-redundant database. We find the ability to limit the search space using taxonomic criteria to be an extremely useful feature of BLAST, especially given the current rapid growth of sequence databases. Another useful way to limit the search space is to include the expression 'srcdb_refseq[PROP]' in the 'Limit results by Entrez query' window. This will limit the search to NCBI's RefSeq database (♦3.4), thus preserving most of the sequence diversity, while avoiding redundancy.

4.4.2. Choosing BLAST parameters

Composition-based statistics and filtering

As noted above, low-complexity sequences (e.g., acidic-, basic- or proline-rich regions) often produce spurious database hits in non-homologous proteins. Currently, this problem is addressed by using composition-based statistics (see ♦4.2.3) as the default for NCBI BLAST; filtering with SEG is available as an option, but is turned off by default. As shown in large-scale tests [748] and confirmed by our own experience, composition-based statistics eliminates spurious hits for all but most severe cases of low sequence complexity [878].

Table 4.6 shows an example of how filtering low-complexity regions in a protein affects the BLAST scores with a true homolog compared to a hit that is due solely to the low-complexity segment and how composition-based statistics solves this problem without filtering. Breast cancer type 1 susceptibility protein (BRCA1 or BRC1_HUMAN) is a 1863-aa protein, which is mutated in a significant fraction of breast and ovarian cancers (see OMIM entry 113705). This protein contains several low-complexity regions. Using the unfiltered BRCA1 sequence in a BLAST search retrieves a number of BRCA1 fragments and its variants from human, mouse, rat, and other vertebrates, which are not particularly helpful for predicting the function(s) of the query protein. The first significant hit beyond BRCA1 variants is dentin sialophosphoprotein, an Asp,Ser-rich protein found in mineralized dentin. There is no genuine homologous relationship between BRCA1 and dentin phosphoprotein and this hit is entirely spurious. The default filtering with SEG decreases the similarity score reported by BLAST from 73 to 49 (and, accordingly, increases the E-value from 3e-11 to 5e-4), which still appears to be a statistically significant similarity. Only more stringent filtering, which increases the number of masked residues from 117 to 172 pushes the E-value of the dentin phosphoprotein hit into the unreliable territory. In contrast, filtering only slightly decreases the similarity scores between BRCA1 and an *Arabidopsis* protein that also contains a BRCT domain [454] or between human and oppossum BRCA1. Of course, when most of the protein is masked, the scores start to drop (and E-values go up). These data were generated using the standard (old) statistics option. The last row of the table shows that using composition-based statistics largely eliminates the problem by recognizing the compositional bias in the dentin phosphoprotein and accordingly decreasing the score for this hit. Thus, this new option makes explicit low complexity filtering largely (albeit not completely) irrelevant for the purposes of routine database searches, but not for exploration of protein structure. Methods like SEG remain important tools for delineating probable globular domains and in that capacity may still be useful for searches, e.g. when a predicted globular domain is used a query. Using composition-based statistics is the only feasible choice for any large-scale automated BLAST searches. However, the same tests have shown that, for some queries, this statistical procedure resulted in artificially high (not significant) E-values [748]. Therefore, for detailed exploration of certain, particularly short, proteins, it is advisable to also try a search with composition-based statistics turned off.

Table 4.6. Removing spurious database hits for the low sequence complexity protein BRCA1 by modifying SEG parameters[a]

Parameter set of SEG[a]	Number of residues		E-values of the BLAST hits		
	Masked	Unmasked	Dentin	Plant BRCA1	Opossum BRCA1
No filtering	0	1,863	3e-11	4e-15	1e-28
12 2.1 2.4	35	1,828	4e-9	4e-15	1e-28
12 2.2 2.5 (default)	117	1,746	5e-4	5e-12	7e-22
12 2.3 2.6	172	1,691	-	5e-11	3e-21
12 2.4 2.7	279	1,584	-	5e-11	1e-14
12 2.5 2.8	487	1,376	-	6e-11	8e-10
12 2.6 2.9	616	1,247	-	2e-10	5e-9
12 2.7 3.0	908	955	-	4e-06	2e-8
12 2.8 3.1	1,164	699	-	0.003	6e-7
Composition-based filtering	0	1,863	-	3e-12	1e-20

[a]SEG parameters are trigger window length, trigger complexity, and extension complexity, (see ♦4.2.3 and [921]).

Expect value, word size, gap penalty, substitution matrix

Expect (E) value can be any positive number; the default value is 10. Obviously, it is the number of matches in the database that one should expect to find merely by chance. Typically, there is no reason to change this value. However, in cases when extremely low similarity needs to be analyzed, the threshold may be increased (e.g. to 100) and, conversely, when it is desirable to limit the size of the output, lower E-values may be used.

Word size (W) must be an integer; the default values are 3 for protein sequences and 11 for nucleotide sequences. This parameter determines the length of the initial seeds picked up by BLAST in search of HSPs. Currently supported values for the protein search are only 3 and 2. Changing word size to 2 increases sensitivity, but considerably slows down the search. This is one of the last resorts for cases when no homologs are detected for a given query with regular search parameters.

We have already discussed the advantages of protein sequence searches over nucleotide sequence searches (♦4.2). The necessarily greater values of the word size parameter in nucleotide searches may lead to curious situations like the following one. Having sequenced the cysteine proteinase gene from alfalfa weevil *Hypera postica* (GenBank accession no. AF157961 [893]), the authors searched for similar sequences using both BLASTX and BLASTN. A BLASTX search with this gene showed that its product is very similar to cathepsin L-like proteinases from a variety of organisms, including the relatively well-characterized proteinase from *Drosophila melanogaster*. In contrast, a BLASTN search with this gene sequence, surprisingly, did not find any homologous genes. This seems surprising because, for example, the cathepsin L gene of the southern cattle tick *Boophilus microplus* (GenBank accession no. AF227957) contains a region with as much as 61% identity (98 identical nucleotides out of 161) to the gene from *Hypera postica*:

```
AF157961 GAATATATGAAGACCAAGATTGCAGTCCTGCTGGCCTGAACCACGCTATTCTTG
AF227957 GAGTGTACGATGAGCCCGAGTGTAGCAGTGAAGATCTGGACCACGGTGTACTCG
Identity ** * ** ** ** *  ** ** **   **  *  *** ****** * * ** *

AF157961 CTGTTGGTTACGGAACCGAGAATGGTAAAGACTACTGGATCATTAAGAACTCCT
AF227957 TTGTCGGCTATGGTGTTAAGGGTGGGAAGAAGTACTGGCTCGTCAAGAACAGCT
Identity  *** ** ** **     **  *** **  * ****** ** * ******  **

AF157961 GGGAGCCAGTTGGGGTGAACAAGGTTATTTCAGGCTTGCTCGTGGTAAAAAC
AF227957 GGGCTGAATCCTGGGGAGACCAAGGCTACATCCTTATGTCCCGTGACAACAAC
Identity ***  *     ***** ** ***** **  **    *  * **** ** ***
```

Figure 4.10. The importance of the "word size" parameter. An alignment of segments of cysteine protease genes from alfalfa weevil (AF157961) and cattle tick (AF227957).

How come BLASTN failed to find this similarity? It happened because BLASTN has the default word size of 11, i.e. reports as an HSP only a run of 11 identical nucleotides. Even decreasing the word size to 7, the lowest word size currently allowed for BLASTN, would not change the result because the longest stretch of identical nucleotides in this alignment is only 6 bases long. This example not only shows once more why protein searches are superior to DNA-DNA searches. It also demonstrates that establishing that two given sequences are ***not*** homologous requires as much caution as proving that they are homologous. Accordingly, the statement that the reported sequence is "novel" and has no homologs in GenBank, often found in scientific literature, should always be treated with a healthy dose of skepticism.

As described above, different amino acid ***substitution matrices*** are tailored to detecting similarities among sequences with different levels of divergence. However, a single matrix, BLOSUM62, is reasonably efficient over a broad range of evolutionary change, so that situations when a matrix change is called for are rare. For particularly long alignments with very low similarity, a switch to BLOSUM45 may be attempted, but one should be aware that this could also trigger an increase in the false-positive rate. In contrast, PAM30, PAM70 or BLOSUM80 matrices may be used for short queries. Each substitution matrix should be used with the corresponding set of ***gap penalties***. Since there is no analytical theory for calculating E-values for gapped alignments, parameters of equation (4.2) had to be determined by extensive computer simulations separately for each combination of a matrix, gap opening penalty and gap extension penalty. Therefore only a limited set of combinations is available for use (Table 4.7). However, there is no indication that substantial changes in these parameters would have a positive effect on the search performance.

Additional aspects of the BLAST setup are discussed in the next subsection because they apply to the output of the search. A useful feature that has been recently added to NCBI BLAST is the ability to save and bookmark the URL with a particular BLAST setup using the 'Get URL' button at the bottom of the page. For a habitual BLAST user, it pays off to save several setups customized for different tasks.

Table 4.7. Substitution matrices and gap penalties

Query length, aa	Substitution matrix	Gap opening cost	Gap extension cost
<35	PAM30	9	1
35-50	PAM70	10	1
50-85	BLOSUM80	10	1
>85	BLOSUM62	10	1

4.4.3. Running BLAST and formatting the output

A BLAST search can be initiated with either a GI number or the sequence itself. In the current implementation at the NCBI web page (http://www.ncbi.nlm.nih.gov/BLAST), the user can run a BLAST search and then try several different ways of formatting the output. The default option involves toggling between two windows, which may become confusing; it may be convenient to switch to a one-window format using the Layout toggle and save the setup as indicated above.

CDD search is run by default in conjunction with BLAST. As discussed above, this search is much faster than regular BLAST and is often more sensitive. The CDD search is normally completed long before the results of conventional BLAST become available. This allows the user to inspect the CDD search output and get an idea of the domain architecture of the query protein, while waiting for the BLAST results. On many occasions, all one really needs from a database search is recognizing a particular protein through its characteristic domains architecture or making sure that a protein of interest ***does not contain*** a particular domain. In such situations, there may be no reason to even wait for the regular BLAST to finish. The CDD search may also be run as a stand-alone program from the main BLAST page. In this mode, it is possible to change the E-value threshold for reporting domain hits (default 0.01), which can be helpful for detecting subtle relationships and new versions of known domains.

The current BLAST setup includes ***limitation on the number of descriptions and the number of alignments*** included in the output; the current defaults are 250 and 100, respectively. With the rapidly growing database size, there is often need to increase these limits in order to investigate a particular protein family. Doing so, however, will likely result in large outputs that are hard to download and navigate. Limiting the search space as outlined above could be a viable and often preferable option.

The ***graphical overview*** option allows the user to select whether a pictorial representation of the database hits aligned to the query sequence is included in the output. Although it slows loading the page, this option is essential for quick examination of the output to get an idea of the domain architecture of the query. Each alignment in the graphical view window is color-coded to indicate its similarity to the query sequence.

The ***Alignments views*** menu allows the user to choose the mode of alignment presentation. The default ***Pairwise alignment*** is the standard BLAST alignment view of the pairs between the query sequence and each of the database hits. All other views are pseudo-multiple alignments produced by parsing the HSPs using the query as a template. ***Query-anchored with***

identities shows only residues that are different from the ones in the query; residues identical to the ones in the query are shown as dashes. ***Query-anchored without identities*** is the same view with all residues shown. ***Flat query-anchored with identities*** is a multiple alignment that allows gaps in the query sequence; residues that are identical to those in the query sequence are shown as dashes. ***Flat query-anchored without identities*** also allows gaps in the query sequence, but shows all the residues. Pairwise alignment is definitely most convenient for inspection of sequence similarities, but the "flat query-anchored without identities" option allows one to generate multiple alignments of reasonable quality that can be saved for further analysis. This option is best used with the number of descriptions and alignments (see above) limited to a manageable number (typically, no more that 50).

The ***Taxonomy Reports*** option allows the user to produce a taxonomic breakdown of the BLAST output. Given that many BLAST outputs are quite large these days, this is extremely helpful, allowing one to promptly assess the phyletic distribution of the given protein family and identify homologs from distant taxa.

Formatting for PSI-BLAST

The output of BLAST can be used as the input for PSI-BLAST. The critical parameter that is typically set before starting the initial BLAST run is ***inclusion threshold***; the current default is E=0.005. This parameter determines the E-value required to include a HSP into the multiple alignment that is used to construct the PSSM. Combined with composition-based statistics, the E-value of 0.005 is a relatively conservative cut-off. Spurious hits with lower E-values are uncommon: they are observed more or less as frequently as expected according to Karlin-Altschul statistics, i.e. approximately once in 200 searches. Therefore, carefully exploring the results with higher E-values set as the inclusion threshold often allows one to discover subtle relationships that are not detectable with the default cut-off. When studying new or poorly understood protein families, we routinely employ thresholds up to 0.1. In the version of PSI-BLAST, which is available on the web, each new iteration has to be launched by the user. New sequences detected in the last iteration with an E-value above the cut-off are highlighted in PSI-BLAST output. PSI-BLAST also has the extremely useful option of manually selecting or deselecting sequences for inclusion into the PSSM. Selecting "hopeful" sequences with E-values below the cut-off may help in a preliminary exploration of an emerging protein family; deselecting sequences that appear to be spurious despite E-values above the cut-off may prevent corruption of the PSSM. The PSSM produced by PSI-BLAST at any iteration can be saved and used for subsequent database searches.

We realize that the above recommendation to investigate results that are not reported as statistically significant is a call for controversy. However, we believe there are several arguments in favor of this approach. First, such analyses of subtle similarities have repeatedly proved useful, including the original tests of PSI-BLAST effectiveness [22]. Second, like in other types of research, what is really critical is the original discovery. Once one gets the first glimpse of what might be an important new relationship, statistical significance often can be demonstrated using a combination of additional methods (more on this in ♦4.3.7). Third, we certainly do not advocate lowering the statistical cut-off for any large-scale, let alone automated searches. This is safe only when applied in carefully controlled case studies.

4.4.4. Analysis and interpretation of BLAST results: separating wheat from the chaff

In spite of the solid statistical foundation, including composition-based statistics, BLAST searches inevitably produce both false positives and false negatives. The main cause for the appearance of false-positives, i.e. database hits that have "significant" E-values but, upon more detailed analysis, turn out not to reflect homology, seems to be subtle compositional bias missed by composition-based statistics or low-complexity filtering. The reason why false-negatives are inevitable is, in a sense, more fundamental: in many cases, homologs really have low sequence similarity that is not easily captured in database searches, and even if reported, may not cross the threshold of statistical significance. In an iterative procedure like PSI-BLAST, both the opportunities to detect new and interesting relationships and the pitfalls are further exacerbated. Beyond the (conceptually) straightforward issues of selectivity and sensitivity, functional assignments based on database search results require careful interpretation if we want to extract the most out of this type of analysis, while minimizing the chance of false predictions. Below we consider both the issues of search selectivity and sensitivity and functional interpretation.

No cut-off value is capable of accurately partitioning the database hits for a given query into relevant ones, indicative of homology, and spurious ones. By considering only database hits with very high statistical significance (e.g. $E < 10^{-10}$) and applying composition-based statistics, false positives can be eliminated for the overwhelming majority of queries, but the price to pay is high: numerous homologs, often including those that are most important for functional interpretation, will be missed.

Now that we have come to practical aspects of database search, specific examples will work best. Let us see what can BLAST analysis tell us about the protein product of human tumor susceptibility gene 101 (TSG101). This protein appears to have important roles in a variety of human cancers and in budding of viruses, including HIV [139,275,514,515,527,721]. We first run the default BLAST search (again, this includes CDD search; composition-based statistics is turned on) using the TSG101 sequence (T101_HUMAN) as the query. Since CDD search is much faster than BLAST, we have the opportunity to examine the potential domain architecture of TSG101 while BLAST is still running. Distinct statistically significant domain hits are reported for the N-terminal and C-terminal regions of TSG101 sequence (Figure 4.11). There is also a low-complexity segment in the middle that probably corresponds to a non-globular domain.

The N-terminal part is similar to the UBCc domain, the catalytic domain of the E2 subunit of ubiquitin-conjugating enzymes. The statistical significance of this similarity is overwhelming, with a E $<10^{-13}$. The domain is unique (the only hit in this part of TSG101 sequence) and UBCc is known to be a globular domain, so there is no reason to suspect a spurious hit due to compositional bias. Therefore we must conclude that the N-terminal portion of TSG101 is a homolog of the catalytic domains of E2 enzymes. Does this mean that this domain of TSG101 has the activity of ubiquitin ligase? The answer comes from a more careful examination of the alignment and is negative. We can easily see that the conserved catalytic Cys residue of E2, which conjugates ubiquitin and is essential for the enzymatic activity [156], is replaced by Tyr in TSG101 (Figure 4.12A). Soon after the discovery of TSG101, similar type of analysis (in the pre-CDD and pre-composition-based statistics days this was much harder!) led to the conclusion that TSG101 was an inactivated homolog of E2 enzymes, probably a regulator of ubiquitination [453,680]. These predictions of TSG101 structure and function have been subsequently confirmed experimentally [516,682].

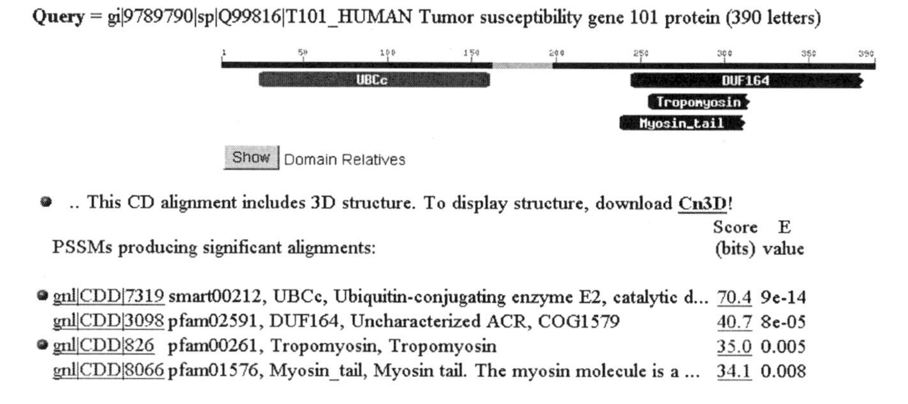

Figure 4.11. Graphical output of the CD search for the TSG101 protein.

The domain hits of the C-terminal region of TSG101 are of a completely different nature. There are three overlapping domains with moderate similarity to the C-terminal portion of TSG101 (Figure 4.11). One of these belongs to an uncharacterized family of bacterial proteins, whereas two others are myosin and tropomyosin, proteins known to contain coiled-coil domains, a distinct non-globular α-helical superstructure. The telltale feature of coiled-coil domains is the periodic distribution of leucines, which tend to occur 7 residues apart. A simple visual examination of the alignment of TSG101 with tropomyosin (Figure 4.12B) reveals this pattern of leucines, suggesting that TSG101 contains a C-terminal coiled-coil domain. This observation can be confirmed by using the COILS2 program (see ♦4.6.2.1). How do we interpret the domain hits for the C-terminal part of TSG101? Coiled-coil domains occur in a huge variety of proteins and it would be ridiculous to consider them all to be homologous. Thus, these hits are not indications of homology and, in all likelihood, do not point to any specific functional similarity between the respective proteins and TSG101. So, in the traditional sense, these domain hits are spurious, i.e. are caused by similar amino acid composition and structure of the query and certain domains present in the CDD collection, rather than by homology. However, these hits are not uninformative. They allow us to predict the existence of a coiled-coil domain in TSG101, and this domain is likely to mediate important protein-protein interactions. Indeed, it has been shown that the coiled-coil domain is responsible for the interaction of TSG101 with stathmin, a phosphoprotein implicated in tumorigenesis [514].

```
                                     *                    A
Consensus      71 IYHPNVD-SSGEICLSILK----EKWS-PA-  94
Query          98 KTGKHVD-ANGKIYLPYLH----E-WKhPQ- 121
2UCZ           77 ILHPNIY-PNGEVCISILH-13-ERWS-PV- 113
2E2C           81 CWHPNVD-QSGNICLDILK----ENWT-AS- 104
1I7K_A        102 CYHPNVD-TQGNISLDILK----EKWS-AL- 125
1AYZ_A         76 MFHPNVY-ANGEICLDILQ----NRWT-PT-  99
                                                          B
                          *      *      *
Query:   254   NALKRTEEDLKKGHQKLEEMVTRLDQ---EVAEVD 285
2TMA_A:   55   DELDKYSEALKDAQEKLELAEKKATDAEADVASLN  89

                     *    *      *      *
Query:   286   KNIEL----LKKKDEELSSALEKMENQSENNDIDE 316
2TMA_A:   90   RRIQLVEEELDRAQERLATALQKLEEAEKAADESE 124
```

Figure 4.12. Results of the CD search for the TSG101 protein. A, alignment of the TSG101 N-terminal domain with ubiquitin-conjugating enzymes. The conserved Cys (replaced by Tyr in TSG101) is shown in bold and indicated with an asterisk. B. Alignment of the TSG101 C-terminal domain with rabbit tropomyosin. The telltale Leu residues of the coiled-coil domain are shown in bold and denoted by asterisks.

We have already learned a lot about the functions of TSG101 from the CD search alone. As mentioned above, in many situations, this information could be all a researcher needs from computational analysis of a sequence. However, for the purpose of this discussion, let us now turn to BLAST. The search returns 184 hits and their distribution along the TSG101 sequence shows the presence of several full-length hits with highly significant similarity to TSG101 (we may immediately suspect these are orthologs) as well as a number of domain-specific hits. Examination of the full-length alignments confirms the conservation of domain architecture in probable TSG101 orthologs not only in distantly related animals (fruit fly and worm), but also in the plant *Arabidopsis thaliana.* Moreover, in all of these alignments, the tyrosine replacing the catalytic cysteine of E2 is conserved, indicating that all these TSG101 orthologs are enzymatically inactive. This leads us to the remarkable conclusion that inactivation of a E2 paralog and its fusion with a coiled-coil domain leading to the regulatory function of TSG101 have already occurred prior to the divergence of animals and plants. Thus, this regulatory mechanism involved both in cellular transformation and in viral budding from animal cells is at least 800 million years old! Let us note that this conclusion, which was easily reached through the analysis of BLAST output, was not immediately obvious from the CDD search result.

What about the domain-specific hits? There are only three of these for the N-terminal part of TSG101 and all appear to be inactivated homologs of E2 enzymes. This is, of course, something that we recognize in retrospect; note that there is no mention of E2 or ubiquitin or anything else in this BLAST output that would even hint at the homology of TSG101 and ubiquitin-conjugating enzymes. The rest of the hits, located in the central and C-terminal portions of the TSG101 sequence, are to low-complexity sequences, including those of coiled-coil domains, with some E-values below 10^{-4}. Does it make sense to run a second PSI-BLAST iteration starting from this output? We think not (an unwieldy proliferation of low-complexity sequences in the output is fully expected), although the reader might want to attempt the experiment.

We will now try a different approach: low complexity filtering instead of composition-based statistics (the two cannot be used together because masking low-complexity regions precludes correct estimation of search parameters). The search with low complexity filtering turned on results in 637 hits. A region of 41 amino acid residues in the middle of TSG101 sequence is masked eliminating the low-complexity hits in this region. However, the C-terminal coiled-coil region is not masked and an even greater number of hits with apparently significant E-values are reported. Thus, filtering does not solve the problem. It does suggest, however, a simple trick we will do next. It is common for low-complexity (non-globular) regions to separate distinct domains in proteins. Therefore we will run BLAST using as the query only the segment of TSG101 preceding the masked region; to do so, we put 1 to 163 in the 'Set subsequence windows' on the BLAST page. The results are dramatically different. We get only 37 hits; all those with significant E-values are (as we know from the results of the CD search) inactivated E2 homologs, but below the threshold, we notice some proteins annotated as ubiquitin-conjugating enzymes. Now we can run PSI-BLAST. The second iteration brings in two more sequences of inactivated E2's; the best hit to ubiquitin conjugating enzyme now has an E-value of 0.009, still below the cut-off. After the third iteration, we get the message "No new sequences were found above the 0.005 threshold!" – the search has converged. Formally, we have not observed a statistically significant sequence similarity between TSG101 and ubiquitin-conjugating enzymes. However, the E-value of 0.009 is suggestive (let us recall: this means that an alignment with this or greater score is expected to be observed in less than one database search in 100 for the given query), and the detected similarity is worth further investigation. Let us run the search using the same 1-161 fragment of TSG101 as the query but with the composition-based statistics turned off. Now, in the second iteration, we detect two ubiquitin-conjugating enzymes with E-values <0.001. In this example, an equivalent result would have been achieved if a less restrictive cut-off (E=0.01) was used with composition-based statistics. However, there are cases when a search with composition-based statistics turned off reveals relationships that are not detectable at all under the default BLAST parameters.

In the third and subsequent iterations, the E-values for ubiquitin-conjugating enzyme sequences become extremely low. Here, a clarifying note on E-values reported by PSI-BLAST is due [21]. When, in a PSI-BLAST search, a sequence crosses the inclusion threshold for the first time, the reported E-value is based on the score of the alignment of this sequence and the PSSM constructed in the previous PSI-BLAST iteration, which did not include the sequence in question. Therefore this E-value accurately reflects the similarity between the given sequence and the rest of the protein family involved in PSSM construction. However, for the next iteration, the sequence(s) in question, e.g. those of E2 enzymes in the above example, become part of the alignment used for PSSM construction and their E-values get unduly inflated. Thus, inferences of homology and functional predictions should take into account only the ***first E-value below the threshold*** reported for the sequence of interest.

In order to introduce another crucial aspect of PSI-BLAST analysis, we will now leave TSG101 and turn to other examples. In an ideal world, PSI-BLAST could be expected to be symmetrical with respect to protein family members. In other words, each member used as the query would retrieve from the database the entire family. In practice, the effect of query choice on the outcome of the search can be dramatic [43]. Consider how different queries fare in retrieving members of the ATP-grasp superfamily of enzymes, which includes primarily ATP-dependent carboligases (Table 3.2). The biotin carboxylase domain of human acetyl-CoA carboxylase initially retrieves other biotin carboxylases. After ~200 sequences of biotin carboxylases, PSI-BLAST starts finding carbamoyl phosphate synthetases, ~100 of them. Only after these, other members of the superfamily (D-alanine-D-alanine ligase, phosphoribosylamine-glycine ligase and others; Table 3.2) start appearing in the output. It takes four iterations and almost 1,000 database hits before this search finds synapsin, a diverged member of the ATP-grasp superfamily, which is involved in regulating exocytosis in synaptic vesicles and has been shown to contain the ATP-grasp domain by crystal structure analysis (PDB entry 1auxA [216]). Using synapsin as a query in PSI-BLAST search with default parameters fares even worse: only other members of the synapsin family itself are retrieved. Only changing the inclusion threshold to 0.01 results in retrieval of other members of this superfamily in the fourth and subsequent iterations. In contrast, using *M. jannaschii* protein MJ1001 as a query in PSI-BLAST search with default parameters results in retrieval of most members of the ATP-grasp superfamily, including the synapsins, in the second iteration. A regrettable personal anecdote belongs here: because of this remarkable asymmetry, we initially failed to predict the structure of synapsin! [262]. This and other examples (see [43]) show that, in order to characterize a protein (super)family, it is never sufficient to run a single PSI-BLAST search. Rather, either all known members or at least a representative subset should be used as queries. The search then needs to be repeated with newly detected members in "superiteration" mode. Just like PSI-BLAST, this process (run either automatically or manually) stops only when no new sequences with credible similarity to the analyzed family are detected. As a potentially useful tip for query choice, it is interesting to note that, in our experience, prokaryotic, and particularly archaeal, proteins are often better queries than their eukaryotic homologs. This conclusion has been reached in the early days of genomics [262,467] and still holds true. Among the obvious reasons for this trend are the generally shorter length and lower content of compositionally biased sequences in prokaryotic proteins.

The final crucial aspect of BLAST searches that we must mention here is the importance of protein domain architecture for inferring the function of the query protein from the functions of homologs. Assigning function on the basis of one homologous domain, while overlooking the multidomain architecture of the query, the database hit or both, is one of the most common sources of regrettable errors in sequence analysis. We will discuss some specific examples in the next chapter when discussing genome annotation.

This brief discussion certainly cannot cover all "trade secrets" of sequence analysis; some more case studies are presented in the next section. However, the above seems to be sufficient to formulate a few rules of thumb that help a researcher to extract maximal amount of information from database searches while minimizing the likelihood of false "discoveries".

- Searching a domain library is often easier and more informative than searching the entire sequence database. However, the latter yields complementary information and should not be skipped if details are of interest.
- Varying the search parameters, e.g. switching composition-based statistics on and off, can make a difference.
- Using subsequences, preferably chosen according to objective criteria, e.g. separation from the rest of the protein by a low-complexity linker, may improve search performance.
- Trying different queries is a must when analyzing protein (super)families.
- Even hits below the threshold of statistical significance often are worth analyzing, albeit with extreme care.
- Transferring functional information between homologs on the basis of a database description alone is dangerous. Conservation of domain architectures, active sites and other features needs to be analyzed (hence automated identification of protein families is difficult and automated prediction of functions is extremely error-prone).

4.5. The Road to Discovery

The notion of punctuated equilibrium introduced by Niles Eldredge and Steven Jay Gould [304] applies not only to the evolution of life but also to the history of science: epochs of relative quiet are punctuated by bursts of discoveries. The event that sets off the avalanche is usually the development of a new method. This phenomenon reproduces itself with an uncanny regularity in all areas of science – and computational biology is no exception. The authors of this book have personally experienced the excitement of two such groundbreaking developments. The first one came in 1994 with the algorithm for iterative motif search implemented in the MoST program [826]. A variety of (relatively) subtle relationships between protein domains that previously were completely unknown suddenly were within our reach.

Arguably, the most remarkable of these new findings was the discovery of the BRCT domain, so named after BRCA1 (hereditary Breast Cancer Associated protein 1) C-terminus [454]. The initial discovery of this domain took something like a leap of faith: the first seed for MoST was derived from an alignment between the C-terminal part of BRCA1 and a p53-binding protein that was produced by BLAST and was not statistically significant at all. Subsequently, of course, some level of statistical significance and, more importantly, the uniqueness of the new domain and the coherence of its diagnostic pattern of amino acid residue conservation were established through MoST searches and multiple alignment analysis. Furthermore, the BRCT domain superfamily was greatly expanded by combining MoST with other profile search tools, Pfsearch and Gibbs sampler [100], and these results were simultaneously corroborated with a completely independent method [136]. In the years since, numerous experimental studies led to the characterization of the BRCT domain as one of the most important adaptors that mediate protein-protein interactions in eukaryotic cell cycle checkpoints (see [407] and references therein).

The BRCT domain also served as the training ground for the next-generation iterative search program, PSI-BLAST. The findings that were originally made quite painstakingly using MoST and other methods were reproduced using PSI-BLAST with minimal human intervention [22]. PSI-BLAST and HMM methods that came of age more or less simultaneously quickly enabled an unprecedented rise of new findings. The discovery of the 3'-5' exonuclease domain in the Werner syndrome protein and the unification of histidine kinases, type II topoisomerases, HSP90 molecular chaperones and MutL repair proteins in a single distinct superfamily of ATPase domains deserve to be mentioned as two of the very first ones [593].

Much of the new knowledge obtained using these methods, along with additional structural and functional information, was soon encapsulated within the new generation of profile search tools combined with domain databases, of which Pfam, SMART and CDD seem to be the most practically important ones (see above). These tools, which employ growing and, for the most part, carefully tested libraries of predefined PSSMs or HMMs, have dramatically simplified and, to a large extent, trivialized the process of identification of already known domains in protein sequences, even in cases of limited sequence conservation. This is, of course, a natural and welcome development: what starts as an exciting investigation of uncharted territories ends up being a routine methodology, and the sooner the better. Nevertheless, it is equally clear that fringe cases, in which, even with these new tools, expert exploration is required to validate and interpret domain finding, still abound, particularly with the rapid progress of genome sequencing. And, of course, many new domains that are still missing in the existing collections remain to be discovered, and for this, the libraries of already known domains are of little use.

Nothing helps capturing the flavor of a relatively complex research approach better than real-life examples, so let us explore some interesting proteins, for which, to our knowledge, there is no annotation in either published literature or available databases.

Phage λ gene Ea31

First, let us consider phage λ gene Ea31, which was briefly mentioned in Chapter 1. We start by using the sequence of this protein in a PSI-BLAST search, with the CDD search option turned on, but the composition-based statistics turned off (see above). The CDD search immediately results in a provocative finding: a hit to the so-called HNH nuclease domain. The similarity seems to be statistically significant, but the E-values are not particularly impressive, about 0.001. Can we conclude that Ea31 contains a nuclease domain of this particular family and predict that this phage protein is a nuclease? Let us first note that there are two distinct questions here because the presence of a conserved domain does not automatically imply the corresponding activity. There are numerous cases of the same domain present in both enzymatically active and inactivated proteins (as we have seen in the TSG101 case). Furthermore, without a more detailed investigation, even the conclusion on the presence of an HNH domain in Ea31 would be premature. Experience shows that spurious hits with a similar level of statistical significance are not particularly uncommon in CDD searches (in other words, they appear somewhat more often than once in 1000 searches as suggested by the E-value), probably due to subtle low complexity present in some of the multiple alignments behind the PSSMs. So let us investigate farther. First of all, by clicking on the HNH domain icon in the CDD search, we obtain a multiple alignment of our query (Ea31) with a diverse set of 10 HNH domains (Figure 4.13).

We should note that nearly all of these proteins, including Ea31, share a conserved pattern of two pairs of cysteines and the aspartate-histidine (DH) doublet. Examination of the relevant literature immediately shows that the cysteines and the histidine in this pattern are the metal-coordinating residues that comprise a Zn-finger-like domain in the HNH family nucleases and are required for the nuclease activity [129]. The structural elements of the HNH nuclease domain known from the available 3D structure (PDB entry 1EMV) are also well conserved in Ea31. At this point, we are in a position to conclude that the CDD hits pointed us in the right direction and Ea31 indeed contains a HNH domain and is, in all likelihood, an active nuclease. As mentioned in Chapter 1, it seems likely that this predicted nuclease forms a two-subunit, ATP-dependent enzyme with the product of the adjacent (and probably coexpressed) gene Ea59. The HNH family includes several restriction enzymes, e.g. McrA, and other enzymes involved in defense function, such as colicins. Therefore it is possible that such an ATP-dependent nuclease functions as a phage-specific restriction-modification enzyme.

Our little discovery is made (or, more precisely, reproduced and expanded here because the HNH nuclease domain of Ea31 has already been described in the course of a comprehensive evolutionary analysis of Holliday junction resolvases [50]) and, hopefully, will be eventually tested experimentally. But let us not leave the HNH nuclease family just yet and see what happens if we do not rely on the CDD search but rather continue with the regular PSI-BLAST analysis. The original PSI-BLAST search with the default cut-off E-value of 0.005 converges after the first iteration. We will try to lower the cut-off and run PSI-BLAST, carefully monitoring the results. A search with E=0.1 brings in numerous proteins in successive iterations, among which proteins annotated as putative HNH family endonucleases start appearing in the second iteration.

After running five PSI-BLAST iterations, we detect about 100 sequences with E-values above the chosen cut-off. Examination of all these sequences shows the conservation of the metal-binding motif, suggesting that even using such a liberal cut-off in this case does not result in spurious hits (let us note parenthetically that this very approach of lowering the cut-off was used for detecting BRCT domains in the original PSI-BLAST work).

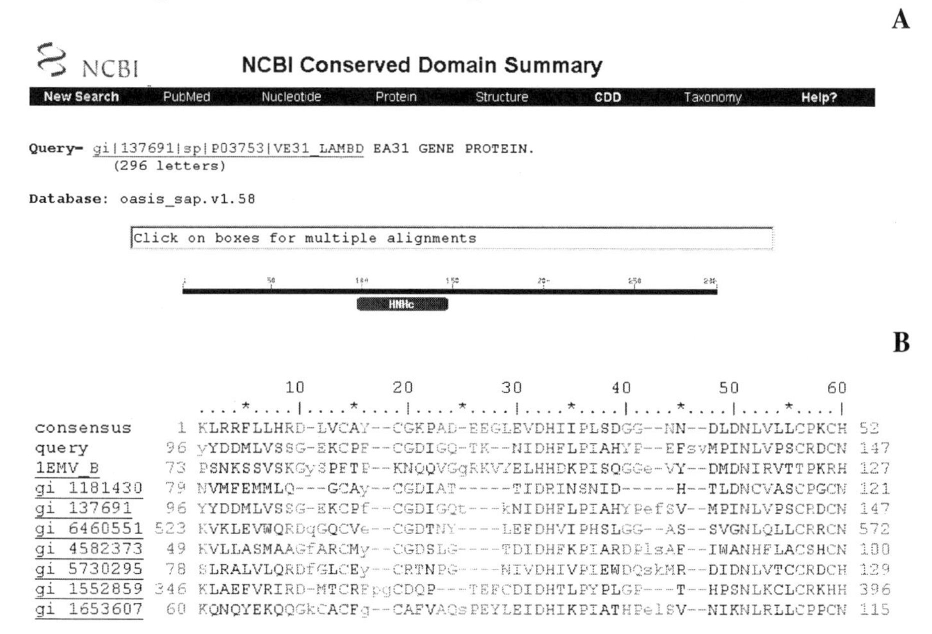

Figure 4.13. Results of the CD search with the Ea31 protein.

Continuing running iterations one by one on the web becomes cumbersome with a large number of retrieved sequences, as well as time-consuming, so, at this point, we switch to a local search on a UNIX workstation, which we run to convergence. The search does indeed converge after the 16th iteration and inspection of the results suggests at least two important conclusions. First, the great majority of the sequences detected in this search and containing the conserved metal-binding site are annotated in the database as "hypothetical proteins" (or, even worse, as "predicted transmembrane proteins") and only a minority are labeled as (predicted) nucleases. Thus, although this family, its conserved motifs, structure and enzymatic activity have been described in a considerable number of publications, its representation in current databases does not even approach completeness and researchers have no direct way to become aware of its scope. Second, our search detected two members of the HNH family from *Arabidopsis thaliana* and one from rice, the latter one already annotated as a predicted nuclease. Therefore, plants also encode nucleases of this family, which so far has been considered to be exclusively prokaryotic.

COG1518

In the above example, we identified a diverged version of an already well-known domain in a previously uncharacterized protein. This was a pretty straightforward observation. Discovering new domains and predicting their function(s) is a much more challenging task. However, we have an excellent source of such potential new domains: the functionally uncharacterized COGs or Pfam families (see ♦3.2.2). The COGs are most convenient because they allow selection of potential "targets" for discoveries using phyletic patterns (♦2.2.6). Figure 4.14 shows the conserved portion of the multiple alignment of proteins from COG1518, which is represented in many archaeal and several bacterial species, primarily thermophiles. These proteins show striking conservation among themselves, but extensive searches for possible homologs using PSI-BLAST, CD search, and even threading methods (♦4.6.3) fail to show any credible relationships (the readers may try these searches again: perhaps the databases change by the time this book is out). However, a little alignment gazing immediately tells us that these proteins have several charged residues (two glutamates, an aspartate and a histidine) that are conserved in all sequences of this family (Figure 4.14). This is a telltale sign of an enzyme: polar amino acids that are invariant in a diverse protein family typically are part of a catalytic center.

We have repeatedly observed that, when a domain is capable of "merely" binding a protein, nucleic acid or a small molecule, there would be no invariant polar residues. Strangely, we found it hard to cite any particular publication for this generalization. Apparently, formal rules for distinguishing enzymes from non-enzymes by analysis of multiple alignments have not been developed so far, although it seems that we more or less know an enzyme when we see the alignment of the respective protein family (provided the aligned sequences are diverse enough). Given the nature of the conserved residues, it appears likely that the activity of this putative new enzyme requires metal coordination. It seems to be dangerous to further speculate on the nature of this activity. We will leave it at that for now. In the next chapter, when discussing the use of genomic context for functional annotation, we will apply a different method of inference to arrive to specific hypotheses on the biological function and biochemical activity of COG1518 proteins.

RNA-dependent RNA polymerase

The final case in this section is a real wild goose chase. Our subject will be the RNA-dependent RNA polymerase (RdRp), an enzyme that attracted enormous attention in the last few years as the apparent amplifier of small RNAs involved in post-transcriptional gene silencing [6]. RdRp is a large protein that is highly conserved in most eukaryotes, including some invertebrate animals, fungi, plants and, importantly, the early-branching protozoon *Giardia*. It is missing in yeast *S. cerevisiae*, the microsporidion *E. cuniculi* and, interestingly, in insects and vertebrates, which is attributed to lineage-specific gene loss during eukaryotic evolution [55]. As noted in the literature and confirmed in our database searches, RdRp does not show significant similarity to any other proteins; in particular, there seems to be no relationship whatsoever with enzymes of RNA viruses that have the same activity. Thus, the origin of this unique enzyme at the dawn of eukaryotic evolution remains a mystery. However, inspection of the alignment of the RdRp's revealed a striking signature present in all these sequences: three invariant aspartates flanked by a few other conserved residues. Negatively charged residues coordinating metal ions are a fixture in all types of DNA and RNA polymerases [176]. A tantalizing hypothesis therefore emerged that this motif was likely to be a part of the active site of RdRp's. Clearly, any similarity between RdRp's and other polymerases, if it exists, must be subtle, otherwise it would have been picked by PSI-BLAST (if not regular BLAST). We ran a PHI-BLAST search using the most conserved segment of the RdRp from the slime mold *Dictyostelium discoideum* as the query and the following delimiting amino acid pattern:

D[ILVMF]DGDx[ILVMFYW]x[ILVMFYW]

```
AF2435    ELMGAEAAARNAYYTKFDE 15 ENEVNAMISFGNSLLYSAVLSEIYHTQLNPAISYLHEPSERRFSLALDIAELFKPVIVDR 238
AF1878    RLLGIEGKASKHYWDAISL 22 KDIVNAMLNYGYSILLAECVKAVELAGLDPYAGFLHVDVSGRSSLAIDLMENFRQQVVDR 263
BH0341    SLRGWEGQAAINYNKVFDQ 19 KDNVNAMLSFAYTLLANDVAAALETVGLDAYVGFMHQDRPGRASLALDLMEELRGLYADR 258
YgbT_Ec   QLRGIEGSRVRATYALLAK 18 GDTINQCISAATSCLYGVTEAAILAAGYAPAIGFVHT--GKPLSFVYDIADIIKFDTVVP 230
PM1126    DKENVEAQAALIYFQTL--  9 ENNINAHLNYAYTVLRSAIARALVLYGWLPQLGIFHRSEVNPFNLADDFVEPFRP-LVDL 268
PH1245    EIMNVEARIRQEYYAKWDE 16 KNEMNALISFLNSRLYATIITEIYNTQLAPTISYLHEPSERRFSLSLDLSEIFKPIIADR 238
SPy1562   SLRGIEGQAANQYFRIFND 19 LDCVNALLSFGYSLLTFECQSALEAVGLDSYVGFFHTDRPGRASLALDLVEEFRSYIVDR 257
MJ0378    EVMNVEGRVRTEYYRLWDE 16 KNEMNALISFLNSRLYPAIITELYNTQLTPTVSYLHEPHERRFSLALDLSEIFKPMIADR 238
MTH1084   DVMNVEGRIRSDYYNAIDS 16 ENMTNAMISFGNSLLYSTVITELYNTQLNPTISYLHEPFERRYSLALDLSEIFKPTLIDR 250
NMA0630   DTGNREAQAAALYFQAL--  9 NNAVNAALNYTYAVLRAAVARALTLYGWLPALGLFHRSELNPFNLADDFIEPLRP-LADL 227
TVN0106   KILGVEGNIWSTYYSAFPF 15 KDELNAMISYGNALLYATVLTKIFITGLNPSISFLHEPSERSFSLALDIADIFKPVIVER 240
Aq_369    ELMSVEAEFRKLCYKKLEE 14 QNPLNALISFGNSLTYAKVLGEIYKTQLNPTVSYLHEPSTKRFSLSLDVAEVFKPIFVDN 230
Rv2817c   ELNGFEGNAAKAYFTALGH 16 LDAFNSMVSLGYSLLYKNIIGAIERHSLNAYIGFLHQDSRGHATLASDLMEVWRAPIIDD 244
Cj1522c   DSKNIEAVAAALYFKTL--  9 LCFENSALNYGYAIIRACIIRAVCISGLLPWLGIKHDNIYNSFALCDDLIEVFRA-SVDD 231
TM1797    ELMGIEGNAREEYYSMIDS 17 KNFANTLISFGNSLLYTTVLSLIYQTHLDPRIGYLHETNFRRFSLNLDIAELFKPAVVDR 238
APE1240   KLLSIEARASRRYWQCIAE 15 LDPFNAALNYGYGMLYSIVEKSLLLVGLDPYLGVFHSEKSGKPSLTLDAIEPFRAPIVDR 263
SSO1450   LLDKDEPAAARVYWQNISQ 15 TDQFNMALNYSYAILYNTIFKYLVIAGLDPYLGFIHKDRPGNESLVYDFSEMFKP-YIDF 229
SSO1405   EVMQKEAEAAKVYWRGVKS 18 LDPFNRALNIGYGMLRKVVWGAVISVGLNPYIGFLHKFRSGRISLVFDLMEEFRSPFVDR 298
cons/95%  ..thEsth...hat.h..    .s.hNthhshh.s.h...h.t.h...th.s.hshhHt...t..shs.Dh.-.hp..hh..
```

Figure 4.14. Alignment of a subset of COG1518 proteins: a predicted new enzyme. Sequence conservation is indicated by shading and by the consensus pattern. Most conserved residues are shown in reverse colors. See ♦5.3 for more on this.

This search retrieved from the database all available sequences of eukaryotic RdRp's and, in addition, another alignment that is shown in Figure 4.15A. Strikingly, although there is no statistical significance at all to this alignment (E-value of nearly 6), our signature fits the known active site located in the second-largest subunit of DNA-dependent RNA polymerase (DdRp), the enzyme that catalyzes transcription in all cellular life forms [586,587]. Furthermore, there seems to be some similarity around this signature that might reflect conservation of the structural elements supporting the polymerase active site (Figure 4.15B). Given the lack of statistical support, these observations should be regarded with utmost caution and perhaps even outright skepticism.

Nevertheless, it seems likely that the resemblance between the most conserved regions of the two classes of polymerases is not fortuitous. Rather, it might be explained either by divergence from a common ancestor or by convergence driven by similar functional requirements to the polymerase active sites. The latter alternative is not unrealistic: in Chapter 2, we discussed rather similar cases of apparent convergence, e.g. among the lysozymes. If the two polymerases are homologous, the evolutionary scenario becomes more or less obvious, with a duplication of the DdRp catalytic subunit at the onset of the evolution of eukaryotes, followed by extreme, rapid divergence, giving rise to the RdRp. We will know the answer once the 3D structure of RdRp is determined. Regardless of the outcome, there seems to be an excellent chance that the conserved region shown in Figure 4.15 is an important part of the catalytic site of RdRp.

We presented here three case studies of uncharacterized proteins that range from straightforward to openly speculative. We believe that these are illustrative of the way computational analysis of proteins works, by drawing from several lines of evidence to produce predictions whose levels of confidence vary to a great extent.

A

(NC_003450) COG0085:DNA-directed RNA polymerase beta subunit/140 kD subunit (*C. glutamicum* AP005279)
Length = 2169, Score = 5.8 bits (19), Expect = 5.9, Identities = 26/91 (28%), Positives = 36/91 (38%), Gaps = 19/91 (20%)

```
Pattern                                               **********
Query:    3 MVIKNPCTHPGDVRYLKAVDNLRLRHLRNVLVFSTKGDVPNFKEISGSDLDGDRYFFCYDKSLIGNRSKS   72
            MV ++P     G VRYL+ V N       L  V V        V +F      D DGD              +GN  K
Sbjct:1556 MVWRDPVIRDGGVRYLRVVIN---DDLHGVAVNPVS--VKSFD----GDFDGDSVGL------VGNLPKK 1610
```

B

```
                                                               * * *
RdRP_Ddis   VMVIKNPCTHPGDVrylkavdnlrlrHLRNVLVFSTkgdv-PNFKEISGSDLDGDRYFFCYDKSLIGNRS
RdRp_Atha   VAIAKNPCLHPGDVrileavdvpqlhHMYDCLIFPQkgdr-PHTNEASGSDLDGDLYFVAWDQKLIPPNR
RdRp_Cele   VLLTKNPCIVPGDVrifeavdipelhHMCDVVVFPQhgpr-PHPDEMAGSDLDGDEYSVIWDQELLLERN
RdRp_Ncra   CVVGRNPSLHPGDIrvveavdvpalrHLRDVVVFPLtgdr-DVPSMCSGGDLDGDDFFVIWDPLLIPKER
RdRp_Spom   CIVARNPSLHPGDVrvckavrcdelmHLKNVIVFPTtgdr-SIPAMCSGGDLDGDEYTVIWDQRLLPKIV
RdRp_Gint   rreASRQNLKPvlpsfvaqeairgm-YRGVICFPKYyegr-PMTDCLSGSDLDGDIYWVSWDASLLIQRE
DdRp_Cglu   -mvwRDPVIRDGGVRY----------LRVVInddlhgvavnPVSVKSFDGDFDGDSVGLVGNLPKKAHEE
DdRp_Ecol   VLLNRAPTLHRLGIQA----------FEPVLIEGKAiqlh-PLVCAAYNADFDGDQMAVHVPLTLEAQLE
DdRp_Aful   IVLFNRPSLHrmsima----------HYVRVLPYKTfrln-PAVCPPYNADFDGDEMNLHVPQSLEAQAE
DdRp_Aper   VLFNRQPSLHRMSIMG----------HIVRVMPGKTfrln-LLVCPPYNADFDGDEMNLHVPRLEEAQAE
DdRp_Scer   VLFNRQPSLHRLSILS----------HYAKIRPWRTfrln-ECVCTPYNADFDGDEMNLHVPQTEEARAE
```

Figure 4.15. **Similarity between the putative catalytic site of RdRp and the known catalytic site of DdRp**. A. The PHI-BLAST hit between RdRp and DdRp. B. Multiple alignment of the conserved domains in RdRps and DdRps. The conserved motif containing the pattern used for the searches described in the text is highlighted. The three aspartates implicated in metal coordination and catalysis are denoted by asterisks.

4.6. Protein Annotation in the Absence of Detectable Homologs

Analysis of various features of protein molecules, which helps predict the type of structure and cellular localization of a given protein or an entire protein family is an indispensable complement to homology-based analysis. Furthermore, identification of certain structural features of proteins, such as signal peptides, transmembrane segments or coiled-coil domains, may provide some functional clues even in the absence of detectable homologs. It is not our intention here to discuss these methods in detail, but a brief summary is necessary to develop a reasonably complete picture of computational approaches that are important in genomics.

4.6.1. Prediction of subcellular localization of the protein

4.6.1.1. Signal peptides

The extensive studies of properties of signal peptides by Gunnar von Heijne and colleagues [149,873,874] resulted in the development of the SignalP program [617], which has become the *de facto* standard for signal peptide prediction and made their identification a relatively straightforward process. SignalP is a neural network method that has been trained separately on experimentally characterized sets of signal peptides from eukaryotes, Gram-positive bacteria and Gram-negative bacteria. Thus, the appropriate version of the program needs to be selected according to the origin of the analyzed protein (the Gram-positive version can be used for other prokaryotes with single-membrane cells, i. e. all other then Proteobacteria). SignalP is available at the web site of the Technical University of Denmark at http://www.cbs.dtu.dk/services/SignalP. The only other widely used algorithm for prediction of signal peptides was developed by Kenta Nakai and Minoru Kanehisa [601] and is included in the PSORT (http://psort.nibb.ac.jp) suite of programs ([600], see below).

4.6.1.2. Transmembrane segments

There is a variety of methods for predicting transmembrane segments or, more precisely, transmembrane α-helices in proteins (Table 4.8). All those programs rely to some degree on the hydrophobicity profiles of the polypeptide chains; several of them use experimentally determined transmembrane segments as training sets.

Table 4.8. Software tools for prediction of transmembrane segments

Program	Author, WWW site	Output			Ref.
		TM borders	Hydrophobicity plot	Topology prediction	
TMHMM	Anders Krogh, http://www.cbs.dtu.dk/services/TMHMM	Yes	Yes	Yes	[474]
TopPred2	Manuel Claros, http://www.sbc.su.se/~erikw/toppred2/	Yes	Yes	Yes	[151]
PhDhtm, PhD topology	Burkhardt Rost, http://cubic.bioc.columbia.edu/pp/	Yes	No	Yes	[715, 716]
PSORT, PSORT II	Kenta Nakai, http://psort.nibb.ac.jp	Yes	No	Yes	[600, 601]
DAS	Miklos Cserzo, http://www.sbc.su.se/~miklos/DAS/	Yes	Yes	No	[164]
TMpred	Kay Hofmann, http://www.ch.embnet.org/software/TMPRED_form.html	Yes	Yes	Yes	[352]
HMMTop	Gabor Tusnady, http://www.enzim.hu/hmmtop/	Yes	No	Yes	[857]
TMAP	Bengt Persson, http://www.mbb.ki.se/tmap/	Yes	No	Yes	[670, 671]
SOSUI	Shigeki Mitaku, http://sosui.proteome.bio.tuat.ac.jp/	Yes	Yes	Yes, with a graph	[349, 577]
Memsat2	David T. Jones, http://bioinf.cs.ucl.ac.uk/psipred/	Yes	No	Yes	[402]

In addition to predicting the positions of transmembrane segments, some of these programs predict membrane topology of the protein, typically using the "positive-inside" rule [875]. While most of the programs listed above (Table 4.8) accept only single sequences, TMHMM, HMMTop and SOSUI would also accept FASTA libraries. PHD additionally accepts multiple alignments, and TMAP works only with multiple alignments.

4.6.1.3. Protein targeting

Prediction of mitochondrial and chloroplast targeting signals allows one to differentiate between cytosolic and organellar proteins (Table 4.9). The PSORT server (http://psort.nibb.ac.jp), developed and maintained by Kenta Nakai at the University of Tokyo, additionally predicts protein targeting to the nucleus, endoplasmic reticulum, lysosomes, vacuoles, Golgi complex, and peroxisomes, and identifies probable GPI-anchored proteins.

Table 4.9. Software tools for prediction of protein targeting

Program	Author, WWW site	Comment	Refs.
PSORT PSORTII iPSORT	Kenta Nakai, http://psort.nibb.ac.jp	A comprehensive set of programs for analysis of protein targeting in prokaryotic and eukaryotic cells.	[600-602]
ChloroP	Olof Emanuelsson, http://www.cbs.dtu.dk/services/ChloroP	Searches for chloroplast transit peptides, can process several sequences at a time	[212]
TargetP	Olof Emanuelsson http://www.cbs.dtu.dk/services/TargetP	Searches for chloroplast transit peptides, mitochondrial targeting peptides, and signal peptides	[212]
MitoProt	Manuel Claros, http://mips.gsf.de/cgi-bin/proj/medgen/mitofilter	Predicts mitochondrial targeting signals; associated with the MitoP database of mitochondrial genes, proteins, and diseases	[150]
Predotar	Ian Small, http://www.inra.fr/predotar/	Predicts mitochondrial and chloroplast targeting peptides, aims at distinguishing between the two classes of targeting signals.	——

4.6.2. Prediction of structural features of proteins

Prediction of structural features of a protein does not directly lead to functional prediction, but is a prerequisite for most functional assignments. These methods contribute to protein analysis both in themselves and in conjunction with sequence-based and structure-based methods for homology detection. As discussed above, identification of low-complexity regions is the standard preliminary step in sequence similarity searches, whereas prediction of the secondary structure elements is a prerequisite for some methods of threading introduced below.

4.6.2.1. Coiled-coil domains

Because of the critical importance of appropriate treatment of low-complexity sequences for similarity searches, these were already introduced and briefly discussed in ♦4.2.3. Coiled-coil domains comprise a distinct class of non-globular protein structures; coiled coils are α-superhelices that are characterized, at the sequence level, by a 7-mer periodicity of hydrophobic residues, usually leucines. Search for such periodicity is the basis of the coiled-coil recognition algorithm in the COILS program by Andrei Lupas (http://www.ch.embnet.org/software/COILS_form.html [531-533]. Peter Kim and colleagues at MIT developed two more programs for coiled-coil prediction, Parcoil (http://nightingale.lcs.mit.edu/cgi-bin/score) and Multicoil (http://gaiberg.wi.mit.edu/cgi-bin/multicoil.pl) [908].

4.6.2.2. Secondary structure

Prediction of the secondary structure of a protein in itself gives little indication of its function(s) or homologous relationship, but nevertheless is important in conjunction with results obtained by other methods. Protein evolution proceeds largely through insertions and deletions in unstructured parts of a domain (loops), whereas secondary structure elements tend to be conserved. Therefore reliable prediction of these elements helps correctly aligning distantly related proteins and revealing subtle sequence similarities that otherwise might have been missed. Sequence-based secondary structure prediction is a well-developed area with a large number of competing methods (Table 4.10). The classic early methods, such as those of Chow-Fasman and Garnier and coworkers, used amino acid residue propensities calculated from 3D structures to predict the most likely structural state for each sequence segment. The modern methods, such as PHD, employ neural networks, which, again, train on the available database of protein structures.

Most of the prediction methods partition the protein sequence into one of the three states: α-helix, β-sheet, and loop and achieve a 70-75% accuracy. Several popular programs, such as PHD and PREDATOR, can accept a multiple alignment as the input, which facilitates identification of conserved structural motifs and notably increases the prediction accuracy. Other programs, such as PSIPRED and Jnet, would take a single sequence as an input, run PSI-BLAST with this sequence as query, and use the alignment generated after three iterations of PSI-BLAST for secondary structure prediction. In addition to generating the structural assignment (α-helix, β-sheet, or a loop) for each amino acid residue, some programs also provide numerical measures of the confidence of prediction. Different secondary structure prediction programs utilize different approaches and often generate conflicting predictions. Therefore it is advisable to attempt prediction with two or three different tools and compare the results. To simplify this task, several servers (Table 4.10) offer simultaneous submission of the given protein sequence to several prediction programs.

The JPred server (http://jpred.ebi.ac.uk), developed by Geoff Barton and colleagues at the EBI (currently at the University of Dundee), simultaneously runs PHD, Predator, DSC, NNSSP, ZPred, Mulpred, Jnet, COILS, and MultiCoil programs [166]. Another convenient site is the PredictProtein server (http://cubic.bioc.columbia.edu/pp/), maintained by Burkhard Rost at the Columbia University and mirrored around the world. In addition to running PHD, PredictProtein also submits the input sequence to Prof, PSIpred, PSSP, SAM-T99, and SSPro programs (see Table 4.11) for secondary structure prediction and to DAS, TMHMM, and TopPred programs (Table 4.8) for prediction of transmembrane segments.

Table 4.10. Portals for secondary structure prediction

Server	Author(s), WWW site	Comment	Ref.
Jpred	James Cuff, Geoff Barton, http://jpred.ebi.ac.uk	Provides access to seven different programs and presents the results in a convenient and easy-to-compare format	[165]
Predict Protein	Burkhard Rost, http://cubic.bioc.columbia.edu/predictprotein	Allows sending the sequence to other servers, which all report the results separately	[717, 718]
NPS@	http://pbil.univ-lyon1.fr	Provides convenient access to several programs	

While the Jpred server runs different programs locally and generates a consensus prediction, PredictProtein does not run other programs by itself. As a result, the user receives separate E-mail messages with outputs from each server used and has to combine their results. The NPS@ server at Pôle Bio-Informatique Lyonnais in Lyon, France (http://pbil.univ-lyon1.fr), provides access to PHD and Predator, as well as several other prediction tools, such as GORI, GORIII, GORIV, and SIMPA96 by Jean Garnier and colleagues [273,274,282,509,510], HNN and MLRC by Yann Guermeur and colleagues [321,322], and SOPM and SOPMA by Geourjon and Deleage [277,278]. Like Jpred, the NPS@ server runs all these programs on its own and allows generation of a single consensus prediction [155].

Despite their convenience, the "meta-servers" do not cover the entire diversity of the existing methods for secondary structure prediction. Servers that run individual prediction programs are listed in Table 4.11.

Table 4.11. Software tools for secondary structure prediction

Program	Author(s), WWW site	Comment	Ref.
PHDsec	Burkhard Rost http://cubic.bioc.columbia.edu/predictprotein	Probably the most popular current program, part of the PredictProtein server.	[717, 718]
PSIpred	David T. Jones http://bioinf.cs.ucl.ac.uk/psipred/	Runs three iterations of PSI-BLAST with the submitted sequence. Fails if there are no homologs in the database.	[402]
Predator	Dmitry Frishman http://bioweb.pasteur.fr/seqanal/structure/	Also available through Jpred, PredictProtein, and NPS@ servers	[246, 247]
Target99	Kevin Karplus http://www.cse.ucsc.edu/research/compbio/ HMM-apps/	Performs iterated search against a library of HMMs of proteins with known 3D structures	[423, 424]
SSP, NNSSP	Victor Solovyev http://genomic.sanger.ac.uk/pss/pss.shtml	Both programs accept single sequences or user-defined multiple alignments	[728, 729, 791]

Table 4.11 – continued

Jnet	James Cuff, Geoff Barton http://jpred.ebi.ac.uk	A neural network-based algorithm, part of the Jpred server	[165]
SSpro, SSpro8	Gianluca Pollastri, Pierre Baldi http://promoter.ics.uci.edu/BRNN-PRED	SSpro8 is the only web-based program that, in addition to α-helices and β-strands, also seeks to predict 3_{10}-helix, π-helix, β-bridge, turn, and bend structures	[70, 677]
nnPredict	Donald Kneller http://www.cmpharm.ucsf.edu/~nomi/nnpredict.html	A neural network-based program that tries to improve performance by detecting periodicities in the input sequence	[446]
PSA	James White, Collin Stultz http://bmerc-www.bu.edu/psa	Useful for the analysis of potential WD repeats	[811, 890]
PSSP, APSSP	G. P. S. Raghava http://imtech.ernet.in/raghava/apssp	Accessible through the PredictProtein server	[690]
DSC	Ross King, MJE Sternberg http://bioweb.pasteur.fr/seqanal/structure/	Was used in CASP2, has been succeded by the Prof program	[441]
Prof	Mohammed Ouali, Ross King http://www.aber.ac.uk/~phiwww/prof	A successor to DSP, accessible through the PredictProtein server	[636]
HNN	Yann Guermeur, http://npsa-pbil.ibcp.fr/NPSA/npsa_hnn.html	Accessible through the NSP@ server	[321]
MLRC	Yann Guermeur http://npsa-pbil.ibcp.fr/NPSA/npsa_mlr.html	Accessible through the NSP@ server	[322]
ZPred MulPred	Markéta Zvelebil http://kestrel.ludwig.ucl.ac.uk/zpred.html	Accessible through the PredictProtein server	[947]

4.6.2.3. Combination and hierarchy of prediction methods

Obviously, to characterize a protein, it is necessary to combine predictions of a variety of structural features, for which purpose the methods outlined above are used. When applying them, one needs to take into account that different predictions may overlap, with a more specific one being subsumed and obscured by a more general one. Specifically, signal peptides may mask as transmembrane segments, whereas both transmembrane segments and coiled coil domains may be given the inappropriately general label of low-complexity sequences. To avoid such clashes between different predictions, it is necessary to establish the hierarchy, in which they apply, with the more specific predictions given higher priority. The following priority order appears reasonable:

signal peptide > transmembrane segment > coiled coil > low complexity.
Secondary structure, which provides a complementary description of a protein sequence (a transmembrane segment is, at the same time, an α–helix), is usually predicted separately. This priority system has to be kept in mind when analysis is done manually, but ordering of prediction methods is also implemented (among software tools familiar to us) in the SMART server and in the UNIPRED program of the SEALS package.

Just like sequence similarity analysis methods, structural prediction needs to be scaled up for the purpose of genome analysis, and this requires local implementation. Most of the researchers who support the servers mentioned above will readily provide their code and/or executables.

4.6.3. Threading

Protein sequence-structure threading (usually simply referred to as threading) is a family of computational approaches that, given a protein sequence, attempt to select, among all known 3D structures, the structure that is best compatible with this sequence [535,775,783]. Metaphorically, the sequence is "threaded" through a variety of structures and the method determines which one fits better than the others [125,405,575,649,651]. The underlying idea is already well familiar to us: protein structure is more conserved in evolution than sequence. Insertions and deletions that change the substrate specificity, thermal stability, and other properties of the protein mostly occur in the loop regions without changing the core set of α-helices and β-strands. Therefore, a comparison of the (predicted) secondary structure of the new protein against a library of known 3D structures could potentially identify distant homologs even in the absence of statistically significant sequence similarity. Generally, threading methods involve calculating residue contact energy for the analyzed sequence superimposed over each structure in the database and ranking the structures by decreasing energy; the structure with minimal energy is the winner (e.g. [125,405,580,939]). Several statistical models to estimate the probability of "native" fold detection have been developed (e.g. [125,575,812]). It has been consistently reported that combining the traditional contact-potential-based threading with the use of sequence profiles and secondary structure alignment leads to a substantially greater success rate of fold recognition than either threading or profile searches or secondary structure comparisons alone [406,559,649,651].

Further discussion of threading is beyond the scope of this book; a detailed review of the physical theory behind threading methods has been published recently [574]. However, before ending this brief section with a list of threading software tools, which are available on the web, we must add a cautionary note based on our own research experience. Despite the well-documented success of threading approaches, using several different threading methods in the analysis of a variety of protein families, we usually failed to detect any demonstrably relevant relationships beyond what we could have identified by extensive PSI-BLAST-based sequence analysis. In contrast, we faced a considerable number of false leads that were associated with apparently statistically significant scores. We are relating this experience not to question the impressive performance of threading methods in fold recognition, but to caution the reader that current threading approaches still might not be robust enough for routine use in large-scale genome analysis.

Even in case-by-case manual analysis, before gleaning any far-reaching conclusions from the threading results, one has to be aware of the complexities of the approach and its potential pitfalls. It is important to carefully analyze the outputs to make sure that the reported secondary structure similarity is genuine and is not caused, for example, by "fatal attraction" of long (>15 aa) helices in two proteins that are otherwise entirely different. It also helps to check the list of hits and make sure that they belong to the same SCOP and/or CATH fold or at least align well in VAST and/or DALI analysis. One should be extremely suspicious when the query protein produces high-scoring alignments with proteins known to have different folds.

Some of the popular threading tools accessible on the web are listed in Table 4.12. A convenient portal to most of these methods is maintained by Leszek Rychlewski and colleagues at BioInfoBank (http://bioinfo.pl/meta) in Poznan, Poland [128]. In addition, this server offers a consensus prediction through the Pcons tool (http://www.sbc.su.se/~arne/pcons [530]).

Table 4.12. Software tools for protein threading and related methods for protein structure prediction

Program	Author(s), WWW site	Comments[a]	Ref.
SAM-T99	Kevin Karplus http://www.cse.ucsc.edu/research/compbio/HMM-apps/	Performs an iterated search against a library of HMMs, then builds a new HMM to search PDB. Not a threading approach in the classic sense, but rather a fold recognition procedure.	[423, 424]
InBGU	Daniel Fischer, http://www.cs.bgu.ac.il/~bioinbgu/	Compares sequence profiles and secondary structure profiles for the query and for proteins of different folds. Combines five different methods to produce a consensus prediction	[227]
UCLA/DOE Fold Server	Tom Holton, David Eisenberg, http://fold.doe-mbi.ucla.edu/	Same as the server at InBGU, but uses a different fold library; linked to a motif-based fold recognition server	[543, 734]
GenThreader	David Jones, http://bioinf.cs.ucl.ac.uk/psipred/	Runs three iterations of PSI-BLAST with the submitted sequence and uses the resulting profile as query. Fails if there are no homologs in the database.	[402]
3D-PSSM	Lawrence Kelley, Robert MacCallum, Michael Sternberg http://www.sbg.bio.ic.ac.uk/~3dpssm/	Compares 1D and 3D profiles coupled with secondary structure and solvation potential.	[434]
FFAS	Lukasz Jaroszewski, Weizhong Li, Adam Godzik, http://bioinformatics.ljcrf.edu/FFAS/	Uses PSI-BLAST to find homologs, then creates a sequence profile and compares it with sequence profiles of protein families in PDB	[725]

FUGUE	Jiye Shi, Tom Blundell, Kenji Mizuguchi http://www-cryst.bioc.cam.ac. uk/~fugue/	Searches the query sequence against the library of structure-based sequence alignments. Creates either a global or a global-local alignment, depending on the difference in sequence length. Features environment-specific substitution tables and variable gap penalties	[765]
SUPER FAMILY	Julian Gough, Cyrus Chothia http://stash.mrc-lmb.cam.ac.uk/SUPERFAMILY/	Compares the query sequence against a library of HMMs for SCOP superfamilies (each superfamily can be covered by one or more HMMs). Not a threading approach in the classic sense, but rather a fold recognition procedure.	[300, 301]
LOOPP	Jarek Meller, Ron Elber http://ser-loopp.tc.cornell.edu/loopp.html	Performs a comparison of pairwise and profile-based alignments with and without gaps to design optimal scoring functions for each particular case	[565]
123D, 123D+	Nickolai Alexandrov http://genomic.sanger.ac.uk/123D/123D.html or http://123d.ncifcrf.gov/	Threads the NNSSP-generated secondary structure prediction through a library of backbone assignments. Allows the user to choose the type of alignment (local or global), the substitution matrix, and gap penalty	[13,14]
RPFOLD	G.P.S. Raghava, http://www.imtech.res.in/ raghava/rpfold/	Combines a sequence similarity search with a Clustal-based alignment of the secondary structure elements	–

4.7. Conclusions and Outlook

The methodological armory of computational biology has been evolving at a substantial rate for only 20 years, but has already reached impressive diversity, such that it would be impossible to present it properly, even if we devoted the entire book to this subject alone. In this chapter, we attempted to briefly discuss only those methods that, in our understanding, are central to comparative genomics. These are: (i) gene prediction, (ii) sequence similarity analysis and, in particular, database search, including iterative methods based on PSSMs and HMMs, and (iii) prediction of protein structural features. We believe that, at the time of this writing (middle of 2002), sequence similarity analysis, along with comparison of 3D structures, remains the main source of biologically important discoveries and predictions made by computational biologists. We tried to show that some of these finding are distinctly non-trivial. In the next chapter, we discuss how these methods work when applied to genome comparison and also present some new approaches that more explicitly rely on the information that can be extracted only from (nearly) complete genome sequences.

4.8. Further Reading

1. Doolittle RF. 1986. *Of Urfs and Orfs: A primer on how to analyze derived amino acid sequences.* University Science Books, San Diego.

2. Baxevanis AD and Ouellette BFF (eds). 2001. *Bioinformatics: a practical guide to the analysis of genes and proteins.* John Wiley & Sons, New York.

3. Mount DW. 2000. *Bioinformatics: Sequence and genome analysis.* Cold Spring Harbor Laboratory Press, Cold Spring Harbor, NY. Chapter 1.

4. Durbin R, Eddy SR, Krogh A and Mitchison G. 1997. *Biological Sequence Analysis: Probabilistic models of proteins and nucleic acids.* Cambridge University Press, Cambridge, UK.

5. Waterman MS. 1995. *Introduction to Computational BiologyMaps, Sequences and Genomes.* CRC Press, Boca Raton, FL.

6. Gusfield D. 1997. *Algorithms on Strings, Trees, and Sequences: Computer Science and Computational Biology.* Cambridge University Press, Cambridge, UK.

CHAPTER 5.
GENOME ANNOTATION AND ANALYSIS

In the preceding chapter, we gave a brief overview of the methods that are commonly used for identification of protein-coding genes and analysis of protein sequences. Here, we turn to one of the main subjects of this book, namely how these methods are applied to the task of primary analysis of genomes, which often goes under the name of "genome annotation". Many researchers still view genome annotation as a notoriously unreliable and inaccurate process. There are excellent reasons for this opinion: genome annotation produces a considerable number of errors and some outright ridiculous "identifications" (see ♦3.1.3 and further discussion in this chapter). These errors are highly visible, even when the error rate is quite low: because of the large number of genes in most genomes, the errors are also rather numerous. Some of the problems and challenges faced by genome annotation are an issue of quantity turning into quality: an analysis that can be easily and reliably done by a qualified researcher for one or ten protein sequences becomes difficult and error-prone for the same scientist and much more so for an automated tool when the task is scaled up to 10,000 sequences. We discuss here the performance of manual, automated and mixed approaches in genome annotation and ways to avoid some common pitfalls. Mostly, however, we concentrate in this chapter on the so-called context methods of genome analysis, which are the recent excitement in the annotation field. These approaches go beyond individual genes and explicitly take advantage of genome comparison.

5.1. Methods, Approaches and Results in Genome Annotation

5.1.1. Genome annotation: data flow and performance

What is genome annotation? Of course, there hardly can be any exact definition but, for the purpose of this discussion, it might be useful to define annotation as a subfield in the general field of genome analysis, which includes more or less anything that can be done with genome sequences by computational means. In simple, operational terms, annotation may be defined as the part of genome analysis that is customarily performed before a genome sequence is deposited in GenBank and described in a published paper. We say "customarily" because the annotations available through GenBank and particularly the types of analysis reported in the literature for different genomes vary widely. For instance, the reports on the human genome sequence [488,870] clearly include considerable amount of information that goes beyond typical genome annotation. The "unit" of genome annotation is the description of an individual gene and its protein (or RNA) product and the focal point of each such record is the function assigned to the gene product. The record may also include a brief description of the evidence for this assigned function, e.g. percent identity with a functionally characterized homolog or the boundaries of domains detected in a domain database search, but there is no room for any details of the analysis.

Figure 5.1 shows a rough schematic of the data flow in genome annotation, starting with the finished sequence; we leave finishing of the sequence out of this scheme but indicate the possibility of feedback resulting in correction of sequencing errors. Of these procedures, which must be integrated for predicting gene functions, statistical gene prediction and search of general-purpose databases for sequence similarity are central in the sense that this is done comprehensively as part of any genome project. The contribution of the other approaches in the scheme in Figure 5.1, particularly specialized database search, including domain databases, such as Pfam, SMART and CDD (♦3.2.2), and genome-oriented databases, such as COGs, KEGG or WIT (♦3.4), and genomic context analysis, varies greatly from project to project. So far, these relatively new methods and resources remain ancillary to traditional database search in genome annotation, but we argue farther in this chapter that they can and probably will transform the annotation process in the nearest future.

Before we consider several aspects of genome annotation, it may be instructive to assess its brutto performance, i.e. the fraction of the genes in a genome, to which a specific function is assigned. Table 5.1 lists such data for several genomes sequenced in 2001 and annotated using relatively up-to-date methods. This comparison shows notable differences between the levels of annotation of different genomes. Some genomes simply come practically unannotated, such as, for example, *Sulfolobus tokodaii*, which is a crenarchaeon closely related to *S. solfataricus*, and represented in the COGs to the same extent as the latter species. In most genomes, however, functional prediction has been made for the majority of the genes, from 54% to 79% of the protein-coding genes. Obviously, these differences depend both on the taxonomic position of the species in question (e. g. it is likely that, for Crenarchaea whose biology is in general poorly understood, the fraction of genes, for which functional prediction is feasible, will be lower than for bacteria of the well-characterized *Bacillus-Clostridium* group, such as *C. acetobutylicum* or *L. lactis*) and on the methods and practices of genome annotators.

Table 5.1. Microbial genome annotation 2001

Species	Total no. of genes[a]	Genes with assigned function	"Conserved hypothetical" proteins	"Hypothetical" proteins	Assigned to COGs	Ref.
Agrobacterium tumefaciens	5,419	3,475 (64%).	1,236 (22%)	708 (13%)	4490 (83%)	[917]
Caulobacter crescentus	3,737	2,030 (54%)	725 (19%)	1,012 (27%)	3,514 (93%)	[618]
Clostridium acetobutylicum	3,672	2,888 (79%)	187 (5%)	597 (16%)	2,941 (80%)	[622]
Lactococcus lactis	2,310	1,482 (64%)	465 (20%)	363 (16%)	1,849 (80%)	[97]
Listeria innocua	3,052	1920 (63%)	757 (25%)	375 (12%)	2,444 (80%)	[286]
Mycobacterium leprae[b]	2,720	1802 (66%)	776 (29%)	142 (5%)	1,231 (45%)	[153]
Nostoc (*Anabaena*) sp. PCC7120	5,368	45%	27%	28%	4,002 (75%)	[416]
Pasteurella multocida	2,014	1,814 (64%)	531 (26%)	200 (10%)	1,881 (93%)	[554]
Sinorhizobium meliloti	6,204	3,704 (60%)	1,991 (32%)	509 (8%)	5298 (85%)	[255]
Staphylococcus aureus	2,595	63%	23%	14%	2,126 (82%)	[481]
Streptococcus pyogenes	1,752	1137 (65%)	145 (8.2%)	470 (27%)	1,390 (79%)	[223]
Sulfolobus solfataricus	2,977	1,624 (57%)	619 (21%)	734 (25%)	1,910 (64%)	[764]
Sulfolobus tokodaii	2,826	14 (0.5%)	920 (33%)	1,892 (67%)	1,778 (63%)	[426]
Yersinia pestis	4,012	76%	13%	9%	3,669 (91%)	[656]

[a] In contrast to Table 1.4, the total gene numbers, as well as the numbers of genes with assigned function, "conserved hypothetical" and "hypothetical" genes, were taken from the original publications.

[b] The low fraction of *M. leprae* genes, assigned to COGs, is due to the large number of pseudogenes in this genome [153].

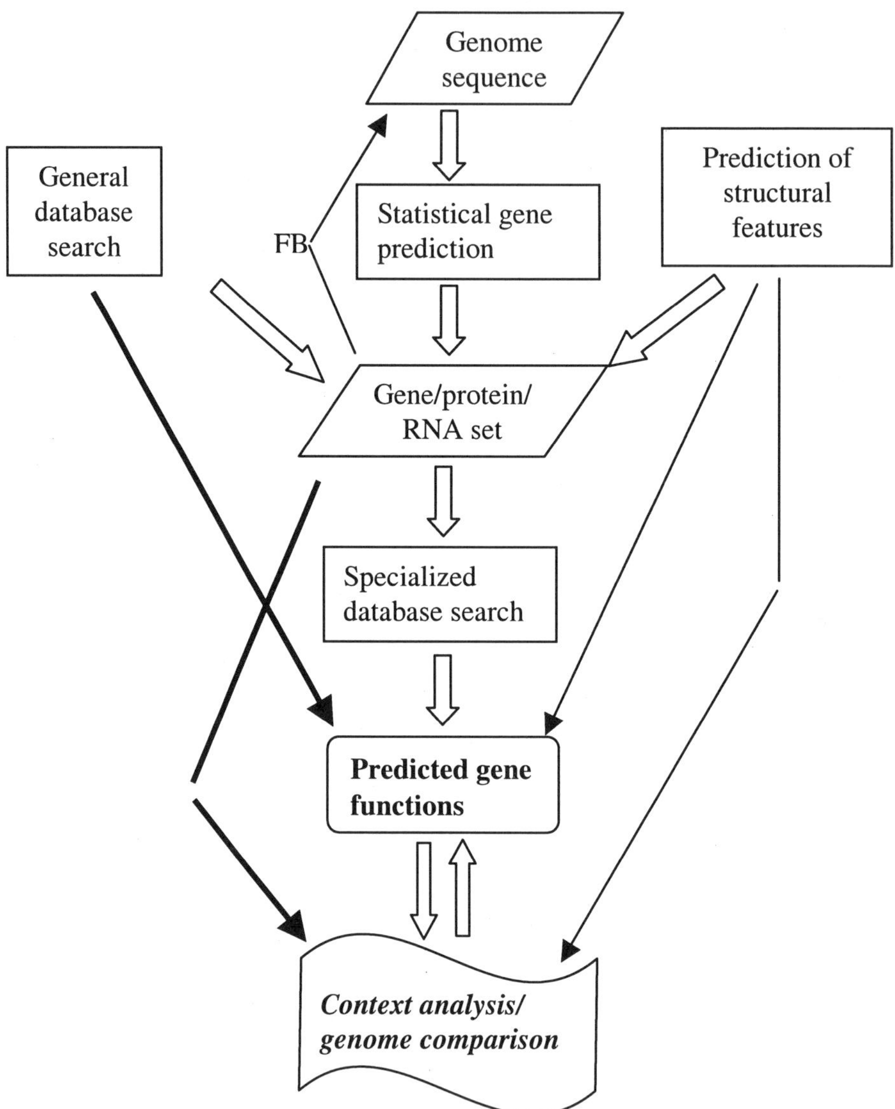

Figure 5.1. A generalized flow chart of genome annotation. FB, feedback from gene identification for correction of sequencing errors, primarily frameshifts. General database search: searching sequence databases (typically, NCBI NR) for sequence similarity, usually using BLAST; specialized database search: searching domain databases, such as Pfam, SMART and CDD, for conserved domains, genome-oriented databases, such as COGs, for identification of orthologous relationship and refined functional prediction, metabolic databases, such as KEGG for metabolic pathway reconstruction, and possibly, other database searches. Statistical gene prediction: use of methods like GeneMark or Glimmer to predict protein-coding genes. Prediction of structural features: prediction of signal peptide, transmembrane segments, coiled domain and other features in putative protein functions.

Even in better-characterized genomes, for hundreds of genes (those encoding "conserved hypothetical" and "hypothetical" proteins), there is no functional prediction whatsoever. Furthermore, among those proteins that formally belong to the annotated category, a substantial fraction of the predictions are only general and are in need of major refinement. Some of these problems can be solved only through experiment, but the above numbers show beyond doubt that there is ample room for improvement in computational annotation itself; farther in this chapter, we touch upon some of the possible directions.

Genome annotation necessarily involves some level of automation. No one is going to manually paste each of several thousand-protein sequence encoded in a genome into the BLAST window, hit the button and wait for the results to appear on screen. For annotation to be practicable at all, software is necessary to run such routine tasks in a batch mode and also to organize the results from different programs in a convenient form, and each genome project employs one or another set of tools to achieve this. After that point, however, genome annotation is still mostly "manual" (or, better, "expert") because decisions on how to assign gene functions are made by humans (supposedly, experts). Several attempts have been made to push automation beyond straightforward data processing and to allow a program to actually make all the decisions. We briefly discuss some of the automated systems for genome annotation in the next section.

5.1.2. Automation of genome annotation

Terry Gaasterland and Christoph Sensen once estimated that annotating genomic sequence by hand would require as much as one year per person per one megabase [253]. We now believe, on the basis of our own experience of genome annotation (e.g. [622,779,805]), that this estimate is exaggerated perhaps by a factor of 5 or 6. Nevertheless, there is no doubt that genome annotation has become the limiting step in most genome projects. Besides, humans are supposed to be inconsistent and error-prone. Hence the incentives for automating as much of the annotation process as possible.

GeneQuiz (http://www.sander.ebi.ac.uk/genequiz/) project was the first automatic system for genome analysis, which performed similarity searches followed by automatic evaluation of results and generation of functional annotation by an expert system based on a set of several predefined rules [749]. Several other similar systems have been created since then, but GeneQuiz remains the only such tool that is open to the general public [350].

GeneQuiz runs automated database searches and sequence analysis by taking a protein sequence and comparing it against a non-redundant protein database, generated by automated cross-linking and cross-referencing of PDB, SWISS-PROT, PIR, PROSITE, and TREMBL databases, with the addition of human, mouse, fruit fly, zebrafish, and *Anopheles gambiae* protein sets obtained from the Ensemble project (http://www.ensembl.org) and a *C. elegans* protein set (http://www.sanger.ac.uk/Projects/C_elegans/wormpep). This comparison is done by running BLAST and FASTA programs and is used to identify the cases with high similarity, where function can be predicted. Additionally, searches for PROSITE patterns are performed. Predictions are also made for coiled-coil regions using COILS2 [533], transmembrane segments using PHDhtm [715] and secondary structure elements using PHDsec [718]. The system further clusters proteins from the analyzed genome by sequence similarity [822] and constructs multiple alignments. The results are presented in a table that contains information on the best hits (including gene names, database identifiers, and links to the corresponding databases), predictions for secondary structure, coiled-coil regions etc., and a reliability score for each item. The functional assignment is then made automatically on the basis of the functions of the homologues found in the database. At this level, functional assignments are qualified as clear or as ambiguous.

The effectiveness and accuracy of such fully automated system has been the subject of a rather heated discussion but still remains uncertain. While the authors originally estimated the accuracy of their functional assignments to be 95% or better [638,749], others reported that only 8 of 21 new functional predictions for *M. genitalium* proteins made by GeneQuiz could be fully corroborated [466]. A similar discrepancy between the functional predictions made by the GeneQuiz team [31] and those obtained by mostly manual annotation [466] was reported for the proteins encoded in the *M. jannaschii* genome ([264], see http://www.bioinfo.de/isb/1998/01/0007). It appeared that GeneQuiz analysis suffered from the usual pitfalls of sequence similarity searches (see ♦3.1.3, the next section and [99,104,264]).

PEDANT, MAGPIE, ERGO, IMAGENE

While GeneQuiz seems to be the only fully automated genome annotation tool that is open to the public for new genome analysis, there have been reports of similar systems developed by other genome annotation groups. These include Dmitrij Frishman's PEDANT (http://pedant.gsf.de, [245,248], Terry Gaasterland's MAGPIE and its sister programs (http://genomes.rockefeller.edu, [252,253]), Ross Overbeek's ERGO (http://ergo.integratedgenomics.com/ERGO, [642,643]), Alan Viari's Imagene (http://wwwabi.snv.jussieu.fr/research, [561]), and some others. Although none of these systems is freely available to outside users, many of the genome annotation results they produced are accessible on the web and can be used to judge the performance.

The PEDANT web site contains by far the most information open to the public and can be used as a good reference point for automated genome analyses (see also ♦2.4).

SEALS

In addition to completely automated systems, some tools that greatly facilitate and accelerate manual genome annotation are worth a mention. System for Easy Analysis of Lots of Sequences (SEALS) developed by Roland Walker at the NCBI is, for obvious reasons, the one most familiar to the authors of this book (available for downloading at http://www.ncbi.nlm.nih.gov/CBBresearch/Walker/SEALS, [878]). The SEALS package consists of ~50 simple, UNIX-based tools (written in PERL), which follow consistent syntax and semantics. SEALS combines software for retrieving sequence information, scripting database searches with BLAST, viewing and parsing search outputs, searching for protein sequence motifs using regular expressions, and predicting protein structural features and motifs. Typically, using SEALS, a genome analyst first looks for structural features of proteins, such as signal peptides (predicted by SignalP), transmembrane domains (predicted by PHDhtm), coiled-coil domains (predicted by COILS2), and large non-globular domains (predicted using SEG). Once these regions are identified and masked, database searches are run in a batch mode using the chosen method, e.g. PSI-BLAST. The outputs can be presented in a variety of formats, of which filtering with taxonomic queries implemented in the SEALS script TAX_COLLECTOR is among the most useful. SEALS has been extensively used in the comparative studies of bacterial, archaeal and eukaryotic genomes (e.g. [52,55,540].

5.1.3. Accuracy of genome annotation, sources of errors and some thoughts on possible improvements

Benchmarking the accuracy of genome annotation is extremely hard. It has been shown on numerous occasions that more advanced methods for sequence comparison, such as gapped BLAST and subsequently PSI-BLAST, sometimes used in combination with threading, as well as various forms of motif analysis and careful manual integration of the results produced by all these approaches, substantially improve detection of homologs (e.g. [168,401,434,466,585]). At the end, however, genome annotation is not about detection of homologs but rather about functional prediction, and here, the problem of a standard of truth is formidable. By definition, functional annotation (more precisely, functional prediction) deals with proteins whose functions are unknown, and the rate of experimental testing of predictions is extremely slow. We believe that it is possible to design an objective test of the accuracy of genome annotation in the following manner. The protein set encoded in a newly sequenced genome is analyzed and specific active centers and other functionally important sites are predicted for as many proteins as possible. When a new, preferably phylogenetically distant genome becomes available, orthologs of the proteins from the first genome are identified and the conservation of the predicted functional sites is assessed. Lack of conservation would count as an error; this is, of course, a harsh test that would give the low bound of accuracy because, first, functional site prediction may be partly wrong but the function of the protein still would be predicted correctly and, second, some active sites might be disrupted in the new genome. In this way, the accuracy of the prediction could be assessed quantitatively and, in principle, even a "tournament" analogous to the CASP competition in protein structure prediction [869] could be arranged.

However, so far, evaluation of the accuracy of genome annotation has been largely limited to the assessments of consistency of annotations of the same genome generated by different groups and various "sanity checks" and expert judgments. Steven Brenner published an interesting comparison of three independent annotations [242,467,639] of the smallest of the sequenced bacterial genomes, *Mycoplasma genitalium* [116]. Without attempting to determine which annotation was "better", he manually examined all conflicting annotations eliminating trivial semantic differences and counting the apparent irreconcilable ones as errors (in at least one of the annotations). His conclusion was that there was at least 8% error rate among the 340 genes annotated by at least two of the three groups. In a similar exercise that we have done on the basis of the COG database, we found that, of 786 COGs that did not include paralogs (the number for the end of 1999), members of 194 had conflicting annotations in GenBank [267]. This suggests, more pessimistically, an annotation error rate of at least 25% using the same criterion as applied by Brenner. Clearly, even the lower of these estimates represents a serious problem for genome annotation, bringing up the specter of error catastrophe [89,104]. We first briefly discuss the most common sources of errors and than some ideas regarding the ways out. Manual and automated genome annotation encounter the same typical problems, which we already mentioned in the discussion of the reliability of sequence database records (see ♦3.1.3). Inevitably, even partial automation of the annotation process tends to increase the likelihood of all these types of errors.

In order to examine various kinds of errors that are common in genome annotation, it is convenient to re-examine four cases of discrepancies in the annotation of *M. genitalium* proteins that were specifically highlighted in the aforecited article of Steven Brenner (Table 5.2). Although one of the authors was involved in one of the compared annotations, we think we can be completely impartial in the spirit of Brenner's article, especially since six years have passed, an eternity for genomics.

The protein MG302 was not annotated in the original genome publication by Fraser and colleagues and was assigned conflicting annotations by the other two groups. Ouzounis and coworkers notably characterized this protein as a "mitochondrial 60S ribosomal protein L2", whereas Koonin and coworkers annotated it is a permease, perhaps specific for glycerol-3-phosphate. A database search performed in 2002 leaves no doubt whatsoever that the protein is a permease; this is, of course, readily supported by transmembrane segment prediction. However, the glycerol-3-phosphate specificity is not supported at all. Instead, these searches, particularly the CDD search, unequivocally pointed to a relationship between MG302 and a family of cobalt transporters. Nevertheless, since the similarity between MG302 and the cobalt transporters is not particularly strong and transporters switch their specificity with relative ease during evolution, caution is due and the annotation as "probable Co transporter" seems most appropriate. This single case nicely covers several common problems of genome annotation. The most benign but also apparently most widespread of these is ***over-prediction*** or, more precisely, ***overly specific prediction***. Even with the methods available in 1996 (ungapped BLAST, FASTA, various alignment methods and transmembrane segment prediction), the conclusion that MG302 was a permease was quite firm. However, glycerol-3-phosphate permease turned up as the most similar functionally characterized protein just by chance (Co^{2+} transporters have not been characterized at the time). Transferring functional information from this unreliable best hit, however tentatively, was a typical error of over-prediction; the appropriate annotation at the time would have been, simply, "predicted permease". The annotation of MG302 as "mitochondrial 60S ribosomal protein L2" is, of course, much more conspicuous. At face value, this does not even pass a "reality check": there certainly can be no mitochondria and no 60S ribosomes in mycoplasmas.

Such semantic snafus are pretty common in genome annotation, especially those that are either produced fully automatically or manually but non-critically (e.g. the "discovery" of head morphogenesis in bacteria mentioned in Chapter 3). However, these are probably the least serious annotation errors.

Table 5.2 Different types of errors in genome annotation

Protein	Annotation				
	Fraser and coworkers	Ouzounis and coworkers	Koonin and coworkers	GenBank 2002	Conclusion 2002
MG085	Hydroxymethyl-glutaryl-CoA reductase (NADPH)	NADH-ubiquinone oxidoreductase	ATP(GTP?)-utilizing enzyme	HPr (Ser) kinase, putative	HPr kinase
MG225	Hypothetical	Histidine permease	Amino acid permease	Hypothetical	Amino acid permease
MG302	No database match	Mitochondrial 60S ribosomal protein L2	(Glycerol-3-phosphate?) permease	Hypothetical	Probable cobalt transporter
MG448	Pilin repressor (pilB)	PilB protein	Putative chaperone-like protein	Hypothetical/ Peptide methionine sulfoxide reductase	Peptide methionine sulfoxide reductase B

Let us just assume that the authors of this annotation meant "homolog of mitochondrial 60S ribosomal protein L2". What is worse, the search result that presumably gave rise to this annotation is impossible to reproduce at this time, at least not without detailed research, which we are not willing to undertake. It is most likely that this blatantly wrong annotation was due to ***a spurious database hit*** to a ribosomal protein that was not critically assessed. It is not clear, in this particular case, how could this spurious hit pass the significance threshold, but in general, this happens most often because of the lack of proper filtering for low complexity (or alternative approaches, such as composition-based statistics, which are available in 2002, but have not been developed in 1996; see Chapter 4). Alternatively or additionally, the problem might lie in non-critical transfer of annotation from ***an unreliable database record***, i.e. a low-complexity sequence erroneously labeled as a ribosomal protein. Notably, our re-analysis shows that the annotations assigned by each of the three groups were not completely correct: one was an outright error, another one involved over-prediction and the third one an under-prediction. Although less notorious than false predictions (false-positives, in statistical terms), lack of prediction, where a confident one is feasible with available methods, is still an error (a false-negative).

The case of the MG225 protein is quite similar except that there was no clear false prediction involved. Once again, the original genome project gave no annotation (a false-negative), whereas one of the remaining groups annotated the protein as "histidine permease" and the other one stopped at a "amino acid permease" annotation without proposing specificity. Today's searches support the latter decision because no convincing, specific relationship between this protein and transporters for any particular amino acid could be detected (in fact, given the small repertoire of transporters in mycoplasmas, this one might have a broad specificity). Notably, both MG302 and MG225 remain "hypothetical proteins" in GenBank to this day, although closely related orthologs from *M. pneumoniae* are correctly annotated as permeases [168].

The MG085 protein was annotated as an oxidoreductase (of different families) in the original genome report and by Ouzounis and coworkers, whereas Koonin and coworkers predicted that it was an ATP(GTP?)-utilizing enzyme on the basis of the conservation of the P-loop motif in this protein and its homologs. In 2002, database searches immediately identify this protein as HPr kinase (this annotation is now correctly assigned to MG085 in GenBank), a regulator of the sugar phosphotransferase system, which indeed is a P-loop-containing, ATP-utilizing enzyme [723]. Back in 1996, this was the only informative annotation that could be derived for this protein; HPr kinase genes have not been identified at the time. Once again, the specific source of the oxidoreductase assignments is hard to determine; spurious hits, non-critical use of incorrect database annotations or a combination thereof must have caused this.

The case of MG448 is of particular interest. This protein was annotated as "pilin repressor" or simply PilB protein by Fraser and coworkers and Ouzounis and coworkers and, somewhat cryptically, as "chaperone-like protein" by Koonin and coworkers. This protein remains "hypothetical" in GenBank but became a peptide methionine sulfoxide reductase (PMSR) in SWISS-PROT. A database search detects highly significantly similarity with numerous proteins that are annotated primarily as PMSR and, in some cases, as PilB-related repressors. In reality, this protein is indeed a recently characterized, distinct form of PMSR, MsrB [476,526], which is evolutionarily unrelated to, but is often associated with the classic PMSR, MsrA, either as part of a multidomain protein or as a separate gene in the same operon [267]. These fusions resulted in the annotation of MG448 as PMSR, which, ironically, turned out to be correct, but mostly (except for the recently updated SWISS-PROT description), for a wrong reason, because it was the MsrA domain that was recognized in the fusion proteins. Furthermore, in several bacteria, these two domains are fused to a third, thioredoxin domain. The three-domain protein of *Neisseria gonorrhoeae* has been characterized as a regulator of pili operon expression, and this is what caused the annotation of MG448 as PilB, which was reproduced by two groups. This annotation is outright wrong and does not even pass a "reality check" because there are no pili in mycoplasmas (parenthetically, latest reports appear to indicate that even the original functional characterization of the *Neisseria* protein was erroneous [776]).

Unrecognized multidomain architecture of either the analyzed protein or its homologs or both is a common cause of erroneous annotation. The "chaperone-like protein" annotation was based on the notion that the PMSR function could be interpreted as a form of chaperone action and, accordingly, the associated domain was also likely to have a chaperone-like activity. In retrospect, this looks like over-prediction combined with insufficient information included in the annotation. A straightforward annotation of MG448 as a PMSR-associated domain, perhaps with an extra prediction of redox activity on the basis of conservation of cysteines in this domain, the way it has been done in a subsequent publication [267], would have been appropriate. We revisit this interesting set of proteins when discussing context analysis in ♦5. 2.

While considering only four proteins with contradictory annotations, we encountered all the main sources of systematic error in genome annotation. We list them here again, more or less in the order of decreasing severity, as we see it: (i) spurious database hits, often caused by low-complexity regions in the query or the database sequence, (ii) non-critical transfer of functional prediction from an unreliable database record, (iii) incorrect interpretation (lack of recognition) of multidomain architecture of the query and/database sequences, (iv) overly specific functional prediction, and (v) under-prediction.

We believe that this brief discussion highlights more general problems beyond these specific causes of errors. Even the apparently correct database annotations are insufficiently informative. Typically, the records do not include the evidence behind the prediction or include only minimal data that may be hard to interpret, such as E-values of the hits to particular domains. In this situation, any complicated case will not be represented adequately (e.g. the PMSR-associated domain discussed above). In addition, there is no controlled vocabulary for genome annotation, which creates numerous semantic problems, although an attempt to correct this situation is being undertaken in the form of the Genome Ontology project [60,513].

The above discussion shows that the general state of genome annotation is far from being satisfactory. What can be done to improve it? In his paper on genome annotation errors, Steven Brenner noted that, "to prevent errors from spreading out of control, database curation by the scientific community will be essential." [116]. Curation, however, implies that databases other than GenBank will have to be employed because GenBank, by definition, is an archival database (Chapter 3). It appears that the future and, to some degree, already the present of genome annotation lies in specialized databases that actually function as annotation tools. The beginnings of such tools can be seen in databases like KEGG, WIT and COGs complemented by tools for domain identification, such as CDD and SMART (see Chapters 3 and 4).

Conceptually, the advantage of this approach may be viewed as reduction and structuring of the search space for genome annotation. Thus, when using COGs, a genome analyst compares each protein sequence not to the unstructured set of more than a million proteins (the NR database), but instead to a collection of ~5,000 mostly well-characterized protein sets classified by orthology, which is the appropriate level of granularity for functional assignment. Already today genome annotation is starting to change through the use of the new generation of databases and tools. However, smooth integration of these and development of new, richer formats for annotation are things of the future. In the next subsection, we turn to a specific example to illustrate how the use of COGs helps genome annotation.

5.1.4. A case study on genome annotation: the crenarchaeon *Aeropyrum pernix*

Aeropyrum pernix was the first representative of the Crenarchaeota (one of the two major branches of archaea; see Chapter 6) and the first aerobic archaeon whose genome has been sequenced [427]. *A. pernix* was reported to encode 2694 putative proteins in a 1.67 Mbase genome. Of these, 633 proteins were assigned a specific or general function in the original report on the basis of sequence comparison to proteins in the GenBank, SWISS-PROT, EMBL, PIR, and Owl databases. Given the intrinsic interest of the first crenarchaeal genome and also because of the unexpectedly low fraction of predicted genes that were assigned functions in the original report, *A. pernix* was chosen for a pilot annotation project centered around the COG database [605].

Figure 5.2 (see the color plates) shows the protocol employed for the COG-based genome annotation. This procedure was not limited to straightforward COGNITOR analysis but also explicitly drew from the phyletic patterns. Whenever *A. pernix* was unexpectedly not represented in a COG (e.g. a COG that included all other archaeal species), additional analysis was undertaken. To identify possible diverged COG members from *A. pernix,* PSI-BLAST searches were run with multiple members of the respective COGs, and to detect COG members that could have been missed in the original genome annotation, the translated sequence of the *A. pernix* genome was searched using TBLASTN. Conversely, unexpected occurrence of *A. pernix* proteins in COGs that did not have any other archaeal members were examined case by case to detect likely HGT events and novel functions in the crenarchaeal genome.

Proteins were assigned to COGs through two rounds of automated comparison using COGNITOR, each followed by curation, that is, manual checking of the assignments. The first round attempts to assign proteins to existing COGs; typically, >90% of the assignments are made in this step. The second round serves two purposes: first, to assign paralogs, that might have been missed in the first round, to existing COGs; and, second, to create new COGs from unassigned proteins.

The results of COG assignment for *A. pernix* are shown in Table 5.3. Manual curation of the automatic assignments revealed a false-positive rate of less than 2% (23 of 1123 proteins). Even if the less severe errors, when a protein was transferred from one related COG to another, are taken into account, the false-positive rate was 4%, which is not negligible, but substantially lower than the estimates cited above for more standard genome annotation methods. The number of identified false-negatives was even lower, but in this case, of course, it is not possible to determine how many proteins remain unassigned. It is further notable that the great majority of assigned proteins belonged to pre-existing COGs, which facilitates a (nearly) automatic annotation.

Altogether, 1102 *A. pernix* proteins were assigned to COGs. Some of these proteins (154) were members of functionally uncharacterized COGs. Subtracting these, annotation has been added to 315 proteins, which is an increase of about 50% compared to the original annotation. These newly annotated *A. pernix* proteins included, among others, the key glycolytic enzymes glucose-6-phosphate isomerase (APE0768, COG0166) and triose phosphate isomerase (APE1538, COG0149), and the pyrimidine biosynthetic enzymes orotidine-5'-phosphate decarboxylase (APE2348, COG0284), uridylate kinase (APE0401, COG0528), cytidylate kinase (APE0978, COG1102), and thymidylate kinase (APE2090, COG0125). Similarly, important functions in DNA replication and repair were confidently assigned to a considerable number of *A. pernix* proteins, which, in the original annotation, were described as "hypothetical". Examples include the bacterial-type DNA primase (COG0358), the large subunit of the archaeal-eukaryotic-type primase (COG2219), a second ATP-dependent DNA ligase (COG1423), three paralogous photolyases (COG1533), and several helicases and nucleases of different specificities.

The case of the large subunit of the archaeal-eukaryotic primase is particularly illustrative of the contribution of different types of inference to genome annotation. COGNITOR failed to assign an *A. pernix* protein to the respective COG (COG2219). However, given the ubiquity of this subunit in euryarchaea and eukaryotes and the presence of a readily detectable small primase subunit in *A. pernix* (COG1467), a more detailed analysis was undertaken by running PSI-BLAST searches against the NR database with all members of COG2219 as queries. When the *A. fulgidus* primase sequence (AF0336) was used to initiate the search, the *A. pernix* counterpart (APE0667) was indeed detected at a statistically significant level.

An interesting case of re-annotation of a protein with a critical function, which also led to more general conclusions, is the archaeal uracil DNA glycosylase (UDG; COG1573). The members of this COG were originally annotated (and still remain so labeled in GenBank) as a "DNA polymerase homologous protein" (APE0427 from *A. pernix*) or as a "DNA polymerase, bacteriophage type" (AF2277 *from A. fulgidus*) or as a hypothetical protein. However, UDG activity has been experimentally demonstrated for the

COG1573 members from *T. maritima* and *A. fulgidus* [740,741]. The reason for the erroneous annotation of these proteins as DNA polymerases is already well familiar to us: independent fusion of the uracil DNA glycosylase with DNA polymerases was detected in bacteriophage SPO1 and in *Yersinia pestis* [44]. Although these fusions hampered the correct annotation in the original analysis of the archaeal genomes, they seem to be functionally informative, suggesting that this type of UDG functions in conjunction with the replicative DNA polymerase.

The 1102 COG members from *A. pernix* comprise 41% of the total number of predicted genes. This percentage was significantly lower than the average fraction of COG members (72%) for the other archaeal species. It seems most likely that this was due to an overestimate of the total number of ORFs in the genome. Many of the *A. pernix* ORFs with no similarity to proteins in sequence databases (1538, or 57.1%) overlap with ORFs from conserved families, including COG members. Based on the average representation of all genomes in the COGs (67%) and the average for the other archaea (72%), one could estimate the total number of *A. pernix* proteins to be between 1550 and 1700. This range is also consistent with the size of the *A. pernix* genome (1.67 Mb), given the gene density of about one gene per kilobase, which is typical of bacteria and archaea. More conservatively, 849 ORFs originally annotated as probable protein-coding genes, significantly overlapped with COG members and could be confidently eliminated, which brings the total number of protein-coding genes in *A. pernix* to a maximum of 1873. Unfortunately, the spurious ORFs still remain in the NR database, polluting it and potentially even leading to emergence of ghost "protein" families once new related genomes are sequenced. Evidence has been presented that spurious "proteins" have been produced by other microbial genome products also [777], although probably not at the same scale as with *A. pernix*. This regrettable pollution emphasizes the value of specialized, curated databases that are free of apparitions.

Despite this over-representation of ORFs in *A. pernix*, we nonetheless added 28 previously unidentified ORFs that were detected by searching the genome sequence translated in all six frames for possible members of COGs with unexpected phyletic patterns. These newly detected genes represent conserved protein families, including functionally indispensable proteins, such as chorismate mutase (APE0563a, COG1605), translation initiation factor IF-1 (APE_IF-1, COG0361), and seven ribosomal proteins (APE_rpl21E, COG2139; APE_rps14, COG0199; APE_rpl29, COG0255; APE_rplX, COG2157; APE_rpl39E, COG2167; APE_rpl34E, COG2174; APE_rps27AE, COG1998).

This pilot analysis, while falling far short of the goal of comprehensive genome annotation, highlights some advantages of specialized comparative-genomic databases as annotation tools. In this particular case, the original annotation probably had been overly conservative, which partly accounts for the large increase in the functional prediction rate. However, the employed protocol is general and, with modifications and addition of some extra procedures, has been used in primary genome analysis [622,779]. In other genome projects, the WIT system has been employed in a conceptually similar manner [179,418]. As shown above, this type of analysis yields reasonable accuracy of annotation even when applied in a fully automated mode (Table 5.3). However, additional expert contribution, particularly in the form of context analysis discussed in the next section, adds substantial value to genome annotation.

Table 5.3. Assignment of predicted *Aeropyrum pernix* proteins to COGs.

Protein category	No. of proteins
Assigned by COGNITOR automatically	1123
Included in COGs after validation	1102
True positives	1062
Preexisting COGs	1035
New COGs	27
False positives	44
Rejected	21
Re-assigned to a related COG	21
Re-assigned to an unrelated COG	2
False negatives (added during manual checking)	17
Proteins in COGs:Update 2001	1178
Proteins in COGs:Update 2002	1242

5.2. Genome Context Analysis and Functional Prediction

All the preceding discussion in this chapter centered on prediction of the functions of proteins encoded in sequenced genomes by extrapolating from the functions of their experimentally characterized homologs. The success of this approach depends on the sensitivity and selectivity of the methods that are used for detecting sequence similarity (see Chapter 4) and on the employed rules of inference (see ♦ 5.1). There is no doubt that homology analysis remains the central methodology of genomics, i.e. the one that produces the bulk of useful information. However, a group of recently developed approaches in comparative genomics go beyond sequence or structure comparison. These methods have become collectively and, we think, aptly known as genome context analysis [267,368,369,372]. The notion of "context" here includes all types of associations between genes and proteins in the same or in different genomes that may point to functional interactions and justify a verdict of "guilt by association"[36]: if gene A is involved in function X and we obtain evidence that gene B functionally associates with A, then B is also involved in X. More specifically, context in comparative genomics pertains to phyletic profiles of protein families, domain fusions in multidomain proteins, gene adjacency in genomes and expression patterns. Indeed, genes whose products are involved in closely related functions (e.g., form different subunits of a multisubunit enzyme or participate in the same pathway) should all be either present or absent in a certain set of genomes (i.e. have similar, if not identical phyletic patterns) and should be coordinately expressed (i.e. are expected to be encoded in the same operon or at least to have similar expression patterns). This simple logic gives us a potentially powerful way to assign genes that have no experimentally characterized homologs to particular pathways or cellular systems. Although context methods usually provide only rather general predictions, they represent a new and important development in genomics that explicitly takes advantage of the rapidly growing collection of sequenced genomes.

5.2.1. Phyletic patterns (profiles)

Genes coding for proteins that function in the same cellular system or pathway tend to have similar phyletic patterns [259,828]. Numerous examples for a variety of metabolic pathways are given in Chapter 7. These observations led to the suggestion that this trend could be used in the reverse direction, i.e. to deduce functions of uncharacterized genes [665]. However attractive this idea might be, the real-life phyletic patterns are heavily affected by such major evolutionary phenomena as partial redundancy in gene functions, non-orthologous gene displacement, and lineage-specific gene loss. As a result, there are thousands different phyletic patterns in the COGs, most of them represented only once or twice. Moreover, examination of a variety of multi-component systems and biochemical pathways (http://www.ncbi.nlm.nih.gov/cgi-bin/COG/palox?sys=all) shows that, despite the tendency of the components of the same complex or pathway to have similar patterns, there is not even one pathway, in which ***all*** members show exactly the same pattern. Even the principal metabolic pathways, such as glycolysis, TCA cycle, purine and pyrimidine biosynthesis, show considerable variability of phyletic patterns due to non-orthologous gene displacement ([265,270,370], see Chapter 7).

Because of this variability, the predictive power of the observation that two genes have the same phyletic pattern is, in and by itself, limited. However, when supported by other lines of evidence, such observations prove useful. Somewhat counterintuitively, the universal pattern is one of the most strongly indicative of gene function: among the 63 universal COGs, at least 56 consist of proteins involved in translation. The functions of those few proteins in the universal set that remain uncharacterized can be predicted with considerable confidence through combination of this phyletic pattern with other line of evidence. For example, the uncharacterized protein YchF, which belongs to the universal set (COG0012), is predicted by sequence analysis to be a GTPase; in addition, this protein contains a C-terminal RNA-binding TGS domain [909]. Taken together with the ubiquity of this protein and with the fact that, in phylogenetic trees, the archaeal members of the COG clearly cluster with eukaryotic ones, this strongly suggests that YchF is an uncharacterized, universal translation factor [267]. This is supported by the juxtaposition of the *ychF* gene with the gene for peptidyl-tRNA hydrolase (*pth*) in numerous proteobacteria. The discussion of this protein made us run ahead of ourselves and invoke other context methods, which are considered in the next subsections, namely analysis of domain fusions and gene juxtaposition. This situation is quite typical: context methods are at their best when they complement one another. Although statistical significance estimates for a combination of context methods do not currently seem feasible, in a case like YchF, the evidence appears to be, for all practical purposes, irrefutable.

Another similar case involves the predicted ATPase or (more likely) kinase YjeE from *E. coli* [256] and its orthologs from a majority of bacterial genomes that comprise COG0802. Domain analysis identified this protein as a likely P-loop ATPase but failed to give any indications as to its cellular role. The phyletic pattern of this COG shows that YjeE is encoded in every bacterial genome, with the exception of *M. genitalum*, *M. pneumoniae*, and *U. urealyticum*, the only three bacterial species in the COG database that do not form a cell wall. Since other conserved proteins with the same phyletic pattern (MurA, MurB, MurG, FtsI, FtsW, DdlA) are enzymes of cell wall biosynthesis, it can be predicted that YjeE is an ATPase or kinase involved in the same process. Again, this prediction is supported by the adjacency of the *yjeE* with the gene for N-acetylmuramoyl-L-alanine amidase, another cell wall biosynthesis enzyme.

There is more to phyletic pattern analysis then prediction based on identical or similar patterns. Guilt by association can be established also through identification of sets of genes that are ***co-eliminated*** in a given lineage; this approach exploits the widespread phenomenon of lineage-specific gene loss. A systematic analysis of the set of genes that have been co-eliminated in the yeast *S. cerevisiae* after its divergence from the common ancestor with *S. pombe* led to the prediction that a particular group of proteins, including one that contained a helicase and a duplicated RNAse III domain, was involved in post-transcriptional gene silencing [55]. This protein turned out to be the now famous dicer nuclease, which indeed has a central role in silencing [365,436].

On many occasions, non-orthologous gene displacement manifests in ***complementary***, rather than identical or similar, phyletic patterns, like we have seen for phosphoglycerate mutase in ♦2.2.6. The complementarity is rarely perfect because of partial functional redundancy: some organisms, particularly those with larger genomes, often encode more than one protein to perform the same function. This can be illustrated by the case of the recently discovered new type of fructose-1,6-bisphosphate aldolase, referred to as FbaB or DhnA [257]. The two well-known variants of this enzyme, class I (Schiff-base forming, metal-independent) and class II (metal-dependent), have long been considered to be unrelated (analogous) enzymes until structural comparisons revealed their underlying similarity (see Figure 1.9) [95,187,257,549]. These enzymes are generally limited in their phyletic distribution to eukaryotes (class I) and bacteria (class II); some bacteria, however, have both variants and yeast has the bacterial (class II) form of the enzyme [549]:

```
-------------c----s--j----   COG3588  Class I FBF aldolase
------yqvdrlbcefgh-nuj--tw   COG0191  Class II FBF aldolase
```

Sequencing of archaeal genomes revealed the absence of either form of the fructose-1,6-bisphosphate aldolase. The same was the case with chlamydiae, which were predicted to have a third form of this enzyme [412,805]. Indeed, investigation of the metal-independent fructose-1,6-bisphosphate aldolase activity in *E. coli* led to the discovery of another metal-independent Schiff-base-forming variant [844] whose sequence, however, was more closely related to those of class II enzymes than to typical class I enzymes [257]. Highly conserved homologs of this new, third form of fructose-1,6-bisphosphate aldolase were found in chlamydial and archaeal genomes:

```
aom-kz-q------e--h---j-i--  COG1830 DhnA-type FBF aldolase
```

As with phosphoglycerate mutase, combining these phyletic patterns shows almost perfect complementarity, with aldolase missing only in *Rickettsia*, which does not encode any glycolytic enzymes, and in *Thermoplasma*, which appears to rely exclusively on the Entner-Doudoroff pathway (see ♦7.1.1):

```
-------------c----s--j----  COG3588 Class I FBF aldolase
------yqvdrlbcefgh-nuj--tw  COG0191 Class II FBF aldolase
aom-kz-q------e--h---j-i--  COG1830 DhnA-type FBF aldolase
```

Other interesting examples of complementary phylogenetic patterns include lysyl-tRNA synthetases, pyridoxine biosynthesis proteins PdxA and PdxZ [256], thymidylate synthases [267], and many others. The case of thymidylate synthases is particularly remarkable. Thymidylate synthase is a strictly essential enzyme of DNA precursor biosynthesis, and its apparent absence in several bacterial and archaeal species became a major puzzle as their genome sequences were reported.

```
a-m---y--drlb-efghsn-j---w  COG0207 Thymidylate synthase
-o-pkz-qv-r--c------u-xit-  COG1351 Predicted alternative
                                    thymidylate synthase
```

The alternative thymidylate synthase was predicted [267] on the basis of a phyletic pattern that was nearly complementary (with just one case of redundancy) to that of the classic thymidylate synthase (ThyA) and the report that the homolog of the COG1351 proteins from *Dictyostelium* complemented thymidylate synthase deficiency [206]. Just before this book went to print, a new issue of *Science* reported the confirmation of this prediction: not only was it shown that the COG1351 member from *H. pylori* had thymidylate synthase activity, but also the structure of this proteins has been solved and turned out to be unrelated to that of ThyA [589,598].

5.2.2. Gene (domain) fusions: "guilt by association"

It is fairly common that functionally interacting proteins that are encoded by separate genes in some organisms are fused in a single polypeptide chain in others. This has been confirmed by statistical analysis that demonstrated general functional coherence of fused domains [930]. The advantages of a multidomain architecture are that this organization facilitates functional complex assembly and may also allow reaction intermediate channeling [546].

The basic assumption in the analysis of domain fusions is that a fusion will be fixed during evolution only when it provides a selective advantage to the organism in the form of improved functional interaction between proteins. Thus, finding fused proteins (domains) in one species suggests that they might interact, physically or at least functionally, in other species. In and by itself, this notion is trivial and has been employed for predicting protein and domain functions on an anecdotal basis for years (see [100], just as an example). However, with the rapid growth of the sequence information, the applicability of this approach widened and two independent groups proposed, in well-publicized papers, that analysis of domain fusions could be a general method for systematic and, moreover, automatic, prediction of protein functions [213,546]. In one of these studies [546], domain fusions are referred to as "Rosetta Stone" proteins – clues to deciphering the functions of their component domains, and this memorable name stuck to the whole approach. (The Rosetta Stone metaphor is quite loose: the notorious stone used by François Champollion to decipher the Egyptian hieroglyphs and now on public display in the British Museum, is a tri-lingua, i.e. a monument that has on it the same text in three different languages. There is nothing exactly like that about domain fusions, it is just possible to say vaguely that the "language" of domain fusions is translated into the "language" of functional interactions. The "guilt by association" simile [36] seems much more apt if less glamorous).

In his comment on the "Rosetta Stone" excitement, Russell Doolittle pointed out that cases that establish a link between two well-known domains or those that link two unknown domains are not likely to lead to any scientific breakthroughs [188]. Only those "Rosetta Stone" proteins, in which an unknown domain is linked to a previously characterized one, can be used to infer the function(s) of the uncharacterized domain. Analysis of domain fusions in complete microbial genomes indicates that they are a complex mixture of informative, uninformative and potentially misleading cases, which certainly provide many clues to functions of uncharacterized domains. However, interpretations stemming from domain fusion seem to require case-by-case examination by human experts and, most of the time, become really useful only when combined with other lines of evidence.

One of the advantages of the guilt by association approach is that, at least in principle, it allows transitive closure, i.e. expansion of functional associations between transitively connected components. In other words, detection of domain combinations AB, BC, and CD suggests that domains A, B, C and D form a functional network. This approach has been successfully applied to the analysis of prokaryotic signal-transduction systems, resulting in the prediction of several new signaling domains. Participation of these domains in signaling cascades has been originally proposed solely on the basis of their conserved domain architectures and subsequently confirmed experimentally [269].

In Figure 5.3, we illustrate the "guilt by association" approach using the peptide methionine sulfoxide reductase example discussed in the previous section as a case of annotation complicated by domain fusion. As in the examples above, the logic of the analysis does not allow us to use domain fusions only; we also have to invoke phyletic patterns and organization of genes in the genome.

In most organisms, protein methionine sulfoxide reductase A (MsrA) is a small, single-domain protein. However, in *H. influenzae*, *H. pylori* and *T. pallidum,* it is fused with another, highly conserved domain (MsrB) that is found as a distinct protein in all other organisms that encode MsrA. In other words, the two fusion components show the same phyletic patterns:

```
-om---y--drlbcefghsnuj--tw  COG0225 MsrA
-om---y--drlbcefghsnuj--tw  COG0229 MsrB
```

In *B. subtilis*, the genes for MsrA and MsrB are not fused, but are adjacent and may form an operon. In contrast, in *T. pallidum*, MsrA and MsrB are fused, but in reverse order, compared to *H. influenzae* and *H. pylori* (Figure 5.3). The *H. influenzae* and *H. pylori* "Rosetta Stone" proteins are most closely related to each other, but the one from *T. pallidum* does not show particularly strong similarity to any of them, suggesting two independent fusion events in these two lineages.

In *Neisseria* and *Fusobacterium*, a third, thioredoxin-like domain joins the MsrAB fusion (Figure 5.3). In *H. influenzae*, the ortholog of this predicted thioredoxin is encoded two genes upstream of MsrAB. The gene in between encodes a conserved integral membrane protein, designated CcdA for its requirement for cytochrome c biogenesis in *B. subtilis*. Its ortholog is encoded next to MsrAB in *H. pylori* and next to thioredoxin in several other genomes (Figure 5.3).

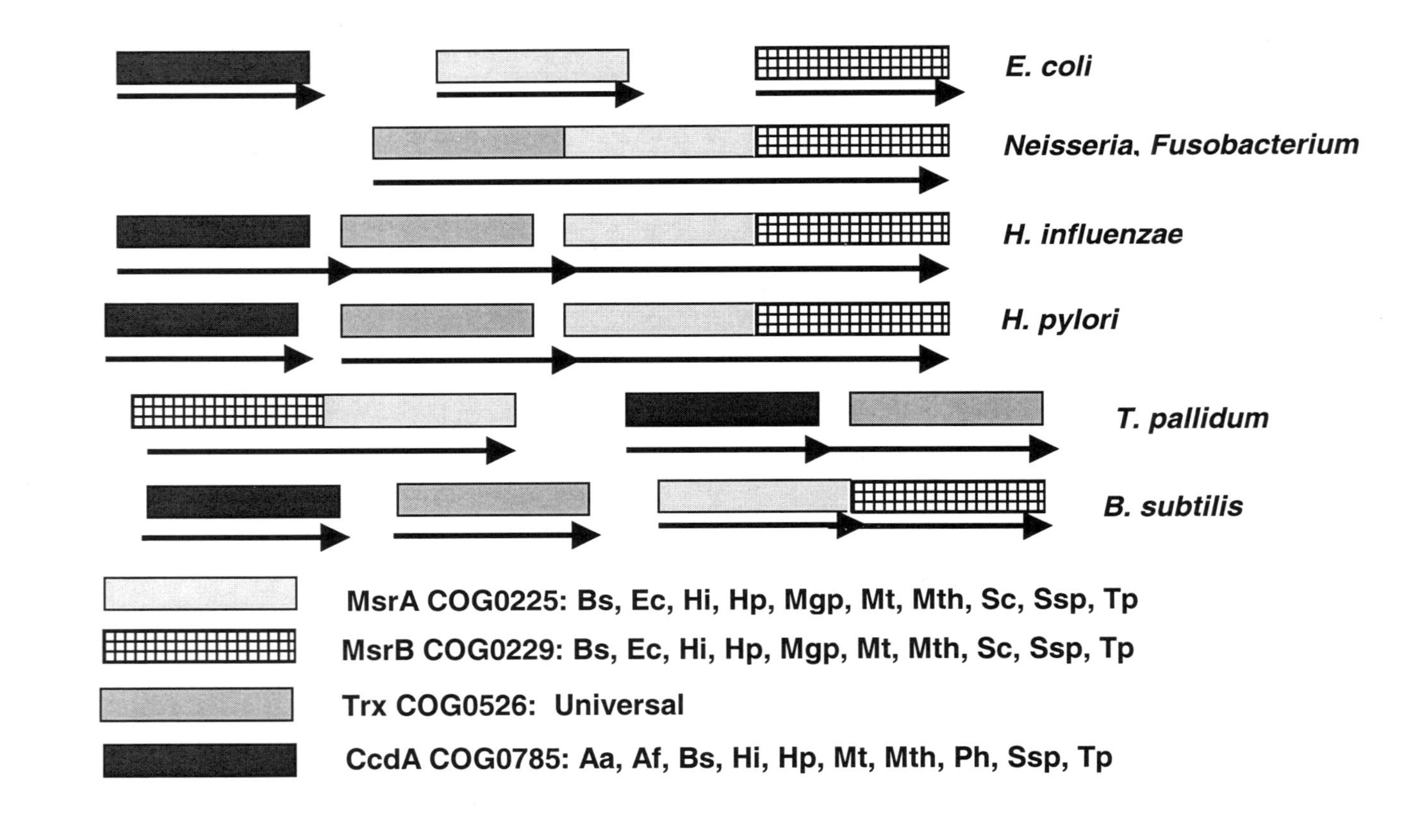

Figure 5.3. A Rosetta Stone case: domain fusions and gene clusters that involve peptide methionine sulfoxide reductases.

Combining all this evidence from the guilt by association approach, gene adjacency data, phyletic profiles, and sequence analysis, it has been predicted that the MsrA, MsrB and thioredoxin form an enzymatic complex, which catalyzes a cascade of redox reactions and is associated with the bacterial membrane via CcdA. However, this is probably not the only complex in which MsrAB is involved, because not all genomes that have this gene pair also encode CcdA (Figure 5.3). Since the publication of this prediction, it has been largely confirmed by the demonstration that MsrB is a second, distinct, thioredoxin-dependent peptide methionine sulfoxide reductase, which cooperates with MsrA in the defense of bacterial cells against reactive oxygen species [316,526,776]. However, the CcdA connection remains to be investigated.

This case study demonstrates both the considerable potential of domain fusion analysis as a tool for protein function prediction, particularly when combined with other context-based and homology-based approaches, and potential problems. One could be tempted to extend the small network of domains shown in Figure 5.3 by including other domains that form fusions (or are encoded by adjacent genes) with the thioredoxin domain. It appears, however, that such an extension would have been ill-advised. Firstly, orthologous relationships among thioredoxins are ambiguous, and secondly, although thioredoxins are not among the most "promiscuous" domains, the variety of their "guilt by association" links still is sufficiently large to make any predictions regarding potential functional connections between the respective domains and MsrAB dubious at best. These two issues, identification of orthologs and "promiscuity" characteristic of certain domains, are the principal problems encountered by the "guilt by association" approach. Domain fusions often are found only within a specialized, narrow group of orthologous protein domains, and translating their functional interaction into a general prediction for the respective domains is likely to be grossly misleading. A relatively small number of "promiscuous" domains, particularly those involved in signal transduction and different forms of regulation (e.g. CBS, PAS, GAF domains), combine with a variety of other domains that otherwise have nothing in common and therefore significantly increase the number of false-positives among the Rosetta Stone predictions. Although it is possible to simply exclude the worst known offenders from any Rosetta Stone analysis [546], other domains also have the potential of showing "illicit" behavior and compromising the results. Manual detection of such cases is relatively straightforward, but automation of this process may be complicated.

5.2.3. Gene clusters and genomic neighborhoods

As already mentioned in Chapter 2, comparisons of complete bacterial genomes have revealed the lack of large-scale conservation of the gene order even between relatively close species, such as *E. coli* and *H. influenzae* [595,829] or *E. coli* and *P. aeruginosa* (Figure 2.6B). Although these pairs of genomes have numerous similar strings of adjacent genes (most of them predicted operons), comparisons of more distantly related bacterial and archaeal genomes have shown that, at large phylogenetic distances, even most of the operons are extensively rearranged [461,884]. The few operons that are conserved across distantly related genomes typically encode physically interacting proteins, such as ribosomal proteins or subunits of the H^+-ATPase and ABC-type transporter complexes [169,385,461,595].

It should be noted that only a relatively small number of operons have been identified experimentally, primarily in well-characterized bacteria, such as *E. coli* and *B. subtilis* [363,732]. However, analysis of gene strings that are conserved in bacterial and archaeal genome strongly suggested that the great majority of them do form operons [916]. This conclusion was based on the following principal arguments: (i) as shown by Monte Carlo simulations, the likelihood that identical strings of more than two genes are found by chance in more than two genomes is extremely low; (ii) most of those conserved strings that include characterized genes either are known operons or include functionally linked genes and can be predicted to form operons; (iii) typical conserved gene strings include 2 to 4 genes, which is the characteristic size of operons; (iv) conserved gene strings that include genes from adjacent, independent operons are extremely rare; (v) nearly all conserved gene strings consist of genes that are transcribed in the same direction [916]. As a result, one can usually assume that conserved gene strings are co-regulated, i.e. form operons, even if they contain additional promoters.

Pairwise genome comparisons showed that, on average, ~10% of the genes in each genome belong to gene strings that are conserved in at least one of the other available genomes [385,916]. These numbers vary widely from <5% for the cyanobacterium *Synechocystis* sp. to 23-24% in *T. maritima* and *M. genitalium*; the fraction of genes that belonged to predicted operons in the archaeal genomes was only slightly lower than that in bacterial genomes [916].

These observations indicate that conserved gene strings are under stabilizing selection that prevents their disruption. For functionally related genes (e.g. those encoding proteins that function in the same pathway or multimeric complex), this selective pressure probably comes from the

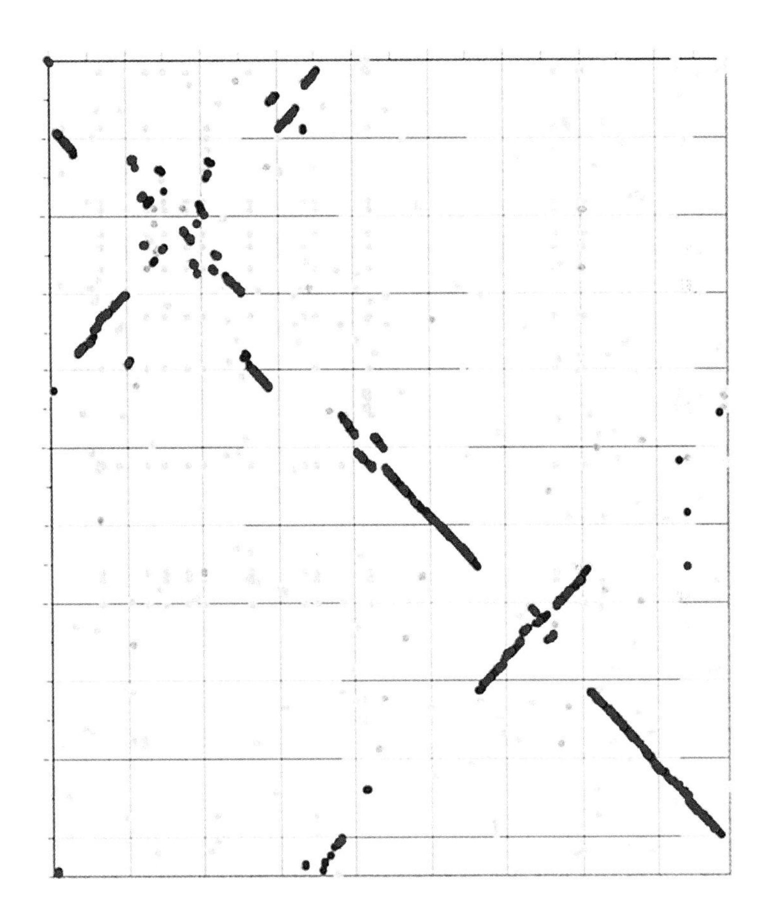

Figure 2.6. Gene order comparison plots.
A (This page). *Chlamydia trachomatis* (X axis) vs *Chlamydophila pneumoniae* (Y axis)
B (Next page). *Escherichia coli* (X axis) vs *Pseudomonas aeruginosa* (Y axis)
Each dot represents a pair of genes with the level of similarity between the encoded proteins sequences indicated by color: red - >1.3 bits/position; blue - from 0.8 to 1.3 bits/position; grey - from 0.3 to 0.8 bits/position; light blue <0.3 bits/position. The similarity scores are expressed in bits/position, rather than in total scores per protein, to remove the bias caused by variation in the protein length.

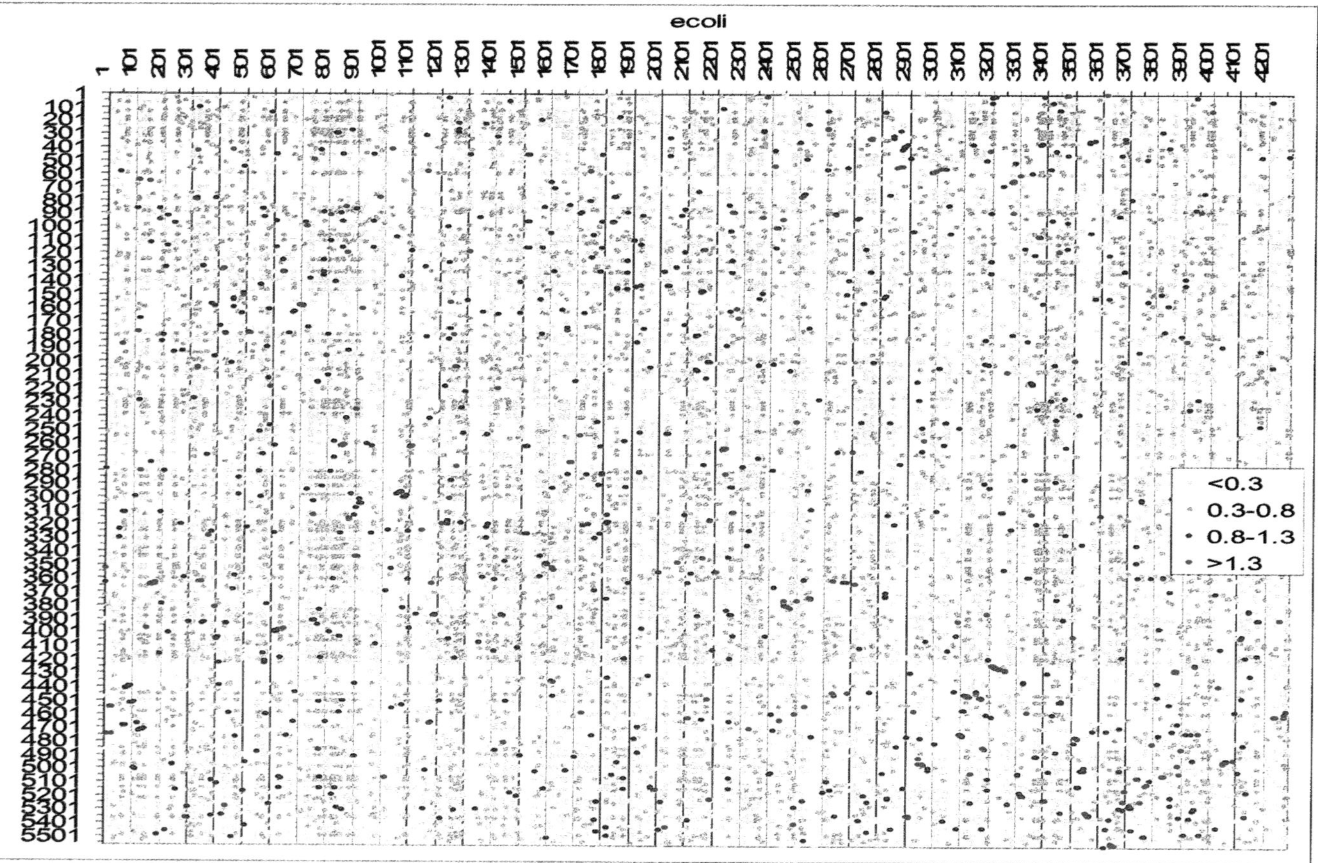
ecoli
paar
<0.3
0.3-0.8
0.8-1.3
>1.3

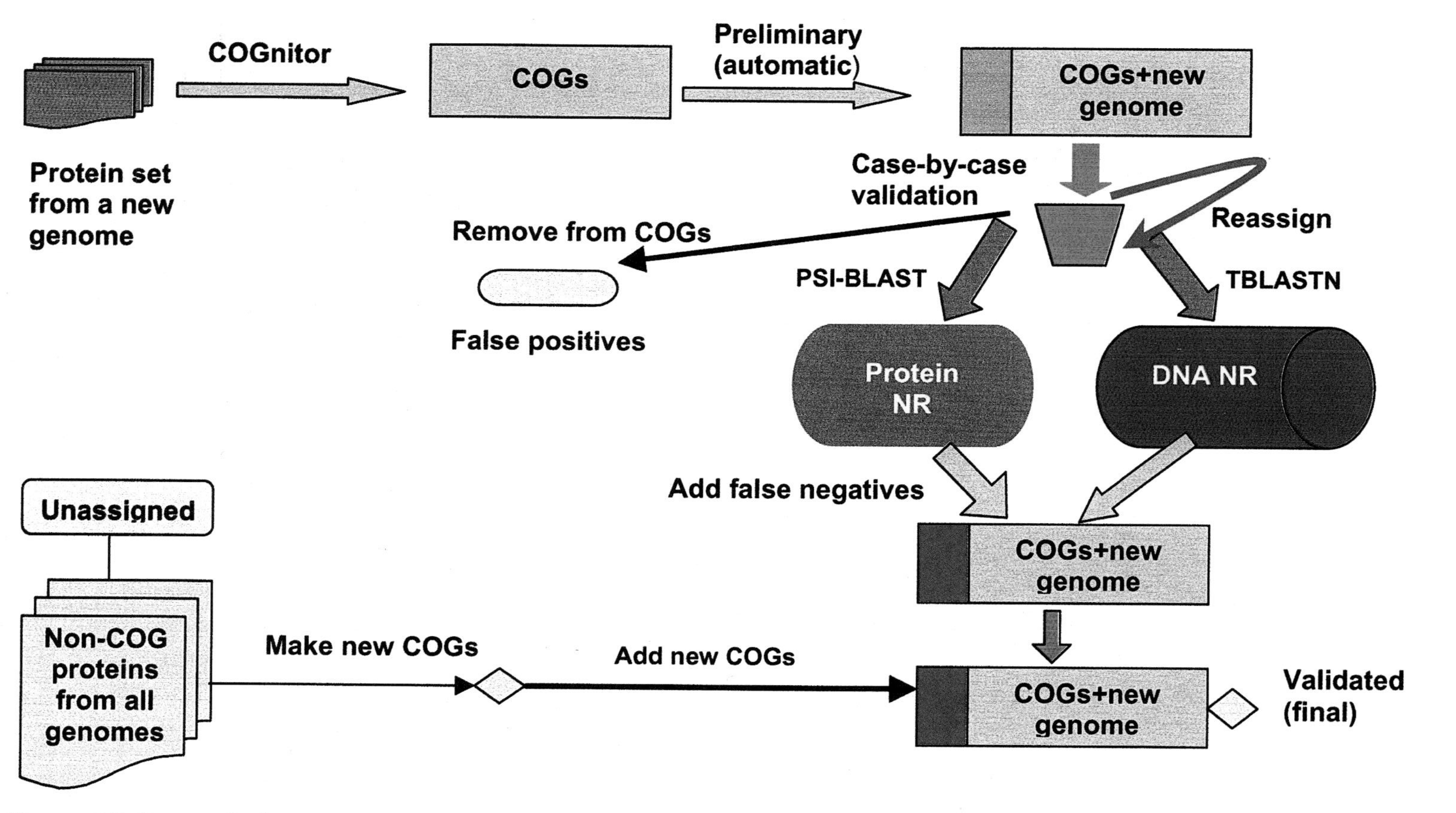

Figure 5.2. Protocol of genome annotation using the COG database

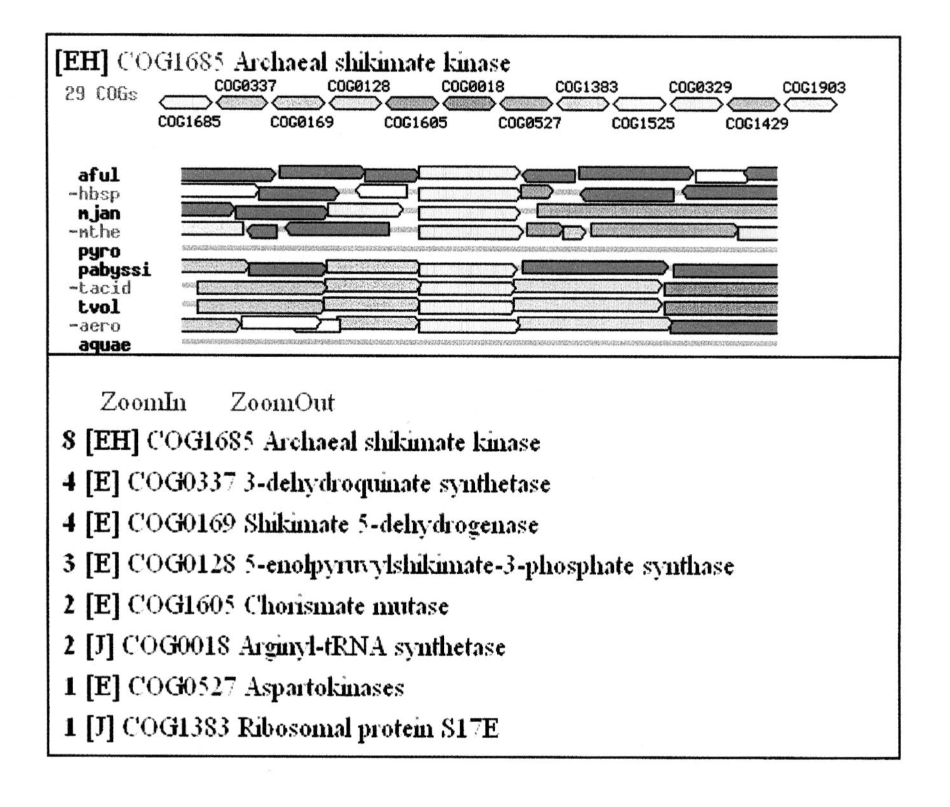

Figure 5.4. Genome context of COG1685 "Archaeal shikimate kinase".
Each line corresponds to an individual genome: aful, *Archaeoglobus fuligidus*; hbsp, *Halobacterium* sp.; mjan, *Methanococcus jannaschii*; mthe, *Methanobacterium thermoautotrophicum*; pyro, *Pyrococcus horikoshii*; pabyssi, *Pyrococcus abyssi*; tacid, *Thermoplasma acidophilum*; tvol, *Thermoplasma volcanumi*; aero, *Aeropyrum pernix*; aquae, *Aquifex aeolicus*. The genes encoding members of COG1685 are shown in the middle in light yellow. Genes encoding members of the same COG are indicated by the same color. The genomes that do not encode a member of COG1685 are indicated by empty lines. The names of all COGs represented in the picture are listed starting from the most common ones. Note that in *Halobacterium* sp. (second line) and *M. thermoautotrophicum* (fourth line), COG1685 genes are followed by the genes encoding chorismate mutase (*tyrA*_1, COG1605), shown in dark green. In *Thermoplasma* spp. and *A. pernix* (lines 6-9), COG1685 genes are sandwiched between the genes encoding shikimate-5-dehydrogenase (*aroE*, COG0169), shown in purple, and genes encoding 5-enoylpuruvoylshikimate-3-phosphate synthetase (*aroA*, COG0128), shown in purple. See Figure 7.7 (page 331) for the chart of the complete pathway of phenylalanine and tyrosine biosynthesis.

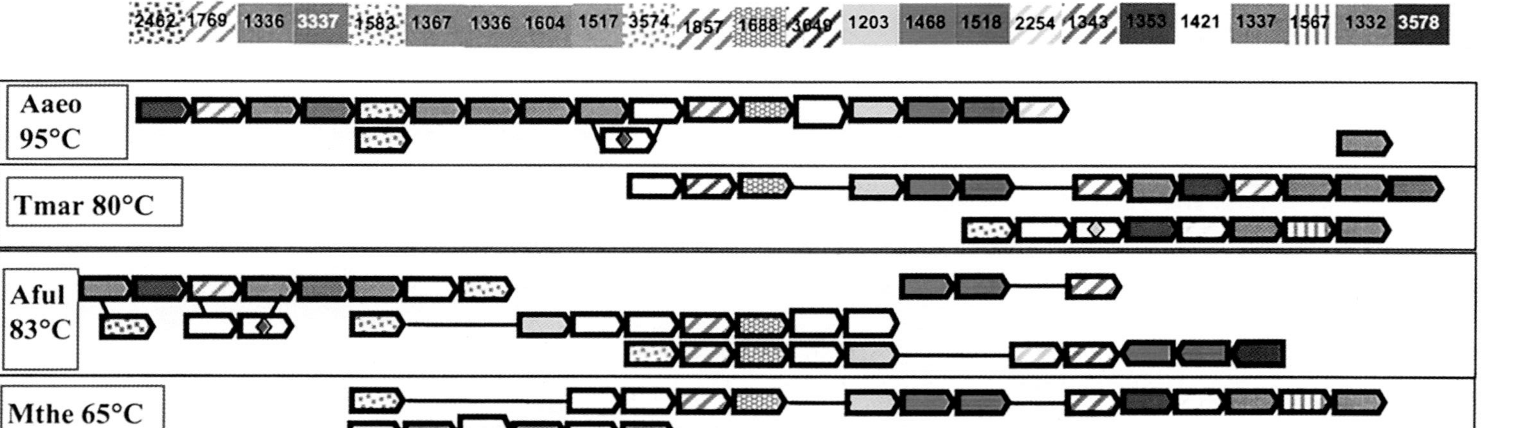

Figure 5.5. Predicted DNA repair system in hyperthermophiles.
The pink boxes show optimal growth temperatures for each of the analyzed species (*A. aeolicus*, *T. maritima*, *A. fulgidus*, *M. thermoautotrophicum*, *M. jannaschii*). The genes are not drawn to scale; arrows indicate the direction of transcription. The uppper row shows the COG numbers for the corresponding proteins. Some of the newly predicted COG functions are: COG2462, helix-turn-helix transcriptional regulator; COG1203, helicase; COG1468, RecB family exonuclease; COG2254, nuclease of the HD superfamily; COG1353, novel DNA polymerase;.

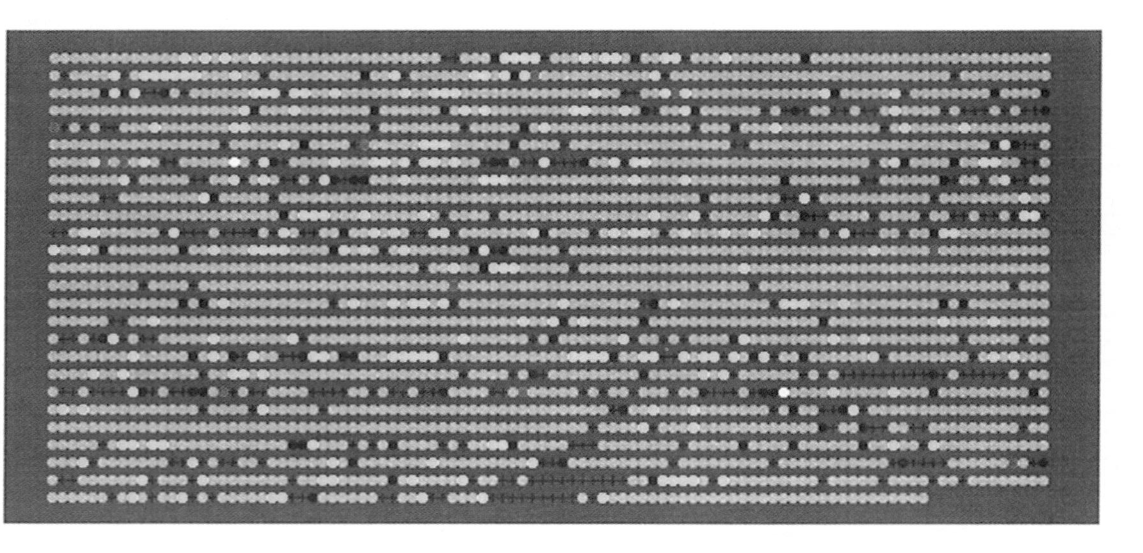

Figure 6.5. Genomic maps of apparent phylogenetic affinities for two bacterial genomes.
A (This page): the hyperthemophile *Thermoanaerobacter tencongenesis*: 258 of 2588 proteins (10%) with significantly greater similarity to archaeal than to bacterial homologs.
B (Next page): The mesophile *Bacillus subtilis*: 174 of 4112 proteins (4.2%) with significantly greater similarity to archaeal than to bacterial homologs.
The genomes are shown as wrapping strings of genes. Yellow circles indicate genes whose products have significantly greater similarity to archaeal proteins than to their bacterial homologs; blue circles indicate genes that encode proteins that are most similar to proteins from other bacteria; purple circles indicate genes whose products are most similar to eukaryotic proteins.

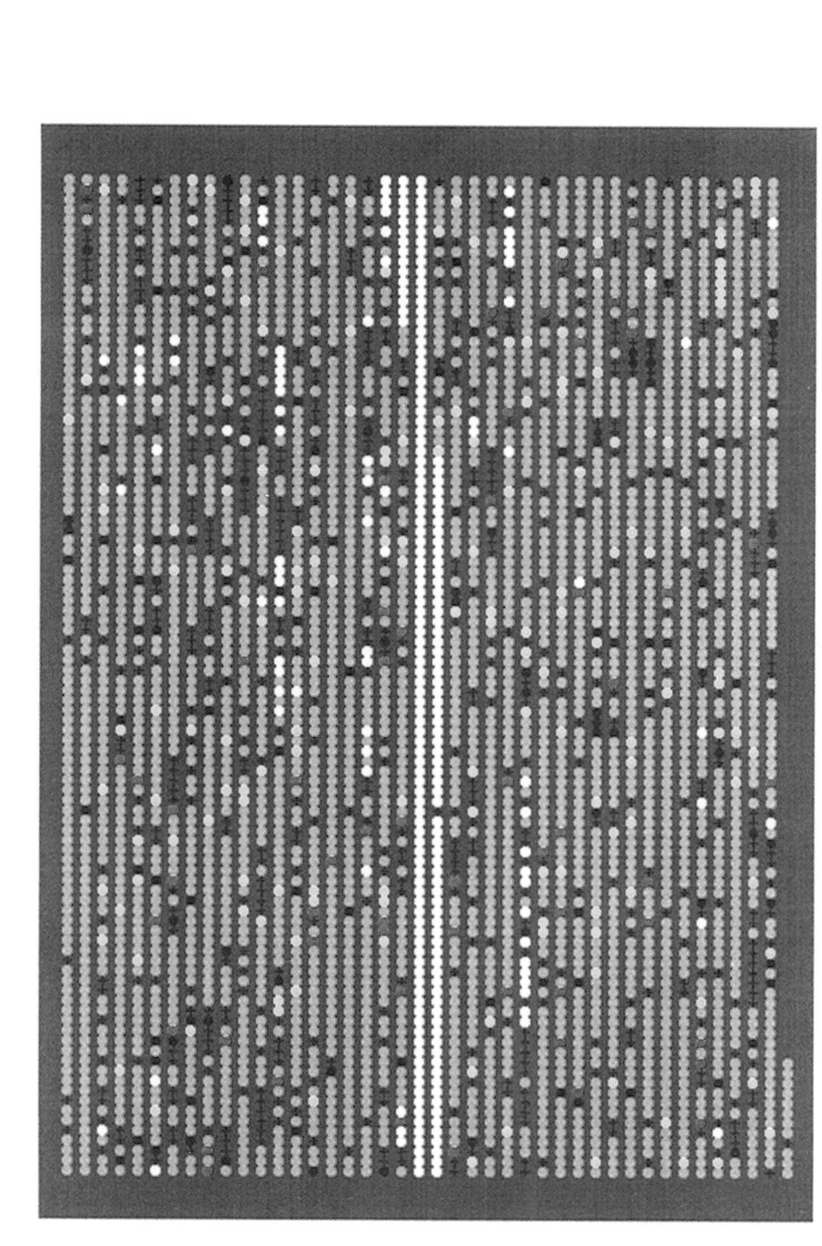

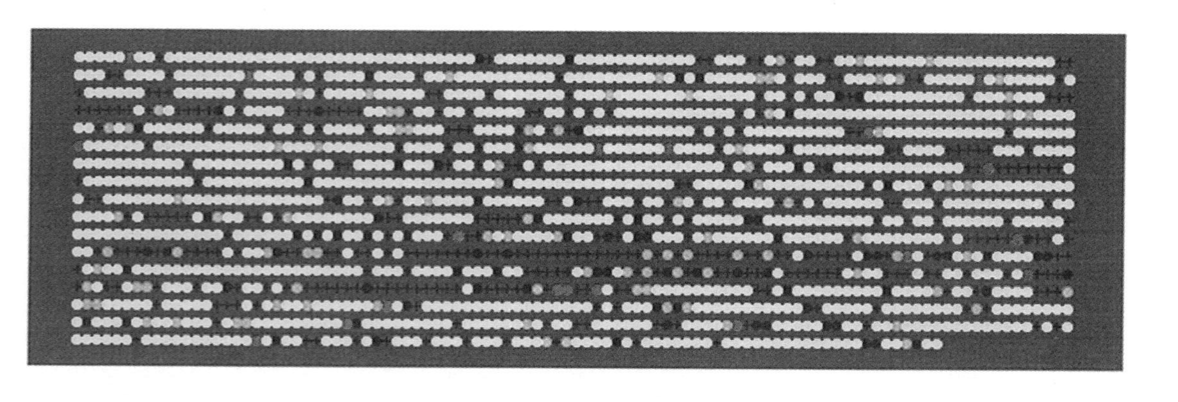

Figure 6.6. Genomic maps of apparent phylogenetic affinities for two archaeal methanogens.
A (This page): The hyperthermophile *Methanopyrus kandleri*. 98 of 1687 proteins (6%) with significantly greater similarity to bacterial than to archaeal homologs.
B (Next page): The mesophile *Methanosacrina acetivorans*. 1453 of 4540 proteins (32%) with significantly greater similarity to bacterial han to archaeal thomologs. The coloring of genes in the genomes is as in Figure 6.5.

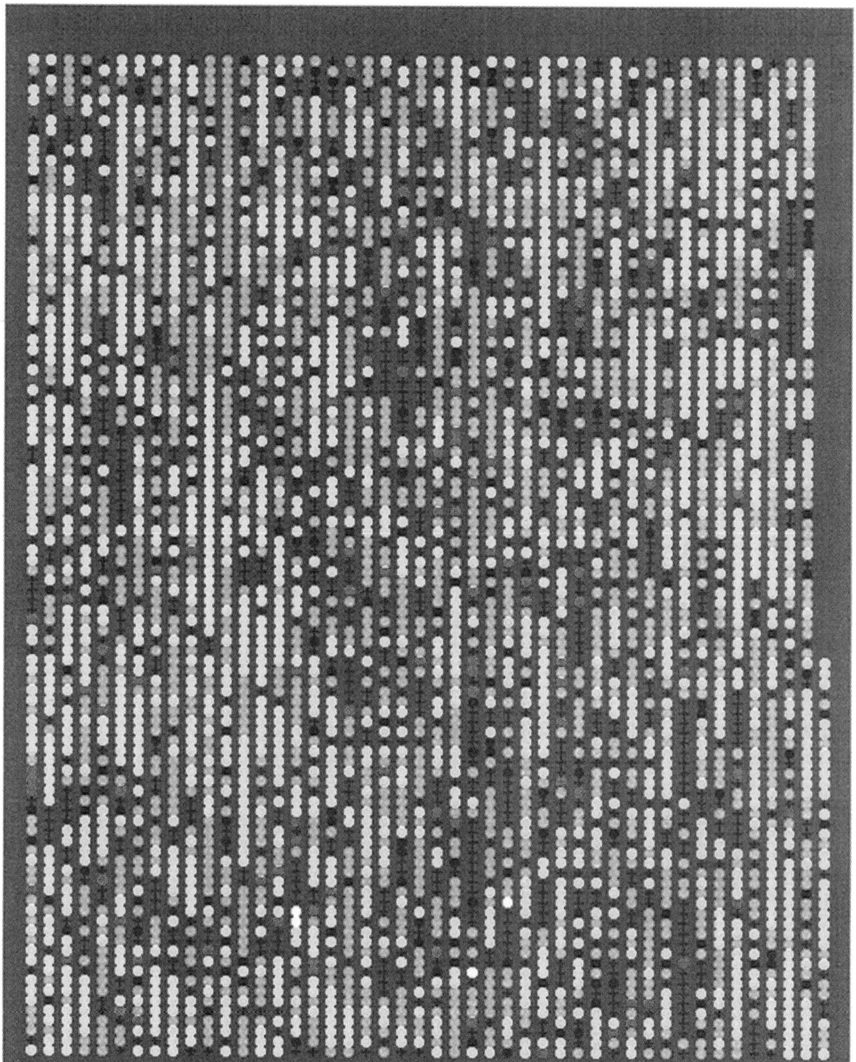

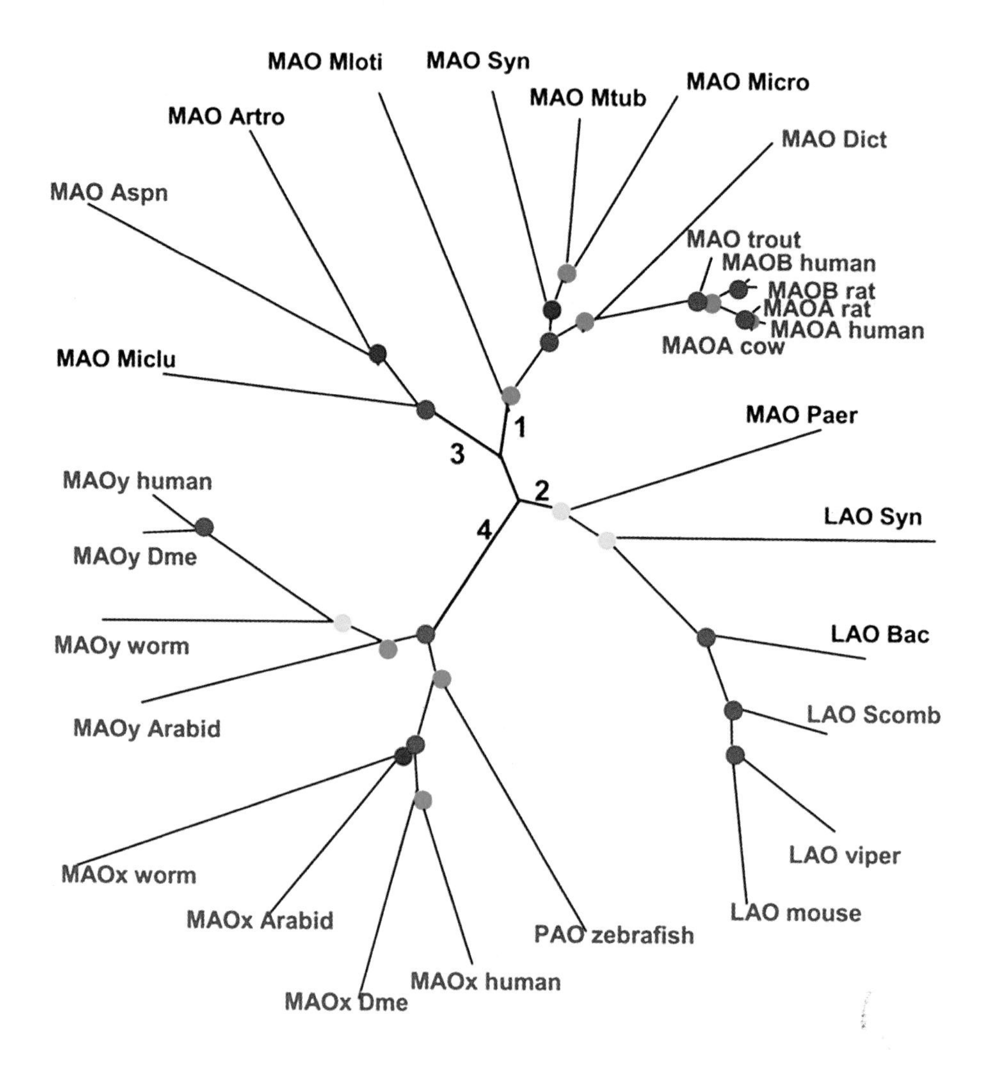

Figure 6.8. Phylogenetic tree of monoaminoxidases. (
MAOx and MAOy, uncharacterized predicted monoaminooxidases; LAO, L-amino oxidase; PAO, polyamine oxidase. The names of eukaryotic species in the MAOA-MAOB-Dictyostelium group are colored red, other eukaryotic species names are colored purple, bacterial species names are in black. The bootstrap probabilities for internal nodes calculated using the Resampling of Estimated Log-likelihoods method of Kishino and Hasegawa (Methods Enzymol. 1990, 183: 550-570) are color-coded as follows: red – 91-100%, gold – 81-90%, green – 71-80%, blue – 51-70%. Abbreviations: Arabid, *Arabidopsis thaliana*; Artro, *Arthrobacter nicotinovorans*; Aspn, *Aspergillus niger*; Bac, *Bacillus subtilis*; Dict, *Dictyostelium discoideum*; Dme, *Drosophila melanogaster*; Mloti, *Mezorhisobium loti*; Miclu, *Micrococcus luteus*; Micro, *Micrococcus rubens*; Mtub, *Mycobacterium tuberculosis*; Paer, *Pseudomonas aeruginosa*; Scomb, *Scomber japonicum*; Syn, *Synechocystis* sp.

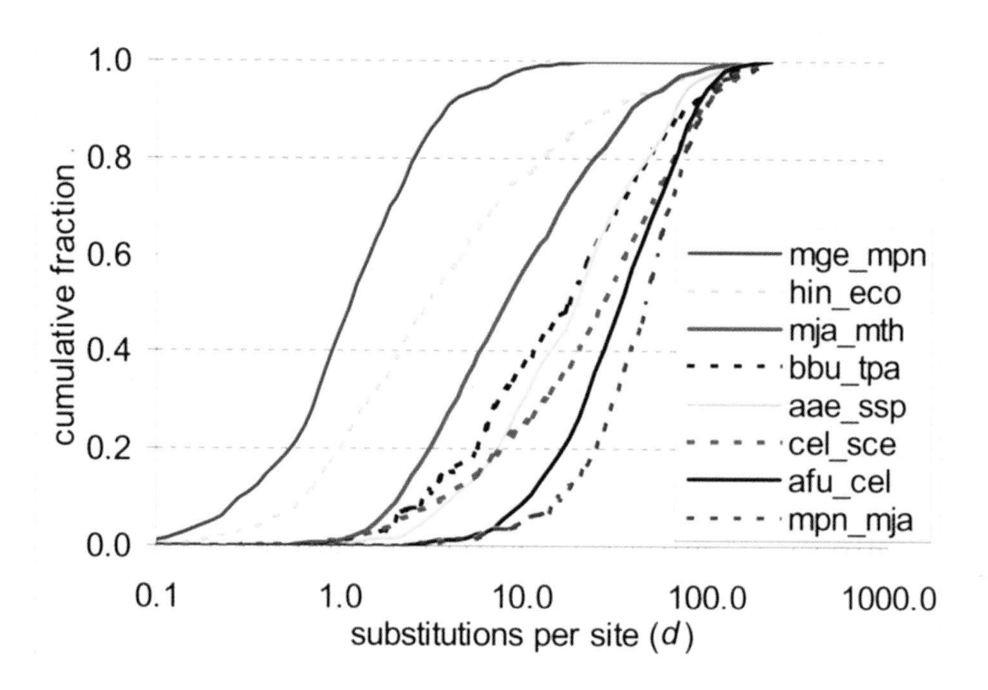

Figure 6.12. Cumulative distributions of evolutionary distances between orthologs for different genome pairs.

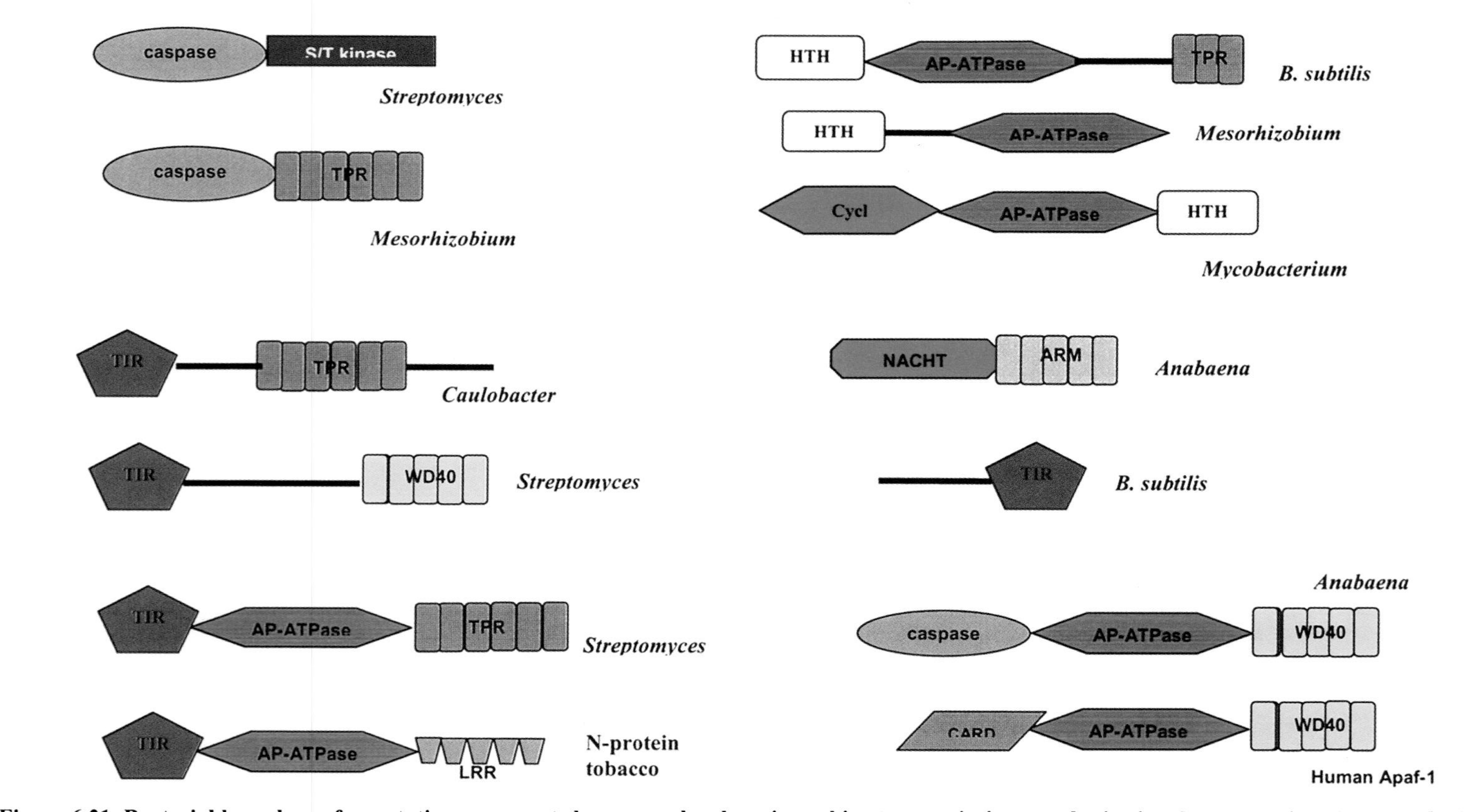

Figure 6.21. Bacterial homologs of apoptotic components have complex domain architectures pointing to roles in signal transduction. Apparently they interact even in bacteria. TIR, Toll-interleukin-receptor domain; TPR, tetratricopeptide repeats; LRR, leucine-rich repeats; Cycl, adenylate cyclase; ARM, Armadillo repeats; for the rest of the designations, see Chapter 6.

necessity to synchronize their expression. This conclusion holds even in the face of the "selfish operon" hypothesis, which posits that operons survive during evolution ***because*** they are disseminated via HGT [494,495]. We believe that the selfish operon hypothesis seems to put the cart ahead of the horse: operons certainly do spread via HGT, but their transfer leads to fixation more often than transfer of individual genes because of the selective advantage conferred to the recipient by the acquired operon. In contrast, for functionally unrelated genes, there would be no selection towards coexpression. Therefore, an observation of similar operons found in phylogenetically distant species can be considered an indication of a potential functional relationship between the corresponding genes, even if these genes are scattered in other genomes. Because of the simplicity and elegance of this approach to functional analysis of complete genomes, there are several web sites that offer slightly different approaches to delineation of the conserved gene strings.

WIT/ERGO

The operon comparison tool in the WIT database (http://wit.mcs.anl.gov), the first of the genome context-based tools, was developed by Ross Overbeek in 1998 [640,641]. This tool identifies conserved gene strings by searching for pairs of homologous proteins that are encoded by genes located no more than 300 bp apart on the same DNA strand in each of the analyzed genomes. Each of these pairs is then assigned a score based on the evolutionary distance between the respective species on the rRNA-based phylogenetic tree. It is expected that chance occurrence of pairs of homologous genes in distantly related species is less likely than in closely related ones, so such pairs are more likely to be functionally relevant. Homologous genes are defined as bidirectional best hits in all-against-all BLAST comparisons, which is similar to the method used in constructing the COG database [828].

Because the number of potential gene linkages grows exponentially with the number of the analyzed genomes [640], the sensitivity of methods based on the detection of conserved gene strings can be significantly improved by taking into consideration even unfinished genome sequences. For this reason, WIT and ERGO databases include many incomplete genome sequences from the DOE Joint Genome Institute and other sequencing centers. This approach was used in the successful reconstruction of several known metabolic pathways and led to the correct prediction of candidate genes for some previously uncharacterized metabolic enzymes [82,171,641]. Unfortunately, while this book was in preparation, the ERGO database has been closed for the public, while WIT was still missing some of the useful functionality. We will therefore illustrate the use of the method by exploiting a somewhat similar tool in the COG database.

COGs

The COG database (http://www.ncbi.nlm.nih.gov/COG) allows a simple and straightforward search for conserved operons. Because all proteins in the same COG are presumed to be orthologs, the "Genome context" view, available from each COG page, shows the genes that encode members of the given COG together with the surrounding genes. Genes whose products belong to the same COG are identically colored. This provides for easy identification of sets of COGs that tend to be clustered in genomes. Of course, this tool only works for the genes whose products belong to COGs, so the relationships between genes that are found in only two complete genomes and hence do not belong to any COG would be missed. An exhaustive matching of the co-localization of genes encoding members of the same two COGs allowed new functional predictions for almost 90 COGs, which comprised ~4% of the total set [469,916].

For a practical example of the use of this method, let us consider the search for the archaeal shikimate kinase, the enzyme that is not homologous to the bacterial shikimate kinase (AroK) and hence was not found by traditional sequence similarity searches [171]. Reconstruction of the aromatic amino acids biosynthesis pathway in archaea showed that genomes of *A. fulgidus*, *M. jannaschii*, and *M. thermoautotrophicum* encoded orthologs of bacterial enzymes for all but three reactions of this pathway ([540], see Figure 7.6).

Two of these missing enzymes catalyze first and second reactions of the pathway, indicating that aromatic acids biosynthesis in (most) archaea uses different precursors than in bacteria, whereas the third reaction, phosphorylation of shikimate, was attributed to a non-orthologous kinase, encoded only in archaea [540]. Daugherty and coworkers made a list of the genes involved in aromatic amino acid biosynthesis in archaea and looked for potential neighbors of the *aroE* gene whose product, shikimate dehydrogenase, catalyzes the reaction immediately preceding the phosphorylation of shikimate (Figure 7.6). In *P. abyssi* genome, the *aroE* gene (PAB0300) was followed by an uncharacterized gene (PAB0301) encoding a predicted kinase, which is distantly related to homoserine kinases. This was also the case in *A. pernix* and *T. acidophilum* genomes, where the PAB0301-like gene (COG1685, Figure 5.4) was found sandwiched between the *aroE* gene and the *aroA* gene, whose product catalyzes the next step of the pathway after shikimate phosphorylation [171]. Genes encoding PAB0301 orthologs (COG1685) were also found in other archaeal genomes, but not in any of the bacterial genomes that contain the typical *aroK* gene (Figure 5.4). Given this connection, Daugherty et al. expressed MJ1440, the COG1685 member from *M. jannaschii* and demonstrated that it indeed had shikimate kinase activity [171].

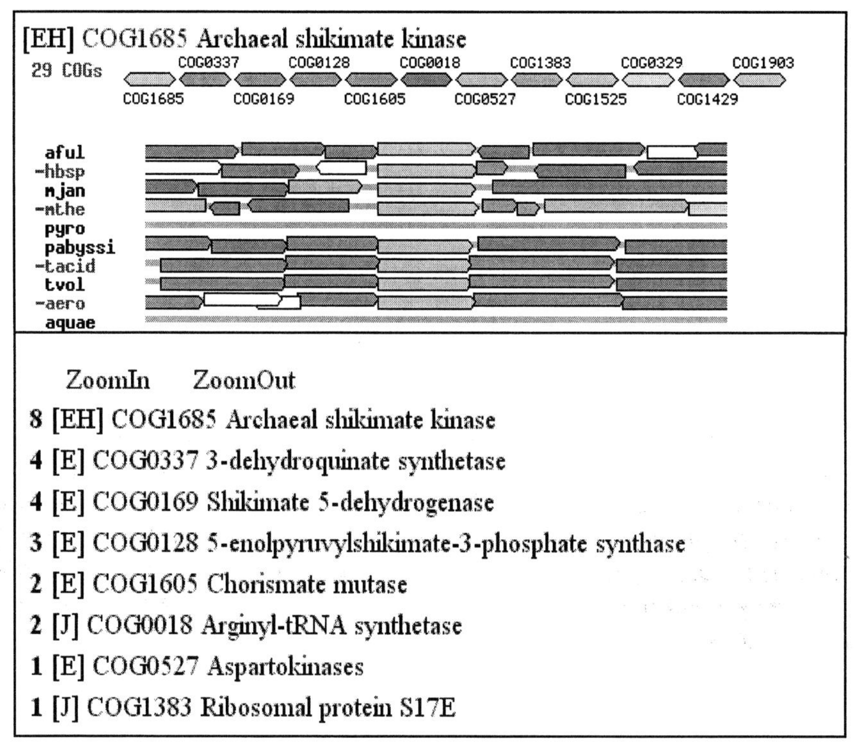

Figure 5.4. Genome context of COG1685 "Archaeal shikimate kinase". Each line corresponds to an individual genome: aful, *Archaeoglobus fuligidus*; hbsp, *Halobacterium* sp.; mjan, *Methanococcus jannaschii*; mthe, *Methanobacterium thermoautotrophicum*; pyro, *Pyrococcus horikoshii*; pabyssi, *Pyrococcus abyssi*; tacid, *Thermoplasma acidophilum*; tvol, *Thermoplasma volcanium*; aero, *Aeropyrum pernix*; aquae, *Aquifex aeolicus*. The genes encoding members of COG1685 are shown in the middle. Genes encoding members of the same COG are indicated by the same shade of grey. Genomes that do not encode a member of COG1685 are indicated by empty lines. The names of all COGs represented in the picture are listed starting from the most common ones. Note that in *Halobacterium* sp. (second line) and *M. thermoautotrophicum* (fourth line), COG1685 genes are followed by the genes encoding chorismate mutase (*tyrA*_1, COG1605). In *Thermoplasma* spp. and *A. pernix* (lines 7-9), COG1685 genes are sandwiched between the genes encoding shikimate-5-dehydrogenase (*aroE*, COG0169), and genes encoding 5-enoyl-puruvoylshikimate-3-phosphate synthetase (*aroA*, COG0128). See Figure 7.7 (page 331) for the chart of the complete pathway of phenylalanine and tyrosine biosynthesis. This figure is also available in color plates.

STRING

The Search Tool for Recurring Instances of Neighbouring Genes (STRING, http://www.bork.embl-heidelberg.de/STRING), developed by Peer Bork and colleagues, is based on a similar approach [788]. Gene clusters are defined by STRING the same way as in WIT, namely as strings of genes on the same strand located no more than 300 bp from each other. Orthologs are identified as bidirectional best hits using Smith-Waterman comparisons. The STRING search starts from a single protein sequence that can be entered as a FASTA file or just by its gene name in the complete genome. The sequence entered in FASTA format is compared against the database of all proteins encoded in complete genomes so that the user could choose one of the best hits for further examination. Like COGs, STRING contains information only on completely sequenced genomes. The default option in STRING further reduces the number of analyzed genomes by eliminating closely related ones (this option can be switched off by the user). Additionally, STRING features a useful tool that allows the user to perform an "iterative" analysis of gene neighborhoods. After the nearest neighbors of a gene in question are identified, the next "iteration" of STRING would look for their neighbors and record if any of these were found previously. If no new neighbors are found, STRING reports that the search has "converged". If this does not happen even after five consequent search cycles, the program would just tabulate how many times was each particular gene found in the output. Combined with impressive graphics, this approach makes STRING a fast and convenient tool to search for consistent gene associations in complete genomes.

SNAPper

The SNAP (Similarity-Neighbourhood APproach) tool at MIPS (http://mips.gsf.de/cgi-bin/proj/snap/znapit.pl, [447]) is similar to STRING, but instead of precomputed pairs of orthologs, it simply looks for BLAST hits with user-defined E-values. In addition, SNAP does not require the related genes to form conserved gene strings, they only need to be in the vicinity of each other. SNAPper looks for the homologs of the given protein, than takes neighbors of the corresponding genes, looks for their homologs, and so on [447]. The program then builds a similarity-neighborhood graph (SN-graph), which consists of the chains of orthologous genes in different genomes and adjacent genes in the same genome. The hits that form a closed SN-graph, i.e. recognize the original set of homologs, are predicted to be functionally related. The advanced version of SNAPper offers the choice of several parameters, which allow fine-tuning the performance of the tool depending on the particular query protein.

KEGG

In contrast to the tools described above, identification of gene strings in the KEGG database (http://www.genome.ad.jp/kegg-bin/mk_genome_cmp_html) is geared toward an analysis of the operon conservation. It allows one to find all genes in any two selected complete genomes whose products are sufficiently similar to each other and are separated by no more than five genes. The user can specify the desired degree of similarity between the proteins in terms of the minimal pairwise BLAST score (or maximal E-value), the minimal length of the alignment, and the type of BLAST hits (bidirectional or unidirectional hits, or just any hits with the specified BLAST score). The user can also specify maximum allowable distances between the genes in either organism, limiting it to any number of genes from zero to five. This option allows one to retrieve much more distant gene pairs than those detected by the ERGO tool. The downside of this richness is that unless one uses fairly strict criteria for protein similarity and the intergenic distances, he or she will end up with dozens or even hundreds of reported gene pairs, few of which would have predictive power. Nonetheless, a sensible use of this tool can bring some very interesting results [268].

Genome context tools in genome annotation

To evaluate the power of gene order-based methods for making functional predictions, we have isolated those cases where a substantial functional prediction did not appear possible without explicit use of gene adjacency information [916]. In spite of the inherent subjectivity of such assessments, the result was instructive: such unique predictions were made for ~90 genes (more precisely, COGs) or ~4% of all COGs analyzed. Given that, as noted above, homology-based approaches already allow functional predictions for a majority of the genes in each sequenced prokaryotic genome, this places gene-string analysis in the position of an important accessory methodology in the hierarchy of genome annotation approaches. Other genome context-based methods may also be useful but are clearly less powerful. This is, of course, a pessimistic assessment because more subtle changes in prediction for gene already annotated by homology-based methods were not taken into account.

These limitations notwithstanding, some of the predictions made on the basis of gene order conservation combined with homology information seem to be exceptionally important. Perhaps the most straightforward case is the prediction of the archaeal exosome, a complex of RNAses, RNA-binding proteins and helicases that mediates processing and 3'->5' degradation of a variety of RNA species [469]. This finding was made by examination of archaeal genome alignments, which led to the detection of a large superoperon, which, in its complete form, consists of 15 genes. This full complement of co-localized genes, however, is present in only one species, *M. thermoautotrophicum*, whereas, in all other archaea, the superoperon is partially disrupted and, in some cases, certain genes have been lost altogether. Remarkably, the predicted exosomal superoperon also includes genes for proteasome subunits. According to the logic outlined above, this points to a hitherto unknown functional and possibly even physical association between the proteasome and the exosome, the machines for controlled degradation of RNA and proteins, respectively.

Gene order-based functional prediction seems to be impossible for eukaryotes because of the apparent lack of clustering of functionally linked genes. However, several operons that have been identified in *C. elegans* [645,894,944] comprise the first exceptions to this rule and suggest that gene order analysis could be eventually used for eukaryotes, too. Besides, the above prediction of proteasome-exosome association might potentially extend to eukaryotes, offering yet another example of the use of prokaryotic genome comparisons for understanding the eukaryotic cell.

Given the fluidity of gene order in prokaryotes, detection of subtle conservation patterns requires fairly sophisticated computational procedures that search for ***gene neighborhoods***, sets of genes that tend to cluster together in multiple genomes, but do not necessarily show extensive conservation of exact gene order [447,491,640,641,709]. One of the interesting findings that have been made possible through these approaches is the prediction of a new DNA repair system in archaeal and bacterial hyperthemophiles [541]. As shown in Figure 5.5 (see color plates), the gene neighborhood predicted to encode this system forms a complex patchwork, with very few conserved gene strings. However, the overall conservation of the neighborhood is obvious (once the analysis is completed and the results are summarized as in Figure 5.5) and statistically significant [541,709]. In an already familiar theme, prediction of this repair system involved a combination of genomic neighborhood detection with fairly complicated protein sequence analysis and structure prediction. One of the notable findings was the identification of a novel family of predicted DNA polymerases (COG1353). Finally, this is where we encounter, once again, COG1518, the protein family already discussed in ♦4.5. When we first analyzed those proteins, we were inclined to predict that they were novel enzymes, perhaps with a hydrolytic activity. Context analysis allows us to make a much more specific prediction: these proteins mostly likely are nucleases involved in DNA repair.

5.3. Conclusions and Outlook

In this chapter, we discussed both traditional methods for genome annotation based on homology detection and newer approaches united under the umbrella of genome context analysis. We noted that, although functions can be predicted, at some level of precision, for a substantial majority of genes in each sequenced prokaryotic genome, current annotations are replete with inaccuracies, inconsistencies and incompleteness. This should not be construed as any kind of implicit criticism of those researchers who are involved in genome annotation: the task is objectively hard and is getting progressively more difficult with the growth of databases (and accumulation of inconsistencies). Fortunately, we believe that the remedy is already at hand (see ♦3.1.3). Specialized databases, designed as genome annotation tools, seem to be capable of dramatically improving the situation, if not solving the annotation problem completely. Prototypes of such databases already exist and function and their extensive growth in the near future seems assured.

The context-based methods of genome annotation are quite new: the development of these approaches started only after multiple genome sequences became available. These approaches have a lot of appeal because they are, indeed, true ***genomic*** methods based on the notion that the genome (and, especially, many compared genomes) is much more than the sum of its parts. The results produced by these methods are often very intuitive and even visually appealing as in gene string analysis. Objectively, however, these methods yield considerably less information on gene function than homology-based methods, at least for the foreseeable future. Nevertheless, different genome context approaches substantially complement each other and homology-based methods. In fact, homology-based and context-based methods often produce different and complementary types of functional predictions. The former tend to predict ***biochemical*** functions (activities), whereas the latter result in ***biological*** predictions, such as involvement of a gene in a particular cellular process (e.g. DNA repair in the example above), even if the exact activity cannot be predicted.

We would like to end this chapter on an upbeat note by stating, in large part on the basis of personal experience, that genome annotation is not a routine, mundane activity as it might seem to an outside observer. On the contrary, this is exciting research, somewhat akin to detective work, which has the potential of teasing out deep mysteries of life from genome sequences.

5.4. Further Reading

1. Brenner S. 1999. Errors in genome annotation. *Trends in Genetics* 15: 132-133.

2. Galperin MY, Koonin EV 2000. Who's your neighbor? New computational approaches for functional genomics. *Nature Biotechnology* 18: 609-613.

3. Huynen, MA, Snel B. 2000. Gene and context: integrative approaches to genome analysis. *Advances in Protein Chemistry* 54: 345-379.

4. Huynen MA, Snel B, Lathe W, Bork P. 2000. Predicting protein function by genomic context: quantitative evaluation and qualitative inferences. *Genome Research* 10: 1204-1210.

5. Wolf YI, Rogozin IB, Kondrashov AS, Koonin EV. 2001. Genome alignment, evolution of prokaryotic genome organization and prediction of gene function using genomic context. *Genome Research* 11: 356-372.

6. Makarova KS, Aravind L, Grishin NV, Rogozin IB, Koonin EV. 2002. A DNA repair system specific for thermophilic Archaea and bacteria predicted by genomic context analysis. *Nucleic Acids Research* 30: 482-496.

7. Ouzounis CA, Karp PD. 2002. The past, present and future of genome-wide re-annotation. *Genome Biology* 3, COMMENT2001.

CHAPTER 6.
COMPARATIVE GENOMICS AND NEW EVOLUTIONARY BIOLOGY

The affinities of all beings of the same class have sometimes been represented by a great tree. I believe this simile largely speaks the truth.
Charles Darwin, 1859, The Origin of Species, Chapter IV.

I should infer from analogy that probably all organic beings which have ever lived on this earth have descended from some one primordial form, into which life was first breathed.
ibid, Chapter XIV

In Chapter 2, we primarily focused on the foundations of comparative genomics that come from evolutionary theory and only briefly summarized the evolutionary implications of genome comparisons. In this chapter, we address the connection between comparative genomics and evolution from a different angle. The question we ask is: how does comparative genomics affect our understanding of major aspects of the evolution of life? We believe that the effect is (or at least has the potential to be) truly profound. Perhaps most importantly, comparative genomics has already led to the reappraisal of the central trends of genome evolution. Instead of the classic concept of relatively stable genomes, which evolve through gradual changes spread through vertical inheritance, we now have the new notion of "genomes in flux" [787]. According to this concept, evolution involves gene loss and horizontal gene transfer as major forces shaping the genome, rather than isolated incidents of little consequence.

This new picture of the evolutionary process is incomparably more complicated than the classic one but, in addition to revealing the true complexity of the phenomena than need to be analyzed to understand evolution, genomics provides the data that are required for this analysis. The genomes threaten to uproot the Tree of Life [667], but in the end, they may help build a better, more realistic tree. The new methods taking full advantage of the wealth of information contained in genome sequences are only starting to emerge. Most of the theoretical and algorithmic developments clearly lie ahead, which makes the field of evolutionary genomics particularly exciting. The availability of genome sequences from many diverse phylogenetic lineages provides for the possibility of reconstructing genomes of ancestral life forms, including the Last Universal Common Ancestor (LUCA) of all extant life forms. Furthermore, even deeper reconstruction becomes feasible and we are starting to glimpse some aspects of the primordial RNA world.

6.1. The Three Domains of Life

In the mid-1970s, while studying some unusual groups of bacteria, thermophilic methanogens and halophiles, Carl Woese and colleagues came to the revolutionary conclusion that these organisms were not really bacteria, but should be assigned to a separate domain (also called primary kingdom or superkingdom or urkingdom) of life with the same status as bacteria and eukaryotes. This group was originally referred to as archaebacteria and later renamed archaea [901,906]. The uniqueness of the archaea was apparent even from some of their biochemical features, such as the unusual structure of lipids, but what really clinched the case was the topology of phylogenetic trees of 16S rRNA, which were first built using oligonucleotide catalogs (we are talking here about the pre-genomic and even pre-sequencing era!) and subsequently derived from complete RNA sequence alignments. These trees clearly indicated that archaea comprised a unique branch of life, distinct from both bacteria and eukaryotes. Furthermore, although, phenotypically, archaea are obviously prokaryotes, like bacteria, i.e. have small cells without nuclei or organelles, they are, in some important respects, to eukaryotes than to bacteria.

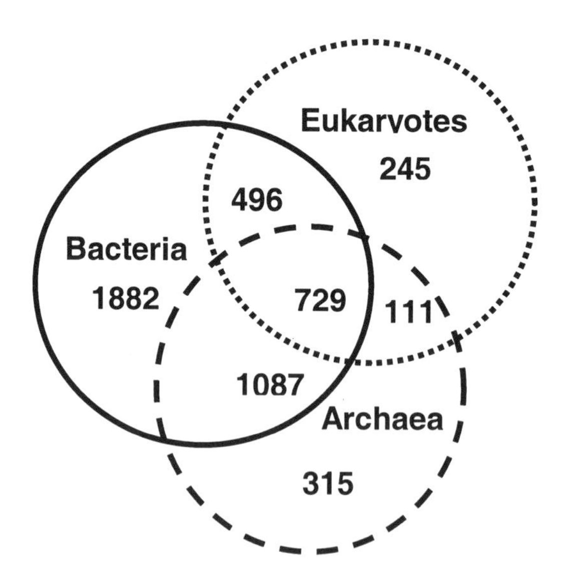

Figure 6.1. Distribution of the COGs in the three domains of life.
The data were obtained using the Phyletic Pattern Search tool of the COG system. The partitions of the Venn diagram are not to scale.

These eukaryote-like features of archaea include the structure of the ribosomes, which have a numbers of proteins shared with eukaryotes, but not with bacteria, the presence of histones (in one of the two major branches of archaea), the organization of the basal transcriptional apparatus, with several transcription factors of the eukaryotic variety, and the organization of the DNA replication apparatus, which is also conserved in archaea and eukaryotes, but not in bacteria [122].

Comparative analysis of the complete genomes of 13 archaea now available strongly supports their uniqueness as a distinct domain of life [540]. This becomes immediately apparent from a simple taxonomic classification of the COGs (Figure 6.1). As many as 315 COGs are unique to Archaea (~14% of the total number of COGs in which archaea are represented) and may be considered to comprise the archaeal "genomic fingerprint" [307,540]. We should note that only 16 of these COGs are found in ***all*** archaea, a fact that will become important in the next section when we discuss horizontal gene transfer and gene loss.

A complementary analysis shows that the sequence similarity of archaeal proteins to homologs from other archaea is, on average, much greater than their similarity to homologs from bacteria or eukaryotes (Figure 6.2).

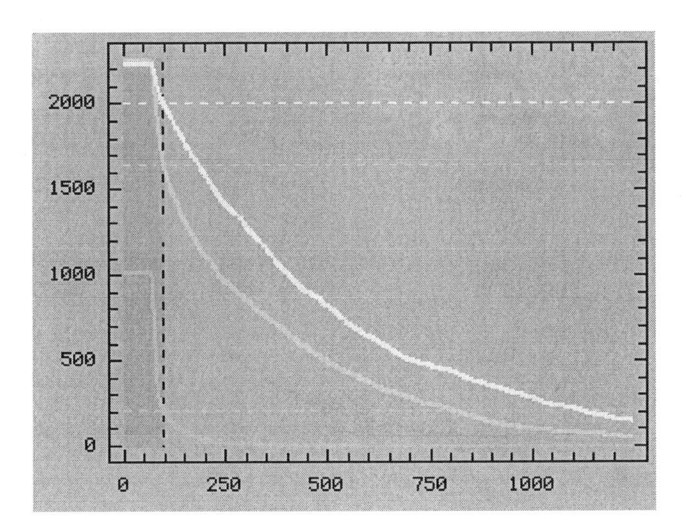

Figure 6.2. Taxonomic breakdown of the database hits for the proteins of the archaeon *Archaeoglobus fulgidus*. The vertical axis shows the number of hits with a score greater than the value indicated on the horizontal axis (from the Entrez genomes division web site: http://www.ncbi.nlm.nih.gov/cgi-bin/Entrez/taxik?gi=131). Top curve: hits to archaeal proteins, middle curve: hits to bacterial proteins, lower curve: hits to eukaryotic proteins.

In a memorable phrase of W. Ford Doolittle and colleagues, archaea are "bacterial in shape and eukaryotic in content" [950]. "Shape" here means primarily the prokaryotic features of the cellular organization, particularly the small size and the absence of nuclei and cytoskeleton. Genomic comparisons, however, show that much of the archaeal content, i.e. the apparent phylogenetic affinities of the genes, is bacterial, too [466]. This becomes clear already from the inspection of the data in Figure 6.2: there are many more bacterial similarities and they are, on average, considerably stronger, than eukaryotic similarities, among archaeal proteins. This notion of a rift between the "bacterial" and "eukaryotic" components of the archaeal gene complement is supported in a dramatic fashion by COG analysis. The number of COGs that include archaeal and bacterial proteins but not eukaryotic ones is almost 10 times greater than the number of archaeal-eukaryotic COGs without bacterial members (Figure 6.1)! This result might be biased because the COG set we analyzed includes only three eukaryotes with small genomes (the yeasts *Saccharomyces cerevisiae* and *Schizosaccharomyces pombe* and the microsporidian *Encephalitozoon cuniculi*), but the bias could not be too significant because there are very few genes uniquely shared by archaea and muticellular eukaryotes.

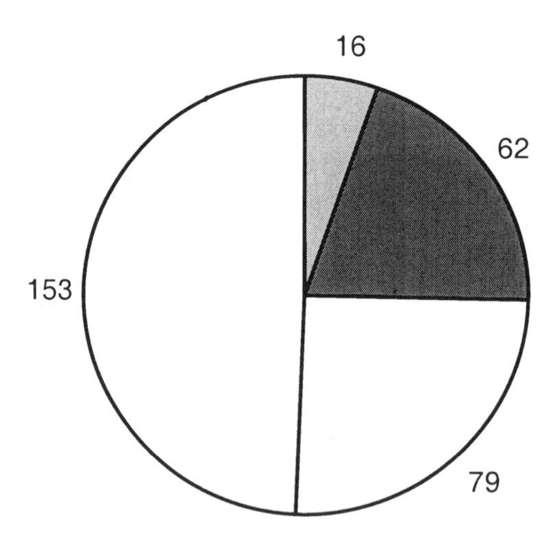

Figure 6.3. Taxonomic breakdown of the conserved archaeal COGs.
The analysis was done for the 310 COGs represented in all 13 archaeal genomes. The gray sector on top represents 16 COGs that include exclusively archaeal proteins, other sectors show, clockwise: COGs with archaeal and bacterial, but not eukaryotic proteins; COGs with archaeal and eukaryotic, but not bacterial proteins; and COGs with proteins from all three domains.

This prevalence of archaeo-bacterial COGs seems to emphasize the existence of what might be called a common gene pool shared by these two domains of life; this has much to do with the prevalence of horizontal gene transfer as discussed in the next section.

It is also notable that, when a similar breakdown is produced for the set of 310 COGs that are represented in all 13 available archaeal genomes and thus comprise the conserved core of archaeal genes, the number of COGs shared exclusively with eukaryotes becomes somewhat greater than the number of archaeo-bacterial COGs (Figure 6.3). Thus, as already noticed in the early days of archaeal genomics, there is a major "eukaryotic" component in the conserved core of the archaeal genomes, whereas the "variable shell" is overwhelmingly bacterial [540].

Of course, the distinction between the "eukaryotic" and "bacterial" components of the archaeal genomes is not only quantitative. The distributions of (predicted) functions in the archaeo-eukaryotic and the acrhaeo-bacterial subsets of COGs are strikingly different as shown here for the conserved archaeal core of 310 COGs (Figure 6.4).

The great majority of archaeo-eukaryotic COGs in the conserved core represent information processing functions, i.e. DNA replication, transcription, and translation (in reality, this fraction is probably even greater because some of the poorly characterized proteins in this set are most likely to function in translation as discussed in Chapter 5).

In a stark contrast, the archaeo-bacterial subset is enriched in metabolic enzymes; those proteins in this subset that are implicated in information processing are largely transcription regulators: this part of archaeal biology is predominantly bacterial [42]. Thus, it appears that archaea are "eukaryotic" in their basal information processing systems and "bacterial" in metabolism and much of cell biology [540,703].

There are at least two major inferences that seem to follow from these complex relationships between the protein sets encoded in the genomes from the three domains of life: (i) archaea and bacteria share a substantial gene pool, part of which is ancient heritage of the common ancestor of these two domains, and part is the result of HGT, and (ii) there is a small but critically important core of proteins, primarily involved in information processing, that reflect shared history of archaea and eukaryotes. In the remaining sections of this chapter, we provide more perspective on each of these issues.

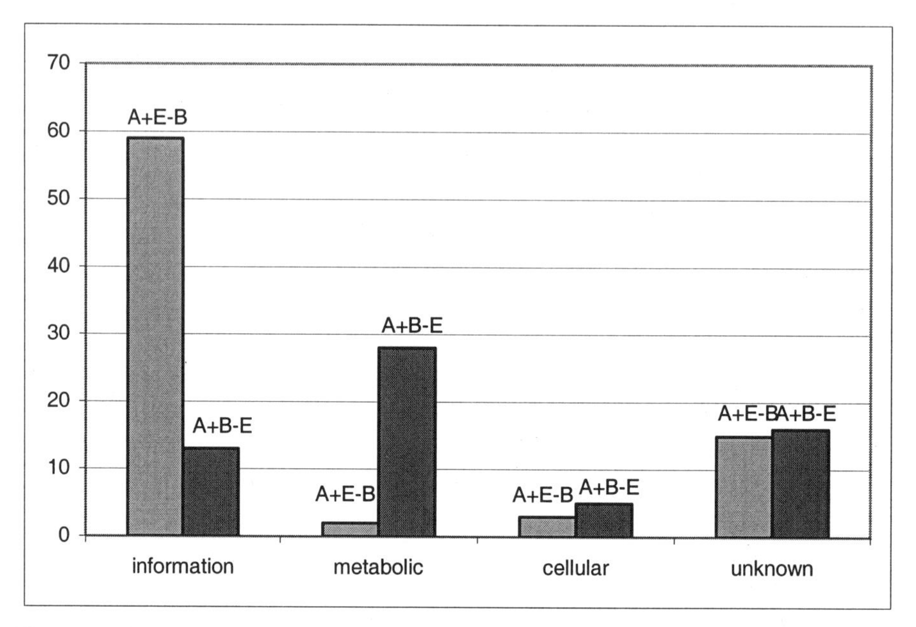

Figure 6.4 Protein functions in the archaeo-eukaryotic and archaeo-bacterial subsets of the conserved archaeal core (310 COGs total, Figure 6.3B).

A: archaea, B: bacteria, E: eukaryotes. Information: proteins involved in replication, repair, transcription and translation; metabolic: metabolic enzymes and transporters; cellular: proteins involved in various cellular functions, such as molecular chaperones and membrane biogenesis; unknown: poorly characterized and uncharacterized proteins.

6.2. Prevalence of Lineage-Specific Gene Loss and Horizontal Gene Transfer in Evolution

Horizontal gene transfer (HGT) and lineage-specific gene loss are tightly linked phenomena. As we will see shortly, any observable phyletic pattern could be potentially explained by gene loss, by HGT or through a combination of both types of events. However, the status of the two types of events in the molecular evolutionary literature is quite different. While gene loss is fully accepted a common evolutionary phenomenon, widespread occurrence of HGT is still contested. No one denies that it occurs, in principle, but there is serious and sometimes heated debate on its extent [479]: is the frequency of HGT comparable to that of gene loss or is HGT several orders of magnitude less frequent? There is, indeed, a rather good reason to assess the two phenomena differently because there are well-defined situations in evolution where gene loss cannot be reasonably questioned. These include evolution of parasites, which obviously have lost large parts of their original gene sets, and evolution of free-living heterotrophs as well [55,114,425,581] (see also ♦6.4 and Chapter 7). There are no such obvious smoking guns for massive HGT. Besides, gene loss presents much less of a problem from the point of view of evolutionary theory because no special selective advantage needs to be postulated for a gene loss event: as soon as a gene becomes dispensable, it may be as well eliminated. In contrast, the fixation of a gene acquired by HGT in the genome of its new host is very much in need of an adaptive explanation.

We believe, however, that strong evidence of large-scale HGT between phylogenetically distant species does exist. Such evidence seems to be provided by clear correlations between similarity in organisms' lifestyles and the apparent number of genes they exchange via HGT. This notion first came to fore when it was shown that the hyperthermophilic bacterium *Aquifex aeolicus* (the first genome of a bacterial thermophile to be sequenced) contained significantly (with good statistical support) more "archaeal" genes (that is, genes that are either missing in bacteria altogether or are more closely related to archaeal than to bacterial orthologs) than any other bacterial genome available at that time [52]. Subsequently, similar observations have been reported for another bacterial hypethermophile, *Thermotoga maritima* [610]. The genomes of these bacterial hyperthemophiles encode ~15-20% proteins of probable archaeal descent compared to 1-5% in mesophiles [462]. A lineage-specific gene loss explanation has been proposed even for these observations, under the notion that *Aquifex* and *Thermotoga* appear to be early-branching bacteria (see next section) and might have retained ancient thermophilic heritage that had been lost in the rest of bacteria as a result of only one loss event [483]. This argument does not appear to be particularly strong because it fails to account for the sharp divide, in terms of the apparent phylogenetic affinities, between these "archaeal" genes and the rest of the genomes of the bacterial hyperthermophiles, which consist of "garden variety" bacterial genes [53]. What seems to really clinch the case, however, is the fact that the same trend is seen in the recently sequenced *Thermoanaerobacter tengcongensis* [73]. This thermophilic bacterium belongs to the Bacillus-Clostridium group of Gram-positive bacteria but, nevertheless, has many more archaeal genes than its mesophilic cousins (Figure 6.5, see color plates). In this case, the gene loss explanation would necessarily require multiple, independent elimination of the same set of genes in many bacterial lineages and does not look plausible at all. The connection between an organism's lifestyle and HGT has been confirmed in the most dramatic fashion by the recent sequencing of the genomes of mesophilic archaeal methanogens *M. acetivorans* and *M. mazei* [181]. In *M. acetivorans*, nearly 30% of the genes seem to be of bacterial origin, an order of magnitude more than in phylogenetically related hyperthermophilic archaeal methanogens (Figure 6.6 in color plates). As in the previous case, an explanation based on gene loss seems to be unrealistic given the position of the methanogens in the archaeal tree [779].

Using the significantly greater sequence similarity to homologs from a distant taxon compared to homologs from the "native" taxon, to which the given species belongs, as an argument for HGT is one of the so-called surrogate criteria for HGT detection [688]. Other such criteria include unusual phyletic patterns (more about this below), unexpected conservation of local gene order between distant species that might be indicative of transfer of entire operons, and anomalous nucleotide composition and codon usage [462,498,625,687,688]. These approaches have been dubbed surrogate because the "real" method for detecting HGT is supposed to be phylogenetic tree analysis. Suppose we have a set of orthologs (COG), which is represented in all archaea and in only one bacterial species, and, furthermore, the bacterial species is not equidistant from all archaeal orthologs but, when a tree is built, specifically cluster with a particular archaeal branch. This would have been an apparently irrefutable case for HGT. Such perfect situations are extremely rare. In the current COG database, there are only four COGs with that exact phyletic pattern, and only two of them seem to produce the desirable tree topology (EVK, unpublished observations). Thus, under a more relaxed criterion, any statistically supported "paradoxical" clustering in a tree would strongly suggest the HGT case. The general difficulty with this approach is that phylogenetic trees, even those constructed with the most powerful modern methods (see ♦6.3), are often ambiguous (claims of "rigor" often found in the literature notwithstanding). This becomes much more of a problem when attempts are made to construct trees (and underlying alignments) automatically and on a large scale. A recent attempt of a genome-scale phylogenetic study, aimed at the characterization of what the authors called the "phylomes" of several prokaryotic species (i.e. the sets of phylogenetic trees for all genes that are sufficiently conserved in evolution to allow tree construction), revealed many instances of unexpected clustering, which suggests widespread HGT [769]. However, it is hard to assess the reliability of this result.

The confidence that an unusual tree topology actually reflects HGT may be bolstered when there is support from an independent line of evidence. In a much smaller benchmarking study, Itai Yanai, Yuri Wolf and one of the authors (EVK) investigated the evolutionary scenarios for gene fusions, aiming to distinguish cases of dissemination of fused genes via HGT from vertical inheritance accompanied by fission in some lineages and from independent evolution of the same fusion on two or more occasions [931]. To this end, fusions (two-domain proteins) were split into the component domains and phylogenetic trees were constructed for each of the corresponding orthologous sets, including both fusion components and products of stand-alone genes from other species. The topologies of the resulting trees were compared to each other and to the topology of a tree made from a concatenated alignment of ribosomal proteins, which was treated as the species tree (see next section). The distribution of the fusion components in the phylogenetic trees for orthologous clusters would follow the phylogeny of the species that have the fusion if the fusion events occurred more than once independently or were vertically inherited, perhaps followed by fission in some lineages. In contrast, if the fusion gene disseminated via HGT, fusion components are expected to form odd clusters different from those in the species tree. When the trees for the two fusion components agree, the case for HGT becomes strong. The conclusion: of the ~50 analyzed fusion proteins that are present in both bacteria and archaea, ~2/3 have spread via HGT.

Anecdotal studies that support HGT by means of phylogenetic analysis have been sufficiently numerous to conclude that this phenomenon had a major role in evolution, at least as far as prokaryotes are concerned. It seems that HGT cuts throughout the range of biological functions, although, among genes coding for core proteins of translation, transcription and replication, transfers probably are less common. Aminoacyl-tRNA synthetases (aaRS) are a notable exception: numerous cases of probable HGT have been detected for these essential enzymes involved in translation [190,330,907,909]. Although aaRS generally follow the "standard model" of evolution, with the original split leading to the separation of the bacterial and archaeo-eukaryotic lines of descent, strong evidence of HGT has been detected for at least 17 of the 20 aaRS specificities [909]. Figure 6.7 illustrates HGT for two aaRS families. The evolution of glutamate and glutamine aaRS (Figure 6.7A) involves one of the most spectacular cases of non-orthologous gene displacement (see also Table 2.3). Most of the bacteria and archaea do not have an aaRS for glutamine (Q-RS); instead, glutamine is formed by transamidation of glutamate-tRNA^{Gln} whose formation is catalyzed by the so-called non-discriminating glutamate-RS (E-RS) [854,855]. Eukaryotes and at least some gamma-proteobacteria lack the enzymatic complex responsible for transamidation and instead encode Q-RS.

This is one of the rather rare situations where the entire evolutionary scenario seems to be clear from the tree. Q-RS apparently evolved via a duplication of E-RS at an early stage of eukaryotic evolution and was subsequently acquired by gamma-proteobacteria, which was followed by obliteration of the ancient transamidation machinery (according to the general scenario for non-orthologous gene displacement discussed in Chapter 2). This sequence of events is dictated by the tree topology in Figure 6.7A, where the gamma-proteobacterial branch is within the archaeo-eukaryotic cluster. The opposite direction of HGT, from gamma-proteobacteria to eukaryotes, would have put eukaryotes in the midst of the bacterial cluster. The direction of the single HGT event for tryptophanyl-RS (W-RS) is equally certain from Figure 6.7B: the archaeon *P. horikoshii* acquired W-RS from a eukaryote via HGT; the alternative, namely acquisition of W-RS by an early eukaryote from this particular archaeal species, is unrealistic, even if only because the divergence of eukaryotes certainly predates the divergence of pyrococcal species. This case is a clear manifestation of the phenomenon of ***xenologous gene displacement***, the variant of HGT when an ortholog from a distant lineage (xenolog) displaces the "native" gene in a given genome [462]. For essential genes, xenologous gene displacement must involve acquisition of the "alien" gene followed by a period of co-existence of the "native" and "alien" forms and then elimination of the native one. For non-essential genes, an alternative is conceivable whereby the "native" gene is lost first and the "alien" gene is acquired subsequently.

The distribution of apparent HGT events among different functional categories of genes has been interpreted in terms of the so-called complexity hypothesis, which posits that genes coding for protein subunits of macromolecular complexes or, more generally, proteins involved in a wide range of interactions, are less subject to HGT [390].

There is indeed strong logic behind this concept because, unless subunits of a complex are encoded in the same operon and are transferred together, a gene coding for just one subunit of a complex is unlikely to get fixed in the recipient genome. In particular, there seems to be relatively little HGT involved in the evolution of the genes for ribosomal proteins, components of the utmost molecular machine of the cell. However, even in this sanctum of vertical evolution, detailed phylogenetic analysis revealed several instances of HGT, some of which involved the gene for S14, an essential protein located "in the heart of the ribosome" [119,120,542].

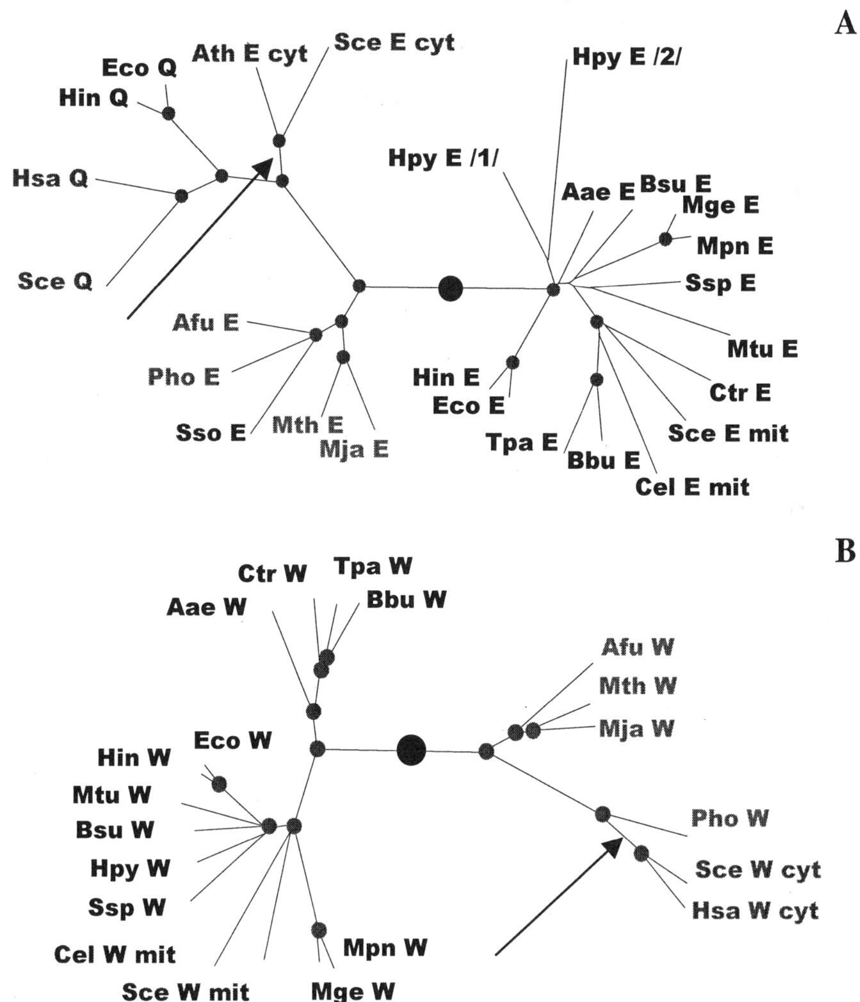

Figure 6.7. Phylogenetic trees for two families of aminoacyl-tRNA synthetases
A (Top panel): glutamate and glutamine; B (Bottom panel): tryptophan. The small gray circles show bootstrap support >70%; the large black circle indicates the likely root position [909]. The nodes that are indicative of HGT are shown by arrows.

It seems that, as far as HGT is concerned, everything is for sale, at least in the prokaryotic world; it is just the price (i.e. the likelihood of HGT) that differs for different genes.

The apparent major role of HGT in the early evolution of eukaryotes is discussed below in ♦6.4. The possibility of acquisition of bacterial genes via HGT relatively late in eukaryotic evolution, namely after the emergence of vertebrates, has become a subject of controversy after it was brought up in the report of the International Human Genome Consortium on the draft sequence of the human genome [463,488,678]. In this report (with a direct contribution from one of the authors of this book, EVK), it was noticed that proteins encoded by 113 human genes either had no detectable eukaryotic homologs outside the vertebrate lineage or showed significantly higher similarity to the bacterial homologs than to homologs from non-vertebrate eukaryotes. The history of these genes was proposed to have involved either lineage-specific gene loss or HGT or both, with HGT considered to be more likely because multiple losses were required to explain the observed phyletic patterns. Moreover, the direction of transfer was thought to be from bacteria to vertebrates because most of these genes were widespread in bacteria. This hypothesis was sharply criticized by three independent groups who observed that many of the genes in question were present in lower eukaryotes, in particular the slime mold *Dictyostelium discoideum* (as shown by searching the slime mold EST database that was not screened in the original study). Moreover, all eukaryotic members of the respective families tended to cluster together in phylogenetic trees ([706,737,799], see also discussion in [29]). All these authors concluded that gene loss, even in multiple lineages, was a much more likely explanation for the observed patterns, whereas acquisition of bacterial genes by vertebrates via HGT was extremely rare, if it occurred at all. These results certainly emphasize caution that is needed in the interpretation of indications for HGT in the absence of a representative set of genomes from the major lineages of eukaryotes (and, we should note parenthetically, the importance of database integration; see Chapter 3). However, we believe that the jury is out on the evolutionary history of these genes and the data can be explained by several alternative scenarios. These include multiple HGT events and a combination of a relatively early gene acquisition from bacteria, e.g. by the common ancestor of *Dictyostelium* and animals, if a phylogeny including such an ancestor is accepted, with subsequent multiple gene losses. It is worth mentioning that analysis of the sequence of *Dictyostelium* chromosome 2, which appeared after the bulk of this manuscript had been completed, showed specific, high similarity to vertebrate homologs for numerous slime mold proteins [288]. We cannot fully assess the significance of these observations because much more analysis is required, but it seems that evolution of eukaryotes has profound mysteries in store.

Figure 6.8 (see color plates) shows the phylogenetic tree for one of the genes identified as possible vertebrate-specific horizontal transfers in the human genome report. This gene attracted special attention because its product, monoaminoxidase (MAO), is an enzyme involved in the metabolism of an essential neuromediator and a drug target in psychiatric disorders [2]. With regard to the origin of the particular subfamily that includes the classic human MAO, the tree topology is indeed compatible with acquisition of a bacterial gene by a common ancestor of slime mold and animals. It is striking, however, that four distinct branches of this family of enzymes each include both bacterial and vertebrate proteins (branches 1-4 in Figure 6.8). This suggests that both multiple HGT events and multiple losses contributed to the evolution of this single family of enzymes. Such complex scenarios might not be uncommon in evolution and are likely to come up repeatedly as the collection of sequenced genomes becomes increasingly more representative.

Perhaps one of the most interesting directions in the analysis of the new reality of genome evolution, which includes numerous gene loss and HGT events, involves development of algorithms for explicit reconstruction of evolutionary scenarios and ancestral gene sets. The simplest of such algorithms take into account only phyletic patterns and a species tree. The principle is illustrated in Figure 6.9, which shows a tentative species tree based on concatenated alignments of universal ribosomal proteins ([915] see next section) and the phyletic distribution of two COGs superimposed upon it.

Consider, first, COG1747, which is represented in four species, two chlamydia and two spirochetes. Given this particular tree topology (again, see next section), there is no need to postulate any gene loss or HGT events to explain the evolution of this COG. The obvious, simplest evolutionary scenario will just hold that the COG emerged at the base of the chlamydia-spirochete clade and was inherited vertically ever since.

Consider, in contrast, COG2810. This COG does not map to a single, compact cluster of species (clade) in the tree and, accordingly, gene loss and/or HGT have to be invoked to explain its phyletic pattern given the species tree. Suppose this protein first emerged (what exactly do we mean by "emergence" of a protein family will be discussed in ♦6.4) in the bacterium *Deinococcus radiodurans* (or, more precisely, in the *Deinococcus* lineage). The observed distribution then could be accounted for by postulating two independent horizontal transfers into different archaeal lineages. The alternative, based solely on gene loss, would require elimination of this gene

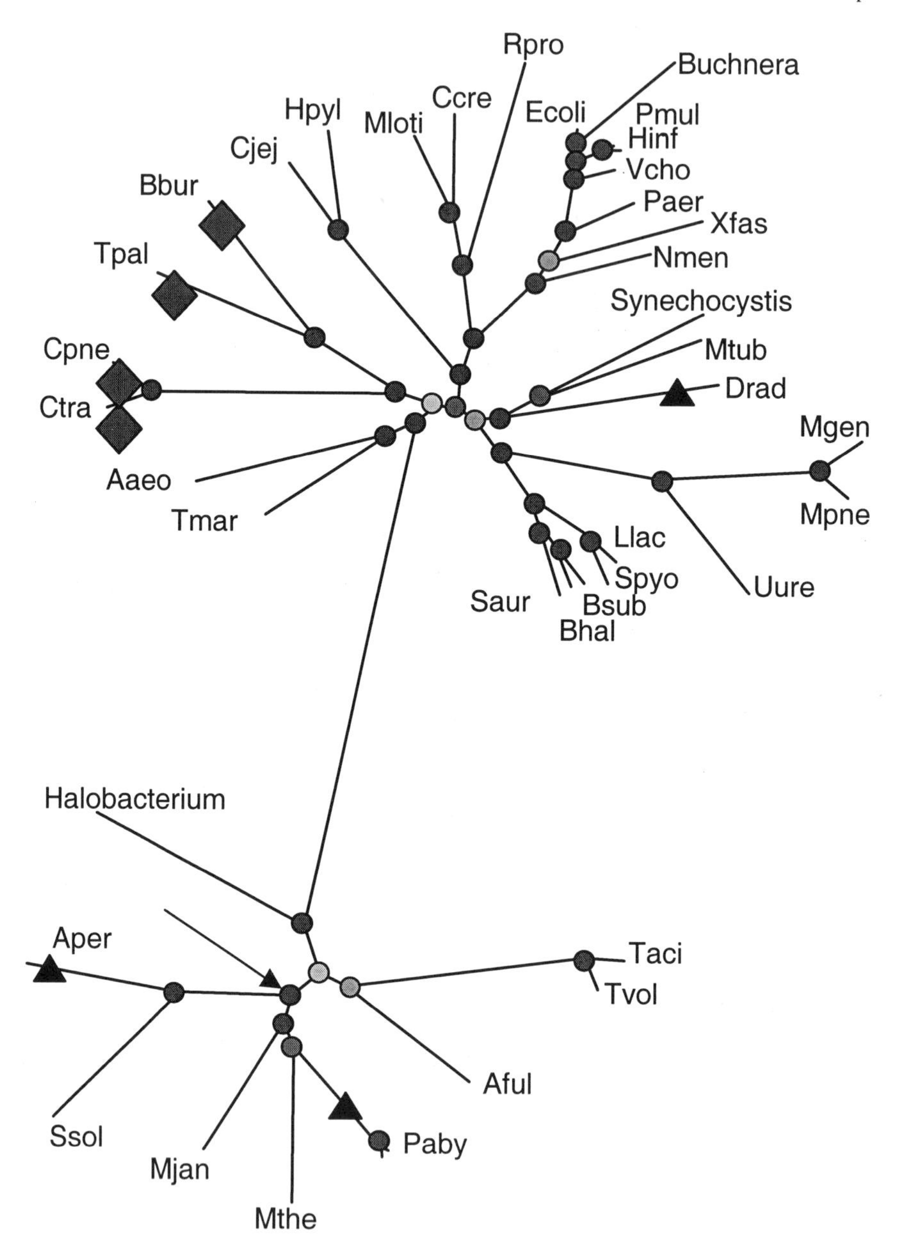

Figure 6.9. Distributions of two COGs on a tentative species tree.
The species with completely sequenced genomes (Table 1.4) are indicated by the first letter of the genus name and first three letters of the species name. COG1747 (indicated by a rhomboid): uncharacterized N-terminal domain of the transcription elongation factor GreA (separate protein in *T. pallidum*); COG2810 (indicated by a triangle): predicted type IV restriction endonuclease.

in five bacterial and five archaeal lineages, 10 evolutionary events altogether (the reader may want to reproduce this exercise if interested). Mixed scenarios are also imaginable. For example, the COG might have appeared first in the ancestral archaeal node indicated by arrow in Figure 6.9. Then, it would take three losses in archaea and one HGT event (four events total) to account for the observed phyletic pattern.

In this fashion, it is possible to compute, for each COG, the minimal number of events that is required to reconcile the phyletic pattern with a given species tree. We designate this measure the Incompatibility Quotient. For a given COG *i*,

$$I_i = l_i + gh_i \tag{6.1}$$

where l is the number of losses, h is the number of HGT events in the minimal (most parsimonious) evolutionary scenario for the given COG, and g is "HGT penalty". Assuming g=1, we will obtain the global minimum of I for each COG given a species tree.

Such calculations for the current COG collection lead to a striking result: the history of a COG, on average, apparently included 3-5 loss or HGT events; only ~14% of the COGs seem to have evolved without these events [951]. We have already seen that most of the COGs include a rather small number of species (Figure 2.8). The I-value calculations illustrated in Figure 6.10 show that this pattern, to a large extent, has been shaped by gene loss and HGT, rather than simply by late emergence of COGs.

Using this approach, fairly straightforward algorithms can be developed to reconstruct the ancestral gene sets for each internal node of the species tree and to assign a series of events to each branch, thus effectively producing the evolutionary scenario for the genomes themselves [787, 951]. The difficulty lies in determining the correct value of the HGT penalty (the g parameter in equation 6.1). It appears plausible that gene loss is more likely to be fixed in evolution as briefly discussed in the beginning of this section, i.e. we should assume $g > 1$; however, we have no data to determine, even approximately, just how much more likely one type of events is compared to the other. This seems to be a critical parameter for the study of genome evolution.

One way to approach the problem is to use the feedback from the results of the reconstruction of the ancestral gene sets. Snel and colleagues [787] found that, with g=1, the reconstructed genomes of early ancestors, in particular, the Last Universal Common Ancestor (so-called LUCA), which corresponds to the root of the species tree, would include unrealistically few genes. In contrast, at high *g* values, which push HGT out of the picture, the hypothetical ancestors (counter-intuitively) grow in size as one moves toward the tree root, making LUCA an "omnipotent" organism with a huge number of genes (see also discussion in the next section). Thus, in principle, one could attempt to obtain an estimate of the number of genes in ancestral forms form independent, biological considerations, and this could lead to reasonable estimates of the *g* value. Of course, as soon as "biological considerations" come into play, the speculative element of the reconstruction becomes worrisome. Another approach to resolving the loss versus HGT problem is to analyze both the phyletic patterns and the actual phylogenetic trees for each COG in conjunction. This hinges on ambiguities in tree topologies and algorithmic problems. Nevertheless, we tend to believe that eventually these and other, more sophisticated computational approaches will give us a description of genome evolution that will be immensely richer and, at the same time, more definitive than anything that could be imagined in the pre-genomic era.

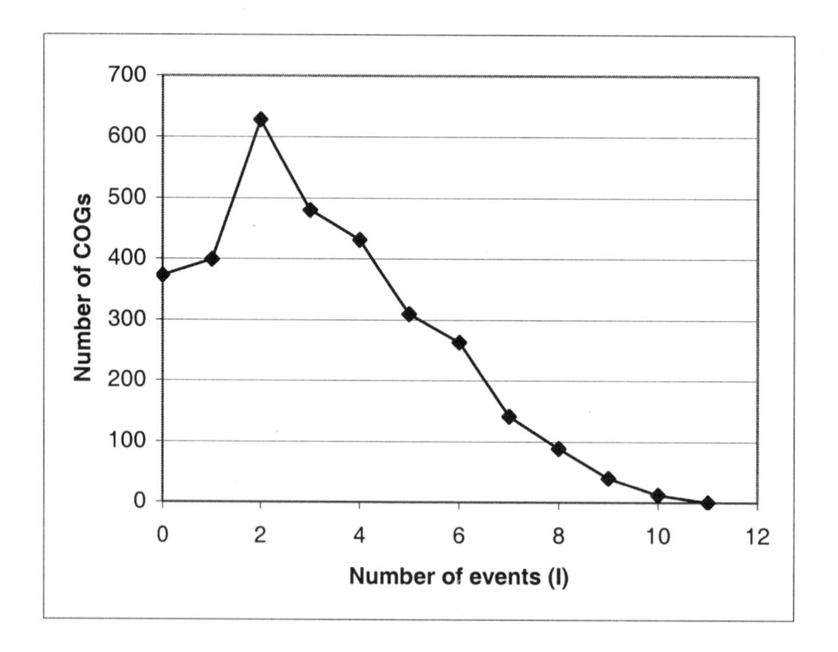

Figure 6.10. Number of gene loss and HGT events in most parsimonious evolutionary scenarios for COGs (*I* values).

6.3. The Tree of Life: Before and After the Genomes

6.3.1. Phylogenetic trees in the pre-genomic era

The concept of a tree of life depicting ***phylogeny*** (rather than simply showing classification of organisms in a convenient form), with leaves corresponding to extant species and nodes to extinct ancestors, was pioneered by Charles Darwin and is embodied in his famous single illustration of *The Origin of Species* [170]. The earliest attempts to populate the tree with real biological entities are associated with the name of Ernst Haeckel [327]. Haeckel's and other early trees were based on the general notion of a hierarchy of relationships between species and higher taxa. Gradually, quantitative criteria have been developed to measure the degree of morphological difference that was generally assumed to reflect evolutionary distance. Another major direction in evolutionary tree construction is cladistics, founded by Willy Hennig, a methodology that employs inferred shared derived characters to identify monophyletic lineages, or clades [760].

The possibility of using molecular sequences for phylogenetic tree construction was first suggested by Francis Crick in his ground-breaking 1958 paper [163] and realized by Emile Zuckerkandl and Linus Pauling who built their trees using aligned sequences of the only two protein families, for which enough sequence information has been available at the time, the cytochromes c and the globins [946]. This seminal analysis was done under the molecular clock hypothesis, introduced by Zuckerkandl and Pauling. The central postulate of this hypothesis is that a given gene evolves at a constant rate as long as the function of the gene product remains unchanged. With the accumulation of numerous sequences from diverse species, phylogenetic studies showed that, even if molecular clock could be adopted as a useful null hypothesis, it is violated all too often, which may easily result in incorrect tree topologies [310]. This inherent problem of phylogenetic tree analysis, together with the necessity for developing measures of evolutionary distance that take into account multiple substitutions in the same site and rate variation between sites within the same gene, led to a plethora of increasingly sophisticated methods for tree construction. The commonly used phylogenetic approaches, each existing in a number of flavors, are ***distance methods***, such as neighbor-joining and least-square method (Fitch-Margoliash), ***maximum parsimony***, and ***maximum likelihood***. Any detailed discussion of these methods is beyond the scope of this book, especially as several highly informative texts on molecular phylogenetics, of both theoretical and more practical inclination, have been published in the last few years [346,607,647].

In the early days of molecular phylogenetics, a gene tree was generally equated with the species tree. This implies the possibility of finding an optimal molecular marker for deciphering the history of life, and indeed, ribosomal RNA sequences became a *de facto* standard in molecular phylogenetics. As we already indicated in the beginning of this chapter, phylogenetic analysis of rRNA revolutionized our understanding of the history of life by establishing the three-domain "standard model" of evolution [901,906]. In addition, phylogenetic analysis of rRNAs brought "the winds of (evolutionary) change" onto taxonomy by revealing, supporting or correcting many major clades among bacteria, archaea, and eukaryotes [630]. It had been recognized for a long time that the exact tree topology depends on the employed phylogenetic method, but the very validity of the rRNA-based approach to species phylogeny was not seriously challenged in the pre-genomic era. This appeared to be the approach of choice to produce the true Tree of Life.

6.3.2. Comparative genomics threatens the species tree concept

The disturbing signs appeared soon after the number of gene families available for phylogenetic analysis became substantial. The problem was that different genes often yielded different trees. This incongruence between tree topologies invaded even the sacred of sacred of phylogenetic taxonomy, the three-domain standard model. In particular, archaeal genes systematically showed different phylogenetic affinities, the components of information-processing systems typically affiliating with eukaryotes, whereas metabolic enzymes and structural proteins displayed bacterial connections ([292,324]; also see above). Although some of the discrepancies could be explained away by pitfalls in phylogenetic tree construction procedures, it was becoming increasingly clear that important evolutionary reality was lurking behind incompatible topologies. All this was but a foreshadow of things to come once multiple complete genome sequences became available for comparison.

As outlined in the preceding section, systematic comparisons of complete gene sets showed beyond reasonable doubt that there was much more to evolution than vertical inheritance, with lineage-specific gene loss and HGT coming to the fore as major evolutionary phenomena, at least in the prokaryotic world. Phylogenetic tree analysis of multiple gene families sends the same message. A detailed study of 28 protein families from prokaryotes suggested that, after probable HGT cases were removed, there was no reliable phylogenetic signal left in the trees [834]. Similar results were obtained for proteins that comprise the conserved core of archaeal genomes: they all showed greater conservation within the archaeal domain than outside it, but no clear consensus phylogeny for the archaea could be determined [611].

Thus, the possibility has been repeatedly considered that comparative genomics might undermine the very concept of a Tree of Life, at least as far as the prokaryotic life is concerned [193-195,666,667] (since prokaryotes comprise two of the three primary domains, a Tree of Life without them is out of the question). Is it necessary to replace the tree representation of life's history with a network-like scenario? In the strict sense, this is certainly the case because, technically, even one HGT event makes a tree an incomplete depiction of the real course of evolution, and we have seen in the previous section that HGT had been widespread (difficulties with more precise estimates notwithstanding). However, genomics that seems to "uproot" any simple-minded tree of life based on a single gene or a small group of genes, might also offer a way to salvage the concept itself, at least in a "weak" form. Soon after multiple genomes of bacteria, archaea and eukaryotes have been sequenced, the idea emerged that phylogenetic analysis could be based not on a tree for selected molecules, e.g. rRNA, but (ideally) on the entire body of information contained in the genomes or on a rationally selected, substantial part of this information [914]. Below we briefly discuss different genome-based phylogenetic approaches (for the sake of brevity, we designate them "genome-tree" methods) and the first results in large-scale prokaryotic phylogeny brought about by the application of these methods.

6.3.3. Genome-trees – can comparative genomics help build a consensus?

The approaches to genome-tree construction and the main results obtained with each are briefly summarized in Table 6.1. The most obvious criterion for genome comparison is based on the analysis of gene content. Closely related species share a large proportion of genes; in contrast, distantly related species should have lost a substantial fraction of the genes inherited from their last common ancestor, rendering the proportion of shared genes low. If this process carries on in a regular fashion (i.e. inter-genomic distances based on gene repertoires can be mapped to time scale uniformly across lineages), it could be used for phylogenetic reconstruction.

The latter requirement raises an obvious objection. The notorious plasticity of prokaryotic genomes (♦6.2) results in gene content being malleable by selective pressures, both in terms of gene loss (e.g. in parasites) and gene acquisition via HGT (e.g. adaptation to extreme environments as in hyperthermophilic bacteria discussed above). As first presciently noted by Charles Darwin himself, traits that are subject to strong selection are less suitable for phylogenetic reconstruction than neutral traits because of highly non-uniform rates of change and the tendency for convergent evolution among the former. This suggests that gene content comparisons could be a relatively poor tool for studying prokaryotic phylogeny, although, if treated as a means to study lifestyle-related similarities and differences between genomes, rather than evolutionary relationships per se, this approach can produce interesting results.

To build trees, data on representation of genomes in orthologous gene sets are either used directly for different variants of parsimony analysis or are converted to evolutionary distances, which are then employed for building neighbor-joining or least squares trees [230,786,835]. Comparison of trees produced with this approach shows that enough phylogenetic information is retained in gene repertoires to provide reliable classification on both ends of the evolutionary distance scale. Gene content trees show good separation between the three primary domains and also consistently group together closely related species. However, on the intermediate distances, i.e. where the relationships between major lineages are concerned, this approach seemed to be less suitable for phylogenetic inference. In appears that the topology of the gene-content trees is determined largely by the relative amount of gene loss in different genomes. In particular, the main division in the bacterial branch is between the free-living and parasitic forms, which resulted in well-defined major lineages (e.g. Proteobacteria) being broken up [915]. This is readily explained by common trends in genome reduction under the selective pressure during the adaptation to parasitism in different lineages. Attempts to overcome this effect included simple removal of parasites from the species set used for tree construction [230,358] and normalization of the intergenomic distances by the number of genes in the smaller genome in each pair [470,786]. The latter method resulted in reasonable phylogenetic reconstructions, with most of the known major prokaryotic lineages recovered. In general, however, gene content analysis seems to have less resolution power than some of the other genome-tree approaches (see below).

More or less the same logic applies to genome comparisons based on gene order. Rearrangements continuously shuffle the genomes, gradually breaking ancestral gene strings. The operonic organization of a prokaryotic genome makes this a complex process. On the one hand, the selective advantage of physical proximity for co-regulation renders some gene arrays less prone to break-up than others, thus extending the range of evolutionary distances, over which gene order conservation is detectable [491,916]; see also Chapter 5). On the other hand, operons are especially amenable to being transferred as a whole [494,495], which could accentuate the effect of HGT on the tree topology.

Given the limited conservation of gene order between phylogenetically distant genomes, the attempts to build trees on the basis of gene order included identification of shared gene pairs. The data on presence-absence of gene pairs in genomes were then subjected to either parsimony or distance analysis [470,915].

Table 6.1. Genome-trees: methods and principal results

Criterion/ Approach	Phylogenetic method(s)	Principal results	Ref.
Gene content	Parsimony, distance methods	Trees reflect partly phylogeny and partly similar lifestyles. Phylogenetic signal enhanced when distances normalized by genome size, but resolution limited.	[230,358, 373,519, 605,786, 835,915]
Gene order	Parsimony, distance methods	Results similar to gene content; effect of HGT noticeable.	[470, 915]
Mean similarity between orthologs	Distance methods	Trees appear to reflect largely phylogenetic relationships; limited resolution but some putative new lineages detected.	[148,470, 915]
Concatenated alignments of proteins less prone to HGT (e.g. ribosomal)	Maximum likelihood, distance methods	Results largely compatible with the mean similarity approach, but with better resolution; several potential new lineages detected.	[119,332, 552,915]
Consensus of phylogenetic analysis of multiple orthologous sets	Maximum likelihood, distance methods	Used to verify the above approaches. Most of the new lineages strongly suggested by genome trees supported.	[915]

Generally, the results obtained with this approach are similar to those of gene content analysis, with a good separation between Archaea and Bacteria and correct clustering of closely related species, but poor resolution at intermediate distances. The influence of HGT on the topology of the resulting trees was readily noticeable. Given the high rate of intragenomic rearrangements, comparison of gene orders, at least in theory, should work particularly well for resolving the phylogeny of closely related species [813].

Evolutionary distances between different pairs of orthologs in the given two genomes show a broad distribution [318]. In theory, this is due to the variability of mean protein evolution rates caused by the differences in the strength of selective constraints, which act on functionally distinct proteins. In practice, several other factors add to the rate variance, including sampling errors, incorrect identification of orthologs and HGT. Nevertheless, if, for the majority of ortholog pairs, the time of divergence coincides with the divergence of species (i.e. HGT involves a minority of genes), it is reasonable to expect that the distance distribution retains enough phylogenetic information to be used for tree construction. On this premise, parameters of this distribution (preferably the median) for all pairs of compared genomes, can be transformed into intergenomic evolutionary distances, which can then be used to construct neighbor-joining or least-squares trees [148,318,915]. This approach produced trees that were fairly robust in terms of correctly reproducing well-known lineages and also suggested the existence of several new ones.

Traditional sequence-based phylogeny relies on gradual sequence change over time. The three main problems with using single genes (more precisely, orthologous sets) to infer a species tree are insufficient number of informative sites, variability of evolutionary rates in different lineages, and the effect of HGT. The former two factors introduce (sometimes major) uncertainty into phylogenetic reconstructions; the latter one leads to gene phylogenies that are genuinely different from the (hypothetical) species phylogeny. In an attempt to overcome these pitfalls, one can concatenate many sequence alignments into one and use the combined long sequence for tree reconstruction. If the likelihood of HGT is reduced by a careful choice of genes, the trees reconstructed from such an alignment have the potential to provide good resolution. In agreement with this anticipation, phylogenetic analysis of concatenated sequences of ribosomal proteins (in some cases, with the addition of other proteins involved in translation, and/or with some proteins suspected to have undergone HGT removed) performed by three independent groups using different methods, produced trees that seemed to contain a strong phylogenetic signal [119,332,552,915]. The topology of these trees was generally compatible with that of the trees constructed using the median similarity between orthologs, but with a greater resolution, which allowed more confident prediction of several new prokaryotic lineages.

Figure 6.11 shows a proposed phylogenetic tree of prokaryotes, in which we combined the results of the genome-tree analyses discussed above (Table 6.1), in particular, trees made using the median similarity between orthologs and those based on concatenated alignments, and attempted to depict the apparent consensus.

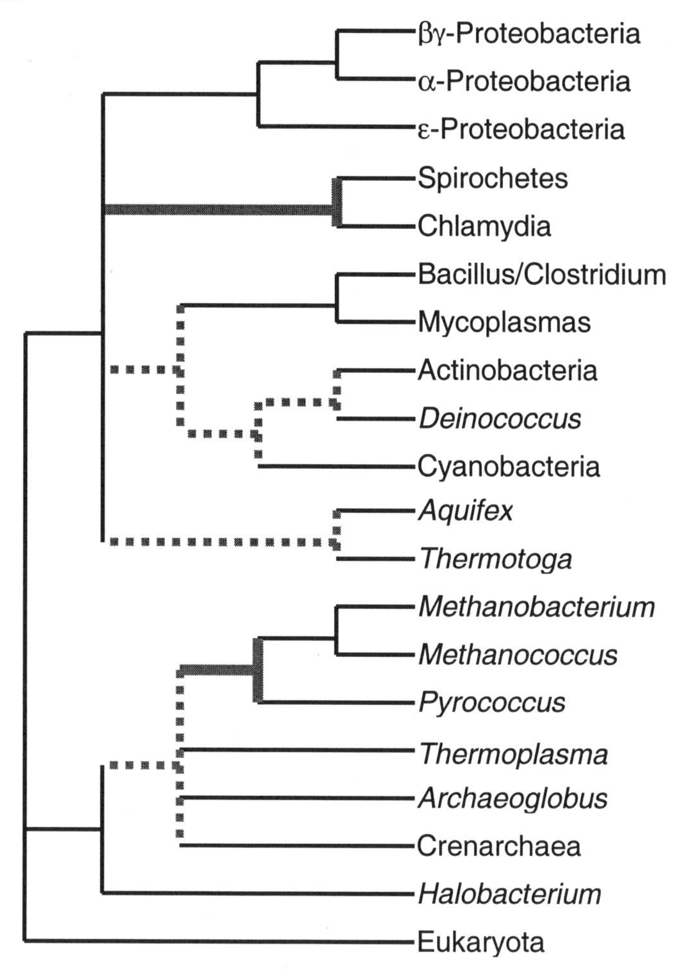

Figure 6.11. A consensus of genome-trees.
Although all genome-tree methods produce unrooted trees, this tree is, for the sake of clarity, shown in a rooted form, with the root position forced between bacteria and archaea. New clades considered to be firmly established are shown by solid thick lines and tentative new clades are shown by dotted thick lines.

At least two major clades that have not been described or, to our knowledge, even suspected to exist prior to the genome-tree studies are strongly supported by different types of analysis and appear reliable: (i) chlamydia-spirochetes among bacteria and (ii) methanogens-pyrococci among Euryarchaeota. In addition, several other major groupings were supported by some but not other approaches and should be considered tentative for the moment. These are the clade comprised of Cyanobacteria, Deinococcales and Actinobacteria, its "grand unification" with the low-GC Gram-positive bacteria (the *Bacillus-Clostridium* group), and the clade including the hyperthermophilic bacteria *Aquifex* and *Thermotoga*.

Furthermore, the genome-tree analysis challenges the traditional phylogeny of archaea, according to which Crenarchaeota (a distinct archaeal lineage, which includes, among the sequenced genomes, *Aeropyrum*, *Sulfolobus*, and *Pyrobaculum*) is traditionally assumed to be a sister group of Euryarchaeota (the rest of the archaea). In addition to the separation of these major archaeal branches in rRNA trees, they differ in certain fundamental aspects of their gene content and biology, including the presence of histones in Euryarchaeota but not in Crenarchaeota. Several of the genome-tree analyses did not support this division and placed the euryarchaeon *Halobacterium* in the archaeal root. However, the euryarchaeal topology emerged as a strong competitor [915] or even the winner [552]. Clearly, the resolution of these major evolutionary puzzles requires many more genomes and more work on the genome-tree approaches.

The results of comparative genomics suggest that the simple notion of a single Tree of Life that would accurately and completely depict the evolution of all life forms is gone forever. Individual genes, especially those of prokaryotes, follow their unique evolutionary trajectories. However, those same comparative-genomic studies that have "uprooted" the Tree of Life provide hope that the concept could be rescued, albeit in a limited sense. Taken together, the results of genome-tree analyses indicate that there is, after all, a phylogenetic signal in the sequences of prokaryotic proteins, but it is weak because of massive gene loss and HGT and, possibly, also because of a punctuated equilibrium mode of evolution, with some of the major transitions having occurred within short time intervals. It seems that, to capture this faint signal, analysis of genome-wide protein sets or carefully selected subsets is required. The concept of the Tree of Life is bound to change in the post-genomic world. It cannot be thought of as a definitive "species tree" anymore, but only as a central trend in the rich patchwork of evolutionary history, replete with gene loss and HGT [903]. Nevertheless, we believe that Darwin's dictum used as one of the epigraphs of this chapter stands: in the epoch of complete genome sequences and "lateral genomics", the tree simile still speaks essential truth about evolution.

6.3.4. The genomic clock

In the preceding section, we concluded that the notion of the Tree of Life as the ultimate depiction of the course of evolution had to be replaced with a softer concept of a "consensus species tree" as a central thread in the patchwork of life's history. Is there a way to similarly reformulate the classic molecular clock concept of Zuckerkandl and Pauling such that, in spite of the fact that changes of the evolutionary rate in individual orthologous gene sets are too common to support the traditional clock (at least at large evolutionary distances), there remains a variable in genome evolution that changes linearly with time? Comparative analysis of the distributions of evolutionary distances between orthologs in pairs of genomes separated by vastly different evolutionary distances suggested a possibility of a positive answer to this question [318]. The shapes of these distributions are very similar, although the characteristic distances are dramatically different, just as expected (Figure 6.12; see color plates). It has been shown that, when multiplied by appropriate scaling factors, most of these distributions become statistically indistinguishable [318]. Thus, however dramatically evolutionary rates for individual sets of orthologous genes may change during evolution (we discuss some anecdotal examples of such drastic changes in the next section), the genome-wide distribution remains largely invariant in shape and only slides along the axis of evolutionary distance toward greater values (Figure 6.12; see color plates). The evolutionary clock concept lives, although it seems to be more like Dali's "Melting Clock" than a proper Swiss timepiece.

6.4. The Major Transitions in Evolution: A Comparative-Genomic Perspective

In their influential book, John Maynard Smith and Eors Szathmary present the history of life as a succession of "major evolutionary transitions" [555]. This view of evolution, clearly parallel (at least in general terms) to the punctuated equilibrium of Eldredge and Gould [304], is helpful in allowing one to concentrate on those relatively brief epochs, during which momentous changes occurred, resulting in the birth of new states of living matter, and which were separated by long periods of relative stasis. Here, we attempt to show how comparative genomics allows us to attain a better and, above all, more concrete understanding of these transitions by considering three of them: (i) from the pure RNA world to RNA-protein life forms, (ii) from RNA to DNA as the substrate of heredity, and (iii) from the prokaryotic to the eukaryotic cell (we should note parenthetically that only the first two are transitions in the strict sense, which represent the succession of different types of living systems replacing one another; the third one is "just" the emergence of a new state, which does not negate the ancestral one; it is, however, of tremendous interest, both fundamental and parochial).

6.4.1. Ancestral life forms and evolutionary reconstructions

One of the most striking features of life on this planet is the surprising unity of the molecular framework of all living things. A human being, amoeba, *E. coli* and *T. acidophilum*, a hyperthermophilic archaeon that lives in nearly boiling acid, may not look like close relatives, but they share highly conserved regions in numerous proteins, particularly those involved in information processing (transcription and translation), and many structural features of key macromolecular assemblies, such as the RNA polymerase, the ribosome, and the plasma membrane. And, of course, minimal variations notwithstanding, they all use the same genetic code to translate information stored in their genomes into proteins. All these common features leave no reasonable doubt that all life forms known to us have evolved from a single common ancestor, which we will call the Last Universal Common Ancestor, or LUCA, an acronym first coined by Patrick Forterre [236]. Remarkably, the conclusion that all life on Earth evolved from a single common ancestor has been presciently reached by Charles Darwin in the Chapter 4 of *The Origin of Species* (see the first epigraph to this chapter), however daring this idea seems to be in the absence of any molecular data. We have dealt with the convergence counter-argument when we discussed the nature of similarities between individual proteins in

Chapter 2, and the implausibility of convergence is only amplified when applied to multiple proteins (and the code itself) shared by all organisms. It is worth noting that the conclusion that all extant life on Earth shares a common ancestor has nothing to do with our specific ideas on the origin of life. In particular, the existence of LUCA is fully compatible with panspermia [359] and even with the somewhat less plausible idea that the first cell had been constructed by an alien genetic engineer. Nor does it exclude multiple origins of life and a primordial (pre-LUCA) diversity of biological systems. All that follows from the impressive repertoire of conserved features shared by the modern organisms is that, at some point, evolution went through a bottleneck, in which LUCA was uniquely positioned to give rise to the entire diversity of life as we know it today.

There are, of course, important, indeed dramatic differences between prokaryotes and eukaryotes, and, within prokaryotes, between bacteria and archaea (♦6.1). Given the firm evidence that all extant life forms evolved from LUCA, it should be possible to study how gene duplication, divergence of functions, gene loss and HGT have shaped the distinct gene repertoire of each major lineage. The goal of evolutionary genomics can be defined as reconstruction of ancestral genomes that existed at different stages of evolution, including LUCA. In ♦6.3, we briefly discussed the algorithmic aspects of this problem. In this section, we touch upon some biological features of the hypothetical ancestors that can be gleaned from comparisons of extant genomes.

6.4.1.1. LUCA and origins of DNA replications

The existence of LUCA seems to be demonstrated beyond reasonable doubt by the unity of many molecular systems of all known cells, above all, the genetic code itself. Given this conclusion and the already considerable collection of sequenced genomes, reconstruction of the LUCA genome inevitably emerges as a fundamental and tantalizing problem. The most naïve approach would simply posit that all genes inherited from LUCA are probably essential and must be represented in all existing life forms. Should that be the case, the task of reconstructing LUCA's gene repertoire would be reduced to the (at least conceptually) trivial problem of finding all universal genes. This approach is quite sensible, but clearly does not work in its simplest form. Indeed, we have already seen, when introducing phyletic patterns, non-orthologous gene displacement and the "minimal genome" concept in Chapter 2, that the number of truly universal genes is almost ridiculously small, only 65 or so. Furthermore, the set of universal genes is strongly functionally skewed: the great majority of these genes encode components of translation machinery, with just a few coding for basal components of the transcription system and molecular chaperones. It is probably indisputable that all universal genes are parts of LUCA's heritage (although, in principle, HGT could have resulted in ubiquitous dissemination of some genes of later origin), but it is equally clear that these proteins alone could not even come close to forming a functional organism. Thus, LUCA encoded many proteins that are not universally present in modern cells. The underlying reasons should be clear from the above discussion (♦6.2 and Chapter 2): firstly, parasites and some free-living heterotrophs undergo substantial loss of genes coding for proteins that are essential in autotrophs and, secondly, extensive non-orthologous gene displacement results in patchy phyletic patterns even for many essential genes. These phenomena, particularly non-orthologous gene displacement, seriously confound the task of reconstructing LUCA's gene repertoire. Only careful biological reasoning, combined with detailed genome comparison, can produce defendable answers to the question, which cases of multiple, non-orthologous solutions for the same function reflect displacement of ancestral proteins that were present in LUCA, and which are later, independent inventions.

A full reconstruction of LUCA's genome is a project far beyond the scope of this chapter. However, a preliminary sketch of what can be inferred of LUCA from comparisons of modern genomes (using the COG database for information on phyletic patterns) is both feasible and instructive. Examination of phyletic patterns suggests that many functional systems of LUCA were only slightly less complex than their counterparts seen in the simplest modern archaea or bacteria. A recent detailed comparative analysis of the entire RNA metabolism machinery suggested that LUCA had not only the basal translation system, but also a considerable repertory of RNA-modifying enzymes, such as methylases and pseudouridine synthases, as well as a rudimentary system for RNA polyadenylation, and the molecular system for translation-coupled protein secretion ([27]; Table 6.2). Although the diversity of RNA modification in LUCA was probably less than that in any modern organism, all principal types of modification were already in place, so that the difference between LUCA and modern life forms seems to be quantitative only.

The classic early concepts of the Origin of Life, starting with Darwin's "little warm pond", put much emphasis on the existence of a primordial (prebiotic) soup rich in all kinds of organic molecules, which allowed the first life forms to enjoy a lavish heterotrophic lifestyle [632]. More recently, this notion has been challenged on the grounds that high concentration of abiogenic organic matter under the conditions of the primitive Earth was

Table 6.2. A tentative reconstruction of LUCA's repertoire of proteins involved in RNA metabolism

Ancient Conserved Families	Activity
RNA modification	
Rossmann-fold methylases 1) KsgA/ERM1/Dim1p methylase 2) HemK methylase 3) MJ0438-like methylase 4) SUN methylase 5) RRMJ (FtsJ-like) methylase, Common ancestor of the Trm2p/ YcbY methylase superfamily	Various RNA methylation and ribosomal protein methylation activities
SPOUT class methylase	Various RNA methylation activities
Thiouridylate synthase (MJ1157-like)	Thiouridylation of tRNA
Pseudouridine synthase type I (TruB)	Pseudouridylation of rRNA and tRNA
Pseudouridine synthase type II (TruA)	Pseudouridylation of rRNA and tRNA
Methylthioadenine synthase (MiaB)	Synthesis of thioadenine derivatives in tRNA
Nucleotide deaminase	Deamination of cytosine, adenine or guanine in RNA
Archaeosine-Queosine Synthase	Synthesis of achaeosine and queuosine in tRNA
Nucleotidyltransferase	PolyA polymerization/CCA addition
CPSF Metallo-β-lactamase fold hydrolase	Cleavage of polyadenylation site
Aminoacyl tRNA synthetases	
Class I Aminoacyl tRNA synthetases: 9 distinct members	Aminoacylation of tRNAs
Class II Aminoacyl tRNA synthetase: 7 distinct members	Aminoacylation of tRNAs
Accessory RNA-binding domains of aminoacyl tRNA synthetases	Recognition of tRNA

Table 6.2 – continued

Translation factors: GTPases 1) YchF 2) OBG/DRG 3) IF2 4) EFG/EF2 5) EF-Tu 6) SelB/EIF2-G 7) YqlF/KRE35p	Various steps of translation initiation, elongation and ribosomal assembly
Non-GTPase translation factors 1) IF-1 / eIF-1A 2) eIF1 / SUI1	Recognition of start codon
Ribosomal proteins 15 families of small subunit proteins 18 families of large subunit proteins	Structural and RNA-binding components of the ribosome
GTPase involved in cotranslational secretion 1) SRP54 2) SR	Part of the Signal Recognition Particle ribonucleoprotein complex and its receptor
RNA polymerase subunits 1) α-(~40K) subunit 2) β-(~140K) subunit 3) β'-(~160k) subunit 4) ω- subunit	Synthesis of RNA using DNA templates
NusG/Spt5	Transcription elongation
Stand-alone Macro-domain protein	Phosphoesterase involved in processing of intermediates in RNA maturation
LigT family 2'-3 phosphoesterase	Phosphoesterase involved in processing of intermediates in RNA maturation
RNAse HII	RNA degradation, mainly of RNA-DNA hybrids in replication
Thermonuclease	RNA Degradation
RNAse PH	3'-5' exonucleolytic RNA degradation
Sm-family protein	RNA-binding in diverse RNP complexes involved in regulatory and RNA processing functions
PIN (PilT Amino terminal domain)	Probable RNA binding domain regulating degradation

highly unlikely [536]. Regardless, even if a primordial soup was available to the very first, primitive life forms, it certainly must have been exhausted by the time an organism with a system for RNA metabolism as complex as outlined here for LUCA has evolved. Therefore, it appears most likely that LUCA was a chemoautotroph resembling, in terms of metabolic capabilities, the simplest modern chemoautotrophs, such as hyperthermophilic archaea (e.g. *Methanococcus*) and bacteria (e.g. *Aquifex*). Thus, LUCA must have had the central metabolic pathways, such as glycolysis and some form of the TCA cycle, as well as the main anabolic pathways, namely those for the biosynthesis of amino acids, nucleotides, lipids and several coenzymes. The corresponding phyletic patterns, a few cases of non-orthologous displacement notwithstanding, are compatible with the notion that all these metabolic pathways in LUCA were equipped with more or less the same set of enzymes as are present in modern organisms, rather than with some radically different primitive forms. Beyond doubt, LUCA had a plasma membrane, into which it inserted proteins. This follows not only from general biological considerations, but also from the universal conservation of the protein components of the signal recognition particle (two paralogous GTPases) and of the SecY protein, which couples translation and secretion in prokaryotes. The rest of the secretory machinery, however, is not conserved and details of this process and the biochemistry of the membrane lipids in LUCA are hard to deduce.

Thus, many functional systems in LUCA might have been nearly as complex as they are in the simplest modern cellular life forms and the organism itself, in many respects, might have resembled modern chemoautotrophic archaea and bacteria. However, the grand mystery that remains is the nature of LUCA's genome and the mode of its replication. In a striking contrast to the other central information processing systems, i.e. translation and transcription, several critical components of the DNA replication machinery are either unrelated or distantly related and apparently not orthologous in archaea-eukaryotes and in bacteria (Table 6.3). Most conspicuously, the apparently unrelated components include the elongating DNA polymerase, the principal replication enzyme. This apparent lack of homology between the DNA replication systems in the two main branches of life has been noticed as soon as the first complete genomes of bacteria and archaea became available and several hypotheses have been proposed by way of explanation [209,237]. One straightforward possibility is that replication components that appear to be unrelated are, in fact, extremely diverged orthologs. However, a detailed analysis of the sequences and structures of the key replication proteins all but refuted this hypothesis. It has been shown, in particular, that archaeo-eukaryotic and bacterial primases have completely different structures with unrelated folds [63,431,676]. Specifically, the structure of the archaeo-eukaryotic primase is distantly related to that of the palm domain that is present in a broad variety of DNA and some RNA polymerases [51]. In contrast, bacterial primases have the so-called TOPRIM catalytic domain shared with several other enzymes (topoisomerases and nucleases) that are not necessarily directly involved in replication [49]. Similarly, the replicative DNA helicases in the two branches appear to have been independently recruited from distinct groups of ATPases, which function in processes other than replication (Table 6.3).

The second explanation for the existence of the two distinct replication systems involves non-orthologous gene displacement and/or differential gene loss. The displacement hypothesis posits that one of the extant replication systems, e.g. the archaeo-eukaryotic version, was present in LUCA and has been displaced in the other branch, in this case bacteria, by the alternative system, whose components perhaps might have been acquired from viruses (bacteriophages) [226,236]. The differential loss scenario postulates that LUCA had two replication systems, one of which might have been used for repair, and that these systems have been differentially eliminated in the two branches of life.

A central aspect of each of these models is that LUCA had a double-stranded DNA genome generally similar to those of modern prokaryotes. In other words, these are continuity hypotheses, which, while postulating displacement or loss of essential genes, picture LUCA essentially as a modern-type cell.

However, these schemes seem to face serious difficulties in interpreting the evolution of the two DNA replication systems. Indeed, it is hard to imagine the evolutionary forces behind the purported displacement of a well-developed, multicomponent replication system. As for the possibility that LUCA had two distinct systems for replication, not only is this without precedent in known organisms, but it would also imply that, at least with respect to replication, LUCA's complexity substantially exceeded that of its descendants, which seems to be an unlikely proposition and poses the problem of the ultimate origin of these systems (but see discussion farther in this section).

Considering these problems, a different, in a sense, more radical proposal regarding LUCA's genome and replication system has been made [504,596]. This hypothesis postulates a sharp discontinuity between LUCA and modern cells, in that the former simply did not have a large double-stranded DNA genome and the system for its replication, which are central to all modern cells. Instead, it is proposed that LUCA's genome consisted of multiple segments of RNA, which replicated via DNA intermediates in a

Table 6.3. Evolutionary relationships between the major components of the DNA replication machinery in archaea-eukaryotes and bacteria

Function	Archaeal/ Eukaryotic protein	Bacterial protein (*E. coli*)	Comment
Apparently unrelated components			
Main replicative polymerase, polymerization domain	B family polymerases	PolIII (DnaE_pol)	No conserved sequence detected despite extensive searches. Structure of PolIII is required to distinguish between lack of homology or a very distant homologous relationship.
Main replicative polymerase, predicted phosphatase domain (subunit)	Calcineurin-type superfamily phosphatase	Predicted PHP superfamily phosphatase (DnaE_PHP)	The PHP domain is predicted to have a TIM-barrel structure unrelated to the calcineurin structure [41]
DNA primase	DNA polymerase α-associated, two-subunit primase	DnaG-type primase	Archaea (but not eukaryotes) encode orthologs of DnaG, but given the presence of the orthologs of the two eukaryotic primase subunits, archaeal DnaG proteins are implicated in repair [54]. The genes for these proteins might have been acquired from bacteria via HGT. The closest eukaryotic homolog is the Toprim domain of topoisomerases.

Table 6.3 – continued

Distantly related, apparently not orthologous components			
Gap-filling DNA polymerase, polymerization domain	DNA polyme rase □ or □	DNA polymerase I (PolA_pol)	Although both polymerase families have the palm domain, the relationship is extremely distant. In contrast, eukaryotic gap-filling polymerase is closely related to other Family B polymerases.
ATPase involved in initiation	ORC1	DnaA	The closest homologs of DnaA are eukaryotic CDC48 ATPases that are involved in cell cycle control, and their archaeal orthologs.
Replicative helicase involved in elongation	MCM	DnaB	Bacterial DnaB helicase evolved from the RecA recombinase [503]. MCM proteins belong to the AAA+ superfamily of ATPases, distantly related to RecA [613].
ssDNA-binding protein	RPA protein (multiple OB-fold domains)	Ssb (OB-fold domain)	These proteins show low similarity detectable only at the structural level.
Main replicative DNA polymerase, 3'-5'-exonuclease domain	3'-5'-exonuclease domain of Family B DNA polymerases	3'-5'-exo domain of polIII	The closest eukaryotic homolog appears to be an RNase involved in splicing.
Gap-filling DNA polymerase, 3'-5'-exonuclease domain	3'-5'-exo domain of Family B DNA polymerases	3'-5'-exo domain of polI	The closest eukaryotic homolog appears to be an RNase involved in splicing.

Table 6.2 – continued

Distantly related orthologous proteins			
Sliding clamp subunit of DNA polymerase	Proliferating cell nuclear antigen (PCNA) and its archaeal orthologs	PolIII β-subunit (DnaN)	Homology was first established via structure comparisons [478], but currently is readily detectable by PSI-BLAST
DNA ligase	ATP-dependent ligase	NAD-dependent ligase	NAD-dependent and ATP-dependent ligases were initially considered unrelated, but homology was established independently by structure and sequence comparisons [43,774].
5'-3' exonuclease (flap nuclease)	Flap nuclease (FEN1, Rad2)	5'-3'-exo domain of PolI	
Highly conserved orthologs			
Clamp-loader ATPase	Replication factor C	PolIII ZX-subunit	
Topoisomerase I/III	Topoisomerase I/III	Topoisomerase I (swivelase)	
Nuclease responsible for removal of RNA primers during lagging DNA strand synthesis	RNAse HII	RNAse HII	
Recombinase	RadA	RecA	

retrovirus-like replication cycle (Figure 6.13). This hypothesis strives to account both for the lack of conservation of several central components of modern replication systems and for the presence of some other conserved components, such as, for example, the sliding clamp, clamp-loader ATPase, and RNAse H, as well as enzymes of DNA precursor biosynthesis, and the basal transcription machinery. The latter group of enzymes, which are ubiquitous in modern life forms, are thought to have been involved in the DNA-based part of the genome replication cycle in LUCA (Figure 6.13; see [504] for details). According to this hypothesis, although many functional systems of modern cells have already evolved in LUCA, the organism itself was not modern, but rather a transitional form on the path from the ancient RNA world to our DNA world.

As discussed above, LUCA must have had at least several hundred protein-coding genes and 30 or so genes for structural RNAs. It appears most likely that the RNA segments of LUCA's genome were of an operon size, i.e. a typical segment carried three to five genes. Comparisons of the gene order in extant bacterial and archaeal genomes show that some operons coding for ribosomal proteins are universally conserved [250,491,916] and (there being no evidence of sweeping HGT of ribosomal protein operons) must have been inherited from LUCA.

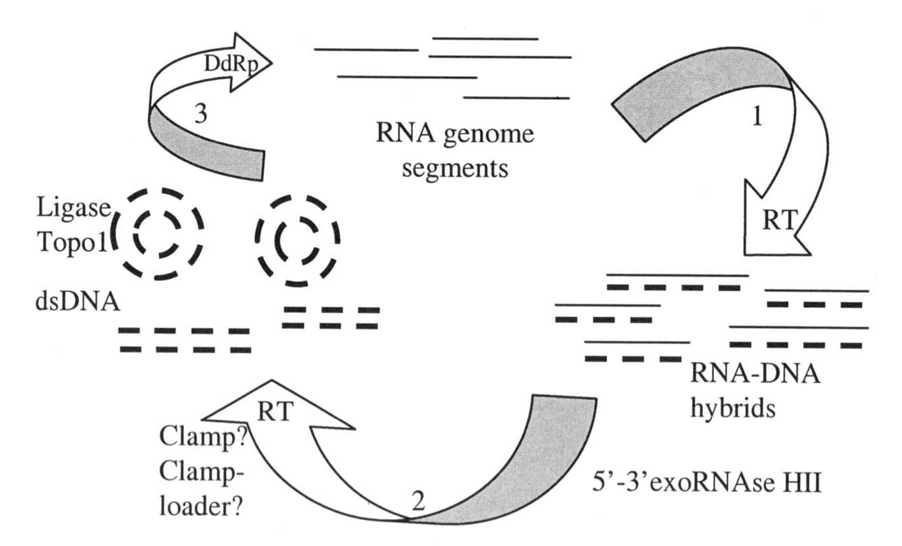

Figure 6.13. The proposed retrovirus-like genome replication cycle of LUCA. 1, reverse transcription; RT, reverse transcriptase; 2, removal of RNA strand, synthesis of second strand of DNA; 3, transcription; DdRp, DNA-dependent RNA polymerase. The points in the cycle where the conserved components of the DNA replication machinery could function are indicated.

Should this be the case, LUCA's genome might have been a collection of 200-400 RNA molecules (and the corresponding DNA intermediates if the scheme in Figure 6.13 is accepted as a working model).

Such a set of genomic segments hardly could segregate with high accuracy into the daughter protocells during division, although miltiploidy could have increased the likelihood that each received the complement of essential genes. Therefore, what we call LUCA inevitably must have been a collection of protocells with similar but not identical sets of genome segments, in a notable agreement with the original notion of the "progenote" and subsequent conceptual developments of this idea [196,900,902]. Furthermore, in such a population, horizontal gene transfer would amount to reassortment of genome segments (not unlike what happens with certain modern viruses, e.g. influenza) and would be extremely common, allowing rapid emergence of new variants and, accordingly, rapid evolution of the LUCA population. However, reassortment could not be allowed to occur without restriction because, should that be the case, natural selection of protocellular entities would have been impossible. It is imaginable that there was a transition during evolution from the early stage of selection at the level of (self-replicating) genome segments, which, in many respects, seem to be quasi-selfish, virus-like entities, to the selection at the level of primitive protocells after reassortment was partially curbed and ensembles of genome segments have become more stable. Within the framework of this concept, the distinction between two hypotheses of origin of DNA genomes and replication, one of which holds that these evolved independently in the bacterial and archaeo-eukaryotic lineages, whereas the other posits that the progenitors of these lineages merely "sorted" among pre-existing replication systems, becomes blurred and perhaps less relevant. Under either of these hypotheses, however, a replication system similar to the one shown in Figure 6.13 seems to be a likely predecessor of the modern DNA replication systems.

In the parlance of evolutionary biology, the crucial transition in the early evolution of cells discussed above could be construed as change of selection agency from replicators (self-replicating, virus-like segments of RNA) to interactors (protocells) [304]. Carl Woese, in his recent theoretical treatise "On the evolution of cells", designated a similar transition the "Darwinian threshold", meaning that speciation has become possible at this point in evolution [904]. This concept seems to be quite compatible with the above notions, but the name could be somewhat disingenuous, in its implication that evolution was "non-Darwinian" before the threshold. The essence of Darwinian evolution is natural selection and it was definitely in place from the earliest stages of biological evolution, as soon as the first replicating and mutating entities emerged, even if the principal agency of selection changed along the way.

6.4.2. Beyond LUCA, back to the RNA world

The ideas of LUCA's gene repertoire and the nature of its genome discussed in the previous section seem to fit within another staple of Carl Woese's general concept of early evolution, the notion of "asynchronous crystallization" of different cellular systems [900]. More specifically, we address questions of the following type: given that LUCA is confidently predicted to have had a well-developed translation system and certain metabolic capabilities, what can be inferred regarding the nature of its genome and replication system? The line of reasoning developed above, albeit not definitive, suggests that the relatively advanced stage of evolution reached by LUCA could still lie on the boundary between the ancient RNA-protein world and the DNA world that replaced it. More generally, this transition is an inevitable step in the evolution of life under any model that accepts the notion of a primordial RNA world. The question then emerges: can we look beyond that transition phase, which perhaps coincides with LUCA, and trace some of the steps of evolution within the RNA world itself?

Even without a full reconstruction of LUCA's gene repertoire, evidence of complex evolution of proteins at earlier epochs is unequivocal. Clearly, paralogous protein families that can be inferred to have been present in LUCA are footprints of ancient, pre-LUCA duplication events. This fact was used in the classic work on paralogous rooting of phylogenetic trees that gave early support to the standard model of evolution [291,386]. For example, within the class of P-loop NTPases, which are the most common protein domains, at least in prokaryotes (see Chapter 8 for some details), perhaps as many as 50 distinct families appear to go back to LUCA, which, in all likelihood, encoded the ancestor of each of them. This includes about 10 families of GTPases alone [505] and about the same number of helicase families [27]. The topology of the evolutionary tree of the P-loop NTPases is fairly well resolved and the presence of many diverse families in LUCA implies that the preceding history involved a long series of duplications, followed by divergence, between the ultimate ancestor of this protein class and the LUCA stage. The same logic applies to other widespread protein domains, such as, for example, Rossmann-fold oxidoreductases or methyltransferases, each of which was represented by multiple paralogs in LUCA [51].

Analysis of each of these domain classes gives us insights into very early stages of life's evolution, but some families are particularly important for aligning protein domain diversification with the timeline of the transition from the primordial RNA world to the modern-type cells. Consider the evolution of aminoacyl-tRNA synthetases (aaRS), crucial enzymes of the modern translation system [376,907,909]. There are 20 specific aaRS altogether, one for each amino acid (eukaryotes have all 20, whereas most bacteria and archaea encode only 17 to 19 because they have specialized mechanisms for incorporation of glutamine and asparagine that do not involve specialized aaRS and archaeal methanogens also lack CysRS [376,907]). This (nearly) ubiquitous presence of aaRS indicates that at least 18 of them were already encoded in LUCA's genome. All aaRS are complex, multidomain proteins, but the principal catalytic domains, which are directly responsible for the formation of aminoacyl adenylates followed by coupling of amino acids to their cognate tRNAs, fall into two distinct classes, each including 10 amino acid specificities. The catalytic domains of all aaRS within each of the classes are homologous, even if some of them show only low sequence conservation, whereas the two classes are unrelated as indicated by sequence and structure comparisons. Since LUCA already had most of the aaRS of both classes, the series of duplications that led from the ancestor of each class to the complete sets of aaRS dates to earlier, pre-LUCA stages of evolution (Figure 6.14). Even more dramatically, each of the catalytic domains of the aaRS is related to several other families of nucleotide-binding domains. Figure 6.14 shows the evolutionary scenario for the HUP domain, so named after the HIGH-domains, UspA and PP-loop NTPases, three protein superfamilies that have this type of nucleotide-binding domain [37]. The HIGH superfamily, designated after the distinct signature in the phosphate-binding loop, unites the catalytic domain of class I aaRS and a family of nucleotidyltransferases, primarily involved in coenzyme biosynthesis, whose relationship with the aaRS is readily demonstrable at the sequence level [101]. The connections between the HIGH superfamily and the rest of the HUP domains were detected primarily through structure comparisons, but the hierarchy of the observed relationships was sufficiently well resolved to infer the tree topology [37]. In this tree, the series of duplication leading to the emergence of the individual aaRS specificities is but a terminal elaboration, which was preceded by multiple ancient duplications (Figure 6.14). The most remarkable aspect of the evolution of the HUP domain class is that multiple HUP-containing proteins, namely aaRS of 9 specificities and tRNA thiouridine synthases (members of the PP-loop ATPase superfamily), are indispensable for translation in its modern form. Since a major diversification of the HUP domains preceded the radiation of the aaRS and the PP-loop ATPases, the conclusion becomes inevitable that ancestors of many modern protein families with well-defined structural and sequence features and distinct biochemical functions, e.g. the ancestral PP-ATPase, HIGH, and USPA domains, ***antedate the modern translation apparatus***. The above analysis leads us to conclude that series of duplications followed by diversification of each of the above domains occurred within a system, in which the specificity of translation was determined not by aaRS as in modern cells, but by RNAs. An important corollary is that such a primitive, partially RNA-based translation system was sufficiently accurate and efficient to allow complex protein evolution. It is easy to argue that this primitive, largely RNA-based translation apparatus should be classified as part of the RNA world.

Very similar conclusions could be reached through analysis of the catalytic domains of Class II aaRS, which are related to the nucleotidyltransferase domain of biotin synthases [59], as well as translation factors. In the modern translation system, the latter are represented by several paralogous GTPases, each of which is indispensable for efficient and accurate translation (Table 6.2). Inevitably, the divergence of these GTPases must have occurred within the confines of a more primitive, at least partly RNA-based translation system [27,505]. It might be harder to directly link the evolution of other protein classes to the evolution of the translation system but, on the whole, there is no doubt that the emergence of most of the major protein folds and the diversification of dozens of protein families within them are evolutionary events that map to the RNA world stage. Strikingly, it might not be an exaggeration to say that the most important stage of protein evolution, the formation of the majority of widespread protein folds, had already concluded by the time of the final "crystallization" of the translation system. Of course, from a purely logical point of view, this conclusion is almost trivial. In retrospect, it seems obvious that, in order to replace an old toolkit with a new one, the new tools had to be made using the old ones.

These reconstructions of the initial radiation of protein classes also give us, with considerable clarity, the teleology of early protein evolution. Consider, once again, the HUP class of nucleotide-binding domains. What becomes apparent when one descends down the tree toward the root is the gradual loss of specificity (Figure 6.14). Indeed, the common ancestor of the HUP class, which, in all likelihood, was an ATP-binding domain with a generic ATP-pyrophosphatase and/or nucleotidyltransferase activity, could not perform the multiple, specific functions assumed by its descendants. In particular, it was impossible for this generic ATP-hydrolase to mediate processes that require highly specific molecular interactions, such as tRNA aminoacylation or thiouridylation of specific bases in tRNA. The solution to this problem, already mentioned above, is that, in the primitive biochemical system, many functions, particularly those related to translation, that are performed by proteins in modern cells, relied on RNA molecules, including the ancestors of rRNA and tRNA (to our knowledge, this idea was first explicated by Francis Crick in 1968 [162]). As shown above, at this early stage of evolution, proteins already had catalytic and ligand-binding capabilities, which suggests that the RNAs were mainly responsible for the specificity. In particular, the ancestor of the HUP class probably interacted non-specifically with proto-tRNAs, thus facilitating aminoacylation, whereas

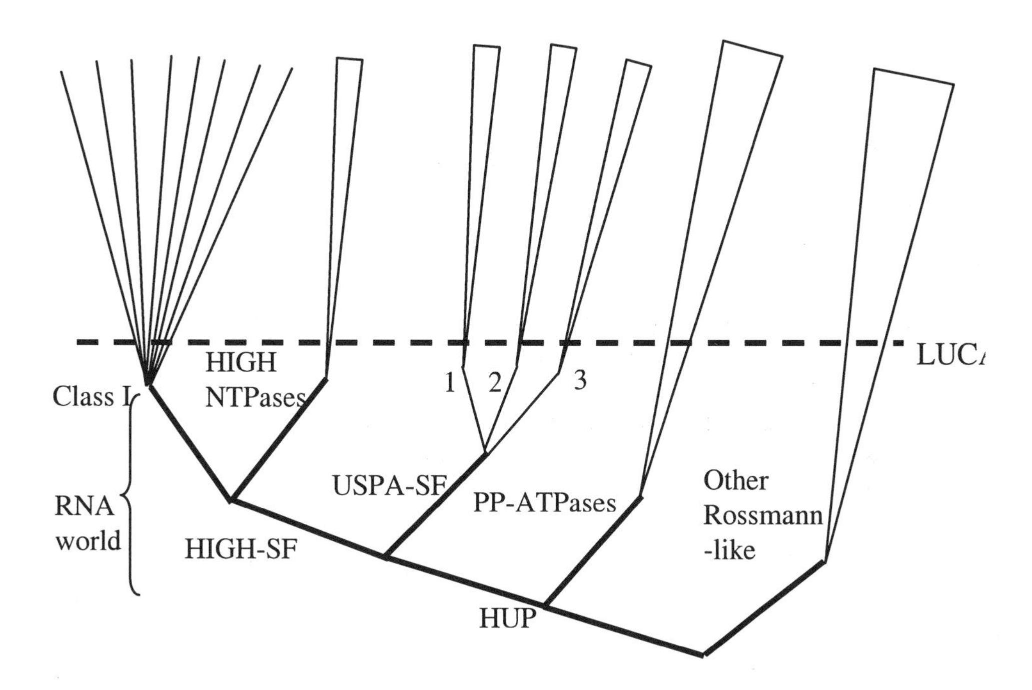

Figure 6.14. A schematic tree of the evolution of the HUP domain class.
The tree was derived through a combination of sequence comparison and cladistic analysis of structural features. As discussed in the text, diversification of aaRS predates LUCA (the LUCA stage is shown by the broken horizontal line). PP-ATPases, a superfamily of ATP-pyrophosphatases, many of which are involved in coenzyme biosynthesis and also in RNA modification [27,102]. USPA-SF, superfamily of nucleotide-binding domains named after the bacterial Universal Stress Protein A; 1, Electron transfer flavoprotein (ETFP) family, 2, DNA photolyase family, 3, USPA family. HIGH-SF, superfamily of ATP pyrophosphatase domains named after the diagnostic signatue in the phosphate-binding loop of HIGH NTPases and class I aaRS.

the specificity of these reactions was conferred bythe cognate tRNA itself, perhaps with an additional contribution from other, accessory RNA molecules. In support of this scenario, specific self-aminoacylation catalyzed by an RNA molecule has been demonstrated experimentally [377]. The same ancestral HUP domain probably functioned, in cooperation with other RNAs, to facilitate additional reactions, which require ATP hydrolysis or nucleotide transfer, including RNA modification and cofactor biosynthesis. After the first duplication that separated the two main branches of the HUP class (Figure 6.14), the ancestor of the HIGH-USPA lineage probably took over the function of a generic nucleotidyltransferase involved in translation and cofactor biosynthesis, whereas the ancestor of the PP-loop ATPases became a generic ATPase involved in RNA modification and some metabolic functions.

6.4.3. A brief history of early life

Generalizing from these analyses, we can now sketch the path from the RNA world to the RNA-protein world to the DNA-RNA-protein world of LUCA (Figure 6.15). Proteins most likely started off as RNA-binding and nucleotide-binding cofactors for the primordial RNA catalysts. Notable relics of this ancient stage of evolution are seen in modern cellular systems, including RNA-protein enzymes, such as RNAse P and, above all, the ribosome itself [143,294,620]. In these systems, to this day, RNA functions as the principal catalyst (ribozyme), whereas RNA-binding proteins are cofactors that stabilize the ribozyme and facilitate the reaction. Further evolution of these primordial proteins led, first, to the origin of generic, non-specific enzymes, such as NTPases, nucleotidyltransferases that gradually became the first polymerases, and nucleases and, subsequently, through series of duplications, to the emergence of specificity. The structural similarity between the palm domain of DNA and RNA polymerases and the RRM-fold RNA-binding domain [51], which might point to the route of origin of first polymerases, supports the vector of evolution from RNA-binding protein cofactors to enzymes.

It appears that a well-developed RNA-protein world with numerous, specific protein enzymes and some protein regulators was a distinct stage in the evolution of life. Under this view, DNA was introduced as an alternative means for storage of genetic information and subsequently as the principal chemical substrate of the genome only at a relatively late stage, after the translation mechanisms and the principal metabolic pathways have been firmly established.

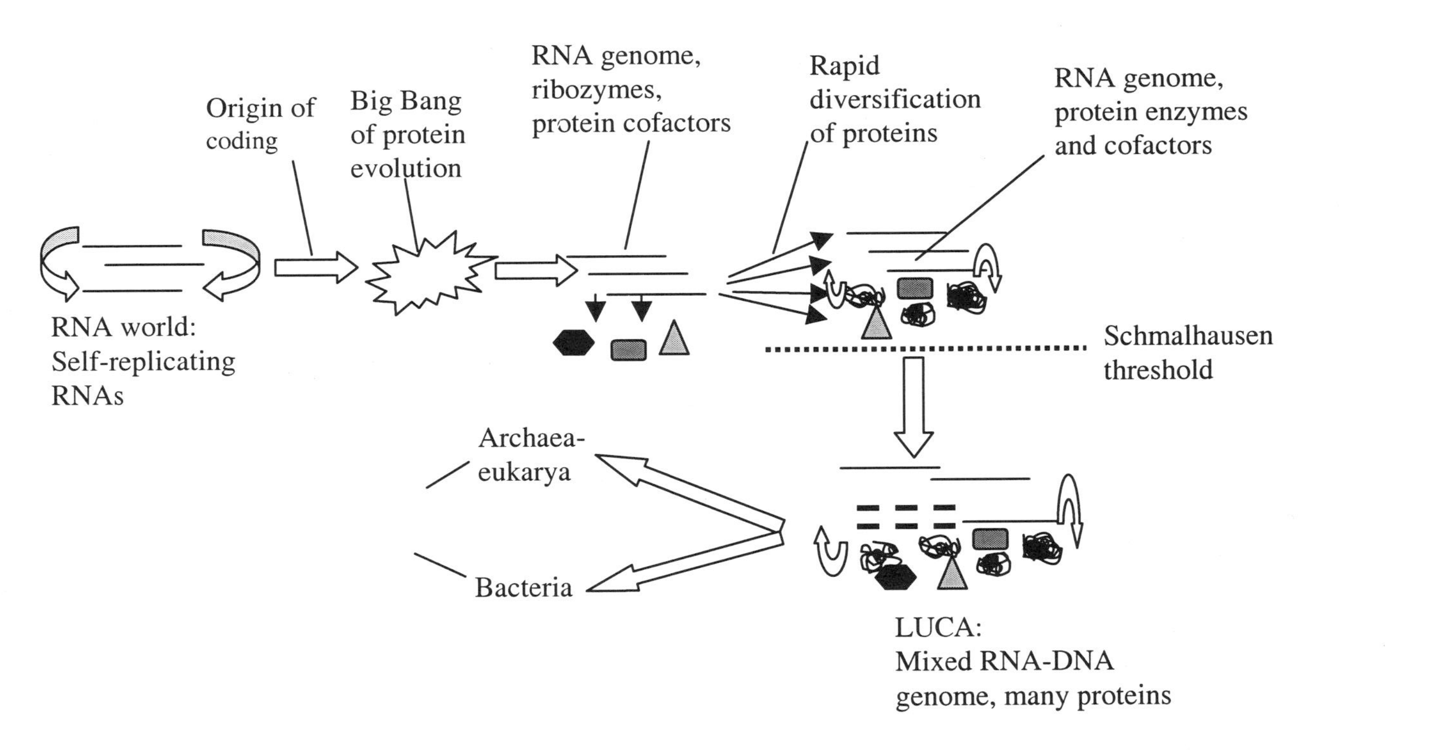

Figure 6.15. A hypothetical sequence of major events in the evolution of life from self-replicating RNA to the emergence of modern-type DNA replication.

A scheme like the one in Figure 6.15 could be easily drawn from first principles at the time when the very idea of a primordial RNA world took shape, i.e. in the late 1960's [162]. However, comparative genomics makes a major difference by providing specific support for each of these hypothetical steps and giving us a fairly clear idea of the protein forms that functioned in these ancient worlds.

LUCA's existence is usually dated at 3.5-3.8 billion of years ago because, by then, diverse prokaryotes seem to have already evolved as indicated by microfossil evidence [753,754], although later dating has been proposed [189]. This leaves perhaps 500 million years between the time the Earth cooled down enough to allow RNA and protein chemistry and the emergence of LUCA. How could it be that, as outlined above, much more dramatic changes in the protein world took place in this relatively short early era than during all subsequent evolution (the logic here will not change much if LUCA is moved, say, 500 million years closer to us)? The bewilderment over this question might lead to the believe that there was simply not enough time for life to emerge on Earth and that the only explanation (barring intelligent design, of course) for its presence is panspermia, seeding of our planet with simple life forms from outer space [359].

While there is nothing intrinsically impossible about panspermia, we believe that, in the absence of direct supporting evidence, this hypothesis fails the Okkam test. Instead, the general solution to the paradox seems to be that the modes of protein evolution radically differed during the early period when the major superfamilies were forming (as discussed above, it appears certain that this happened within the framework of the RNA world, with most catalytic functions performed by RNA molecules) and during the subsequent evolution, after the major protein folds have "crystallized" (Figure 6.15). The latter phase of evolution was and still is dominated by purifying selection and had been generally characterized by a relatively low rate of amino acid replacement; although the classic molecular clock seems to be a gross over-simplification even for this stage of evolution, "soft", genomic molecular clock probably applies (♦6.3). Certainly, this conservative course of life's evolution was punctuated by numerous bursts of positive selection, which were associated with the birth of new protein superfamilies and even new folds. However, the bulk of evolution in the 3.5 billion years or so since LUCA occurred during periods of stasis, i.e. gradual functional adaptation without changes in the basic structure and biochemical activity of proteins (we deliberately adopt the language of the theory of punctuated equilibrium [304], which seems to be a good, if not necessarily precise simile for a description of the fundamentals of molecular evolution). In contrast, the early phase of evolution, during which the major protein folds have emerged in a semblance of biological Big Bang (Figure 6.15), most likely was dominated by positive selection, which caused rapid change of the sequences. Mechanistically, this mode of evolution seems to be compatible with the high error rate of RNA replication, which reaches 10^{-3} even in modern viruses with their evolved RNA-dependent RNA polymerases [882,883,888], and was probably even greater for the early RNA or RNA-protein polymerases. We may call this transition from positive selection to stabilizing selection, arguably one of the pivotal events in the evolution of life, the "Schmalhausen threshold", after the eminent Russian evolutionist who was the first to introduce, in the 1930's, the crucial concept of stabilizing (purifying) selection [750]. It is interesting to note that this view of the earliest phases of evolution is decidedly at odds with Darwin's opinion that "...at the very first dawn of life, when very forms of the simplest structure existed, the rate of change may have been slow in an extreme degree" [170].

6.4.4. The prokaryote-eukaryote transition and origin of novelty in eukaryotes

6.4.4.1. The nature of the transition and origins of eukaryotic genes

As emphasized repeatedly and ever so eloquently by Stephen Jay Gould (e.g. [303]), the iconography of progress from inconsequential prokaryotes to spectacular bipedals, which is commonplace in biology textbooks and popular literature, is demonstrably false if construed as an objective depiction of the evolution of the biosphere on this planet. Indeed, whether we consider the energy flow in the biosphere, the spread of life over diverse habitats, including extreme ones, or the diversity of species themselves, the entire history of life is definitely the age of prokaryotes, which will continue until the ultimate demise of the earthly life. It is equally undeniable, however, that this iconography does represent something important and profoundly interesting in the evolution of life, namely the tremendous increase in the ***maximum organizational complexity*** of life forms observed at a given time. Arguably, the single most important leap in the direction of greater complexity was the origin of the eukaryotic cell. Indeed, the simplest eukaryotic cell is considerably more complex than even the most advanced prokaryote [11]. To begin with, the eukaryotic cell is much larger, on average by about a factor of 1,000 (in volume). Furthermore, the eukaryotic cell has several distinct organelles and intracellular structures without counterparts in prokaryotes, the most important of which are the nucleus, the mitochondria, the endoplasmic reticulum, and the cytoskeleton. There are no candidates for evolutionary intermediates between prokaryotic and eukaryotic cells, which suggests a true "phase transition", a distinct, dramatic evolutionary event leading to the emergence of eukaryotes. The nature of this event is one of the greatest mysteries of life's history.

Rather unexpectedly, recent studies on early-branching eukaryotes have suggested a strong candidate for this event: the mitochondrial endosymbiosis itself. The very idea that endosymbiosis (although not necessarily mitochondrial) underlies the very emergence of eukaryotes has been proposed by Lynn Margulis in her pioneering 1967 paper that reinstated the endosymbiotic theory [727]. This idea, however, had not won wide acceptance, and the field of eukaryotic evolutionary biology was dominated by the Archezoan hypothesis, which contends that the host of the promitochondrial endosymbiont was an amitochondrial unicellular organism that already possessed all the main features of the eukaryotic cell, the hypothetical archezoan [140,659,707]. However, the search for an archezoan branch of eukaryotes so far has been futile. Although numerous protists, such as, for example, *Giardia* or *Trichomonas*, that live under anaerobic conditions, lack regular mitochondria, most of them have hydrogenosomes, small membranous organelles that metabolize malate and pyruvate to produce ATP, H_2 and CO_2. Typical hydrogenosomes have no genome or translation system, but are composed of proteins, which, as shown by phylogenetic analysis, have been derived from the mitochondria and are encoded by former mitochondrial genes transferred to the nuclear genome [10,90,127,203,204,356,713]. Phylogenetic studies on protists (an incredibly fascinating field in itself, but the details of which are beyond the scope of this book) indicated that, in early eukaryotic evolution, secondary loss of mitochondria was a common event, which probably occurred in at least 15 lineages [707.] The hunt for an archezoan continues, but with each failure to discover one, the odds are increasing that the very origin of the eukaryotic cell was triggered by the acquisition of the α-proteobacterial symbiont that became the progenitor of the mitochondrion. The second partner, the recipient, in this event remains an enigma. Certainly, much evidence points to an archaeon (♦6.1), but whether or not this primary ancestor of eukaryotes belonged to one of the currently known archaeal lineages remains unclear.

Even if we accept that endosymbiosis triggered the emergence of the first eukaryotic cell, the nature of the transition is a great mystery. Since the cytoskeleton is one of the critical parts of the eukaryotic cell and, more specifically, the part that allows the eukaryotic cell to maintain its characteristic large size, the unusual evolutionary patterns seen among the principal cytoskeletal proteins might be directly linked to the transition. Actin, the most abundant eukaryotic cytoskeletal protein, is a highly derived ortholog of MreB and FtsA, bacterial proteins involved in cell division.

These proteins are present in almost all bacteria, but only in a few scattered archaeal species, which probably acquired the respective genes via HGT. Actin and MreB/FtsA show very low sequence similarity to each other but share all diagnostic motifs of the HSP70-like ATPase-fold ATPases [106], have very similar structures and form similar filaments [210,864,865]. The absence of these proteins in most archaea suggests that, in eukaryotes, they are an ancient bacterial acquisition, although the alternative, namely that these proteins were already present in LUCA (and then, presumably, lost in archaea), also has been considered [191]. The unusually low sequence conservation between MreB/FtsA and actin, in all likelihood, reflects the abrupt change in

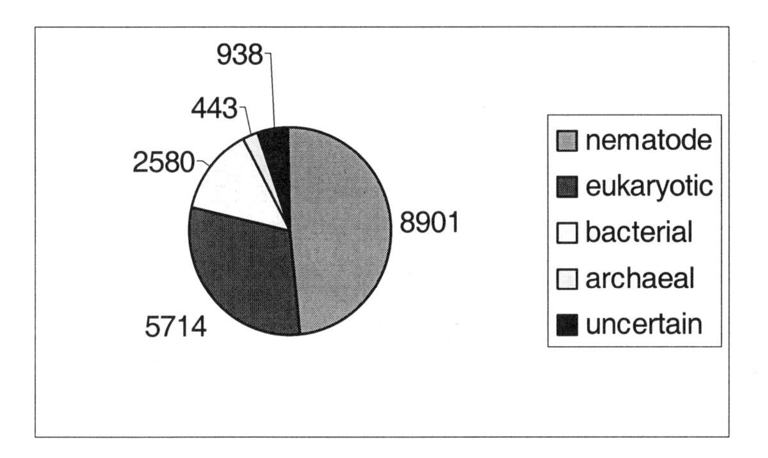

Figure 6.16. Taxonomic breakdown of the BLAST search results for the proteins of the nematode *C. elegans*. 'Nematode' (-specific) are proteins that did not show significant sequence similarity ($E < 10^{-3}$) to proteins from any species other than the nematode itself; 'eukaryotic' are proteins that have detectable (same criterion) homologs only in other eukaryotes, 'bacterial' are proteins that had bacterial homologs and either had no detectable archaeal homologs or had archaeal homologs with much lower similarity (the difference between the bacterial and archaeal homolog in E-values had to be at least 10 orders of magnitude), 'archaeal', similarly defined proteins of apparent archaeal provenance, 'uncertain', proteins with both archaeal and bacterial homologs detectable with comparable levels of similarity. The data for this figure were obtained before the complete fruit fly and human genomes became available, so the number of nematode-specific proteins is an over-estimate; some of these proteins should move to the 'eukaryotic' category. However, the relationships with prokaryotic homologs should not be affected.

Table 6.4. Extreme divergence between some essential eukaryotic proteins and their apparent prokaryotic orthologs

Eukaryotic protein, function	Bacterial ortholog, function. Phyletic pattern	Archaeal ortholog, function. Phyletic pattern	Probable evolutionary scenario	Domain organization, other comments and references
Actin, major cytoskeleton component	FtsA/MreB, cell division. All except mycobacteria and *Synechocystis*	MreB, cell division. Only *Archaeoglobus*, two methanogens and thermoplasmas	HGT from an endosymbiont to protoeukaryotes	HSP70-like ATPases [106,210,864,865]
Tubulin, major cytoskeleton component	FtsZ, cell division. All except chlamydia and *Ureaplasma*	FtsZ, cell division. All Euryarchaeota	Vertical inheritance from LUCA according to the standard model	Distinct GTPase domain [214,215,621]
LC7, small subunit of the dynein motor	MglB, implicated in GTPase regulation for signal transduction and gliding motility. Sporadic distribution.	MglB, implicated in GTPase regulation for signal transduction. Sporadic distribution.	Possible origin in the common archaeo-eukaryotic ancestor; additional HGTs.	[455]
Separase (separin), protease responsible for chromatid separation in mitosis and meiosis	HetF, heterocyst formation regulator in Cyanobacteria; Cyanobacteria, *Streptomyces*, *Caulobacter*	None	HGT from an endosymbiont to protoeukaryotes	The caspase-hemoglobinase protease domain [47,860]

Ddi1, predicted aspartic protease, cell cycle control	Uncharacterized protein in alpha-proteobacteria	None	HGT from promitochondrial endosymbiont	Might be the progenitor of retroviral proteases [87,475]
Ubiquitin, the central mediator of controlled protein degradation	MoaD, ThiS, Molybdenum cofactor and thiamin biosynthesis proteins. Most bacteria except for some parasites	MoaD, ThiS. All except *Aeropyrum*	Vertical inheritance from LUCA according to the standard model	[486]
DNA polymerase small subunit	None	All euryarchaeota	Vertical inheritance from archaeal ancestor	The archaeal ortholog is predicted to be an active phosphatase; the eukaryotic protein is inactivated leading to low sequence conservation [41]
ERCC4, the nuclease required for DNA incision in various forms of repair	None	All euryarchaeota	Vertical inheritance from archaeal ancestor	The archaeal ortholog is predicted to be an active helicase, but the eukaryotic protein is inactivated leading to low sequence conservation [54]

functional constraints that accompanied exaptation (i.e. recruitment of a protein for an entirely new function, which may be mechanistically but not biologically related to the original one [302]) of the bacterial cell division protein for functioning in the emerging cytoskeleton. In a striking parallel, tubulin, the other principal cytoskeleton protein of eukaryotes, also diverged from the prokaryotic ortholog, the cell division protein FtsZ, to the point of near obliteration of sequence similarity (except for the distinct phosphate-binding loop of the GTPase domain), although the structural conservation remains obvious [214,215,621]. This theme can be extended to the large subunit of dynein, a major motor ATPase, which is an extremely diverged AAA+ class ATPase whose prokaryotic progenitor could not be readily identified because of this divergence [613]. Furthermore, a similar trend is seen with one of the small dynein subunits, LC7, whose distant prokaryotic homologs appear to be involved in gliding motility and signal transduction in bacteria and archaea [455]. These dramatic changes in protein sequences that accompanied the origin of eukaryotic cytoskeleton might have been critical in the evolution of the eukaryotic cell. In particular, it seems possible that mutations that resulted in altered properties of the MreB/FtsA protein acquired from the bacterial endosymbiont initiated the entire process of transformation of the ancestral prokaryotic cell into the more complex eukaryotic cell.

Sequence similarity between actin, tubulin, LC7 and their respective prokaryotic orthologs is so weak, presumably due to a radical change in function, that these relationships escape detection in routine genome comparisons. The discovery of these crucial evolutionary connections became possible only as a result of systematic analyses of the sequences of the respective protein families or through direct comparison of protein structures (whereas each of these proteins is highly conserved within prokaryotes and eukaryotes). Case by case sequence and structural studies revealed several additional essential eukaryotic proteins that showed extreme divergence from their apparent evolutionary precursors in prokaryotes (Table 6.4). The fact that several proteins conserved in all eukaryotes and involved in quintessential eukaryotic functions, such as cytoskeleton formation (actin), chromosome separation in mitosis and meiosis (separase), and cell cycle control (Ddi1), show a bacterial provenance and might have been acquired by eukaryotes via HGT emphasizes the potential role of endosymbiosis in the very origin of eukaryotes. A further, focused search for potential prokaryotic counterparts of essential proteins that so far seem to be eukaryote-specific might help in revealing more about eukaryotic origins.

Genome-wide analyses of eukaryotic proteins and studies on evolution of specific functional systems converge on the notion that the eukaryotic proteome is a mix of proteins of apparent archaeal descent, those that seem to originate from bacteria, and eukaryote-specific ones. This is readily illustrated by the taxonomic breakdown of the database search results for proteins encoded in any eukaryotic genome as shown in Figure 6.16 for the nematode *C. elegans* [913]. The eukaryote-specific category is the largest, but, among those proteins that do have prokaryotic homologs, the "bacterial" group dominates. Similar conclusions can be reached by breaking down the COGs by domain-specific phyletic patterns: the number of COGs that consist of bacterial and eukaryotic proteins, to the exclusion of archaea, is more than four-fold that of archaeo-eukaryotic COGs (Figure 6.1).

The most obvious explanation of the presence of a large number of 'bacterial' proteins in eukaryotes is a massive acquisition of genes from the mitochondrial endosymbiont and, in the case of plants, from the chloroplast endosymbiont. Phylogenetic studies of mitochondrial rRNA and proteins have shown beyond reasonable doubt that mitochondria evolved from alpha-proteobacteria [30,313,489]. However, in phylogenetic analyses of eukaryotic proteins encoded in the nuclear genome, only a small number confidently cluster with the alpha-proteobacterial orthologs ([420,480]; K.S. Makarova, M.V. Omelchenko and EVK, unpublished observations). The legacy of the more recent chloroplast symbiosis in plants is easier to detect, but even in this case, only 63 unequivocal plant-cyanobacterial clades out of the 386 examined phylogenies of genes shared by the cyanobacterium *Synechocystis* and the plant *Arabidopsis* were reported [720]. One cause of the problems with proving the influx of genes from endosymbionts is likely to be the accelerated evolution of the acquired eukaryotic genes, perhaps a burst of rapid evolution immediately after gene transfer from the symbiont to the nucleus. Indeed, it is often observed that proteins functioning in the mitochondria cluster within the bacterial branch in phylogenetic trees, but the specific affiliation with alpha-proteobacteria is seen only in the minority of cases; aminoacyl-tRNA synthetases are a good illustration of this trend [909]. This being the case, application (and further refinement) of phylogenetic methods that are minimally sensitive to changing evolutionary rates could clarify the situation.

Another, intriguing explanation that may be called Doolittle's ratchet (after W. Ford Doolittle who proposed this idea in a brief article with the memorable title "You are what you eat" [192]), is that mitochondrial (and to a lesser extent chloroplast) affinity is hard to demonstrate for most of the "bacterial" proteins from eukaryotes simply because they largely do not originate from the mitochondrial ancestor. Under this hypothesis, protoeukaryotes (whatever the nature of these organisms) formed numerous symbiotic relationships with bacteria, the pro-mitochondrial symbiont just being the only one that stayed on, became indispensable for the host and had been passed on to most eukaryotic lineages. The less lucky symbionts might have ended up as food for their host but, in the process, some of them probably contributed genes to the host genome, resulting eventually in the eukaryotic genome becoming "what the protoeukaryotes ate". This idea seems to be compatible both with the behavior of many eukaryotic proteins in phylogenetic studies and with the observations of multiple losses of mitochondria in unicellular eukaryotes (see above).

The characteristic functional distinction between eukaryotic genes that have been vertically inherited according to the standard model and those that seem to have been acquired from bacteria follows from the nature of the shared archaeo-eukaryotic heritage that we already discussed in ♦6.1. The archaeal legacy primarily consists of genes coding for information processing system components (translation, transcription and replication), whereas metabolic enzymes and transporters (what is sometimes collectively called operational systems) seem to be largely of bacterial origin. The core of the DNA replication machinery, which we discussed when examining the probable nature of LUCA, is the most dramatic manifestation of this dichotomy, with eukaryotes using the archaeal system whose key components are unrelated to the bacterial functional counterparts. The distinction, however, is by no means absolute. Thus, detailed computational dissection of the proteins involved in various aspects of DNA repair [54] and RNA metabolism [27] showed that, along with the ancient archaeal core, the eukaryotic systems include a considerable number of proteins that apparently have been acquired from bacteria (see also Table 6.4).

At least three distinct scenarios of bacterial gene integration into the protoeukaryotic genome should be differentiated: (i) displacement of ancestral, archaeo-eukaryotic genes by bacterial counterparts (xenologous gene displacement, see ♦6.2), (ii) acquisition of bacterial genes without elimination of the ancestral archaeo-eukaryotic counterparts so that eukaryotes end up with both versions of a particular protein, and (iii) evolution of new functions by utilization of bacterial proteins (exaptation). The first scenario applies primarily to essential metabolic enzymes whose single eukaryotic forms appear to be of bacterial origin. The most interesting question regarding this displacement phenomenon is what was its underlying cause or, in other words, what was the selective advantage conferred by the acquired bacterial genes onto the recipient that led to displacement of the ancestral versions. One explanation is that bacterial enzymes were generally more efficient than the original archaeo-eukaryotic ones and were selected for that reason. The alternative is Doolittle's ratchet: there might not have been any selective advantage in replacing ancestral archaeo-eukaryotic genes with bacterial ones, but rather plenty of opportunities for this to happen through symbiotic relationships between protoeukaryotes and various bacteria.

The second scenario, acquisition without displacement, applies largely to universal proteins whose distinct forms function in the eukaryotic cytosol and in the mitochondria, particularly translation system components, e.g. the aaRS. Such cases are probably the best available demonstrations of probable gene transfer from the promitochondrial endosymbiont to the nuclear genome, even in cases when phylogenetic trees do not show the alpha-proteobacterial affinity of the respective eukaryotic proteins.

Eukaryotic innovations that appear to have evolved as a result of acquisition of bacterial genes are of special interest because systematic exploration of these cases may shed light on the origins of the unique eukaryotic complexity (e.g. Table 6.4 and the relevant discussion above). Such studies have become truly meaningful only after multiple genomes of both bacteria and eukaryotes have been sequenced, so as of this writing, there simply had not been enough time to conduct the comparative studies that are required to characterize the evolutionary provenance of all unique eukaryotic functional systems. However, below we discuss one example of such a study that has been completed and revealed both prominent bacterial connections and truly unique eukaryotic inventions.

6.4.2.2. Origin and evolution of eukaryotic programmed cell death

The molecular machinery of programmed cell death (PCD, or apoptosis) appears to be a quintessential eukaryotic signaling system. Bacterial cells are known to "commit suicide" under certain circumstances, but the molecular mechanisms of these processes so far identified seem to be unrelated to those of eukaryotic PCD [512,932]. In contrast, PCD appears to be ubiquitous in multicellular eukaryotes and, indeed, should be considered one of the hallmarks of the multicellular state [778]. In any multicellular organism, PCD is indispensable for eliminating cells with impaired division control whose propagation leads to cancer and cells infected with pathogens or damaged by stress [340,701,808,935]. Programmed death of specific cells is also important in normal development [564].

The phyletic patterns for the key components of the eukaryotic PCD system show a clear 'meta-pattern': the enzymes involved in apoptosis tend to have a broad phyletic distribution, with bacterial homologs identifiable, whereas the non-enzymatic components typically have no bacterial homologs and, in some cases, are present in only one eukaryotic lineage (Table 6.5). With the single notable exception of the Apoptosis-Inducing Factor (AIF), the prokaryotic homologs of the proteins involved in PCD are widely represented in bacteria, but not in archaea. This pattern suggests a substantial, perhaps decisive contribution of acquired bacterial genes to the evolution of the eukaryotic PCD system. It is instructive to further assess this hypothesis through a more detailed examination of the bacterial homologs of the eukaryotic proteins involved in PCD and phylogenetic analysis of the respective protein families [38,39,457].

The caspase superfamily proteases. Caspases are the principal proteases that are activated during animal apoptosis and cleave a variety of proteins, ultimately leading to cell death and disintegration [808,845]. The caspases have undergone remarkable proliferation and specialization in vertebrates and function in a cascade, which includes several consecutive cleavage steps. Caspases belong to a distinct class of cysteine proteases (we designate it the CHF-class, after Caspase-Hemoglobinase Fold), which also includes hemoglobinases, gingipains, clostripains and separases, the proteases involved in chromosome segregation in mitosis and meiosis (see also Table 6.4 and discussion above). Recent sequence and structure analyses revealed a much greater diversity of caspase-related proteases than previously suspected [47]. Two families of predicted CHF-proteases were identified and shown to be more closely related to the caspases than to other proteases of this class and hence dubbed paracaspases and metacaspases [861]. A possible regulatory role for the human paracaspase in certain forms of PCD has been demonstrated [861]. More recently, the yeast metacaspase has been shown to mediate programmed cell death upon peroxide treatment and in aged cultures, which not only supports the role of metacaspases in apoptosis, but also indicates that PCD occurs even in (at least some) unicellular eukaryotes via mechanisms related to those in multicellular organisms [537]. A major role for metacaspases in plant PCD is also likely, given the proliferation of the genes coding for metacaspases in plant genomes, the absence of other caspase homologs in plants, and the fusion of some of the plant metacaspases with the LSD1 Zn-finger, a regulator of plant PCD [861]. Paracaspases were detected in animals, slime mold, and one group of bacteria, the *Rhizobiales*, with a notable expansion in *Mesorhizobium loti*, whereas metacaspases are present in plants, fungi, early-branching eukaryotes, and a variety of bacteria (Table 6.5). Phylogenetic analysis of the caspase-like protease superfamily shows a clear affinity of the eukaryotic metacaspases, paracaspases and the classic caspases with the corresponding predicted proteases from the Rhizobia, which belong to the α-subdivision of the Proteobacteria, the free-living ancestors of the mitochondria (Figure 6.17). This topology of the phylogenetic tree is best compatible with the origin of metacaspases from the mitochondrial endosymbiont. The case of caspases-paracaspases is more complicated because this branch of the superfamily is so far has not been detected in eukaryotes other than animals and slime mold. This distribution is compatible with a second, later HGT from α-proteobacteria to eukaryotes or with independent loss of the paracaspase gene in multiple eukaryotic lineages.

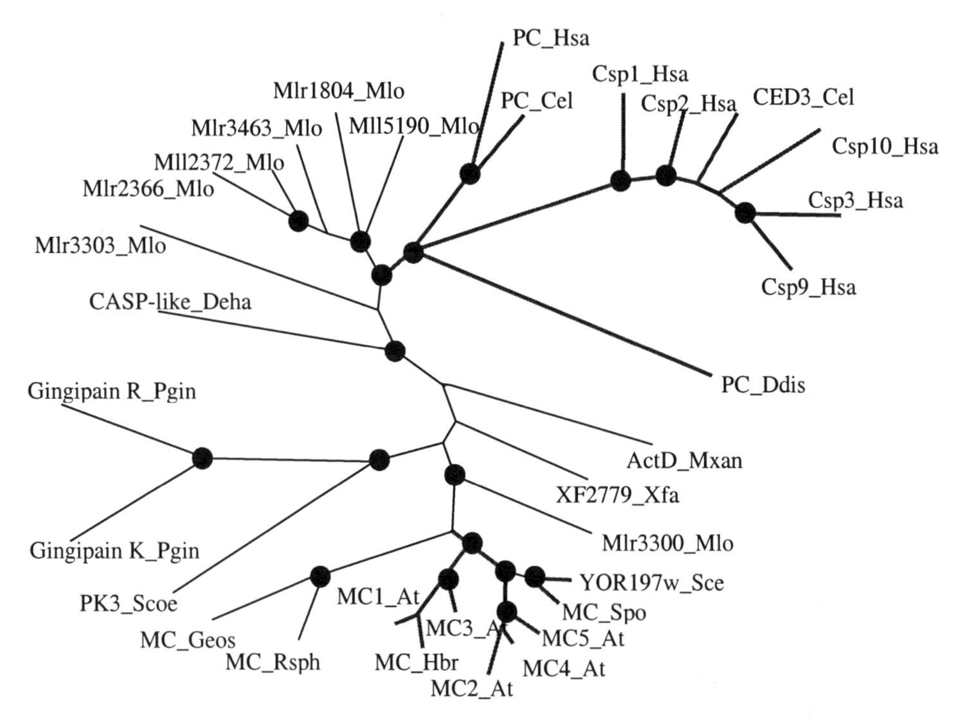

Figure 6.17. Phylogenetic tree of the caspase-like protease superfamily.
The proteins are indicated by their gene names and abbreviated species names. Csp, caspase; PC, paracaspase; MC, metacaspase. Circles show nodes with ≥75% bootstrap support. Thick lines indicate eukaryotic lineages. Species abbreviations on Figures 6.17–6.19 are: Af, *Archaeoglobus fulgidus*; Ap, *Aeropyrum pernix*; At, *Arabidopsis thaliana*; Atu, *Agrobacterium tumefaciens*; Bab, *Brucella abortus*; Bs, *Bacillus subtilis*; Bhen, *Bartonella henselae*, Ca, *Clostridium acetobutylicum*, Ccr, *Caulobacter cresentus*; Cel, *C. elegans*; Cj, *Campylobacter jejuni*; Ddis, *Dictyostelium discoideum*; Deha, *Dehalococcoides* sp.; Dm, *Drosophila*; Dre, zebrafish; Ec, *Escherichia coli*; Geos, *Geosulfurococcus*; Hbr, Hp, *Helicobacter pylori*; Hsa, human; Hvu, barley; Les, tomato; Lla, *Lactococcus lactis*; Lusi, jute; Mj, *Methanococcus jannaschii*; Mlo, *Mesorhizobium loti*; Mmu, mouse; Mta, *Methanobacterium thermoautotrophicum*; Mxan, *Myxococcus xanthus*; Ncr, *Neurospora crassa*; Osa, rice; Pae, *P. aeruginosa*; Pgin, *Porphyromonas gingivalis*, Ph, *Pyrococcus horikoshii*; Pput, *Pseudomonas putida*; Rhiz, *Rhizobium* sp.; Rp-*Rickettsia prowazekii*; Rsph, *Rhodopseudomonas sphaeroides*, Sce, yeast; Scoe, *Streptomyces coelicolor*; Spo, *S. pombe*; Sso, *Sulfolobus sulfotaricus*; Ssp, *Synechocystis* sp.; Sto, *Sulfolobus tokodaii*; Sme, *Sinorhizobium meliloti*; Ta, wheat; Tm, *Thermotoga maritima*; Tp, *Treponema pallidum*; Xfa, *Xylella fastidiosa*.

The OMI (HtrA-like) protease. The OMI protease homologous to the widespread and well-characterized bacterial HtrA family of serine proteases is a recent addition to the repertoire of PCD-associated eukaryotic protein [336,550,815]. This protein, normally located in the mitochondria, is released into the cytoplasm during apoptosis and contributes both to caspase-dependent and caspase-independent PCD. HtrA-like membrane-associated proteases are nearly ubiquitous in bacteria, the sole exception so far being the mycoplasmas, the bacterial parasites with the smallest genomes; in contrast, these proteins are missing in most archaeal genomes sequenced to date.

Phylogenetic analysis of the HtrA-like proteases suggests a major diversification of this family into several distinct lineages in bacteria, with a prominent expansion in α-proteobacteria. This analysis strongly supports the monophyly of the eukaryotic OMI/HtrA2 proteases, which are involved in PCD, with a particular lineage of α-proteobacterial HtrA-like proteases (Figure 6.18). Clearly, this observation is compatible with a mitochondrial origin for OMI.

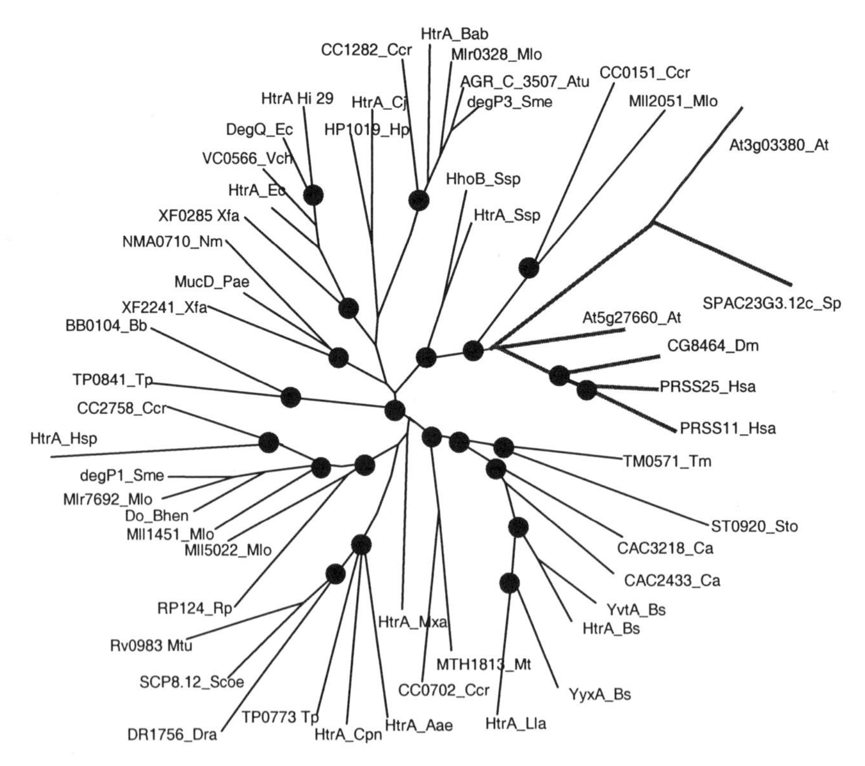

Figure 6.18. Phylogenetic tree of the HtrA family of proteases.
All details are as in Figure 6.17.

Apoptotic (AP) ATPases and NACHT GTPases. AP-ATPases are central regulators of PCD, which interact with caspases to form the so-called apoptosome and are required for caspase activation [142,146]. AP-ATPases are present in animals, plants, in which they are encoded by vastly proliferated pathogen and stress resistance genes, one fungal species (*Neurospora crassa*), many bacteria, and one archaeon, *P. horikoshii*. Among bacteria, AP-ATPase homologs are present in α-proteobacteria, cyanobacteria and Actinomycetes, with a particularly notable proliferation in the latter lineage (Table 6.5). Phylogenetic analysis of the ATPase domain of AP-ATPases strongly supports the monophyly of the plant and animal representatives, but does not group them with any bacterial lineage in particular; in contrast, the *Neurospora* AP-ATPase clusters with those of Actinomycetes as does the only archaeal member of this family (Figure 6.19). The latter two AP-ATPases appear to be obvious cases of HGT from Actinomycetes. The origin of the animal and plant AP-ATPases is less clear. However, a more detailed examination of the alignment of the AP-ATPase domain showed that a large subgroup of these proteins, including those from plants, animals and several bacteria, primarily actinomycetes, contained a distinct C-terminal motif, which was missing in the rest of the bacterial AP-ATPase homologs, including those from alpha-proteobacteria [457]. This feature allows us to tentatively root the AP-ATPase tree and hence establish the connection between eukaryotic AP-ATPases and homologs from Actinomycetes (Figure 6.19). Given these observations and the absence of AP-ATPases from the available genome sequences of yeasts and early-branching eukaryotes, a relatively late, around the time of animal-plant divergence, acquisition of this gene by eukaryotes from Actinomycetes seems to be the most likely evolutionary scenario. In principle, however, transfer of the AP-ATPase from mitochondria cannot be ruled out, assuming that the alpha-proteobacterial progenitor of the mitochondria, unlike rhizobia, had an AP-ATPase with the C-terminal motif and that some eukaryotic lineages have lost this gene.

The NACHT (after NAIP, CIIA, HET- E and TP1) family is another group of NTPases (primarily GTPases) with a eukaryotic-bacterial phyletic pattern [456]. So far, this family is represented in animals, one fungal species (*Podospora anserina*), and several bacteria (Table 6.5). A major proliferation of the NACHT family associated with an involvement in PCD and immune response against diverse viral and bacterial pathogens is observed in vertebrates, whereas other animals have only one NACHT domain that appears to be involved in the telomerase function rather than apoptosis [39]. Typically, the same bacteria that have AP-ATPases tend to encode NACHT NTPases, sometimes multiple ones (Table 6.5). In the general scheme of evolution of P-loop NTPases, the NACHT family appears to be the sister group of AP-ATPases [456]. Given the considerable diversification of each of these families in bacteria, the divergence between them should date to a relatively early stage of bacterial evolution. Although the limited sequence conservation within the NACHT family makes it a poor candidate for phylogenetic analysis, two distinct groups could be discerned within this family. The first group includes the vertebrate-specific expansion of NAIP-like proteins and several bacterial proteins from *Streptomyces* and *Anabaena, Rickettsia;* the second group consists of the animal TP-1-like telomerase subunits and the fungal proteins Het-E-1 from *Podospora anserina* and B24M22.200 from *Neurospora crassa.* Thus, multiple HGT events might have been responsible for the introduction of these proteins in eukaryotes, one occurring early in evolution and resulting in the TP-1-like forms, and the second one occurring much later, perhaps even just prior to the emergence of the vertebrate lineage, and injecting the NAIP-like forms.

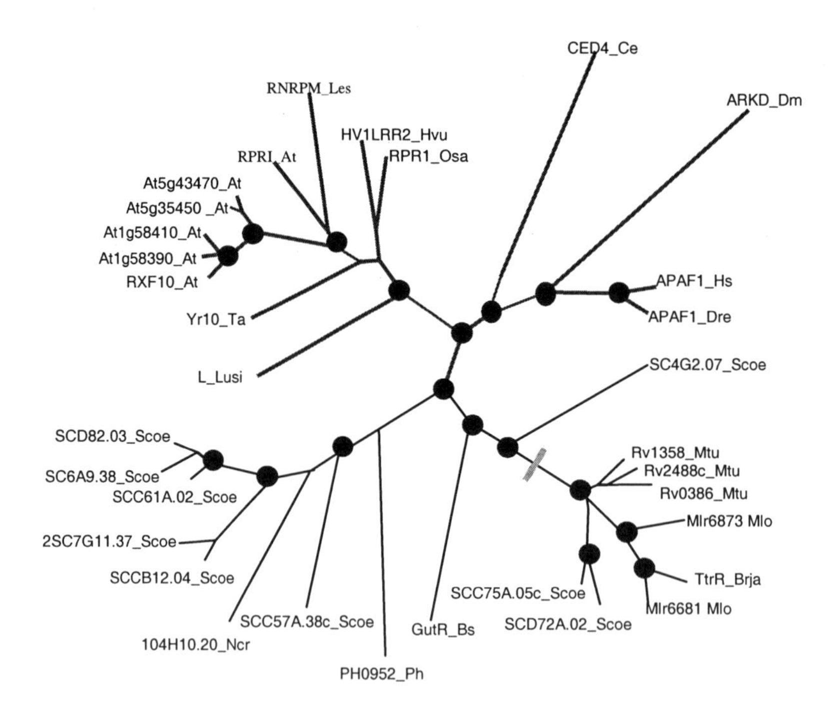

Figure 6.19. Phylogenetic tree of the apoptotic (AP) ATPase family.
The inferred root position is shown by a bar. Other details are as in Figure 6.17.

Table 6.5. Phyletic distribution of domains and proteins involved in eukaryotic apoptosis and related pathways[a]

Protein / domain family	Vertebrates (human)	Arthropods (fruit fly)	Nematodes (worm)	Plants	Fungi (yeasts)	Early-branching eukaryotes	Prokaryotes
Executors of PCD: caspase superfamily proteases							
Caspases	14 (1 inactive)	7	4	0	0	?	0
Paracaspases	1	0	1	0	0	?	7 in *M. loti*, 1 in *Rhizobium*
Metacaspases	0	0	0	~10	1	1 each in *Leishmania*, *Plasmodium*, *Trypanosoma*	2 in *Anabaena*, 1 in *Synechocystis*, 1 in *M. loti*
(Probable) executors and regulators of PCD: other proteases							
A20-family	3	1	1	Several distant homologs of A20; no orthologs			1 distant homolog in *C. pneumoniae*
HtrA-family	At least 4	4	6	At least 5	1	At least 1	1-8 in nearly all bacteria
Regulators of PCD: NTPases							
AP-ATPase	1	1	1	~190	None in yeast, 1 in *Neurospora*	?	9 in *S. coelicolor*, 3 in *Anabaena*, 6 in *M. tuberculosis*

NACHT-GTPases	18 (17 Naip-like, 1 TP1-like)	1 (TP1-like)	1 (TP1-like: 1)	0	None in yeasts but 1 TP-like form in Podospora and Neurospora	?	3 in *S. coelicolor*, 6 in *Anabaena*, 2 in *Synechocystis*, 1 in *Rickettsia conorii*
D-GTPase	2	1	2	1	0	0	0
Regulators of PCD: kinases							
Serine/threonine protein kinases	IKK: 4; DAP:1; NIK:1; IRAK:4	IKK: 2; NIK:1; IRAK:1	DAP:1; IRAK:1	0 (H)	0 (H)	0 (H)	0 (H)
Regulators of PCD: BCL-2-family proteins							
Bcl2	11	2	1	0	0	?	0
PCD adaptors: the 6-helical domains							
Death	30	9	6	0	0	?	0
DED	7	1	0	0	0	?	0
CARD	20	1	2	0	0	?	0
PYRIN	8	0	0	0	0	?	0
PCD Adaptors: other							
TIR	22	10	1	~140	0	?	4 in *S. coelicolor*, 1 each in *Rhizobium*, *Synechocystis*, *B. subtilis*, *Anabaena* and *C. crescentus*

Table 6.5 – continued

MATH	6	3	1	~ 26	1	≥ 1	0
(TRAF-like)	(several additional MATH domains not directly related to TRAF)			MATH domains but none directly related to the TRAFs			
BCL2-family	11	2	1	0	0	0	0
Ligands and receptors involved in PCD							
TNFR	8	H[b]	H	H	0	?	0
IL-1 like	8	0(H)	0(H)	0(H)	0(H)	0(H)	0(H)
Toll-like	10	8	1 (?)	0(H)	0(H)	(H)	0(H)
TNF	17	1	0(H)	0	0	?	0
Cysteine knots	TGF-like: 12; NGF like-3	TGF like: 3 Spaetzle like:3; NGF like 1	H	0	0	?	0
Nuclear factors involved in PCD							
NFKB	5	3	0 (H)	0 (H)	0 (H)	0	0 (H)
NFAT	6	1	0 (H)	0 (H)	0 (H)	0	0 (H)
P53	3	1	0	0	0	0	0
E2F	8	2	3	6	0	0	0
DP1	5	1	1	2	0	0	0
STAT	6	1	4	0	0	0	0
RB	3	1	1	1	0	0	0
CAD	5	4	0	0	0	0	0
BIR	8	4	1	0	1	0	0

[a] H: Indicates the presence of homologous domains but not actual orthologs.

Apoptosis-inducing factor (AIF) is a mitochondrial protein, which is released into the cytoplasm during apoptosis and stimulates a caspase-independent PCD pathway essential for early morphogenesis in mammals [410]. This function of AIF is highly conserved in evolution as indicated by the recent demonstration of the function of the AIF ortholog in PCD in the slime mold *Dictyostelium discoideum* [58]. AIF is a Rossmann-fold, FAD-dependent oxidoreductase, but the redox activity is not required for its pro-apoptotic function [573]. This protein is highly conserved and nearly ubiquitous in bacteria, archaea and eukaryotes. Phylogenetic analysis showed that eukaryotic AIFs cluster with their archaeal orthologs, to the exclusion of bacterial ones, with the sole exception of *T. maritima*, a hyperthermophilic bacterium, which probably acquired this gene from archaea via HGT [457]. Thus, AIF seems to be the only major component of the PCD apparatus that conforms with the standard model of evolution, which is particularly notable because this ancestral protein apparently had been secondarily recruited for a mitochondrial function.

The TIR domain is the only PCD adaptor molecule that has been detected in bacteria, although not so far in fungi or early-branching eukaryotes [38]. The distribution of the TIR domain in bacteria is similar to that of caspase-related proteases, AP-ATPases and NACHT NTPases, with a notable expansion in actinomycetes. The information contained in the TIR domain alignment does not seem to be sufficient to produce a reliable phylogenetic tree. Nevertheless, given that TIR domains seem to be present only in crown-group eukaryotes, possible evolutionary scenarios include a mitochondrial acquisition with subsequent loss in multiple eukaryotic lineages or a later HGT from a bacterial source.

Domain architectures of bacterial homologs of eukaryotic PCD-associated proteins suggest functional interactions. Functional information on bacterial homologs of eukaryotic apoptotic proteins is scarce. For the AP-ATPase homologs, the only available data point to a role of some of these proteins, such as GutR from *B. subtilis* and AfsR from *S. coelicolor*, in transcription regulation [233,681]. The function of GutR, which is a regulator of the glucitol operon, has been shown to be ATP-dependent [681]. Among the numerous caspase-related proteins detected in bacteria, only one, ActD from *Myxococcus xanthus*, has been characterized experimentally. This protein is a regulator of the production of the sporulation morphogen, CsgA, but its mechanism of action remains unknown [319]. Comparative-genomic information partly compensates for the paucity of experimental data: examination of the domain architectures of the bacterial homologs of apoptotic components provides tantalizing functional hints. Firstly, nearly all of these proteins form complex, multidomain architectures (Figure 6.20; see color plates). Secondly, many of them contain repetitive protein-protein interaction modules, such as WD40, TPR and Armadillo repeats, which tend to form scaffolds facilitating the formation of multisubunit complexes. Finally, and most strikingly, some of the bacterial apoptosis-related proteins are fused within multidomain proteins, suggesting functional interactions between them (recall the "guilt by association" or "Rosetta Stone" principle ♦5.2.2). Examples include the fusion of a caspase-like protease with an AP-ATPase and WD40 repeats in the cyanobacterium *Anabaena* and the TIR-AP-ATPase and metacaspase-protein-kinase fusions in Actinomycetes (Figure 6.20; color plates). Some of the domain architectures observed among apoptotic protein homologs in bacteria reflect specifics of prokaryotic signal transduction, e.g., the characteristic fusions of AP-ATPases with helix-turn-helix DNA-binding domains in transcription regulators. These peculiarities notwithstanding, the above observations are sufficient to justify the hypothesis that bacterial homologs of the eukaryotic apoptotic proteins interact functionally and, most likely, also physically in signal-transduction pathways whose exact nature remains to be determined. A bolder speculation is that, in bacteria with complex development and differentiation, such as actinomycetes, cyanobacteria, myxobacteria and some alpha-proteobacteria, the homologs of apoptotic proteins, particularly meta- and paracaspases and AP-ATPases, form large complexes that might be functional analogs or perhaps even evolutionary predecessors of the eukaryotic apoptosome. A search for such complexes in bacteria and elucidation of their potential role in signal transduction and/or an unknown form of PCD seems to be an exciting subject for experimental studies.

Evolution of eukaryotic programmed cell death: the case for multiple infusions of bacterial genes. As discussed above, the principal enzymes and at least one adaptor domain involved in eukaryotic PCD are widespread in bacteria, but are conspicuously missing in archaea. Furthermore, two important lines of evidence support HGT from bacteria to eukaryotes as the principal route of evolution of these proteins. Firstly, in at least two cases, those of OMI and metacaspases, phylogenetic analysis confidently shows a specific affinity of the eukaryotic apoptotic proteins with homologs from α-proteobacteria. These observations strongly suggest a mitochondrial origin for the respective genes (see above). Secondly, in each case, and particularly for caspase-related protease and AP-ATPases, exploration of the bacterial homologs of apoptotic proteins reveals a greater diversity, in terms of phyletic distribution, domain architectures and sequences themselves, than seen in eukaryotes. This points to the probable direction for HGT: from bacteria to eukaryotes.

In principle, all apparent bacterial contributions to eukaryotic PCD could be explained through acquisition of mitochondrial genes. However, this would require multiple losses of the genes for apoptotic proteins in different eukaryotic lineages and, in addition, would be at odds with some phylogenetic analysis results, e.g. those that seem to link eukaryotic AP-ATPases with Actinomycetes (Figure 6.20). Thus, a different scenario, with at least two infusions of bacterial genes contributing to the origin of PCD, appears to be more parsimonious [457]. According to this hypothesis, the first influx of the relevant bacterial genes was part of the domestication of the pro-mitochondrial endosymbiont, whereas the second one probably occurred at the stage of a primitive multicellular eukaryote, perhaps the ancestor of the eukaryotic crown group. Apparently, there were at least occasional subsequent gene transfers, such as the acquisition of an AP-ATPase by the fungus *Neurospora crassa*, and perhaps even the less orthodox acquisition of an additional NACHT NTPase at a late stage of animal evolution. This hypothesis of multiple acquisitions of PCD-related genes by early eukaryotes from bacteria is clearly in line with the Doolittle's ratchet mechanism.

Proteins encoded by scavenged bacterial genes appear to constitute the core of the ancestral eukaryotic apoptotic machinery; a caspase-like protease, probably a metacaspase, an HtrA-like protease and AP-ATPases were principal enzymatic components, whereas the TIR domain might have functioned as the main adaptor. These core components have undergone further, lineage-specific proliferation and specialization, such as expansion of caspases in vertebrates and metacaspases and AP-ATPases in plants. Around this core, the outer layers of the apoptotic machinery have built up gradually from exapted domains that originally might have had different functions, such as MATH or BIR, and of newly "invented" domains, such as the six-helical adaptor domain, which subsequently gave rise to the Death, Death Effector, and CARD domains (Table 6.5).

Returining to the three routes of eukaryotic innovation mentioned in the beginning of this section, we see that routes (ii) and (iii), i.e. exaptation of bacterial and perhaps a few ancestral archaeo-eukaryotic proteins for new functions and "invention" of novel domains, contributed substantially to the evolution of PCD. Whether or not route (i), direct recruitment of a functionally analogous bacterial precursor, was also employed, remains to be established through functional characterization of the bacterial homologs of apoptotic proteins.

The observations described here emphasize the pivotal role of bacterial-eukaryotic HGT in the origin of the eukaryotic PCD system and, by implication, of the eukaryotic multicellularity itself. Indeed, much of the glory of eukaryotic ascension to the ultimate complexity of higher plants and animals might owe to a "lucky" choice of bacteria with complicated differentiation processes as the primary, promitochondrial, and perhaps subsequent symbionts.

Mitochondria appear to be among the principal (if not ***the*** principal) sensors of cell damage that trigger PCD by releasing cytochrome c, which stimulates apoptosome assembly [5,113]. Furthermore, additional proteins, such as AIF and OMI, are also released from mitochondria and contribute to PCD. Is there an intrinsic connection between the role of mitochondria in PCD and the origin of the eukaryotic apoptotic system? This is not immediately obvious, in part, because the involvement of mitochondria in apoptosis has been demonstrated primarily in the vertebrate model system, potentially allowing for the possibility that mitochondria are a late addition to the ancestral repertoire of apoptotic regulators. However, several recent studies suggest that the mitochondrial contribution to PCD is likely to be ancient, e.g. the demonstration of the role of AIF in PCD in the slime mold and the role of mitochondrial endonuclease G in apoptotic DNA degradation in the nematode [58,658]. Indications of a mitochondrial involvement in PCD in plants [487] and of a potential involvement of the metacaspase in mitochondrial biogenesis in yeast [818] add to the growing evidence of an ancient role of mitochondria in eukaryotic PCD. The other side of the problem is that mitochondrial endosymbiosis and the origin of PCD appear to be uncoupled in time because endosymbiosis, a very early event in eukaryotic evolution, apparently was followed by a lengthy age of unicellular eukaryotes, which generally are not known to have PCD. Thus, mitochondrial acquisitions, such as AIF and metacaspase, might have been "pre-adaptations" for PCD, which originally had other roles in primitive eukaryotes, and only later have been exapted for their functions in apoptosis. However, the recent striking experiments demonstrating the role of yeast and possibly even trypanosome (an early-branching protist) metacaspases in PCD might lead to a revision of these views [537,818]. Whether or not early-branching, truly unicellular eukaryotes have PCD is a subject of major interest; the ultimate evidence can be obtained only by direct experiments, but the conservation of the metacaspase in these primitive eukaryotes is suggestive. Regardless of the outcome of such studies, a straightforward hypothesis can connect pro-mitochondrial endosymbiosis with the origin of eukaryotic PCD. The early α-proteobacterial endosymbionts might have been using secreted and membrane proteases, such as metacaspases, paracaspases and HtrA-like proteases, to kill the host cells once the latter became inhospitable, e.g. because of scarcity of nutrients or accumulation of free radicals. Such a mechanism could enable the endosymbionts to efficiently use the corpse of the assassinated host and move to a new one. During subsequent evolution, this weapon of aggression might have been appropriated by the host and made into a means of programmed suicide, with the subsequent addition of regulatory components [240,457].

6.5. Conclusions and Outlook: Evolution Tinkers with Fluid Genomes

In this chapter, we covered a lot of territory, at the inevitable price of sketchiness. Even so, we managed to discuss only a few of the important evolutionary issues brought into focus by comparative genomics. We already touched upon some other aspects, including evolution of genome organization, in Chapters 2 and 5; evolution of metabolic pathways and protein domains and families will be discussed in Chapters 7 and 8, respectively. Nevertheless, we certainly cannot hope to present any comprehensive treatise on "Genomics and Evolution". Not only is the subject vast but, more importantly, the research in this area has started in earnest only three-four years ago (in part, for the obvious and excusable reason that genome data simply were not around before that). Therefore development of new approaches and systematic application of the existing ones are required to tease out answers to evolutionary puzzles hidden in the genomes. The principal notion we tried to convey is that comparative genomics has already made the picture of the evolution of life substantially more complex, but also immensely richer and more interesting than anyone could imagine in the pre-genomic era.

Three interlinked fundamental messages, we believe, are here to stay. Firstly, comparative genomics shows that genomes are much more dynamic, even volatile (on the evolutionary scale) systems when previously thought. Lineage-specific gene loss and horizontal gene transfer can no longer be treated as peripheral evolutionary phenomena that may be involved in important but specialized cases, such as evolution of parasites and antibiotic resistance. Instead, they should be accorded the status of major factors of evolution, which, at least among unicellular life forms, are ubiquitous and as important as vertical inheritance. If HGT (but not necessarily gene loss) is significantly less common in multicellular eukaryotes, they more than compensate with intragenomic mobility, including recruitment of mobile elements for coding and regulatory regions [517,539,609]. We believe that these results of comparative genomics amount to a new view of the evolutionary process. The new picture in no way contradicts what we at least consider to be the cornerstone of Darwinism: the central role of natural selection (in its substantially different forms, we must now add) in evolution. However, just like many modern developments in evolutionary biology itself [304], the new picture promulgated by genomics defies the exclusive emphasis on small, gradual mutational change, which was part of Darwin's message in *The Origin of Species* and had been farther elevated in status by the neo-Darwinian synthesis.

Secondly, comparative genomics reaffirms, through numerous spectacular illustrations, a rather old but so far (we believe) not fully appreciated evolutionary principle captured in a brilliant metaphor of François Jacob [388]: evolution is largely a tinkerer who achieves the best feasible result by combining, sometimes in haphazard ways, whatever materials are at hand. Comparative genomics has shown in abundance how evolution tinkers with protein domains and operons (more about this in Chapters 5 and 8), to produce amazingly diverse, effective and subtle signaling and regulatory systems. Genome flexibility, to a large extent ensured by HGT, provides ample material for this. Tinkerers are not supposed to be particularly good in inventing real new gadgets, and evolution indeed seems to avoid this as much as possible, by reutilizing, modifying and recombining already tried solutions. However, true novelty also emerges, particularly in complex eukaryotes. For sure, on some occasions, what looks completely new is only the ultimate form of tinkering when the original gadget ceases to be recognizable; we have seen such cases when discussing exaptation of prokaryotic proteins for some crucial eukaryotic functions (Table 6.4). In other situations, tinkering turns into invention, which may occur via evolution of a globular domain from a generic non-globular structure, such as coiled coil [45], or even through emergence of coding sequences from non-coding ones. We do not discuss these processes in detail in this book not so much due to a lack of space, but simply because their extent and mechanisms are still not sufficiently clear.

Thirdly, and finally, it seems that comparative genomics not only vastly complicates the picture of life's evolution but also provides the information necessary to resolve the principles and details of this picture. The methods and concrete studies that take us in this direction are starting to appear but much more remains to be done.

6.6. Further Reading

1. Crick FH. 1968. The origin of the genetic code. *Journal of Molecular Biology* 38: 367-379

2. Jacob F. 1977. Evolution and tinkering. *Science* 196: 1161-1166

3. Woese C. 1998. The universal ancestor. *Proceedings of the National Academy of Sciences of the United States of America* 95: 6854-6859

4. Woese CR. 2000. Interpreting the universal phylogenetic tree. *Proceedings of the National Academy of Sciences of the United States of America* 97: 8392-8396

5. Woese CR. 2002. On the evolution of cells. *Proceedings of the National Academy of Sciences of the United States of America* 99: 8742-8747

6. Leipe DD, Aravind L, Koonin EV. 1999. Did DNA replication evolve twice independently? *Nucleic Acids Research* 27: 3389-3401

7. Anantharaman V, Koonin EV, Aravind L. 2002. Comparative genomics and evolution of proteins involved in RNA metabolism. *Nucleic Acids Research* 30: 1427-1464.

8. Snel B, Bork P, Huynen MA. 2002. Genomes in flux: the evolution of archaeal and proteobacterial gene content. *Genome Research* 12: 17-25.

CHAPTER 7. EVOLUTION OF CENTRAL METABOLIC PATHWAYS: THE PLAYGROUND OF NON-ORTHOLOGOUS GENE DISPLACEMENT

One of the central goals of functional genomics is the complete reconstruction of the metabolic pathways of the organisms, for which genome sequences have been obtained. As discussed in Chapter 1, there is no chance that all necessary biochemical experiments are ever done in any substantial number of organisms. Therefore reconstructions made through comparative genomics, combined with the knowledge derived from experiments on model systems, are the only realistic path to a satisfactory understanding of the biochemical diversity of life and to the characterization of poorly studied and hard-to-grow organisms (including extremely important ones, e.g. the syphilis spirochete *T. pallidum* [243,887]).

In the pre-genomic era, metabolic reconstruction might have seemed to be a relatively easy task given the overall similarity of the key metabolic enzymes in several model organisms, such as *E. coli*, *B. subtilis*, yeast, plants, and animals. Although cases of non-orthologous (unrelated or distantly related) enzymes catalyzing the same reaction, such as the two distinct forms of fructose-1,6-bisphosphate aldolases, phosphoglycerate mutases, and superoxide dismutases, have been known for a long time, these cases were generally perceived as rare and, more or less, inconsequential [187,258,271,549]. The availability of complete genomes is gradually changing this perception, making us realize just how common these cases of analogous (as opposed to homologous) enzymes are in nature (♦2.2.5). The phenomenon of non-orthologous gene displacement turned out to be a major complication (but also a major source of unexpected findings) for the analysis of metabolic pathways, making it particularly hard to automate. Indeed, whenever an ortholog of a given metabolic enzyme from the model organisms is not detected in the organism of interest (the initial step of metabolic reconstruction, the identification of orthologs of known enzymes, can be automated almost completely), the process turns into "detective work". The researcher needs to identify a set of gene products that, on the basis of their predicted biochemical activities, potentially could catalyze the reaction in question. Often, there is more than one such candidate, and the choice between these might not be possible without direct experiments. Furthermore, there is always a chance that, however plausible, all candidates detected in such searches are false, whereas the true culprit is a complete unknown. This makes metabolic reconstruction in the era of comparative genomics a less precise, but much more exciting undertaking.

In this chapter, we show how a COG-based reconstruction of bacterial and archaeal metabolism helps organizing the existing data on microbial biochemistry, illuminates the remaining questions, suggests candidates for some of the "missing" enzymatic activities, and predicts the existence of novel enzymes that remain to be discovered. For each metabolic reaction, we list the COGs that are known to catalyze it or can be reasonably predicted to do so. We then compare the phyletic patterns of the corresponding COGs to see if the current set of COGs is sufficient to suggest candidate proteins to catalyze the given reaction in each organism with sequenced genome or still unexplained gaps remain in metabolic pathways.

7.1. Carbohydrate Metabolism

7.1.1. Glycolysis

We have already used the COG approach to demonstrate the complementarity of the phyletic patterns of the three forms of phosphoglycerate mutase (♦2.6). Figure 7.1 shows the COGs that are known or predicted to include glycolytic enzymes and shows their phyletic patterns. This superposition of COGs and metabolic pathways provides a convenient framework for a detailed analysis of the phylogenetic distribution of each of the glycolytic enzymes and the general principles of evolution of carbohydrate metabolism. This figure shows, for example, that *R. prowazekii*, an obligate intracellular parasite and a relative of the mitochondria [30], does not encode a single glycolytic enzyme. In contrast, all other organisms with completely sequenced genomes encode enzymes of the lower (tri-carbon) part of the pathway. This supports the notion that glycolysis is the central pathway of carbohydrate metabolism and makes comparative analysis of variants of this pathway all the more interesting.

Glucokinase (EC 2.7.1.2)

Fermentation of glucose starts with its phosphorylation, which is catalyzed by glucokinase. Although many bacteria bypass the glucokinase step by phosphorylating glucose concomitantly with its uptake by the PEP-dependent phosphotransferase system, some of them, including *E. coli*, encode a glucokinase (COG0837) that shares little sequence similarity with yeast and human enzymes. There is also another bacterial form, found in *S. coelicolor*, *Bacillus megaterium*, and other bacteria [32,795].

Recently, *P. furiosus* has been reported to encode an ADP-dependent glucokinase [435]. This enzyme has no detectable sequence similarity to any other glucokinase, but shows significant structural similarity to enzymes of the ribokinase family [383]. In retrospect, several conserved motifs were detected in this new glucokinase and the ribokinase family proteins, which indicates of a homologous relationship. Thus, a clear-cut case of non-orthologous gene displacement is observed: a ribokinase family enzyme has been recruited to replace the typical glucokinase. So far, the ADP-dependent glucokinase has been found only in *M. jannaschii* and in pyrococci. The existence of at least three distinct forms of glucokinase is remarkable, especially given that this is apparently not an essential component of glycolysis. Moving down the glycolytic pathway, we find similar examples of non-orthologous gene displacement for several other, essential enzymes.

Glucose-6-phosphate isomerase (EC 5.3.1.9).

Bacteria and eukaryotes encode several distinct but homologous forms of glucose-6-phosphate isomerase (phosphoglucomutase) [624]. The classical (*E. coli*) form of the enzyme is found in gram-negative bacteria and in the cytoplasm of the eukaryotic cell. A divergent version of this enzyme is found in gram-positive bacteria including *B. subtilis*, in *T. maritima*, and some archaea, such as *M. jannaschii* and *Halobacterium* sp. [466,761]. The most divergent members of this family of glucose-6-phosphate isomerases were detected in *A. aeolicus* and another subset of archaea, including *M. thermoautotrophicum*, *A. pernix* and *Thermoplasma* spp. No enzyme of this family seems to be encoded in the genomes of *A. fulgidus* or pyrococci. Instead, *P. furiosus* has been shown to encode a novel glucose-6-phosphate isomerase, which has highly conserved orthologs in *P. horikoshii* and in *A. fulgidus*, but so far not in any other organism [331]. Thus, two non-orthologous (in fact, apparently unrelated) versions of this enzyme together account for the phosphoglucomutase activity in all known microbial genomes, with the exception of *R. prowazekii* and *U. urealyticum*. As indicated above, the former does not encode any glycolytic enzymes, whereas the latter apparently obtains fructose-6-phosphate by importing fructose concomitantly with its phosphorylation through the fructose-specific phosphotransferase system, thus bypassing the phosphoglucomutase stage

Phosphofructokinase (EC 2.7.1.11).

The next glycolytic enzyme, phosphofructokinase, offers an even more interesting example of non-orthologous gene displacement. It is also an example of an enzyme where several "missing" enzyme forms have been discovered just in the past year.

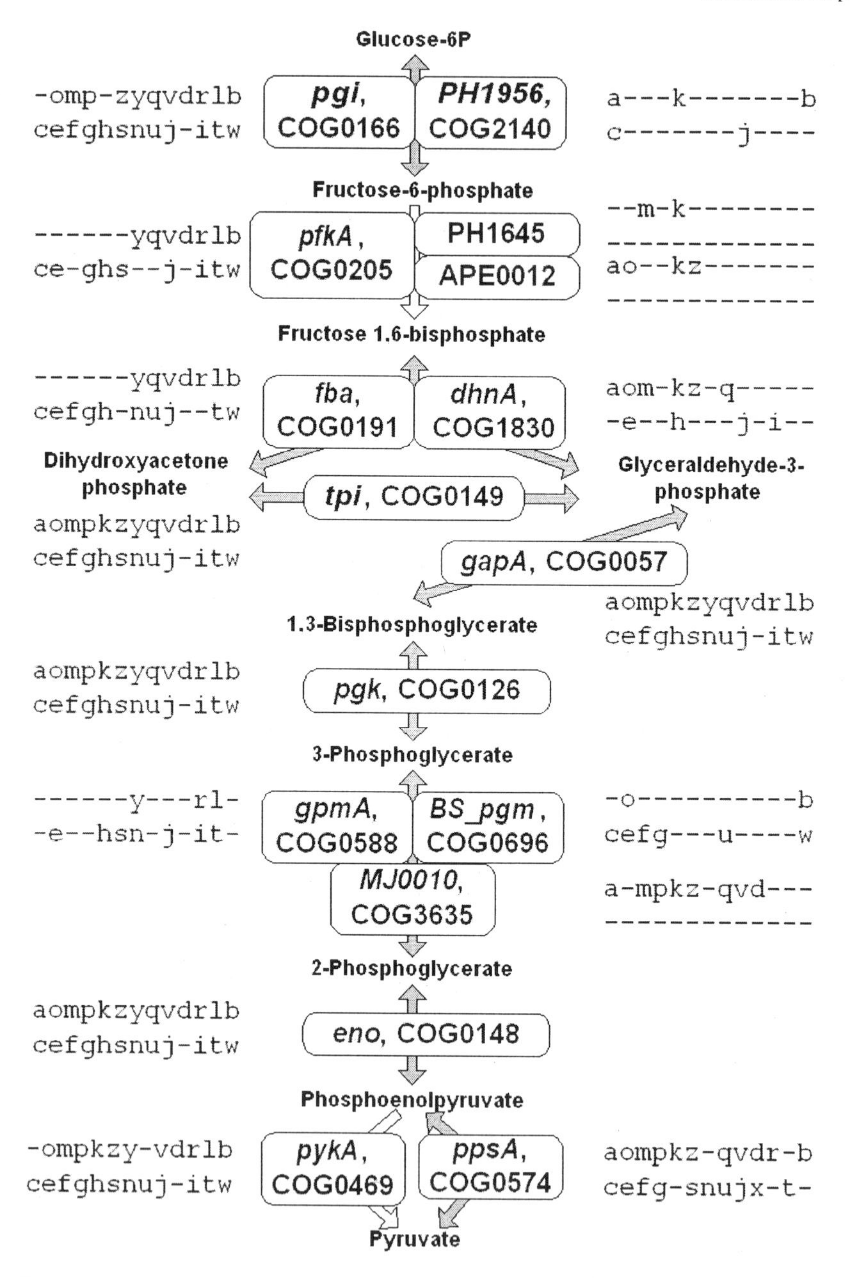

Figure 7.1. Distribution of glycolysis (Embden-Meyerhoff-Parnas pathway) enzymes in organisms with completely sequenced genomes. Each rounded rectangle shows a glycolytic enzyme, denoted by its gene name and the COG number. Alternative enzymes catalyzing the same reaction are shown side-by-side. Each COG is accompanied by its the phyletic pattern (see ♦2.2.6). The species abbreviations are as in Figure 2.7 (page 39).

The most common version of this enzyme, PfkA, is an ATP-dependent kinase of unique structure found in bacteria and many eukaryotes. Plants have a homologous enzyme, which, however, uses pyrophosphate as the phosphate donor. Altogether, homologs of PfkA are found in nearly all bacteria and eukaryotes, but are conspicuously missing in *H. pylori* and in all archaeal genomes sequenced so far. In addition, *E. coli* encodes a second phosphofructokinase, PfkB, which is unrelated to PfkA and instead belongs to the ribokinase family of carbohydrate kinases.

A unique ADP-dependent phosphofructokinase has been described in *P. furiosus* [853]. However, this enzyme appears to have a limited phyletic distribution: so far, it was found only in *M. jannaschii* and in pyrococci. This left the phosphofructokinase activity in other archaea unaccounted for and suggested that additional forms of this enzyme might exist. Very recently, a new ATP-dependent phosphofructokinase, which is a member of the ribokinase family but is not specifically related to PfkB, has been identified in *A. pernix* [712]. Close homologs of this protein (APE0012) were found in *Halobacterium* sp., *A. fulgidus*, *M. thermoautotrophicum*, and several other archaea. Therefore it seems likely that these ribokinase family enzymes function as phosphofructokinases in all these archaea. Finally, *Thermoplasma* does not encode orthologs of any of the four forms of phosphofructokinase described above. This leaves two possibilities: either thermoplasmas lack phosphofructokinase altogether (along with fructose-1,6-bisphosphate aldolase; see below) or they might have yet another, fifth variant of this enzyme.

Fructose-1,6-bisphosphate aldolase (EC 4.1.2.13).

For more than 50 years now, it has been known that fructose-1,6-bisphosphate aldolase exists in two distinct forms, a metal-independent one (class I) in multicellular eukaryotes and a metal-dependent one (class II) in bacteria and yeast [187,549,881]. Certain organisms, such as *Euglena*, seem to have enzymes of both classes. Although these two enzyme forms have similar structures, they do not share any detectable sequence similarity [257].

Sequence analysis of archaeal genomes and those of chlamydia showed that they encode neither a typical class I enzyme, nor a typical class II enzyme. Instead, chlamydia and all archaea, with the exception of thermoplasmas, encode orthologs of the recently described class I aldolase DhnA (FbaB) of *E. coli*, which is only distantly related to the regular class I enzymes and may be considered a third class of aldolases. Recently, fructose-1,6-bisphosphate aldolase activity was demonstrated in the *P. furiosus* homolog of DhnA; this enzyme has been referred to as a class IA aldolase [770]. The phyletic patterns of the bacterial-type class II aldolase (COG0191) and the DhnA-type aldolase (COG1830) are almost complementary, except that both types of aldolases are present in *E. coli* and *A. aeolicus,* and none of them is detectable in *X. fastidiosa* (Figure 7.1). *X. fastidiosa*, a plant pathogen, encodes a eukaryotic class I aldolase, which is specifically similar to the plant class I aldolase, and probably has been acquired from the plant host via HGT. However, typical eukaryotic (class I) fructose-1,6-bisphosphate aldolase is also encoded in several other bacteria, in which cases the underlying evolutionary scenario is less clear.

Although most genomes encode only one type of fructose-1,6-bisphosphate aldolase, different forms of this enzyme do coexist in several organisms. In particular, the relatively large genome of the plant symbiont *M. loti* encodes fructose-1,6-bisphosphate aldolases of all three classes.

The nature of the aldolase, if any, in thermoplasmas remains unclear. The apparent absence in these archaea of both phosphofructokinase and fructose-1,6-bisphosphate aldolase might indicate that these organisms split hexoses into trioses exclusively via the Entner-Doudoroff pathway (see below). Indeed, thermoplasmas encode close homologs of the recently described fructose-6-phosphate aldolase [758].

Finally, given that chlamydiae are important human pathogens and that the unusual class IA fructose-1,6-bisphosphate aldolase is the only aldolase encoded in their genomes, this presumably essential enzyme might be a promising target for anti-chlamydial drug therapy (♦7.6) [257,266].

Triose phosphate isomerase (EC 5.3.1.1).

Triose phosphate isomerase is conserved in all organisms, with the exception of *Rickettsia.* Bacterial-eukaryotic and archaeal isomerases form two clearly separated clusters [239]. This gave rise to the notion that eukaryotic triose phosphate isomerases originated from the promitochondrial endosymbiont whose genes have been transferred into the nucleus of the eukaryotic host [432].

Glyceraldehyde-3-phosphate dehydrogenase (EC 1.2.1.12).

Like triosephosphate isomerases, archaeal glyceraldehyde-3-phosphate dehydrogenases are homologous to those from bacteria and eukaryotes, but form a well-defined cluster, suggesting the mitochondrial origin of this enzyme in eukaryotes. In pyrococci and, probably, in several other archaea, the main glycolytic flow goes through a different enzyme, glyceraldehyde-3-phosphate:ferredoxin oxidoreductase, whereas glyceraldehyde-3-phosphate dehydrogenase appears to be confined to gluconeogenesis [584,867].

In *U. urealyticum*, the typical NADH-dependent glyceraldeldehyde-3-phosphate dehydrogenase is missing and this reaction is apparently catalyzed by a non-phosphorylating, NADP-dependent enzyme, similar to the well-characterized enzymes from plants and *Streptococcus mutans* [326,544]. These enzymes belong to a large superfamily of NADP-dependent aldehyde dehydrogenases and are unrelated to the phosphorylating glyceraldeldehyde-3-phosphate dehydrogenase [568]. Remarkably, an archaeal member of the non-phosphorylating glyceraldeldehyde-3-phosphate dehydrogenase family uses NAD instead of NADP [124].

Phosphoglycerate kinase (EC 2.7.2.3)

Like triose phosphate isomerases and glyceraldehyde-3-phosphate dehydrogenases, phosphoglycerate kinase is conserved in all organisms that have glycolysis, and the sequences from bacteria and eukaryotes are closer to each other than they are to their archaeal counterparts, suggesting the mitochondrial origin of the eukaryotic enzyme.

Phosphoglycerate mutase (EC 5.4.2.1).

The diversity of phosphoglycerate mutases was discussed earlier (♦2.2.6). We would only like to reiterate that there are two unrelated forms of this enzyme, 2,3-bisphosphoglycerate-dependent (animal-type) and 2,3-bisphosphoglycerate-independent (plant-type), either one of which (or both) can be found in various bacteria [138]. Although *E. coli pgm* mutants devoid of its principal (cofactor-dependent) form of phosphoglycerate mutase clearly exhibit a mutant phenotype, a recent study of the second (cofactor-independent) form of this enzyme showed that it accounts for as much as 10% of the total phosphoglycerate mutase activity in *E. coli* [244].

Remarkably, neither form of phosphoglycerate mutase is encoded in any archaeal genome available to date, with the sole exception of *Halobacterium* spp., which has a typical cofactor-independent enzyme, similar to the one in *B. subtilis*. Sequence analysis of archaeal genomes showed that they encode enzymes of the alkaline phosphatase superfamily that are distantly related to the cofactor-independent phosphoglycerate mutase and contain all the principal active-site residues [258]. These enzymes were predicted to have a phosphoglycerate mutase activity [258]. This prediction was supported by the structural analysis of the cofactor-independent phosphoglycerate mutase [261,394] and has been recently confirmed by direct experimental data [308,866]. Thus, like phosphofructokinase and fructose-1,6-bisphosphate aldolase, phosphoglycerate mutase is found in three different (unrelated or distantly related) variants.

Enolase (EC 4.2.1.11).

Enolases encoded in bacterial, archaeal and eukaryotic genomes are highly conserved; phylogenetic trees for enolases show a "star topology", which precludes any definitive conclusions on the evolutionary scenario for this enzyme. Pyrococci and *M. jannaschii* encode additional, divergent paralogs of enolase whose function(s) remains unknown.

Pyruvate kinase (EC 2.7.1.40).

Pyruvate kinase, the terminal glycolytic enzyme, is not encoded in some bacterial (*A. aeolicus*, *T. pallidum*) and archaeal genomes (*A. fulgidus*, *M. thermoautotrophicum*). In these organisms, the pyruvate kinase function is probably taken over by phosphoenolpyruvate synthase, which is capable of catalyzing pyruvate formation by reversing its typical reaction.

Pyruvate kinase, like phosphofructokinase (see above), is also missing in *H. pylori*. Although a ribokinase-like phosphofructokinase and phosphoenolpyruvate synthase could be considered as possible bypasses for these enzymes, it seems more likely that glycolysis is not functional in *H. pylori*. In contrast, this bacterium encodes the complete set of enzymes involved in gluconeogenesis (Figure 8.2). Such organization of metabolism seems to make perfect sense for *H. pylori*, given the challenge of maintaining near-neutral intracellular pH in the highly acidic gastric environment. Sugar fermentation, resulting in intracellular production of acid, would place an additional burden on the pH maintenance mechanism, while gluconeogenesis converts organic acids into sugars and thus removes H^+ from the cytoplasm. For the purposes of energy production, *H. pylori* apparently depends on fermentation of amino acids and oligopeptides that are produced by gastric proteolysis and are transported into the bacterial cells by ABC-type transporters. Amino acid fermentation results in alkalinization of the cytoplasm and could relieve part of the burden of pH maintenance in *H. pylori*. This simple example shows that, even when seemingly plausible candidates for missing steps in a pathway can be suggested, this should be done with caution and the resulting predicted pathways should be assessed against the biological background of the respective organism.

After a string of recent publications [331,383,770], it appears that most glycolytic enzymes have now been accounted for. While there are no clear candidates for phosphofructokinase and fructose-1,6-bisphosphate aldolase in *Thermoplasma* spp., the chances of discovering new enzyme variants in this pathway appear very slim.

7.1.2. Gluconeogenesis

With the exception of reactions catalyzed by phosphofructokinase and pyruvate kinase, glycolytic reactions are reversible and function also in gluconeogenesis (Figure 7.2). The reversal of the latter reaction, i.e. conversion of pyruvate into phosphoenolpyruvate, can be catalyzed by two closely related enzymes, phosphoenolpyruvate synthase and pyruvate,phosphate dikinase. The only other reaction that is specific for gluconeogenesis is the dephosphorylation of fructose-1,6-bisphosphate.

Phosphoenolpyruvate synthase (EC 2.7.9.2)

Phosphoenolpyruvate synthase (pyruvate, water dikinase, EC 2.7.9.2) and pyruvate, phosphate dikinase (EC 2.7.9.1) catalyze two similar reactions of phosphoenolpyruvate biosynthesis

Pyruvate + ATP + H_2O = Phosphoenolpyruvate + AMP + P_i
Pyruvate + ATP + P_i = Phosphoenolpyruvate + AMP + PP_i

and have highly similar sequences. This enzyme is widely present in bacteria, archaea, protists and plants, but is missing in animals, where PEP is synthesized from oxaloacetate in a PEP carboxykinase-catalyzed reaction.

Phosphoenolpyruvate carboxykinase (EC 4.1.1.32 and EC 4.1.1.49)

Phosphoenolpyruvate carboxykinase exists in two unrelated forms, which catalyze ATP-dependent (EC 4.1.1.49) or GTP-dependent (EC 4.1.1.32) decarboxylation of oxaloacetate:

Oxaloacetate + ATP = Phosphoenolpyruvate + ADP + CO_2
Oxaloacetate + GTP = Phosphoenolpyruvate + GDP + CO_2

These forms show remarkably complex phyletic distributions. The GTP-dependent form is found in animals and in a limited number of bacteria, such as *Chlamydia* spp., *Mycobacterium* spp., *T. pallidum*, and the green sulfur bacterium *Chlorobium limicola*. Among archaea, it is encoded only in the genomes of pyrococci, thermoplasmas, and *Sulfolobus*. In contrast, the ATP-dependent form of phosphoenolpyruvate carboxykinase is found in plants, yeast, and many bacteria. The only complete archaeal genome that has been found to encode the ATP-dependent form is that of *A. pernix* (Figure 7.2).

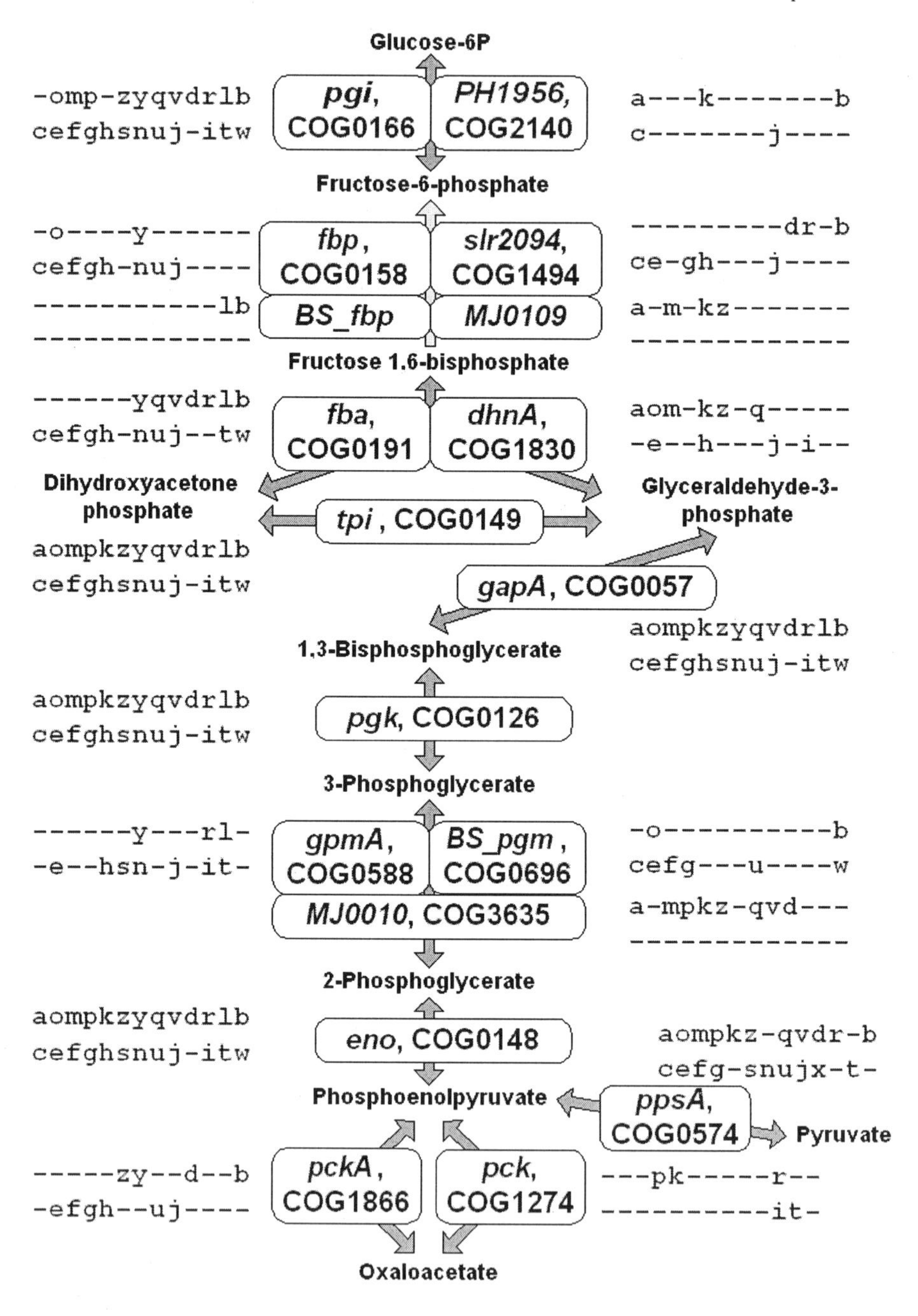

Figure 7.2. Distribution of gluconeogenesis enzymes in organisms with completely sequenced genomes. All details are as in Figure 2.7.

Since the typical bacterial ATP-dependent phosphoenolpyruvate carboxykinase appears to be unrelated to the GTP-dependent form found in humans, this key enzyme of central metabolism might be an interesting drug target for such pathogenic bacteria as *H. influenzae* and *C. jejuni* (see ♦7.6).

There have also been reports of a third, pyrophosphate-dependent, form of phosphoenolpyruvate carboxykinase [819], but they remain unconfirmed and no sequence so far has been identified with this form. Absent this third form, phosphoenolpyruvate carboxykinase appears to be missing in a large number of microorganisms, leaving room for discovery of a new enzyme.

Fructose-1,6-bisphosphatase (EC 3.1.3.11).

The best-studied form of fructose-1,6-bisphosphatase, found in E. coli, yeast, and human (COG0158), has a limited phyletic distribution: it is not encoded in the genomes of chlamydia, spirochetes, gram-positive bacteria, *A. aeolicus* or *T. maritima.* Among archaea, it is present only in *Halobacterium* sp. A second form of this enzyme (COG1494), originally described in cyanobacteria, has been reported to function both as a fructose-1,6-bisphosphatase and as a sedoheptulose-1,7-bisphosphatase [823,824].

This form also has a limited phyletic distribution, being found in a relatively small number of bacteria (Figure 7.2). Although a member of this second family is encoded in *B. subtilis*, this organism also has a distinct form of fructose-1,6-bisphosphatase that is unrelated to the first two and is found only in several other low-GC gram-positive bacteria [251]. Finally, archaea encode yet another, fourth form of this enzyme that belongs to the inositol monophosphatase family and only recently has been shown to possess fructose-1,6-bisphosphatase activity [399,801]. Like *B. subtilis*, several bacterial genomes encode members of more than one protein family, which include known or potential fructose-1,6-bisphosphatases; this makes it hard to predict which of them actually have this function in gluconeogenesis. In contrast, there is no clear candidate for this function in *A. aeolicus*, *T. maritima*, *X. fastidiosa*, *Chlamydia* spp., mycoplasmas, spirochetes, and thermoplasmas. While the first three of these organisms and *B. burgdorferi* encode enzymes of the inositol monophosphatase family, they are not closely related to the archaeal fructose-1,6-bisphosphatase (typified by the MJ0109 protein from *M. jannaschii*) and might represent an independent case of enzyme recruitment. Proteins that function as fructose-1,6-bisphosphatase in *Chlamydia* spp., *Thermoplasma* spp., mycoplasmas, and *T. pallidum*, if any, remain to be identified.

7.1.3. Entner-Doudoroff pathway and pentose phosphate shunt

Alternative pathways for converting hexoses into trioses, the pentose phosphate shunt and the Entner-Doudoroff pathway, are found in many organisms, but cannot be considered universal. Both of these pathways start from the NADP-dependent oxidation of glucose-6-phosphate into phosphogluconolacton and proceed through 6-phosphogluconate (Figure 7.3). Instead of the standard Entner-Doudoroff pathway, some archaea encode the so-called non-phosphorylating variant of this pathway, which starts from glucose and includes unphosphorylated intermediates.

Glucose-6-phosphate 1-dehydrogenase (EC 1.1.1.49)

Glucose 6-phosphate dehydrogenase (Zwischenferment) primarily uses $NADP^+$ as the electron acceptor, although there have been reports of NAD^+-dependent forms. This enzyme is found in many bacteria and eukaryotes, but is not encoded in any of the archaeal genomes sequenced to date. In addition, it is missing in several bacteria, such as *M. leprae*, *B. halodurans*, *S. pyogenes*, *C. jejuni*, and mycoplasmas (Figure 7.3).

6-Phosphogluconolactonase (EC 3.1.1.31)

Although this enzymatic activity had been characterized many years ago, the gene for the lactonase remained unidentified until very recently, which was due, in part, to the inherent instability of its substrate and, in part, to the fact that this activity resides in a protein that is closely related to glucosamine-6-phosphate isomerase/deaminase and might even combine both activities [154,328]. In humans, the lactonase is fused to the glucose-6-phosphate dehydrogenase, forming the C-terminal domain of a bifunctional enzyme. Interestingly, in *Plasmodium falciparum*, the fusion partners switch places, with the lactonase located at the N-terminus [551]. The lactonase is found largely in the same set of species as glucose dehydrogenase, although it appears to be missing, additionally, in *A. aeolicus* and *D. radiodurans*.

7.1.3.1. Pentose phosphate shunt

6-phosphogluconate dehydrogenase (decarboxylating, EC 1.1.1.44)

6-Phosphogluconate dehydrogenase, the product of the *gnd* gene in *E. coli*, is the upstream enzyme specific for the pentose phosphate pathway. Of those organisms that encode phosphogluconate dehydrogenase (COG0362), several (*M. loti, B. subtilis, L. lactis*) also encode its close paralog (COG1023), whose function remains unknown, but which is likely to have the same activity. Phosphogluconate dehydrogenase has an even more narrow phylogenetic distribution than phosphogluconolactonase, being additionally absent from *S. pyogenes*, *X. fastidiosa*, *H. pylori*, and *C. jejuni* (Figure 7.3).

Pentose-5-phosphate-3-epimerase (EC 5.1.3.1)

The next reaction of the pentose phosphate pathway, isomerisation of ribulose 5-phosphate into xylulose 5-phosphate, is catalyzed by phosphoribulose epimerase. In addition to the pentose phosphate pathway, this enzyme also participates in the interconversions of pentose phosphates in the Calvin cycle, which accounts for its wider phyletic distribution than seen for phosphogluconate dehydrogenase.

Ribose 5-phosphate isomerase (EC 5.3.1.6)

Ribose-5-phosphate isomerase, which catalyzes interconversion of ribulose 5-phosphate and ribose 5-phosphate, is found in two apparently unrelated forms, both of which, RpiA and RpiB, have been characterized in *E. coli* [793]. RpiA is found in many bacteria, archaea, and eukaryotes. In contrast, RpiB is limited to certain bacterial species and is the sole form of ribose-5-phosphate isomerase in *B. subtilis*, *M. tuberculosis*, *H. pylori*, and several other bacteria. The phyletic patterns of the two forms of the enzyme are largely complementary:

```
aompkzy--d-l-cefghsn-j-it-   COG0120 RpiA
-------qv-rlbce--h--ujx--w   COG0698 RpiB
aompkzyqvdrlbcefghsnujxitw   RpiA + RpiB
```

Like phosphoribulose epimerase, phosphoribose isomerase participates in the Calvin cycle, which might explain its universal distribution.

Transketolase (EC 2.2.1.1)

In eukaryotes and bacteria, transketolase is a single protein of 610-630 amino acid [821]. In archaea, however, this enzyme is either missing altogether (e.g. *A. fulgidus*, *M. thermoautotrophicum*) or is encoded by two separate genes that may not even be adjacent (in *M. jannaschii*). The hyperthermophilic bacterium *T. maritima* has both types of genes, one full-length and one split gene, the latter probably acquired from archaea via HGT. Transketolase shows high sequence similarity to deoxyxylulose-5-phosphate synthase and other thiamine pyrophosphate-dependent enzymes, which might point to a broad substrate specificity of this enzyme, particularly of in thermophiles.

Transaldolase (EC 2.2.1.2)

Transaldolase is a protein of 310-330 amino acid residues, which is present in eukaryotes and many bacteria and catalyzes the transfer of the tricarbon unit of sedoheptulose-7-phosphate to glyceraldehyde-3-phosphate, producing fructose-6-phosphate and erythrose-4-phosphate [821]. Archaea and some other bacteria encode a closely related but shorter protein, about 210-230 aa long, which has recently been demonstrated to function not as transaldolase, but as fructose-6-phosphate aldolase, which splits fructose-6-phosphate into glyceraldehyde-3-phosphate and dihydroxyacetone [758]. While *E. coli* encodes two paralogous transaldolases (*talA*, *talB*) and two paralogs of the smaller related enzyme (*talC*, *mipB*), many other micro-prokaryotes, including *B. subtilis*, *M. jannaschii*, and *Thermoplasma* spp., encode only the latter protein. Although the exact substrate specificity of these enzymes is not known, enzymes from *B. subtilis* and *T. maritima* have been reported to have transaldolase activity [758]. Thus, different MipB orthologs could have different (primary) activities, which makes complete reconstruction of the pentose phosphate pathway in organisms having these enzymes unrealistic at this time. Clearly, however, the phyletic patterns of the enzymes of this pathway differ significantly, which suggests the existence of still uncharacterized enzyme forms.

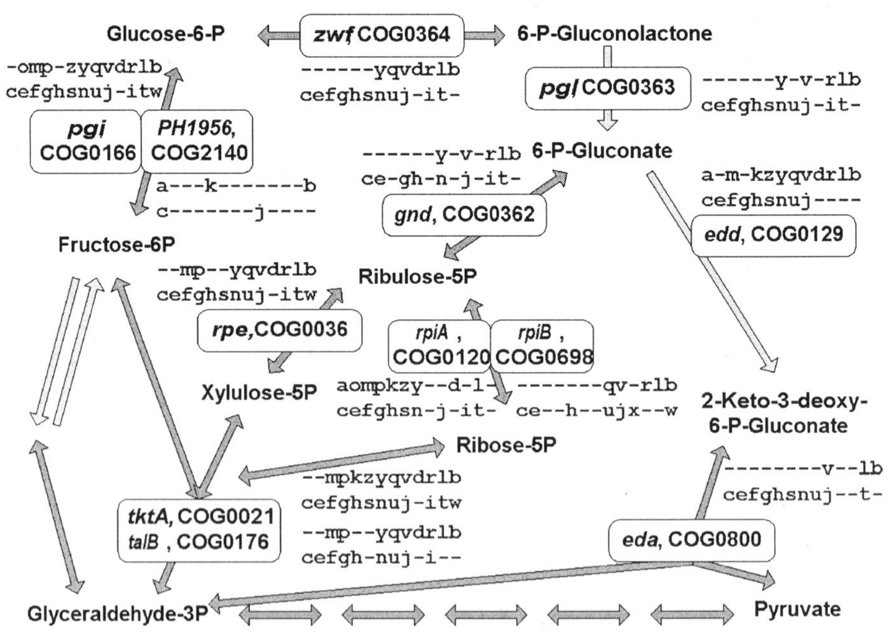

Figure 7.3. Distribution of enzymes of the pentose phosphate and Entner-Doudoroff pathways in organisms with completely sequenced genomes. Details are as in Figure 2.7.

7.1.3.2. The Entner-Doudoroff pathway

Conversion of 6-phosphogluconate into two tri-carbon molecules, 3-phosphoglyceraldehyde and pyruvate, via the Entner-Doudoroff pathway includes only two steps, which are catalyzed by 6-phosphogluconate dehydratase and 2-keto-3-deoxy-6-phosphogluconate aldolase (the products of *E. coli* genes *edd* and *eda,* respectively (Figure 7.3).

Phosphogluconate dehydratase (EC 4.2.1.12)

Phosphogluconate dehydratase is a close paralog of dihydroxyacid dehydratase, an enzyme of isoleucine/valine biosynthesis, which is encoded in almost every genome As a result, it is not easy to decide which organisms encode phosphogluconate dehydratase. In *E. coli* and several other proteobacteria, *edd* and *eda* genes form operons. In other organisms, such as *P. aeruginosa,* even though both these genes are present, they are not adjacent, which complicates the identification of phosphogluconate dehydratase.

2-Keto-3-deoxy-6-phosphogluconate aldolase (EC 4.1.2.14)

KDPG aldolase has a much more narrow phyletic distribution than phosphogluconate/dihydroxyacid dehydratase (Fig 7.3). Assuming that a functional Entner-Doudoroff pathway requires the presence of each of these enzymes, as well as glucose-6-phosphate dehydrogenase and phosphogluconolactonase, the available genomic data suggest that the pathway is limited to certain proteobacteria, *T. maritima,* and some gram-positive bacteria of the *Bacillus/Clostridium* group.

7.1.3.3. Non-phosphorylated variants of the Entner-Doudoroff pathway

While the standard Entner-Doudoroff pathway starts from glucose-6-phosphate and proceeds through phosphorylated sugar intermediates, a variety of bacteria and archaea possess so-called "non-phosphorylated" variants of this pathway, which all start from glucose and delay phosphorylation until later stages. The simplest version of such a modified pathway includes glucose oxidation into gluconate, followed by its phosphorylation into 6-phosphogluconate. The resulting 6-phosphogluconate rejoins the standard Entner-Doudoroff pathway. Another variant of the modified pathway includes an additional non-phosphorylated step, dehydratation of gluconate into 2-keto-3-deoxygluconate, followed by its phosphorylation. In yet another variant of this pathway, phosphorylation is delayed even further, until after splitting of 2-keto-3-deoxygluconate into two tri-carbon molecules, pyruvate and glyceraldehyde. The latter compound is then phosphorylated into 3-phosphoglyceraldehyde. Finally, phosphorylation can be delayed one step further, with glyceraldehyde first oxidized into glycerate and then phosphorylated into 2-phosphoglycerate.

Glucose 1- dehydrogenase (EC 1.1.1.47, 1.1.99.10)

Glucose dehydrogenase, which catalyzes glucose oxidation into glucono-1,5-lactone, is known in several variants, which use different electron acceptors. Two non-orthologous NAD^+-dependent variants of this enzyme (EC 1.1.1.47), typified by enzymes from *T. acidophilum* [397] and *Bacillus megaterium* [929], belong, respectively, to the Zn-containing dehydrogenase family and to the short-chain reductases/dehydrogenases family. One more variant of glucose dehydrogenase, which is present in *E. coli* and several other bacteria, uses pyrroloquinoline quinone as the electron acceptor [637]. Finally, the enzyme from *Drosophila* (DHGL_DROME) is a flavoprotein that can use a variety of electron acceptors [141].

Gluconolactonase (EC 3.1.1.17)

Only a single variant of gluconolactonase has been characterized so far [413]. It has a patchy and relatively narrow phyletic distribution (COG3386), suggesting that alternative versions of this enzyme might exist.

Gluconate kinase (EC 2.7.1.12)

Gluconate kinase is found in two distinct versions, one unique and the other belonging to a large family of sugar kinases. This second form of gluconate kinase has probably evolved from a glycerol kinase or a xylulose kinase via enzyme recruitment (♦2.2.5). Gluconate kinases of the first type are found in yeast, *D. radiodurans*, *E. coli* and several other proteobacteria, whereas the second form is apparently limited to *B. subtilis* and a handful of other gram-positive bacteria.

Gluconate dehydratase (EC 4.2.1.39)

Although gluconate dehydratase activity has been described in bacteria long ago [299] and can be easily detected in archaea [398], the gene(s) for this enzyme has not been identified. *E. coli*, some other bacteria, and *Thermoplasma* spp. encode an enzyme with similar activity, D-mannonate dehydratase (EC 4.2.1.8, the product of *uxuA* gene), which converts mannonate into 2-keto-3-deoxygluconate. It is not known whether or not this enzyme can use gluconate as a substrate. In any case, its narrow phyletic distribution suggests that, even if UxuA functions as gluconate dehydratase in *E. coli*, *M. loti*, *B. subtilis*, and *Thermoplasma* spp., there should exist a different form of this enzyme, which would participate in the non-phosphorylated Entner-Doudoroff pathway in other archaea.

2-keto-3-deoxygluconate aldolase

Although splitting of 2-keto-3-deoxygluconate into pyruvate and glyceraldehyde has been described long ago [16], the first gene for 2-keto-3-deoxygluconate has been identified only recently in the hyperthermophilic crenarchaeon *S. solfataricus*. This enzyme is closely related to N-acetyl-neuraminate lyase and belongs to the same superfamily of Schiff-base-dependent aldolases [126]. Enzymes of this family (COG0329) are present in all archaeal genomes sequenced so far, as well as in most bacteria. Although the exact substrate specificity of each particular member of this family is not yet clear, *Thermoplasma* spp. and *P. abyssi* encode proteins that are highly similar to the enzyme from *Sulfolobus* and can be confidently predicted to catalyze this reaction.

7.1.4. The TCA cycle

The tricarboxylic acid cycle (Krebs cycle) is the central metabolic pathway that links together carbohydrate, amino acid, and fatty acid degradation and supplies precursors for various biosynthetic pathways. Remarkably, the complete TCA cycle, which has been studied in much detail in animal and yeast mitochondria, *E. coli* and *B. subtilis*, is only found in a handful of microorganisms (Figure 7.4). Most organisms with completely sequenced genomes encode only a certain subset of TCA cycle enzymes and, instead of performing the entire cycle, utilize only fragments of it. Another remarkable feature is the diversity of this pathway: cases of non-orthologous gene displacement are detectable for at least five of the eight TCA cycle enzymes. A detailed analysis of the phyletic distribution and evolution of the TCA cycle enzymes has been recently published by Huynen and coworkers [370]. Most of their conclusions remain valid, although the sequences of the genomes of two aerobic archaea, the crenarchaeon *A. pernix* and the euryarchaeaon *Halobacterium* sp., have substantially changed the notions of what can and cannot be found in archaeal genomes. In an impressive confirmation of early biochemical results on halobacterial metabolism [8], both of these organisms have been found to encode the complete set of TCA cycle enzymes as was the microaerophile *Thermoplasma* spp. A reconstruction of the TCA cycle reactions occurring in each organism can be a very interesting project, which we recommend the readers to do on their own (see Problems, page 406). We concentrate here pexclusively on the cases of non-orthologous gene displacement.

Citrate synthase (EC 4.1.3.7)

Citrate synthase is a highly conserved enzyme, which is encoded in most bacterial, archaeal, and eukaryotic genomes (Figure 7.3). It serves as the principal port of entry of acetyl-CoA into the TCA cycle and, in eukaryotes, is exclusively located in the mitochondria. A very similar reaction is catalyzed by ATP:citrate lyase (EC 4.1.3.8), which contains a citrate synthase-like domain at its C-terminus.

Citrate synthase:

Oxaloacetate + Acetyl-CoA + H_2O = Citrate + CoA

ATP-citrate lyase:

Oxaloacetate + Acetyl-CoA + ADP + P_i = Citrate + CoA +ATP

However, ATP:citrate lyase so far has been found exclusively in eukaryotes, where it localizes in the cytoplasm and preferentially catalyzes the reverse reaction, citrate cleavage.

Citrate synthase is missing in spirochetes and mycoplasmas, which do not encode any enzymes of the TCA cycle. It is also missing in pyrococci, *M. jannaschii*, *S. pyogenes* and *H. influenzae,* which encode unlinked branches of the TCA cycle (Figure 7.4). It has been suggested that the TCA cycle has evolved from two separate reductive branches [711], which were subsequently linked by (i) citrate synthase and (ii) either an α-ketoglutarate dehydrogenase or an α-ketoglutarate:ferredoxin oxidoreductase [445,538]. In any case, due to the absence of known displacements, citrate synthase seems to be a good indicator of the presence of a (nearly) complete TCA cycle in a given organism.

Aconitase (EC 4.2.1.3)

There are two distantly related, paralogous aconitases, referred to as aconitase A and aconitase B, both of which are present in *E. coli* and many other proteobacteria (Figure 7.4). Aconitase A has a much wider phyletic distribution and is the form of the enzyme present in α-proteobacteria *M. loti*, *C. crescentus*, and *R. prowazekii*. Accordingly, this is also the form of aconitase found in the mitochondria. Although aconitase B has a much more narrow phyletic distribution, it is the only form of the enzyme encoded in *Synechocystis* sp., *P. multocida*, *H. pylori*, and *C. jejuni*.

Aconitase is closely related to 3-isopropylmalate dehydratase, an enzyme of leucine biosynthesis (♦7.4.4), which sometimes makes its annotation of these enzymes in sequenced genomes not entirely straightforward. However, *leu* genes are usually found in a conserved operon, which helps make the correct assignment.

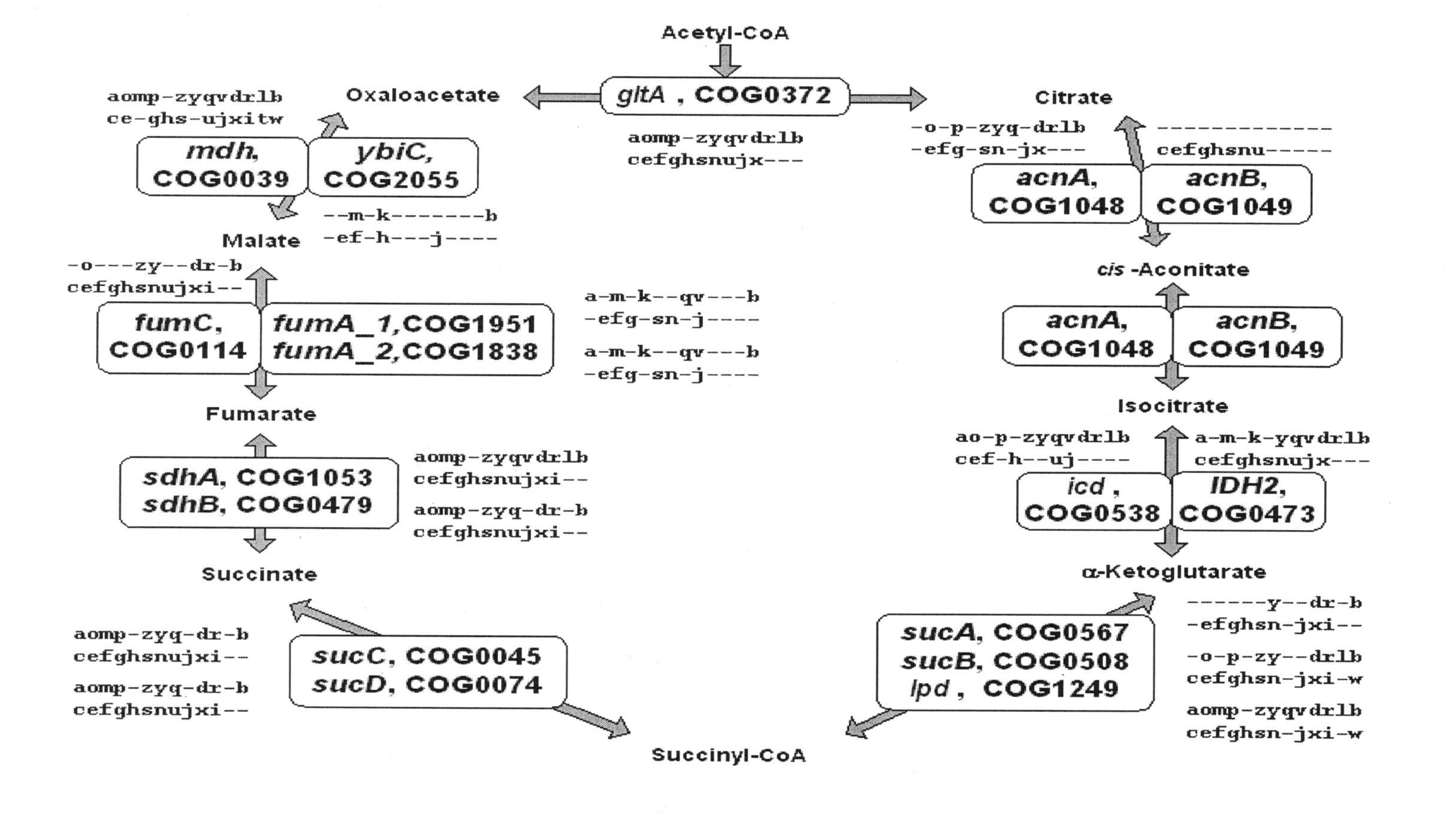

Figure 7.4. Distribution of the TCA cycle enzymes in organisms with completely sequenced genomes. All details are as in Figure 2.7.

Isocitrate dehydrogenase (EC 1.1.1.42)

Like aconitase, isocitrate dehydrogenase is also found in two forms, which, however, appear to be unrelated. The mitochondrial form of this enzyme, also found in *E. coli* and many other bacteria and archaea, is closely related to isopropylmalate dehydrogenase, an enzyme of leucine biosynthesis (♦7.4.4), and it is believed that it could have evolved via a duplication of the *leuB* gene [579]. Again, genome annotation here has to rely on the genetic context, i.e. on the presence or absence of adjacent leu genes. In any case, the product of such a gene is very likely to have both isocitrate dehydrogenase and isopropylmalate dehydrogenase activity. This form is active as a homodimer, which distinguishes it from the second form, referred to as monomeric isocitrate dehydrogenase. This second form was originally found in *Vibrio* sp. [814] and was subsequently discovered in many other bacteria [803]. It is the only form of the enzyme encoded in the genomes of *M. leprae*, *V. cholerae*, and *C. jejuni*.

2-ketoglutarate dehydrogenase (EC 1.2.4.2)

In mitochondria and many aerobic bacteria and archaea, decarboxylation of α-ketoglutarate into the succinyl moiety of succinyl-CoA is catalyzed by the thiamine pyrophosphate and lipoate-dependent α-ketoglutarate dehydrogenase complex. In contrast, many anaerobic bacteria and archaea utilize α-ketoglutarate ferredoxin oxidoreductase, an unrelated enzyme [445,538].

Succinyl-CoA synthetase (EC 6.2.1.4, 6.2.1.5)

Succinyl-CoA synthetases are divided into paralogous, highly similar GTP-dependent and ATP dependent forms. Succinyl-CoA synthetase is a member of a large family of acyl-CoA synthetases (NDP-forming), which also includes acetyl-CoA synthetase found in many archaea and lower eukaryotes. For ATP binding, these enzymes employ the ATP-grasp domain (Table 3.2). Variants of this enzyme with shifted substrate specificities are found in most phylogenetic lineages.

Succinate dehydrogenase/fumarate reductase (EC 1.3.99.1)

Mitochondrial succinate dehydrogenase, which couples the oxidation of succinate to fumarate with the reduction of ubiquinone to ubiquinol, consists of four subunits carrying three iron-sulfur centers, a covalently bound flavin and two b-type hemes (the history of the discovery of these complexes is vividly described in [773]). The fumarate reductase (quinol:fumarate reductase) complex also contains iron-sulfur centers and a covalently bound flavin, but usually consists of only two or three subunits. Succinate dehydrogenase is part of the aerobic respiratory chain, whereas fumarate reductase is involved in anaerobic respiration, with fumarate functioning as the terminal electron acceptor. Accordingly, one or both of these enzymes is found in all organisms, with the exception of pyrococci, spirochetes, and mycoplasmas (Figure 7.4).

Fumarate hydratase (fumarase, EC 4.2.1.2)

Like several other TCA cycle enzymes, fumarase is represented by two unrelated forms. The mitochondrial form of this enzyme (class II) is also encoded in many bacteria and in aerobic archaea, *A. pernix* and *Halobacterium* sp. The second form of fumarase (class I) consists of two subunits that are fused in most bacterial genomes, but are encoded by separate genes in archaea, *T. maritima* and *A. aeolicus*. The two forms of fumarase have largely complementary phyletic patterns:

```
a-m-k--qv---b-efg-sn-j----  COG1951+1838, Fumarase, class I
-o---zy--dr-bcefghsnujxi--  COG0114  Fumarase, class II
--------------------------------------------------------
aom-kzyqvdrlbcefg-snuj----  Fumarase, all forms
```

The only archaeal genome that appears not to encode a fumarase is *T. acidophilum* whose fumarase homolog Ta0258 is much more closely related to aspartate ammonia-lyase (COG1027) than to a typical fumarase (COG0114). The actual activity of this *Thermoplasma* enzyme has not been determined.

Malate dehydrogenase (EC 1.1.1.37)

Malate dehydrogenase is also found in two forms, with the mitochondrial form showing a much wider phyletic distribution. The second form of malate dehydrogenase was originally described in archaea [8,78,320] and is often referred to as the "archaeal" form of the enzyme. However, it is also encoded in certain bacterial genomes, including three paralogous genes in *E. coli* (*ybiC*, *yiaK*, and *ylbC*) and *M. loti*. It is the only form of malate dehydrogenase in pyrococci and in *P. aeruginosa*. Remarkably, *M. thermoautotrophicus, M. jannaschii, B. subtilis, H. influenzae*, and *P. multocida* encode both forms of malate dehydrogenase ([842], Figure 7.3). Why do these organisms, with their relatively small genomes, need two paralogous forms of this enzyme remains unclear. *U. urealyticum* and *T. pallidum* do not encode either of the two forms of malate dehydrogenase, in contrast to their respective relatives *M. genitalium* and *B. burgdorferi*. Therefore, the possibility remains that there exists yet another, third form of this enzyme.

7.2. Pyrimidine Biosynthesis

In contrast to the pathways of carbohydrate metabolism discussed above, enzymes of the pyrimidine biosynthesis pathway show a fairly consistent phyletic pattern, although cases of non-orthologous gene displacement can be found here, too (see Figure 2.7 on page 39). The whole pathway, with the exception of the last three steps, is missing in the obligate parasitic bacteria with small genomes: rickettsiae, chlamydiae, spirochetes, and mycoplasmas, whereas bacteria and archaea with larger genomes encode all or almost all enzymes of pyrimidine biosynthesis.

Carbamoyl phosphate synthase (EC 6.3.5.5)

In bacteria and archaea, carbamoyl phosphate synthase consists of two subunits, which in eukaryotes are fused into a single multifunctional CAD protein that additionally contains dihydroorotase and aspartate carbamoyl-transferase domains. The small subunit, encoded by the *carA* gene, is a typical glutamine amidotransferase of the Triad family [936]. The large subunit consists of two ATP-grasp domains (♦3.3.3) fused in the same polypeptide chain [391,800,841]. In *M. jannaschii* and *M. thermoautotrophicus*, the large subunit is split into two proteins, which are encoded by different, albeit adjacent genes. In addition to the obligate parasites mentioned above, carbamoyl phosphate synthase is missing in *P. horikoshii*, *P. abyssi*, *Thermoplasma* spp., and *H. influenzae* (Figure 2.7). It is present, however, in *Pyrococcus furiosus*, suggesting a relatively recent loss of this enzyme in the other two pyrococci. In *P. abyssi*, carbamoyl phosphate biosynthesis is carried out by an unrelated form of the enzyme, which is closely related to carbamate kinase [683,684]. This second form is also responsible for the carbamoyl phosphate synthase activity in *P. furiosus* [201,692] and might account for this activity in *Thermoplasma* spp. Although both subunits of carbamoyl phosphate synthase belong to large protein superfamilies and are similar to many proteins with different substrate specificities, the sheer size of the large subunit, which typically contains more than 1050 amino acid residues, allows an easy identification of this enzyme in genome analyses. However, caution is due with respect to the annotation of any shorter proteins that give statistically significant hits to the large subunit of carbamoyl phosphate synthase: these are likely to be other ATP-grasp superfamily enzymes (see Table 3.2).

Aspartate carbamoyltransferase (EC 2.1.3.2)

Aspartate carbamoyltransferase, the second enzyme of pyrimidine biosynthesis, has a wide distribution with a phyletic pattern, which is similar to that of carbamoyl phosphate synthase, but additionally includes pyrococci and *Thermoplasma* spp. (Figure 2.7). This enzyme, however, is lacking in *H. influenzae* and in its close relative *P. multocida.* In eukaryotes, aspartate carbamoyltransferase comprises the C-terminal domain of the multifunctional CAD protein [771].

Dihydroorotase (EC 3.5.2.3)

The well-characterized form of dihydroorotase (COG0418), encoded by the *E. coli pyrC* gene [65] and by the *URA4* gene in yeast [325], has a very limited phyletic distribution (Figure 2.7). In contrast, the second form of this enzyme (COG0044) is almost universal, being present in many bacteria, archaea, and eukaryotes [686]. In eukaryotes, this enzyme forms the middle portion of the multifunctional CAD protein [771,943]. In yeast, however, this domain is apparently inactive [794], most likely because of the presence of the alternative form of dihydroorotase. Just as aspartate carbamoyltransferase, neither form of dihydroorotase is encoded in *H. influenzae* or *P. multocida.* Notably, the union of the phyletic patterns for the two forms of dihydroorotase is identical to the phyletic pattern of aspartate carbamoyl-transferase:

```
------y------cefg--nu-----  COG0418   Dihydroorotase
aompkzyqvdrlbcef--s-uj----  COG0044   Dihydroorotase
-----------------------------------------------------
aompkzyqvdrlbcefg-snuj----  All forms Dihydroorotase
aompkzyqvdrlbcefg-snuj----  COG0540   Aspartate carbamoyl-
                                                 transferase
```

Dihydroorotate dehydrogenase (EC 1.3.3.1)

Dihydroorotate dehydrogenase displays the same phyletic pattern as dihydroorotase and aspartate carbamoyltransferase, with the addition of *H. influenzae* and *P. multocida.* Both these bacteria encode the enzymes for all downstream steps of pyrimidine biosynthesis.

Orotate phosphoribosyltransferase (EC 2.4.2.10)

The phyletic pattern of orotate phosphoribosyltransferase differs from that of dihydroorotate dehydrogenase in only one respect, the presence of a *pyrE*-related gene in *C. pneumoniae*. The function of the product of this gene in *C. pneumoniae* is unknown, but given the absence in this organism of the enzymes for the upstream and the downstream steps of the pathway, it is unlikely to function as orotate phosphoribosyltransferase. Rather, this enzyme might be recruited to catalyze a different phosphoribosyltransferase reaction. In eukaryotes, orotate phosphoribosyltransferase is fused to the next enzyme of the pathway, OMP decarboxylase, forming a two-domain UMP synthase. As a result, orotate phosphoribosyltransferase and OMP decarboxylase are occasionally misannotated as UMP synthases and vice versa [264].

Orotidine-5'-monophosphate decarboxylase (EC 4.1.1.23)

Although the phyletic pattern of OMP decarboxylase is identical to that of dihydroorotate dehydrogenase, a closer look at COG0284 shows that it consists of three distantly related families. Two of these include well-characterized enzymes from *E. coli* and other bacteria [856] and from yeast and other eukaryotes [234,570]. The third family includes OMP decarboxylases from archaea and a small number of bacteria, such as *M. tuberculosis*, *M. leprae,* and *Myxococcus xanthus* [12,439]. Mycobacterial OMP decarboxylases seem to be sufficiently distinct from those of eukaryotes and other bacteria to consider them promising targets for antituberculine drugs [266].

Uridylate kinase (EC 2.7.4.-, 2.7.4.14)

There seem to be two distinct forms of uridylate kinase: one specific for UMP and found in bacteria and archaea (COG0528) and another one that phosphorylates both UMP and CMP and is found in eukaryotes [628,739]. The eukaryotic form of the enzyme is closely related to bacterial adenylate kinase, and could have been recruited from an ancestral prokaryotic adenylate kinase. The prokaryotic form of uridylate kinase is encoded in all bacterial and archaeal genomes sequenced to date, including the 'minimal' (see ♦2.2.5) genomes of mycoplasmas and *Buchnera.*

Nucleoside diphosphate kinase (EC 2.7.4.6)

Nucleotide diphosphate kinase (COG0105) is highly conserved in most bacteria, archaea, and eukaryotes. Surprisingly, however, this enzyme is not encoded in *T. maritima*, *L. lactis*, *S. pyogenes*, and mycoplasmas. One could imagine that these organisms employ a different nucleotide diphosphate kinase that might have been recruited, just like the eukaryotic uridylate kinase, from the adenylate kinase family (COG0563).

This, however, would not solve the problem for *T. maritima* and mycoplasmas, which encode only a single enzyme of that family. It therefore seems likely that nucleotide diphosphate kinase in these organisms has been recruited from yet another kinase family. Indeed, a phyletic pattern search for a protein that would be encoded in those four genomes, but not in other organisms with relatively small genomes, such as chlamydiae, spirochetes or *H. pylori*, easily finds an uncharacterized (predicted) kinase related to dihydroxyacetone kinase (COG1461), which appears to be a good candidate for the role of nucleoside diphosphate kinase in these organisms:

```
_   aompkzyqvdrlbcefghsnujxitw Uridylate kinase (all forms)
    aompkzyq-dr-bcefghsnujxit- COG0105 NDP kinase
-----------------------------------------------------------
    --------v--l-------------w Missing NDP kinase
    --------vdrlb------------w COG1461 Predicted kinase
```

CTP synthase (UTP-ammonia ligase, EC 6.3.4.2)

CTP synthase is a two-domain protein, which consists of an N-terminal nucleotide-binding synthetase domain and a C-terminal glutamine amidotransferase domain. This enzyme is extremely highly conserved in bacteria, archaea, and eukaryotes. It is missing only in the genomes of *M. genitalium* and *M. pneumoniae*, which apparently make CTP from CDP or CMP in a salvage pathway, rather than from UTP.

General notes on pyrimidine biosynthesis evolution

Comparison of the phyletic patterns for the enzymes of pyrimidine biosynthesis reveals two important evolutionary trends. First, there appears to be a tendency towards decreasing the genome size by losing genes that have ceased to be essential. Indeed, ample evidence indicates that mycoplasmas evolved from a Gram-positive ancestor by way of massive gene loss associated with their adaptation to parasitism. While bacilli, lactococci, and many other gram-positive bacteria carry the full set of genes of pyrimidine biosynthesis, most of the *pyr* genes have been lost in the mycoplasmal lineage. Similarly, many *pyr* genes apparently have been lost in other parasitic bacteria with small genomes, such as spirochetes, rickettsiae, and chlamydiae (Figure 2.7).

The trend towards gene loss is much more pronounced for the initial steps of the pyrimidine biosynthesis pathway than it is at for the distal steps. Thus, genes for the first three steps of pyrimidine biosynthesis from bicarbonate and ammonia to dihydroorotate (*carA*, *carB*, *pyrB*, and *pyrC*) are missing in *H. influenzae*, but the genes for all the subsequent steps of pyrimidine biosynthesis, from dihydroorotate to CTP, are present (Figure 2.7). This means that, although *H. influenzae* is incapable of *de novo* pyrimidine biosynthesis, it still can synthesize UTP and CTP from dihydroorotate, orotate or OMP. Spirochetes, chlamydiae, rickettsiae, and mycoplasmas show an even deeper loss of pyrimidine biosynthesis genes, but nevertheless retain genes for the last three steps of the pathway, the conversion of UMP into CTP. Thus, while depending on the host for the supply of essential nutrients, this strategy allows the parasite to preserve at least some metabolic plasticity. In particular, every organism seems to encode enzymes to synthesize its own nucleoside triphosphates (NTPs). For thermodynamic reasons, bacteria cannot import NTPs directly, although intracellular bacterial parasites do encode ATP/ADP translocases, which are capable of exchanging ADP generated by the parasite for cytoplasmic ATP [899,910].

7.3. Purine Biosynthesis

Like pyrimidine biosynthesis enzymes, enzymes of the purine biosynthesis pathway follow a consistent phylogenetic pattern, albeit with some inevitable complications (Figure 7.5). With only a few exceptions, enzymes that catalyze the common reactions of the pathway, which leads to the formation of inosine-5'-monophosphate, are missing in parasitic bacteria with small genomes, namely mycoplasmas, rickettsiae, chlamydiae, spirochetes, *Buchnera* sp., and *H. pylori*, and, interestingly, in the aerobic crenarchaeon *A. pernix*. Other bacteria encode the complete set of purine biosynthesis enzymes, whereas the distribution of these enzymes in archaeal genomes is more complex and has to be discussed separately for each enzyme.

Phosphoribosylpyrophosphate synthetase (EC 2.7.6.1)

PRPP synthetase (ribose-phosphate diphosphokinase) is an enzyme that is shared by purine biosynthesis and histidine biosynthesis pathways. This enzyme is found in most completely sequenced genomes, including those of mycoplasmas, spirochetes, and *Buchnera*, which do not encode most purine biosynthesis enzymes (Figure 7.5).

Amidophosphoribosyltransferase (EC 2.4.2.14)

Glutamine phosphoribosylpyrophosphate amidotransferase (PurF) belongs to the N-terminal nucleophile (Ntn) family of glutamine amidotransferases [936]. This enzyme is encoded in every sequenced bacterial genome, with the exception of some obligate parasites, such as rickettsiae, chlamydiae, spirochetes, mycoplasmas, and *H. pylori,* and in every archaeal genome except for *A. pernix*. The same phyletic pattern is seen for the majority of purine biosynthesis enzymes.

Phosphoribosylamine-glycine ligase (EC 6.3.4.13)

Phosphoribosylglycinamide synthetase PurD, an ATP-grasp superfamily (Table 3.2) enzyme, has the same phyletic pattern as amidophosphoribosyltransferase and many other enzymes of this pathway.

Phosphoribosylglycinamide formyltransferase (EC 2.1.2.2)

5'-Phosphoribosyl-N-formylglycinamide synthase (GAR transformylase) exists in two different forms, formate-dependent (PurN) and folate-dependent (PurT), which are unrelated to each other and catalyze entirely different reactions. The folate-dependent form functions as a transferase, catalyzing transfer of the formyl group from formyltetrahydrofolate to phosphoribosylglycinamide. This enzyme is found in many bacteria and eukaryotes, but only in a few archaea, such as *Halobacterium* sp. and *Thermoplasma* spp. The formate-dependent form of the enzyme belongs to the ATP-grasp superfamily (see Table 3.2) and catalyzes an ATP-dependent ligation of phosphoribosylglycinamide with formic acid. This is the only form of GAR transformylase in methanogens and pyrococci. Surprisingly, neither form of the enzyme is encoded in the *A. fulgidus* genome. With the exception of *A. fulgidus*, the combined phyletic pattern of the two forms of GAR transformylase coincides with the patterns for amidophosphoribosyltransferase and phosphoribosylamine-glycine ligase:

```
-o-p--yqvdrlbcefghsnuj---- PurN  Folate-dependent form
--m-k-----r-bcefgh-------- PurT  Formate-dependent form
-----------------------------------------------------
-ompk-yqvdrlbcefghsnuj---- Both forms together
aompk-yqvdrlbcefghsnuj---- PurF
aompk-yqvdrlbcefghsnuj---- PurD
```

Phosphoribosylformylglycinamidine synthase (EC 6.3.5.3)

Like many other amidotransferases, phosphoribosylformylglycinamidine (FGAM) synthase PurL consists of two subunits, a glutamine amidotransferase of the Triad family [936] and a synthetase. The phyletic pattern of both FGAM synthase subunits is the same as that of PurF and PurD. In *E. coli* and many other γ-proteobacteria, as well as in yeast and other eukaryotes, these two subunits are fused in one polypeptide chain, whereas in most other bacteria and in archaea they are encoded by separate genes. In this latter case, FGAM synthase apparently requires an additional 80-aa subunit, referred to as PurS [747].

Phosphoribosylaminoimidazol synthetase (EC 6.3.3.1)

Phosphoribosylformylglycinamidine cycloligase (AIR synthetase) PurM has the same phyletic pattern as PurF, PurD, and PurL.

Phosphoribosylaminoimidazole carboxylase (EC 4.1.1.21)

Phosphoribosylaminoimidazole (AIR) carboxylase (NCAIR synthetase) PurK is, like PurD, an ATP-grasp superfamily enzyme (Table 3.2), which catalyzes ATP-dependent carboxylation of AIR. Unlike other enzyme of purine biosynthesis, PurK is not encoded in the genomes of *A. fulgidus*, *C. jejuni*, methanogens, and pyrococci (Figure 7.5), so that the mechanism of AIR carboxylation in these organisms remains unknown. This reaction can occur spontaneously at elevated temperatures in a CO_2-rich atmosphere, which could explain the absence of this enzyme in hyperthermophilic archaea. This explanation does not seem to work, however, for *C. jejuni*, suggesting the existence of a still unidentified alternative version of PurK (see [270,567] for discussion).

Phosphoribosylcarboxyaminoimidazole mutase

Phosphoribosylcarboxyaminoimidazole (NCAIR) mutase, previously thought to be a subunit of NCAIR synthetase, but recently identified as an individual enzyme [567,583], has the typical phyletic pattern of purine biosynthesis enzymes, identical to the phyletic patterns PurF, PurD, and PurL.

Phosphoribosylaminoimidazolesuccinocarboxamide synthase (EC 4.3.3.2)

Phosphoribosylaminoimidazolesuccinocarboxamide (SAICAR) synthase (PurC) contains a distinct version of the ATP-grasp domain. In addition to the standard set of organisms that are capable of purine biosynthesis, SAICAR synthase is encoded in the genome of *R. prowazekii*. It is hard to imagine what might be the function of this enzyme in an intracellular parasite, which lacks all other enzymes of purine biosynthesis. The sequence of *R. prowazekii* PurC is closely related to the enzymes from other α-proteobacteria but has at least three substitutions of amino acid residues that are otherwise conserved in SAICAR synthases (EVK, unpublished observations). This suggests that rickettsial SAICAR synthase might have lost its enzymatic activity and acquired another, perhaps regulatory function.

Adenylosuccinate lyase (EC 4.3.2.2)

Adenylosuccinate lyase (PurB) has the typical phyletic pattern of purine biosynthesis enzymes, with the addition of *H. pylori*. This is most likely due to the involvement of PurB in the conversion of IMP into AMP, the reaction that appears to occur in *H. pylori*.

AICAR transformylase (EC 2.1.2.3)

Phosphoribosylaminoimidazolecarboxamide (AICAR) formyltransferase (PurH) catalyzes the transfer of the formyl group from formyltetrahydrofolate to AICAR. In every organism studied to date, this protein is fused to the IMP cyclohydrolase in a bifunctional enzyme. AICAR transformylase comprises the C-terminal 300-aa portion of the PurH protein,

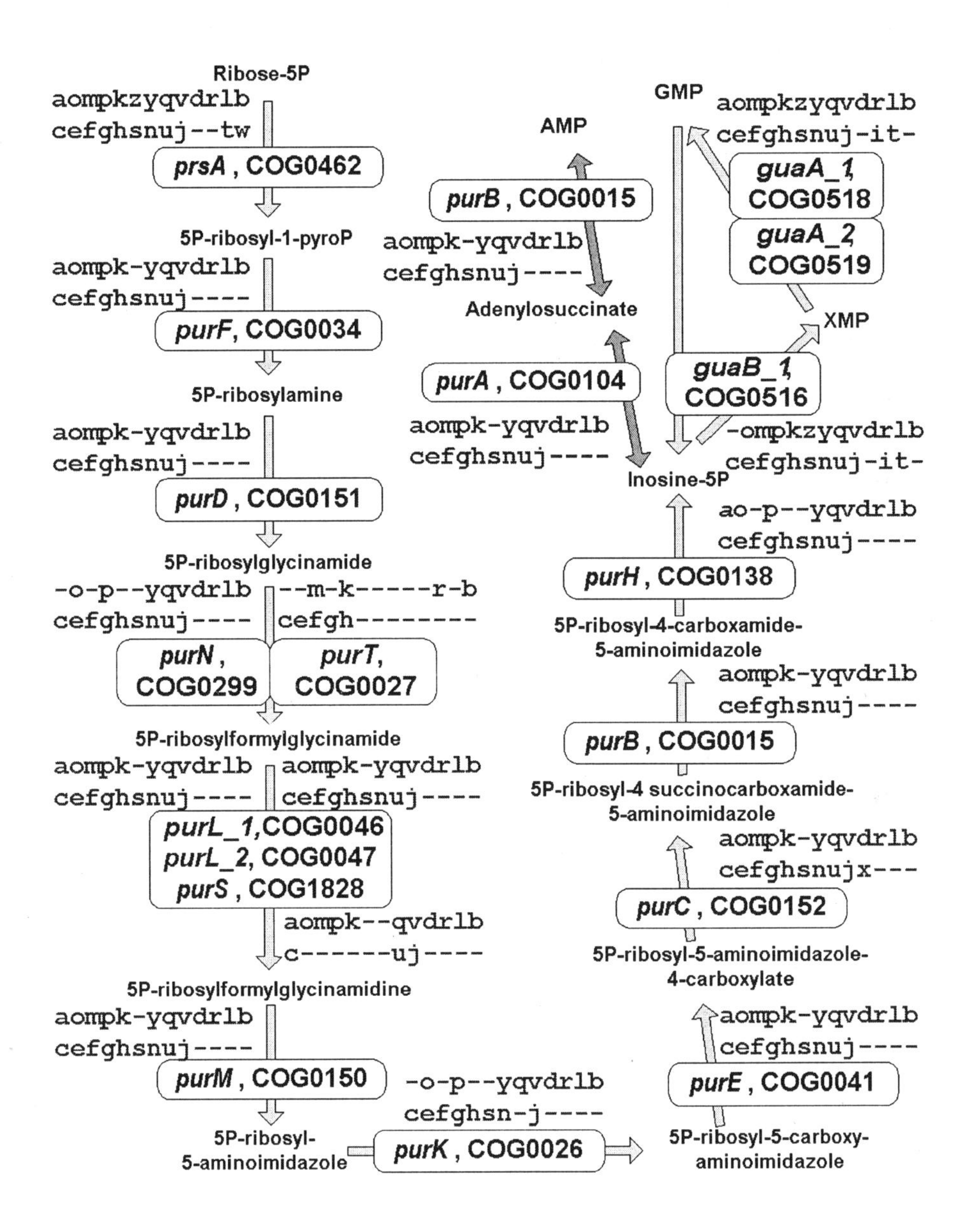

Figure 7.5. Distribution of purine biosynthesis enzymes in organisms with completely sequenced genomes. All details as in Figure 2.7.

whereas IMP cyclohydrolase comprises the N-terminal 200-aa region [693]. AICAR transformylase is encoded in almost the same set of organisms as all other purine biosynthesis enzymes, with the exception of *A. fulgidus*, which encodes only the IMP cyclohydrolase portion of PurH, and methanogens and pyrococci that do not encode either of these enzymes.

IMP cyclohydrolase (EC 3.5.4.10)

IMP cyclohydrolase, which catalyzes the last step of purine biosynthesis, is fused to AICAR transformylase in every organism, except for *A. fulgidus*, which does not have an AICAR transformylase at all, and *Halobacterium* sp., in which the AICAR transformylase domain is fused to PurN, a different folate-dependent GAR transformylase (see above). As noted above, methanogens and pyrococci do not encode a recognizable IMP cyclohydrolase.

Adenylosuccinate synthase (EC 6.3.4.4)

Conversion of IMP into AMP can occur in one step, which is catalyzed by the eukaryote-specific enzyme AMP deaminase (EC 3.5.4.6), or in two steps, as in most bacteria and archaea. First, IMP is converted into adenylosuccinate by adenylosuccinate synthase PurA. In addition to the entire set of organisms that encode enzymes of IMP biosynthesis, PurA is also encoded in *H. pylori*. The second step, the conversion of adenylosuccinate into AMP, is catalyzed by adenylosuccinate lyase PurB (see above), which has the same phyletic pattern as PurA.

IMP dehydrogenase (EC 1.1.1.205)

Although the reverse reaction, catalyzed by GMP reductase (EC 1.6.6.8), occurs in one step, conversion of IMP into GMP takes two steps. First, IMP is oxidized into XMP by IMP dehydrogenase GuaB, a close paralog of GMP reductase, which, however, contains an ~120 amino acid insert comprising two CBS domains involved in allosteric regulation of the enzyme activity [79,941]. Because CBS is a "promiscuous" domain, which is found in association with various proteins [26], it has caused numerous errors in automated genome annotation (see ♦5.2.2) [264]. Thus, at least twelve *A. fulgidus* proteins have been annotated as IMP dehydrogenases or "IMP dehydrogenase-related" proteins [444], whereas, ironically, the real IMP dehydrogenase appears not to be encoded in the *A. fulgidus* genome. With the exception of this archaeon, IMP dehydrogenase is present in almost every bacterial and archaeal genome sequenced to date, including *A. pernix*, *C. pneumoniae*, and *B. burgdorferi*, which do not encode any enzymes of IMP biosynthesis and apparently have to import this nucleotide.

GMP synthase (EC 6.3.5.2)

GMP synthase is another amidotransferase that consists of two subunits, a glutamine amidotransferase of Triad family [936] and a synthetase subunit, which belongs to the PP-loop superfamily of ATP pyrophosphatases [102]. The phylogenetic pattern of both GMP synthase subunits is the same as that of IMP dehydrogenase, with the addition of *A. fulgidus*, i.e. this enzyme is also found in *A. pernix*, *C. pneumoniae*, and *B. burgdorferi,* which lack many other purine biosynthesis enzymes. In bacteria, yeast and other eukaryotes, and in *A. pernix*, these two subunits are fused together in the same polypeptide, whereas, in other archaea, they are encoded by separate genes.

General notes on purine biosynthesis evolution

The phyletic distribution of purine biosynthesis enzymes shows some of the trends noted above for other pathways, i.e. non-orthologous gene displacement and increased loss of enzymes for upstream steps of the pathway as compared to the downstream steps. With the exception of several obligate parasites with very small genomes and *Buchnera* sp., most bacteria encode the entire set of purine biosynthesis enzymes; there is little doubt that they are all capable of IMP formation. Based on their gene content, bacteria *H. pylori, C. pneumoniae,* and *B. burgdorferi* and the archaeon *A. pernix* are only capable of converting IMP into GMP; AMP formation in these organisms probably occurs through the activity of adenine phosphoribosyltransferase or some other mechanism. While *Halobacterium* sp. and *Thermoplasma* spp. encode all the enzymes of purine biosynthesis, other archaea appear to miss at least two *pur* genes. Methanogens and pyrococci lack *purK* and *purH* genes, and *A. fulgidus* additionally lacks *purN/purT* and *guaB*, making it hard to judge whether or not purine biosynthesis pathway is functional in this organism. Purine biosynthesis is much more likely to occur in methanogens and pyrococci, which would then need to harbor alternative versions of AICAR transformylase and IMP cyclohydrolase and, potentially, an alternative version of AIR carboxylase. Thus far, no obvious candidates for these activities have been identified by comparative genome analysis of these organisms. It is amazing that, although purine biosynthesis has been intensely studied for over 50 years, comparative genomics reveals unsuspected gaps in our understanding of this pathway and may eventually lead to the discovery of novel enzymes.

7.4. Amino Acid Biosynthesis

7.4.1. Aromatic amino acids

7.4.1.1. Common steps of the pathway

The biosynthetic pathways for phenylalanine, tyrosine, and tryptophan in bacteria and eukaryotes share common steps leading from phosphoenol-pyruvate and erythrose-4-phosphate to chorismate. Enzymes for most of these steps are encoded also in archaeal genomes.

2-Dehydro-3-deoxy-D-arabino-heptonate 7-phosphate synthase (EC 4.1.2.15)

Although 2-dehydro-3-deoxy-D-arabino-heptonate 7-phosphate (DAHP) synthase is found in *E. coli* in three different versions, AroF, AroG, and AroH, all these enzymes are close paralogs and represent the so-called microbial form of DAHP synthase. A different form of this enzyme was originally described in potato and *Arabidopsis* and designated the plant form. Subsequently, this form has been discovered also in bacteria [205,298,433,879]. This form is encoded in many complete genomes and is the only DAHP synthase in *M. tuberculosis*, *M. leprae*, *H. pylori*, and *C. jejuni* (Figure 7.6). *B. subtilis* and several othei gram-positive bacteria encode a third form of DAHP synthase, referred to as AroA(G), which is homologous to 3-deoxy-D-manno-octulosonate 8-phosphate synthase of *E. coli* [96]. Remarkably, this third form is also found in *T. maritima* and in several archaea, such as *P. abyssi*, *A. pernix*, and *Thermoplasma* spp. Other archaea, such as *A. fulgidus*, *M. jannaschii*, and *M. thermoautotrophicum*, as well as the bacterium *A. aeolicus* encode neither of these three DAHP synthases and appear to synthesize 3-dehydroquinate via a different mechanism that does not include DAHP as an intermediate (see below).

3-Dehydroquinate synthase (EC 4.6.1.3)

3-Dehydroquinate synthase is found in many bacteria and in some archaea. Remarkably, the phyletic pattern of this enzyme exactly corresponds to the overlap of the phyletic patterns for the three forms of DAHP synthase:

```
------y--d-l--efghsn-j----  COG0722  DAHP synthase
----------r----f----uj----  COG3200  DAHP synthase
---pkz--vd-lbc-----n---i--  COG2876  DAHP synthase
--------------------------------------------------
---pkzy-vdrlbcefghsnuj-i--  All forms of DAHP synthase  and
---pkzy-vdrlbcefghsnuj-i--  COG0337 3-dehydroquinate synthase
```

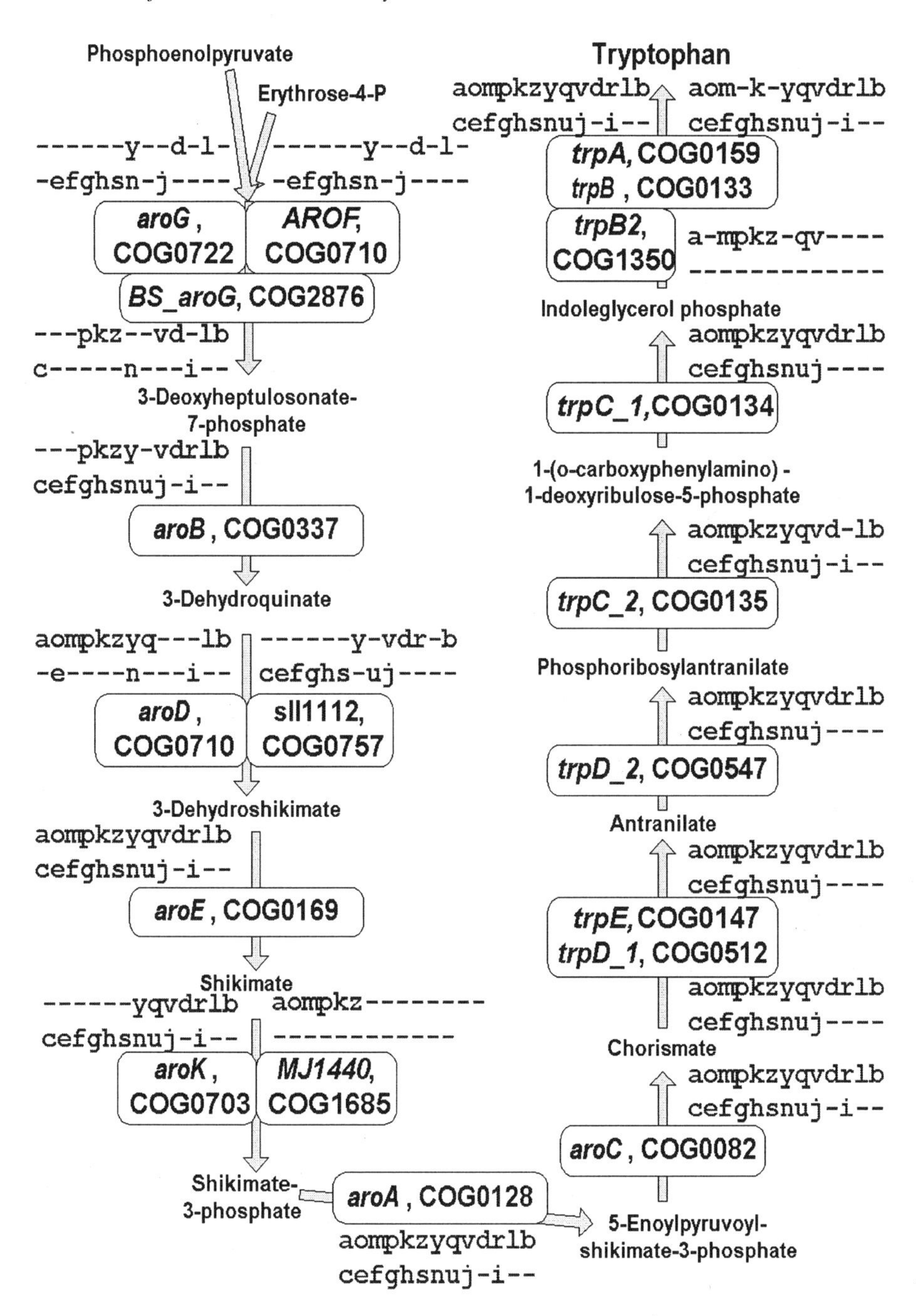

Figure 7.6. Distribution of tryptophan biosynthesis enzymes in organisms with completely sequenced genomes. All details are as in Figure 2.7.

This correlation suggests that a single form of 3-dehydroquinate synthase can account for the conversion of DAHP into 3-dehydroquinate in all the organisms with completely sequenced genomes and probably represents the only form of this enzyme.

3-Dehydroquinate dehydratase (EC 4.2.1.10)

Two forms of 3-dehydroquinate dehydratase have been characterized and designated class I (encoded by *aroD* gene) and class II (encoded by *aroQ* or QUTE genes), respectively. Taken together, these two enzymes completely cover the phyletic diversity of the organisms that encode 3-dehydroquinate synthase:

```
aompkzyq---lb-e----n---i-- AroD COG0710
------y-vdr-bcefghs-uj---- AroQ COG0757
------------------------------------------
aompkzyqvdrlbcefghsnuj-i-- 3-dehydroquinate dehydratase
---pkzy-vdrlbcefghsnuj-i-- 3-dehydroquinate synthase.
```

Notably, dehydroquinate dehydratase (as well as most of the other enzymes of tryptophan biosynthesis) is found in several genomes that do not encode dehydroquinate synthase, indicating the existence of an alternative, still uncharacterized pathway of dehydroquinate formation in *Halobacterium* sp., *A. fulgidus*, *M. jannaschii*, and *M. thermoautotrophicum*, and *A. aeolicus*.

Shikimate 5-dehydrogenase (EC 1.1.1.25)

The last enzyme in the shikimate-producing part of the pathway, shikimate 5-dehydrogenase, is encoded in most of the completely sequenced bacterial (with the exception of rickettsiae, spirochetes and mycoplasmas) and archaeal genomes (with the exception of *P. horikoshii*). The phyletic pattern for shikimate 5-dehydrogenase coincides with the combined pattern of the two forms of 3-dehydroquinate synthase:

```
aompkzyqvdrlbcefghsnuj-i-- All 3-dehydroquinate dehydratases
aompkzyqvdrlbcefghsnuj-i-- COG0169 Shikimate dehydrogenase
```

Shikimate kinase (EC 2.7.1.71)

The typical form of shikimate kinase, found in bacteria and eukaryotes, is not encoded in any archaeal genome sequenced so far. Recently, a shikimate kinase of the GHMP superfamily has been identified and experimentally studied in *M. jannaschii* [171]. This enzyme is encoded in each archaeal genome, except for *P. horikoshii*. Together, these two forms of shikimate kinase have the same phyletic pattern as the combination of the two forms of 3-dehydroquinate dehydratase or shikimate dehydrogenase:

```
------yqvdrlbcefghsnuj-i--  COG0703  Shikimate kinase
aompkz--------------------  COG1685  Shikimate kinase
aompkzyqvdrlbcefghsnuj-i--  Shikimate kinase (all forms).
```

5-enolpyruvylshikimate 3-phosphate synthase (EC 2.5.1.19)

Like shikimate dehydrogenase, 5-enolpyruvylshikimate 3-phosphate synthase (AroA) is found in only one form with the same phyletic pattern as the preceding enzyme.

Chorismate synthase (EC 4.6.1.4)

The only known chorismate synthase (AroC) has the same phyletic pattern as shikimate dehydrogenase and 5-enolpyruvylshikimate 3-phosphate synthase.

7.4.1.2. Tryptophan biosynthesis

After chorismate, the tryptophan biosynthetic pathway deviates from the pathways leading to phenylalanine and tyrosine. In the tryptophan branch, all the remaining enzymes have very similar phyletic patterns.

Anthranilate synthase (EC 4.1.3.27)

Anthranilate synthase and the closely related para-aminobenzoate synthase consist of two components, the synthetase subunit and the glutamine amidotransferase subunit, which, in most organisms, are encoded by separate genes *trpG* (or *pabA*) and *trpE* (or *pabB*). In *E.coli*, the *trpG* gene for glutamine amidotransferase subunit is fused to the *trpD* gene that encodes anthranilate phosphoribosyltransferase, the enzyme catalyzing the next step of the pathway. This sometimes leads to a confusion in nomenclature, with the *trpG* gene being referred to as *trpD* or as *trpD_1*. The phyletic pattern for anthranilate synthase is the same as the patterns described above for shikimate dehydrogenase, 5-enolpyruvylshikimate 3-phosphate synthase, and chorismate synthase, with the exception that anthranilate synthase is missing in chlamydiae.

Anthranilate phosphoribosyltransferase (EC 2.4.2.18)

There is only one form of anthranilate phosphoribosyltransferase that shows almost the same phyletic pattern as anthranilate synthase. The only diversification in the existing set of genomes is the absence of the trpD gene (as well as genes for the remaining steps of tryptophan biosynthesis) in *S.*

pyogenes. This probably means that genes annotated as *trpG* and *trpE* genes in *S. pyogenes* actually encode para-aminobenzoate synthase and have no role in tryptophan biosynthesis.

N-(5'-phosphoribosyl)anthranilate isomerase (EC 5.3.1.24)

Phosphoribosylanthranilate isomerase is also represented by only one form with essentially the same phyletic pattern as anthranilate phosphoribosyltransferase. Here again, the nomenclature is somewhat complicated because of a gene fusion in *E. coli*. The phosphoribosylanthranilate isomerase gene that, in most species, is referred to as *trpF*, in *E. coli* is fused to the *trpC* gene that encodes indole-3-glycerol phosphate synthase, the enzyme for the next step of the pathway. Therefore, in *E. coli*, the *trpF* gene is sometimes also referred to as *trpC*, which can lead to confusion. The phyletic pattern of phosphoribosylanthranilate isomerase is essentially the same as that of other enzymes of tryptophan biosynthesis, with the most notable difference being the apparent absence of *trpF* in *M. tuberculosis* and *M. leprae*. Another peculiarity is the unusual distribution of phosphoribosylanthranilate isomerase in different chlamydial species: while *C. trachomatis* and *C. muridarum* both have the *trpF* gene, *C. pneumoniae* does not. This probably reflects the ongoing gene loss in the evolution of chlamydiae.

Indole-3-glycerol phosphate synthase (EC 4.1.1.48)

Indole-3-glycerol phosphate synthase exists in a single form with the same phyletic pattern as anthranilate phosphoribosyltransferase.

Tryptophan synthase (EC 4.2.1.20)

Tryptophan synthase consists of two subunits, which are encoded by *trpA* and *trpB* genes. Their phyletic patterns are similar to that of anthranilate phosphoribosyltransferase, with the exception that, as seen above for *trpC*, *trpA* and *trpB* genes are found in *C. trachomatis* and *C. muridarum* but not in *C. pneumoniae*.

7.4.1.3. Phenylalanine and tyrosine biosynthesis

Chorismate mutase (EC 5.4.99.5)

Chorismate mutase is involved in both phenylalanine and tyrosine biosynthesis. The best-known version of this enzyme is found in *E. coli* in two paralogous forms fused with prephenate dehydratase in PheA and prephenate dehydrogenase in TyrA. In addition to these two forms, there is (i) a distantly related form of chorismate mutase encoded in yeast, fungi, and in plant cells and (ii) an unrelated monofunctional form found in *B. subtilis*, *Synechocystis* sp., and many other gram-positive bacteria and cyanobacteria.

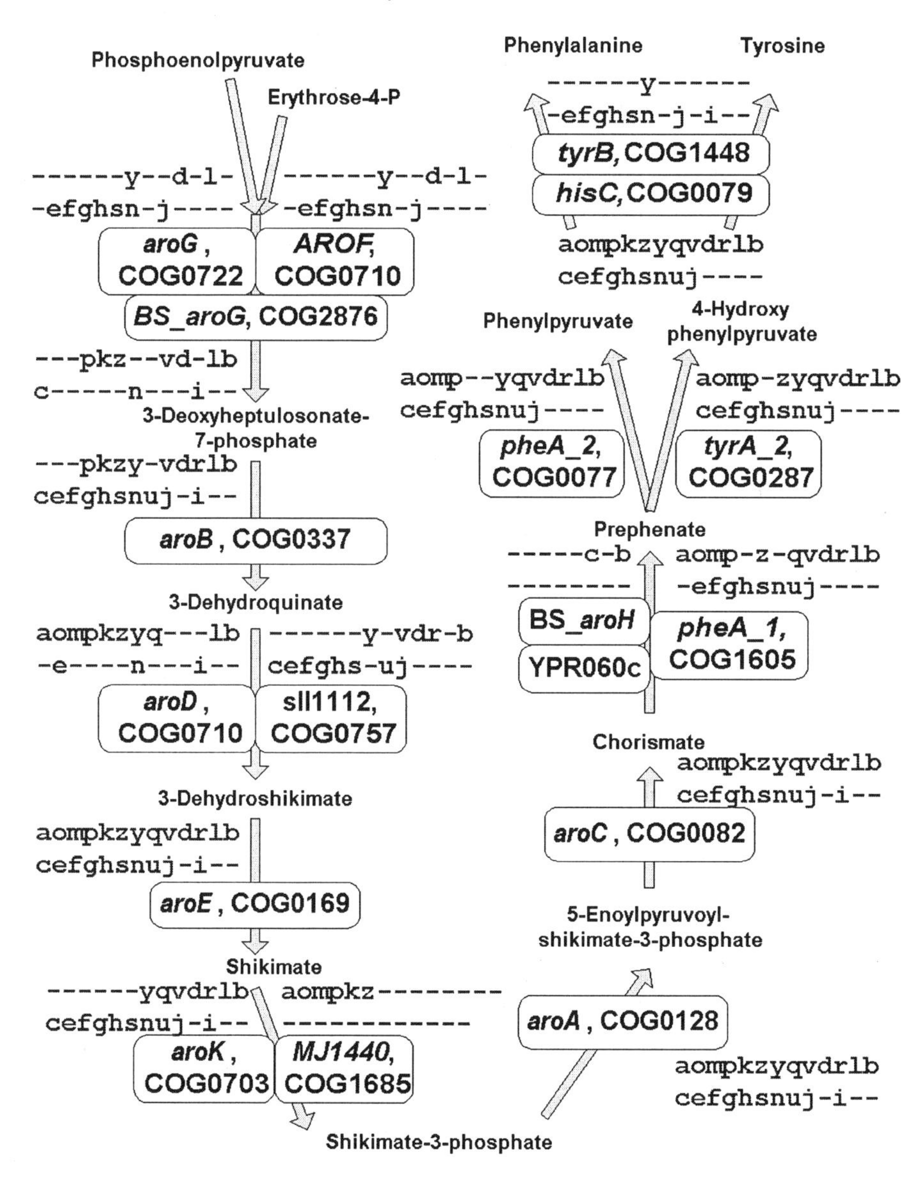

Figure 7.7. Distribution of phenylalanine and tyrosine biosynthesis enzymes in organisms with completely sequenced genomes. All details are as in Figure 2.7.

While these three forms of chorismate mutase show almost no sequence similarity to each other, structural comparisons indicate that the *E. coli* and yeast enzymes are related to each other and unrelated to the form found in *B. subtilis* and *Th. thermophilus*. A comparison of the combined phyletic pattern of all these forms of chorismate mutase with that of chorismate synthase shows that, with the exception of *P. abyssi* and *Chlamydia* spp., all organisms that produce chorismate are capable of converting it to prephenate. The fate of chorismate in *Chlamydia* spp. remains obscure.

Prephenate dehydrogenase (EC 1.3.1.12)

Prephenate dehydrogenase TyrA, an enzyme of the tyrosine biosynthesis branch of the pathway, is found in a single form with almost the same phyletic pattern as chorismate mutase (Figure 7.7). The only exception is *S. pyogenes* that appears not to encode prephenate dehydrogenase.

Prephenate dehydratase (EC 4.2.1.51)

Prephenate dehydratase, an enzyme of the phenylalanine biosynthesis branch of the pathway, is also represented by a single form in all known organisms. However, its phyletic pattern shows the absence of this enzyme in *S. pyogenes*, *H. pylori*, and *A. pernix*, which are all capable of producing prephenate, suggesting that these organisms either lack phenylalanine biosynthesis or have an alternative form of prephenate dehydratase.

Aromatic aminotransferase (EC 2.6.1.1, 2.6.1.5, 2.6.1.9, 2.6.1.57)

There are several families of pyridoxal-phosphate-dependent aminotransferases that are capable of producing tyrosine and phenylalanine from, respectively, 4-hydroxyphenylpyruvate and phenylpyruvate. Although the best-studied tyrosine aminotransferase, *E. coli* TyrB, has a relatively narrow phyletic distribution, homologs of histidinol phosphate aminotransferase and aspartate aminotransferase are encoded in every bacterial and archaeal genome except for spirochetes and mycoplasmas. Thus, once phenylpyruvate and 4-hydroxyphenylpyruvate are synthesized, their transamination into, respectively, phenylalanine and tyrosine can be performed by all organisms whose genome sequences are currently available.

A summary on aromatic amino acid biosynthesis

Because aromatic amino acid biosynthesis shares common steps with biosynthesis of ubiquinone, this pathway displays a stunning variety of alternative enzymes catalyzing the same reaction. This makes analysis of their phyletic patterns rather complicated, but, at the same time, allows one to draw some interesting conclusions. Most bacteria and archaea retain the complete set of genes for tryptophan biosynthesis. The exceptions are the obligate archaeal heterotroph *P. horikoshii* and some obligate bacterial parasites, such as *S. pyogenes*, rickettsiae, chlamydiae, spirochetes and mycoplasmas, which apparently obtain tryptophan, just like many other nutrients, from other microbes and from the host, respectively.

Enzymes of the tyrosine biosynthesis pathway are encoded in almost as many complete genomes, with the conspicuous exception of *P. abyssi*. One could speculate that, while tryptophan is rapidly degraded at 105°C (the optimal growth temperature of this organism), tyrosine is not, which alleviates the requirement for *de novo* synthesis. These considerations could also explain the absence of phenylalanine biosynthesis in *P. abyssi* and *A. pernix*.

The consistency of the phyletic patterns of the enzymes for the downstream stages of aromatic amino acid biosynthesis underscores the remaining problem with the early stages. Indeed, *A. aeolicus* and four archaeal species encode 3-dehydroquinate dehydratase and all the downstream enzymes but do not encode either DAHP synthase or 3-dehydroquinate synthase:

```
aompkzyqvdrlbcefghsnuj-i-- All forms 3-dehydroquinate dehydratase
---pkzy-vdrlbcefghsnuj-i-- All forms DAHP synthase
---pkzy-vdrlbcefghsnuj-i-- COG0337 3-dehydroquinate synthase
—  ——————————————————————————————————————————————————————————————
aom----q------------------ Missing 3-dehydroquinate synthase
```

It appears that these organisms produce 3-dehydroquinate via a different mechanism, which does not include DAHP as an intermediate. Using the COG phyletic pattern search tool, one could search for orthologous protein sets that are represented in those five genomes, but are missing in thermoplasmas, pyrococci, *A. pernix*, and *T. maritima,* all of which encode a DAHP synthase and a 3-dehydroquinate synthase. Such a search identified just four COGs, only one of which, COG1465, consisted of uncharacterized proteins. These proteins, orthologs of *M. jannaschii* MJ1249, can be predicted to function as an alternative 3-dehydroquinate synthases (MYG, unpublished). This prediction seems to be further supported by the adjacency of the genes encoding COG1465 members AF0229 and VNG0310C to the *aroC* gene in the genomes of *A. fulgidus* and *Halobacterium* sp., respectively. However, even if this prediction is correct, the exact nature of the precursor for 3-dehydroquinate and the mechanism of its biosynthesis in these organisms need to be elucidated experimentally.

7.4.2. Arginine biosynthesis

N-acetylglutamate synthase (EC 2.3.1.1, 2.3.1.35),

The first step in arginine biosynthesis from glutamate is its acetylation, with either acetyl-CoA or acetylornithine utilized as donors of the acetyl group (Figure 7.8). In *E. coli* and several other organisms, this reaction is catalyzed by the acetyltransferase ArgA, which employs acetyl-CoA as the acetyl donor. In all proteobacteria that encode this enzyme, the *argA* gene is fused to the gene for N-acetylglutamate kinase, which catalyzes the next step of the pathway. Like in other domain fusion cases, confusion occasionally emerges during genome annotation, especially because the N-terminal kinase domain, which consists of ~300 amino acid residues, can make the C-terminal acetyltransferase domain almost invisible in BLAST outputs (♦2.4.4.5).

A different, unrelated N-acetylglutamate synthase (N-acetylornithine transferase, the *argJ* gene product) is present in *B. subtilis*, yeast, and many other organisms. This enzyme couples acetylation of glutamate with deacetylation of N-acetylornithine, which is the fifth step in arginine biosynthesis. This activity allows recycling of the acetyl group in the arginine biosynthesis pathway.

N-acetylglutamate kinase (EC 2.7.2.8)

Phosphorylation of N-acetylglutamate is catalyzed by the product of the *argB* gene, a kinase with the carbamate kinase fold. This enzyme is found in a wide variety of organisms, such that its phyletic pattern is even broader than the combined patterns of both enzymes that generate N-acetylglutamate:

```
-------q-dr---efgh-n------ ArgA N-acetylglutamate synthase
a-m---yqvdrlbc-f---n-j---- ArgJ N-acetylornithine transferase
---------------------------------------------------------------
a-m---yqvdrlbcefgh-n-j---- Both N-acetylglutamate synthetases
a-m-kzyqvdrlbcefghsnuj---- ArgB N-acetylglutamate kinase
```

However, N-acetylglutamate kinase is not encoded in the genomes of many parasitic bacteria, such as *S. pyogenes*, *H. influenzae*, *H. pylori*, chlamydiae, rickettsiae, spirochetes, and mycoplasmas.

N-acetyl-gamma-glutamyl phosphate reductase (EC 1.2.1.38)

The enzyme that catalyzes the next step of the pathway, ArgC, has the same phyletic pattern as ArgB. In fungi, *argB* and *argC* genes are fused and encode a single bifunctional protein.

N-acetylornithine aminotransferase (EC 2.6.1.11)

N-acetylornithine deacetylase (N-acetylornithinase) belongs to a large family of closely related acetyltransferases (deacetylases), which is represented by two or more paralogs even in the relatively small genomes of *H. influenzae*, *L. lactis*, and *S. pyogenes*. Although proper assignment of substrate specificity in such a case is difficult, if not impossible, the few organisms that produce N-acetylornithine but lack the ArgJ-type N-acetylglutamate synthase offer an ample choice of candidates for the function of N-acetylornithinase.

N- acetylornithine deacetylase (EC 3.5.1.16)

N-acetylornithinase belongs to a large family of closely related acetyl-transferases (deacylases), which is represented by two or more paralogs even in the relatively small genomes of *H. influenzae*, *L. lactis*, and *S. pyogenes*. Although proper assignment of substrate specificity in such a case is difficult, if not impossible, the few organisms that produce N-acetylornithine but lack the ArgJ-type N-acetylglutamate synthase offer an ample choice of candidates for the role of N-acetylornithinase.

Ornithine carbamoyltransferase (EC 2.1.3.6)

Ornithine carbamoyltransferase catalyzes the sixth step of arginine biosynthesis, conversion of ornithine into citrulline. Carbamoyl phosphate that serves as the second substrate of this reaction is provided by carbamoyl phosphate synthetase, which was discussed above (♦7.2). Ornithine carbamoyltransferase has a much wider phyletic distribution than other enzymes of arginine biosynthesis. This is probably due to the fact that it also catalyzes the reverse reaction, i. e. phosphorolysis of citrulline with the formation of ornithine and carbamoyl phosphate, which is part of the urea cycle. Accordingly, ornithine carbamoyltransferase is found in humans and other higher eukaryotes, which have the urea cycle, but are incapable of arginine biosynthesis.

Argininosuccinate synthase (EC 6.3.4.5)

Like ornithine carbamoyltransferase, argininosuccinate synthase participates in the urea cycle. As a result, its phyletic distribution is also wider than that of the enzymes that catalyze early steps of arginine biosynthesis. This enzyme, too, is found in humans and in bacteria, such as *H. influenzae*, which have all the urea cycle enzymes, but lack several enzymes of arginine biosynthesis.

Argininosuccinate lyase (EC 4.3.2.1)

Argininosuccinate lyase, the last enzyme of arginine biosynthesis, splits argininosuccinate into arginine and fumarate. Like the two preceding enzymes, it also participates in the urea cycle, and its phyletic pattern is nearly identical to that of argininosuccinate synthetase.

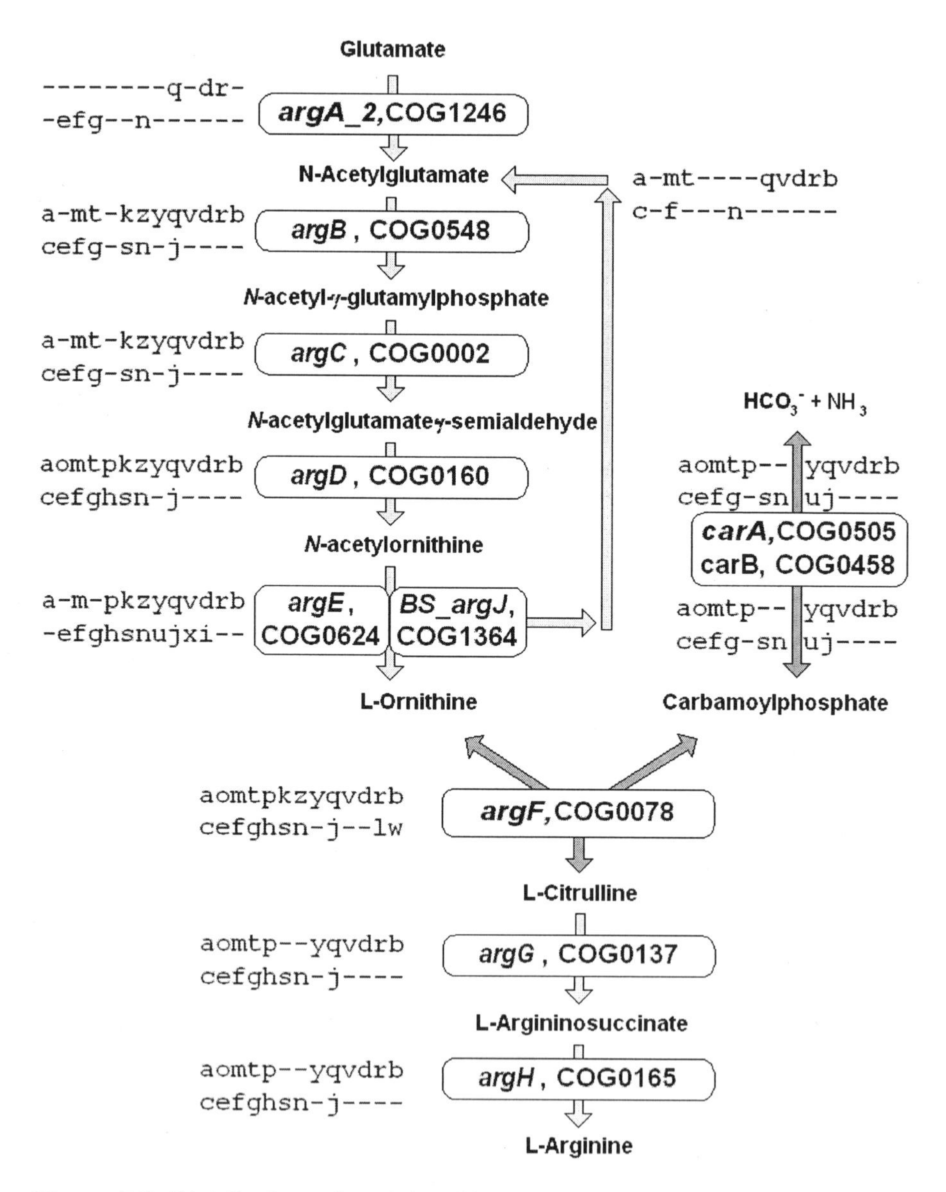

Figure 7.8. Distribution of arginine biosynthesis enzymes in organisms with completely sequenced genomes. All details as in Figure 2.7.

7.4.3. Histidine biosynthesis

In contrast to the pathways of aromatic amino acid and arginine biosynthesis, histidine biosynthesis exhibits remarkable consistency of the phyletic patterns of all the enzymes involved (Figure 7.9). While the first enzyme of the pathway, phosphoribosylpyrophosphate synthetase (EC 2.7.6.1), also participates in the purine biosynthesis pathway (see above), nearly all the committed enzymes of histidine biosynthesis have the same phyletic pattern, indicating that this pathway is encoded in the great majority of complete prokaryotic genomes sequenced to date. The exceptions are the heterotrophic archaea *Thermoplasma* spp., *Pyrococcus* sp., and *A. pernix*, and parasitic bacteria with small genomes, namely rickettsiae, chlamydiae, spirochetes, and mycoplasmas, as well as *S. pyogenes* and *H. pylori* (despite their larger genomes). Remarkably, the aphid symbiont *Buchnera* sp., which has second-smallest genome available to date, encodes the complete set of histidine biosynthesis enzymes.

There are several deviations from this common pattern. First, phosphoribosyl-ATP pyrophosphatase (EC 3.6.1.31) was not detected in *A. fulgidus*. Since this organism encodes genes for all other enzymes of histidine biosynthesis, one should assume that this reaction in *A. fulgidus* is catalyzed by an unrelated pyrophosphatase. Indeed, *A. fulgidus* genome encodes several predicted pyrophosphatases of unknown specificity (COG1694) that could be good candidates for the role of the missing phosphoribosyl-ATP pyrophosphatase

Another deviation from the common pattern is the existence of at least two unrelated histidinol phosphatases (Figure 7.9), one of which has been experimentally characterized in *E. coli* and the other in yeast and *B. subtilis* [15,500]. The latter form of this enzyme (COG1387) belongs to a large superfamily of PHP-type phosphohydrolases [41], which have common sequence motifs but clearly differ in substrate specificity. A closer inspection of COG1387 shows that proteins from yeast, *B. subtilis*, *B. halodurans*, *L. lactis*, *D. radiodurans*, and *T. maritima* comprise a tight orthologous set and can be confidently predicted to possess histidinol phosphatase activity. Other members of this COG are more distantly related to the experimentally characterized histidinol phosphatases from yeast and *B. subtilis* and might have other substrates. In addition, both forms of histidinol phosphatase are missing in *Halobacterium* sp. Therefore it appears likely that there is yet another, so far unrecognized, form of histidinol phosphatase in *Halobacterium* sp., *Thermoplasma* spp., and other organisms. There are plenty of unassigned predicted hydrolases that could potentially have this activity.

Remarkably, the ortholog of the *E. coli* histidinol phosphatase (HisB, COG0241) is encoded in *H. pylori*), which lacks all the other enzymes of histidine biosynthesis. This protein most likely represents a case of enzyme recruitment and functions as a phosphatase that hydrolyzes some other phosphoester

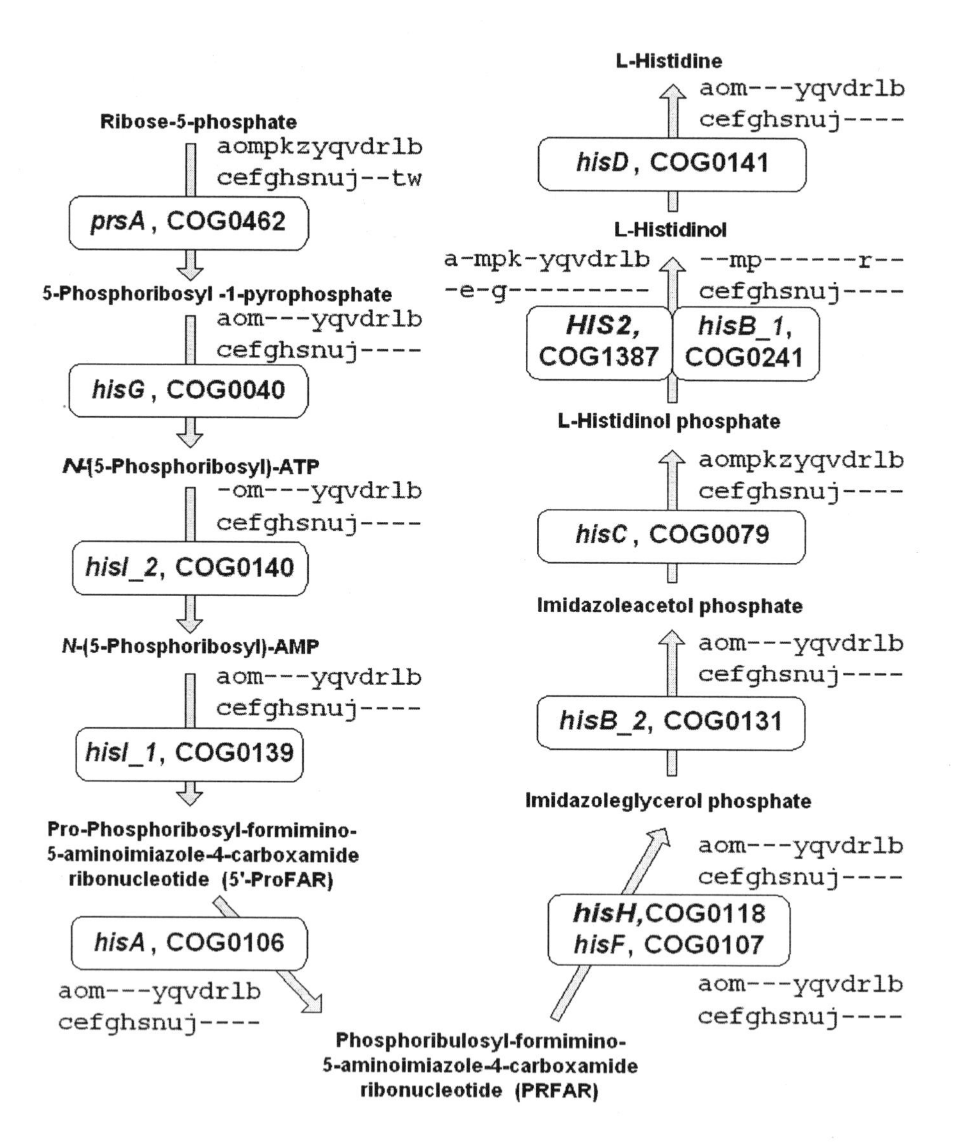

Figure 7.9. Distribution of histidine biosynthesis enzymes in organisms with completely sequenced genomes. All details as in Figure 2.7.

7.4.4. Biosynthesis of branched-chain amino acids

To those readers who are already tired of numerous instances of non-orthologous gene displacement in metabolic pathways, biosynthesis of leucine, isoleucine, and valine offers a well-deserved reprieve. In these pathways, the only instance of alternative enzymes catalyzing the same reaction is the last step, amination of α-ketomethylvaleriate, α-ketoisovaleriate, and α-ketoisocaproate. In addition to the branched-chain amino acid aminotransferase IlvE, this reaction can be catalyzed by alternative aminotransferases (Figure 7.10)

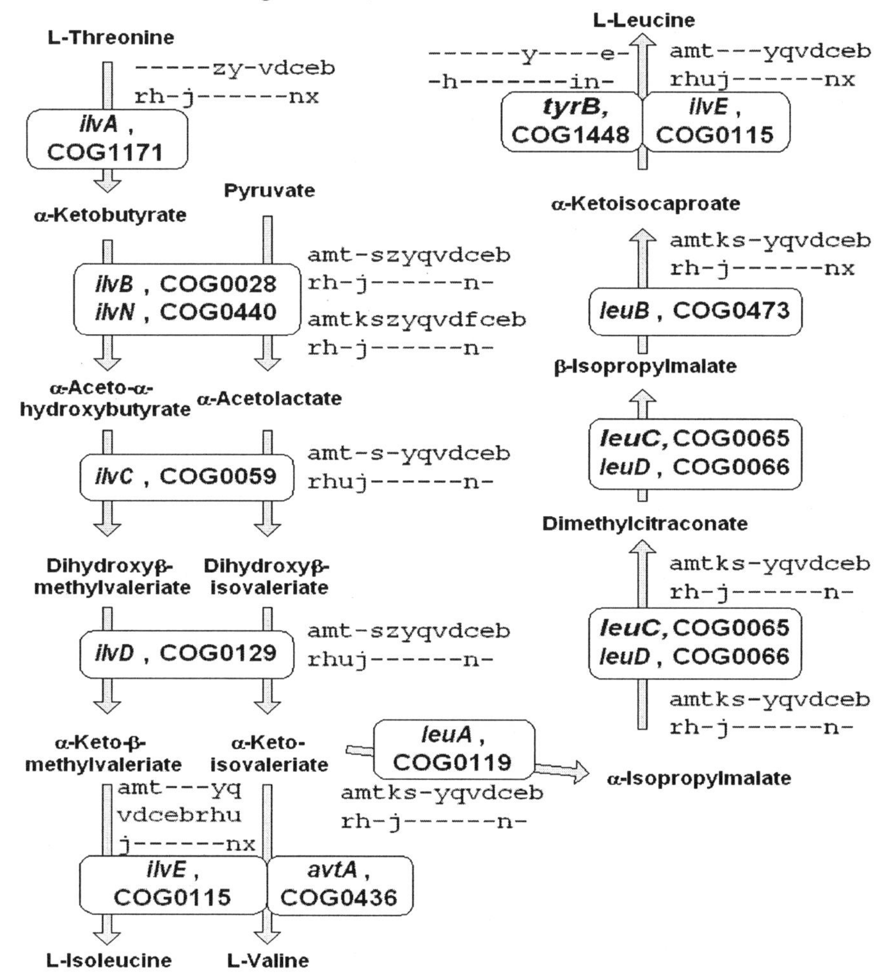

Figure 7.10. Distribution of isoleucine/leucine/valine biosynthesis enzymes in organisms with completely sequenced genomes. All details are as in Figure 2.7.

7.4.5. Proline biosynthesis

The best characterized pathway of proline biosynthesis is a three-step chain of reactions (Figure 7.11) that converts glutamate into proline through consecutive action of glutamate kinase (ProB, EC 2.7.2.11), γ-glutamyl phosphate reductase (ProA, EC 1.2.1.41) and Δ-pyrroline-5-carboxylate reductase (ProC, EC 1.5.1.2). This pathway is encoded in yeast, *E. coli*, *B. subtilis*, and in many other bacteria, including *C. jejuni* (but not *H. pylori*) and *T. pallidum* (but not *B. burgdorferi*). This pathway, however, is not detectable in archaea, except for the two species of *Methanosarcina*, which, in all likelihood, acquired it through HGT. Instead, *Halobacterium* sp., *A. fulgidus*, *M. thermoautotrophicum*, *Thermoplasma* spp., and *A. pernix* encode an unusual enzyme, ornithine cyclodeaminase (EC 4.3.1.12), which directly makes proline from ornithine. This enzyme, first discovered in tumor-inducing (Ti) plasmids of *A. tumefaciens*, was later found in pseudomonads and other bacteria [183,798]. In plants, expression of this interesting enzyme stimulates flowering [850], whereas the mammalian ortholog of this enzyme is expressed in neural tissue, including human retina, and functions as μ-crystallin, a major component of the eye lens in marsupials [438]. Although no such gene was detected in *M. jannaschii*, this archaeon, too, has been reported to possess ornithine cyclodeaminase activity [309].

An interesting aspect of proline metabolism is that its biosynthesis and degradation both proceed through the Δ-pyrroline-5-carboxylate intermediate. As a result, the proline biosynthetic pathway is sometimes confused with proline catabolism. Another complication in the analysis of proline metabolism is that, in *E. coli* and several other bacteria, the genes for proline dehydrogenase (EC 1.5.99.8) and γ-glutamate semialdehyde dehydrogenase (EC 1.5.1.12), the first and second enzymes of proline catabolism, respectively, are fused, forming the bifunctional protein PutA. In the COG database, these two domains of the PutA protein belong to two different COGs, COG0506 and COG1012 (Figure 7.11).

In conclusion, proline metabolism is tightly interlinked with arginine metabolism. Proline biosynthesis from glutamate can be reconstructed in all organisms with completely sequenced genomes with the exception of pyrococci, *H. pylori*, *B. burgdorferi*, chlamydiae, and mycoplasmas. The gene encoding ornithine cyclodeaminase in *M. jannaschii* [309] remains to be identified. It can be expected to be a member of a different enzyme family, unrelated to the known ornithine cyclodeaminases (COG2423).

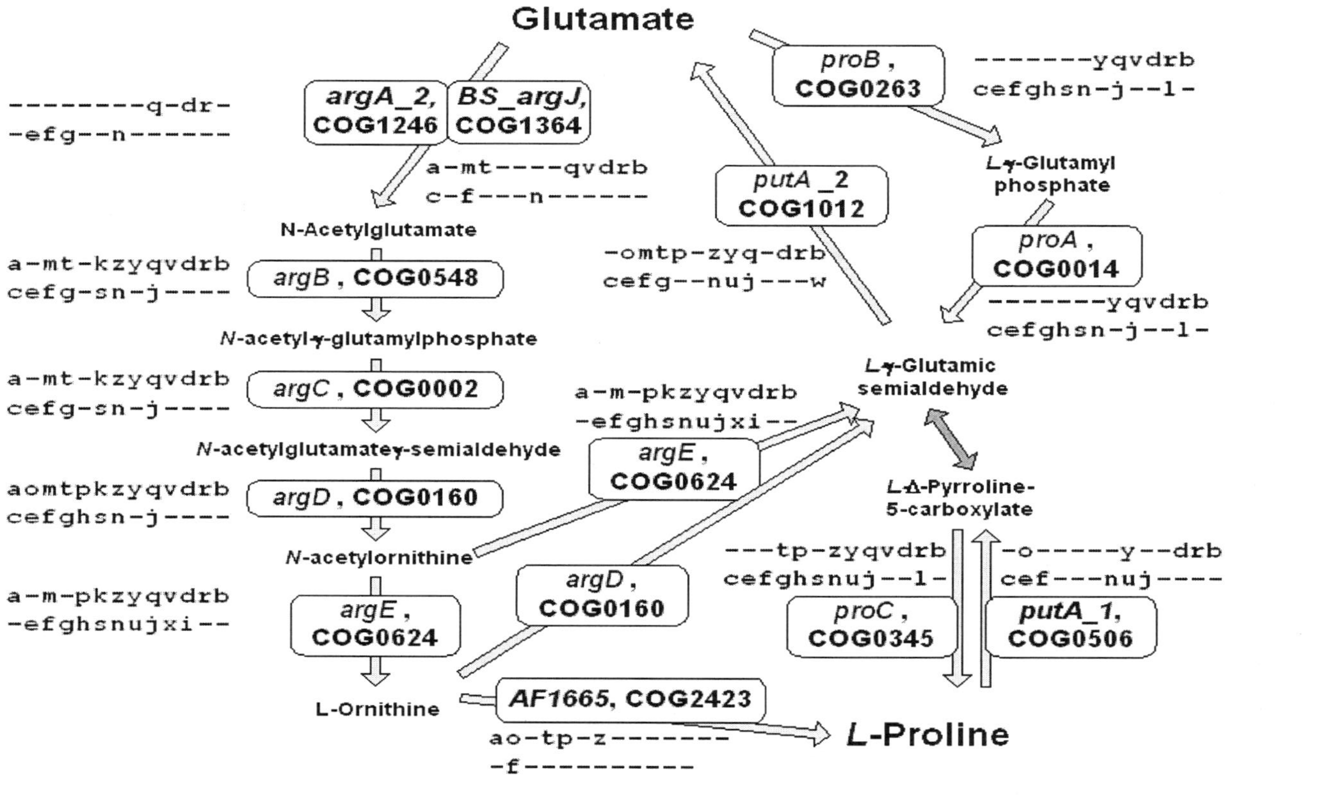

Figure 7.11. Distribution of proline biosynthesis enzymes in organisms with completely sequenced genomes. All details are as in Figure

7.5. Coenzyme Biosynthesis

7.5.1. Thiamine

Biosynthesis of cofactors (coenzymes), particularly thiamine, is a surprisingly poorly studied area of biochemistry. Although first *thi* mutations in *E. coli* have been characterized half a century ago, the complete list of thiamine biosynthesis genes has been determined only in the 1990's [868], and the functions of their products have been characterized only in the last several years [81,83,927]. The scheme for thiamine biosynthesis in Figure 7.12 was drawn using the *E. coli* data. One cannot help noticing that every enzyme on this chart has its own distinct phyletic pattern. This indicates the abundance of non-orthologous gene displacement cases among thiamine biosynthesis enzymes and suggests that different organisms might use different compounds as thiamin precursors. The apparent absence of ThiC in thermoplasmas, *A. pernix*, *H. influenzae*, and *H. pylori*, all of which encode ThiD (Figure 7.12), is a strong indication that some intermediate other than AIR is used as a precursor in these organisms. Thus, although all steps of the thiamine biosynthesis pathway have been resolved for *E. coli* [81], there is still ample opportunity for new discoveries in other organisms.

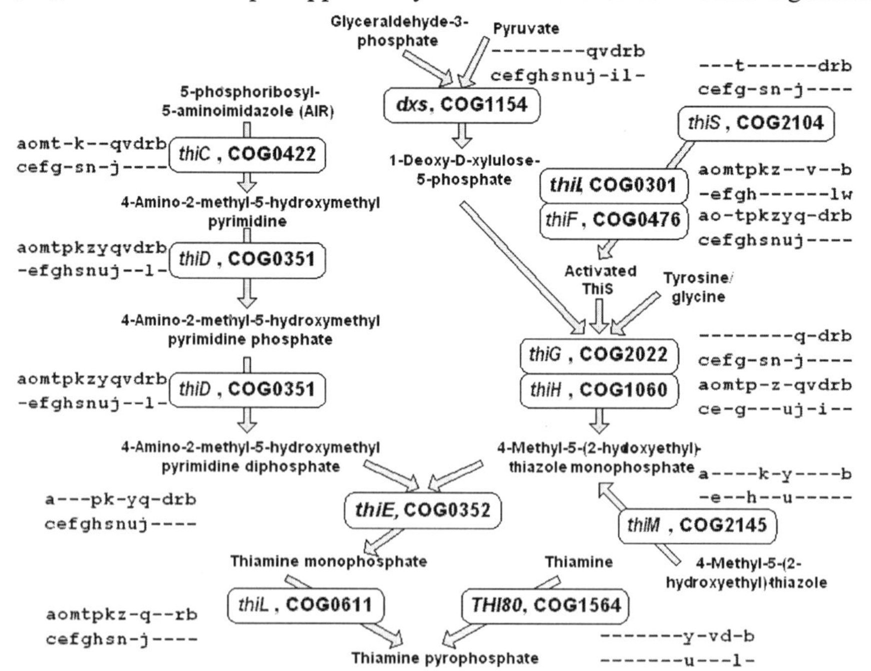

Figure 7.12. Distribution of thiamine biosynthesis enzymes in organisms with completely sequenced genomes. All details are as in Figure 2.7.

7.5.2. Riboflavin

The riboflavin biosynthesis pathway is a challenging case, with three of the seven *rib* genes characterized in *E. coli* and *B. subtilis* having no archaeal orthologs (Figure 7.13). The archaeal variant of riboflavin synthase, the last enzyme of the pathway, has been identified and turned out to be unrelated to the bacterial enzyme [207]. In contrast, the archaeal versions of the first two enzymes of the pathway, GTP cyclohydrolase II (RibA) and pyrimidine deaminase (RibD1), remain unknown, so there is an excellent chance of discovering new enzymes of this pathway.

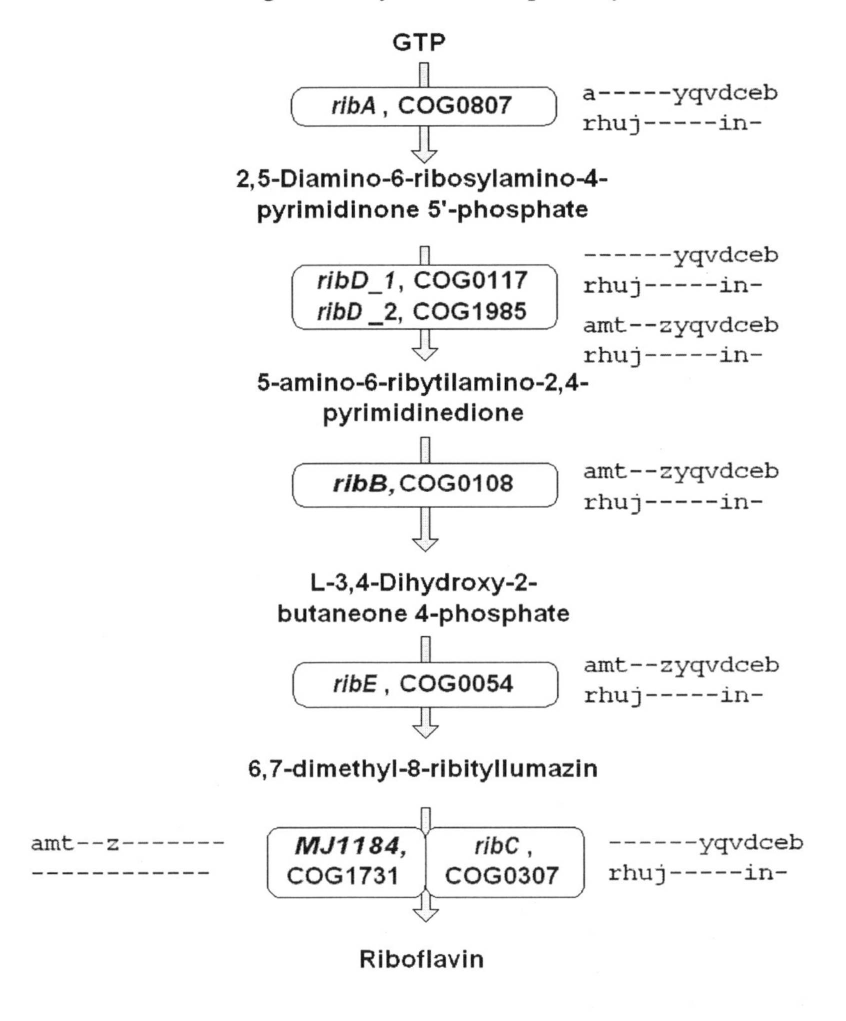

Figure 7.13. Distribution of riboflavin biosynthesis enzymes in organisms with completely sequenced genomes. All details are as in Figure 2.7.

7.5.3. NAD

Nicotinate mononucleotide adenylyltransferase, the last missing enzyme of the NAD biosynthesis pathway, has been characterized only in 2000, thanks in part to the genome-context-based methods [82,563,592]. It turned out that *E. coli* has two distantly related forms of this enzyme, which shows specificity, respectively, for mononucleotides of nicotinic acid and nicotinamide [279]. Most other organisms encode either one or the other form (Figure 7.14).

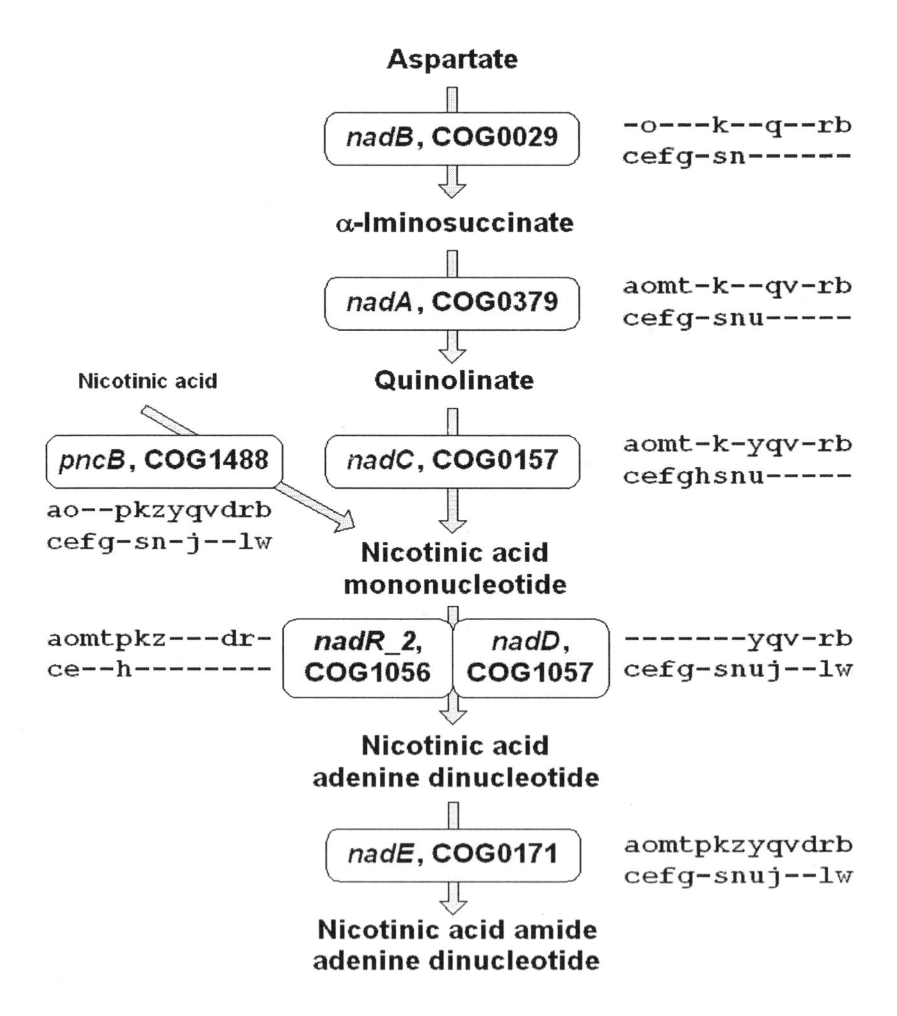

Figure 7.14. Distribution of NAD biosynthesis enzymes in organisms with completely sequenced genomes. All details are as in Figure 2.7.

7.5.4. Biotin

As is the case with many other pathways, the initial steps of biotin biosynthesis are poorly understood. The phyletic patterns of the four enzymes that catalyze the conversion of pimeloyl-CoA into biotin are relatively consistent (Figure 7.15), but the mechanisms of the formation of pimelate (6-carboxyhexanoate) and pimeloyl-CoA are still largely obscure.

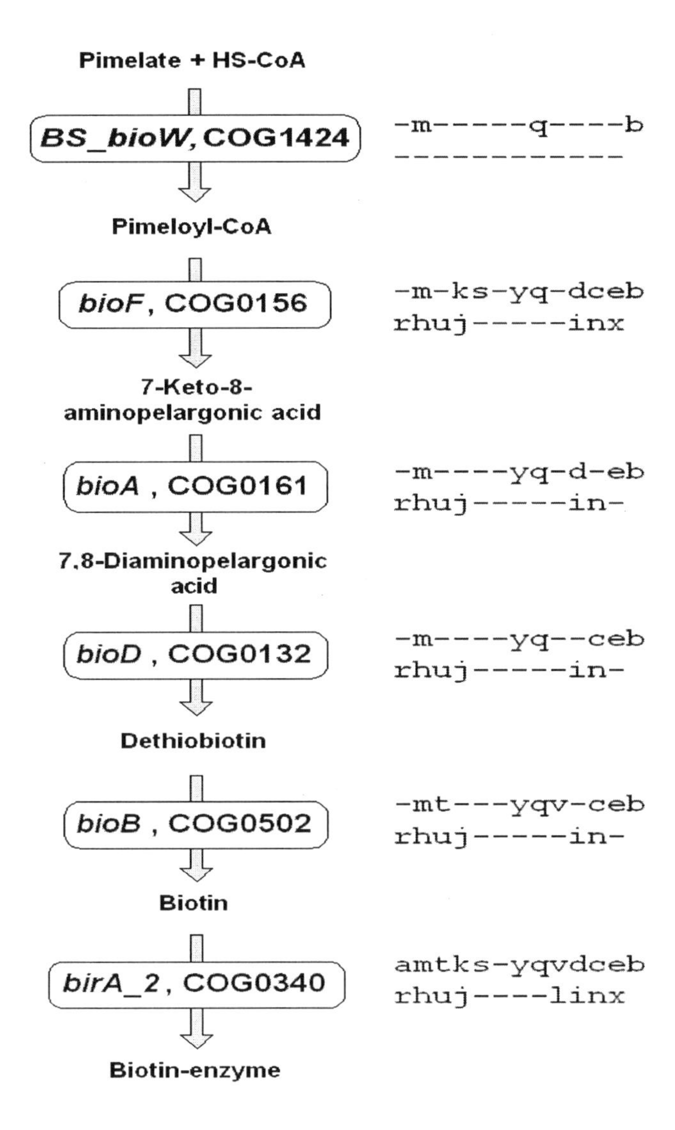

Figure 7.15. Distribution of biotin biosynthesis enzymes in organisms with completely sequenced genomes. All details are as in Figure 2.7.

B. subtilis, *A. aeolicus*, and *M. jannaschii* encode an enzyme that makes pimeloyl-CoA from pimelate and CoA in a reaction that uses the energy of ATP hydrolysis to AMP and pyrophosphate [675]. In contrast, pimeloyl-CoA synthetase from *Pseudomonas mendocina* belongs to the family of NDP-forming acyl-CoA synthetases [91,738]. Neither of these two enzyme families is represented in *Synechocystis* sp., *H. influenzae*, *H. pylori*, *C. jejuni*, and several other bacteria, indicating the existence of yet another enzyme for the synthesis of pimeloyl-CoA (or an entirely different pathway for the formation of 7-keto-8-aminopelargonate).

In spite of the similarity between the phyletic patterns of BioF, BioA, BioD, and BioB, one cannot help noticing that *Synechocystis* sp. lacks the *bioA* gene, suggesting that amination of 7-keto-8-aminopelargonate is catalyzed by a different aminotransferase. The absence of *bioD* and *bioB* genes in *D. radiodurans* makes one wonder whether this bacterium can synthesize biotin at all.

The enzyme catalyzing the last reaction in Figure 7.15, ligation of biotin to the biotin carboxyl carrier protein (or domain), has a much broader phyletic distribution than any of the biotin biosynthesis enzymes. This indicates that *A. fulgidus*, *Halobacterium* sp., *Pyrococcus* spp., *L. lactis*, *S. pyogenes* and many other organisms that do not have known pathway of biotin synthesis still can utilize biotin. Thus, they either have a completely different, unknown biotin synthesis pathway or import biotin from the environment (however, a biotin transport system so far has not been identified).

The paucity of data on the enzymes of biotin biosynthesis and a putative biotin uptake system should encourage active experimentation in this area. There definitely are novel enzymes and transporters yet to be discovered.

7.5.5. Heme

From the comparative-genomic point of view, the heme biosynthesis pathway is characterized by the following trends (Figure 7.16): (i) with the single exception of uroporphyrinogen III synthase (HemD), the enzymes from *R. prowazekii* and yeast (mitochondria) have identical phyletic patterns; (ii) all archaea, including the aerobes *A. pernix* and *Halobacterium* sp., produce siroheme, but not protoheme; (iii) non-orthologous displacement is observed in the downstream steps of the pathway, as opposed to the uniformity of all the upstream steps down to uroporphirinogen III.

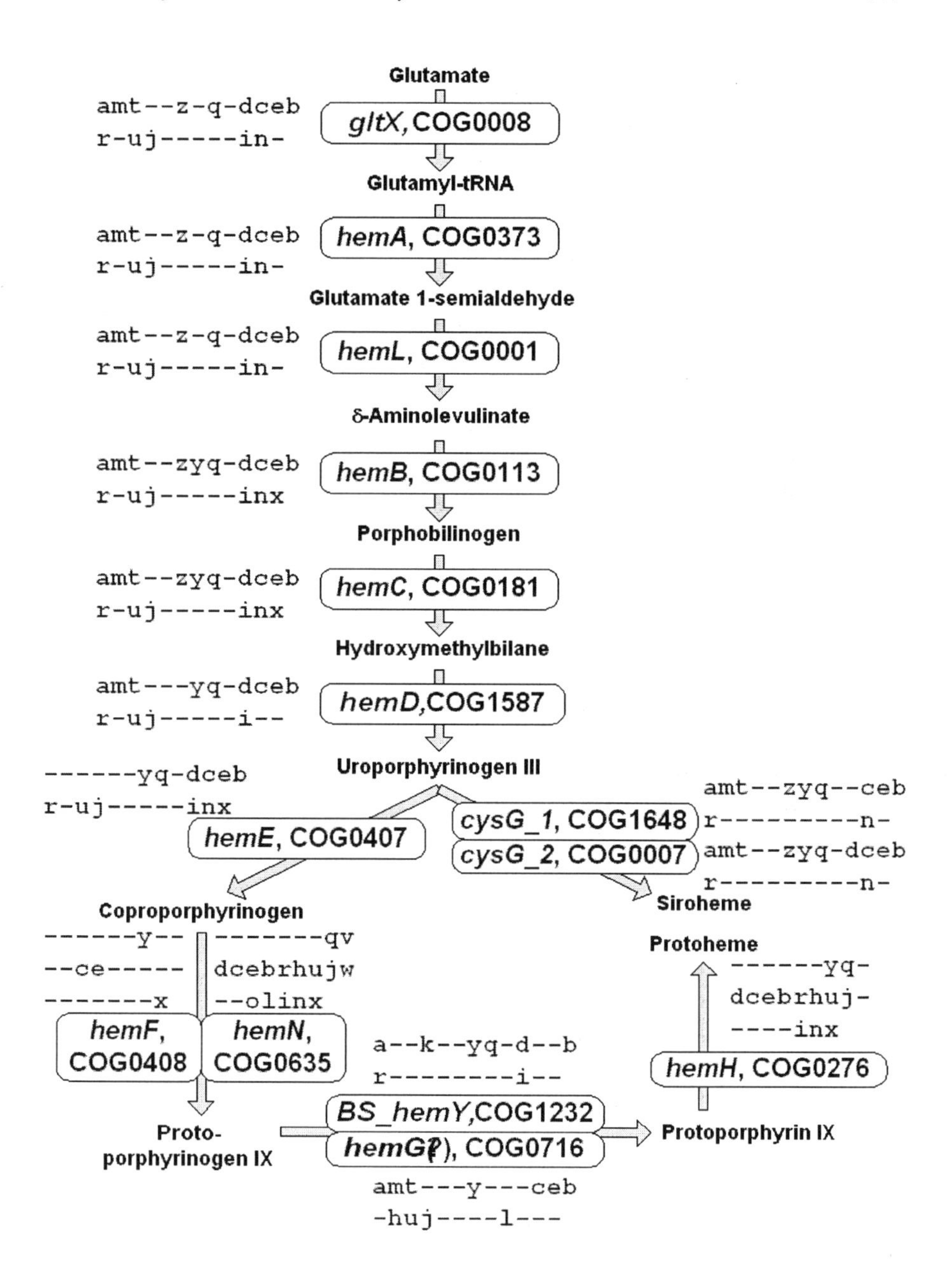

Figure 7.16. Distribution of heme biosynthesis enzymes in organisms with completely sequenced genomes. All details are as in Figure 2.7.

7.5.6. Pyridoxine

We conclude our survey of central metabolic pathways with the pyridoxine biosynthesis pathway, which, despite recent efforts, is still not completely understood. The scheme below is drawn based on the *E. coli* data [198,485]. In other organisms, the carbon backbone of the pyridoxine ring is formed of 4-hydroxythreonine (or its phosphate) and 1-deoxy-D-xylulose (or its phosphate) with the nitrogen supplied by either glutamate (in the PdxAJ-catalyzed reaction), or glutamine (in the PDX1,PDX2-catalyzed reaction) [211,578,635,825,832]. Since 1-deoxy-D-xylulose phosphate synthetase (Dxs, COG1154) so far has been identified only in bacteria, it is possible that archaea and eukaryotes use a different sugar as a pyridoxine precursor. Obviously, new enzymes of this pathway remain to be discovered.

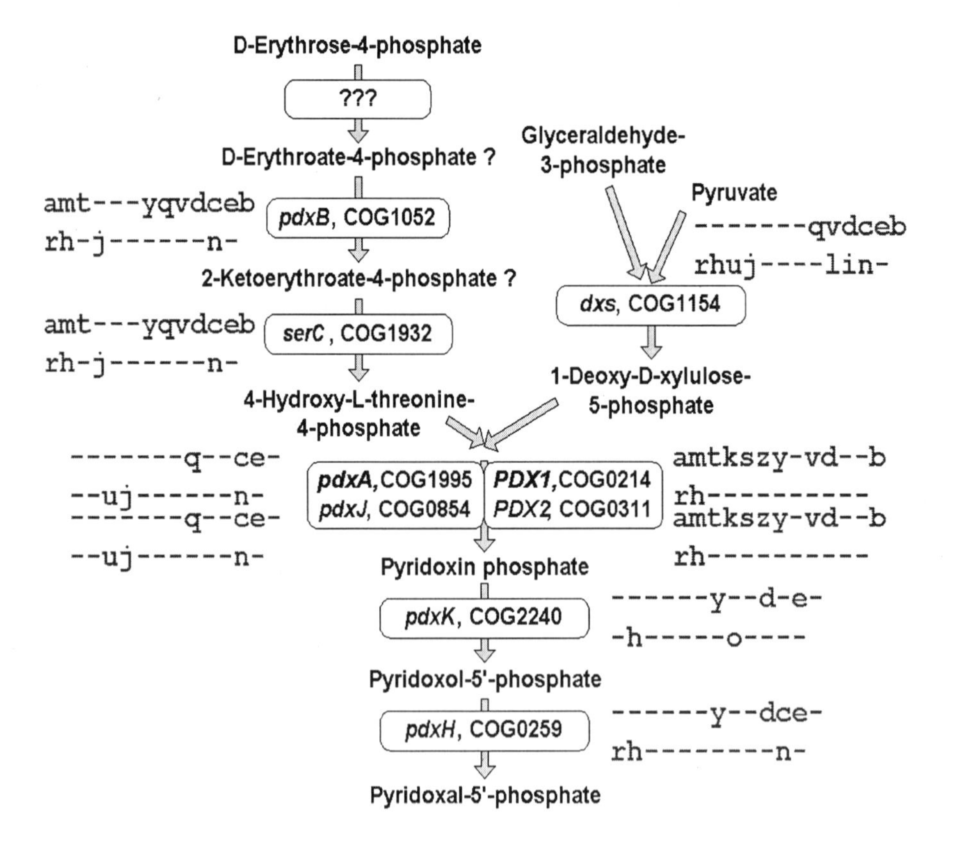

Figure 7.17. Distribution of pyridoxine biosynthesis enzymes in organisms with completely sequenced genomes. All details are as in Figure 2.7.

7.6. Microbial Enzymes as Potential Drug Targets

One of the major incentives behind the genome sequencing of numerous pathogenic bacteria is the desire to better understand their peculiarities and to develop new approaches for controlling human diseases caused by these organisms. This task has become even more urgent with the rapid evolution of antibiotic resistance in many bacterial pathogens, including multidrug-resistant enterococci, pneumococci, pseudomonads, staphylococci, and tuberculosis bacilli. Unfortunately, finding new antibiotics is an extremely laborious process that includes (i) testing numerous compounds for their activity against model organisms (*E. coli*, *P. aeruginosa*, *S. aureus*) that are easy to maintain in culture; (ii) screening these compounds against mammalian cell cultures to eliminate those toxic to humans; (iii) testing the efficacy and safety of each chosen drug in animal models; and (iv) pre-clinical and clinical testing, which alone takes several years. This process ensures that only highly effective and reasonably safe drugs make it to the market. The majority of drug candidates fail the tests, usually because in low concentrations they turn out to be safe but ineffective, whereas at high doses they are effective but show unfavorable side effects.

In spite of what one might have read in popular press, genomics cannot accelerate most steps of the drug development process. What it can do, however, is to increase the success rate by helping to choose drug candidates that are most likely to be effective (being targeted at essential systems of the bacterial cell) and least likely to be toxic (having no targets in the human cell). Indeed, while not all currently used antibiotics have well-characterized targets, those targets that have been characterized comprise bacterial proteins that (i) are essential for bacterial cell metabolism and (ii) are not represented (or represented in a very distinct form) in human cells (see Table 6.2). Microbial genome sequences provide us with complete lists of the proteins encoded in any given pathogen, including all the virulence factors that it could potentially produce. This "parts list" offers a wide selection of potential drug targets.

Comparative analysis of microbial genomes based on the notion of a phyletic pattern, which is discussed throughout this book, allows the identification of gene products that are common to all (or most) pathogenic microorganisms in a chosen group, as well as of those specific for a particular organism. The proteins in the former set are attractive targets for broad-spectrum antibiotics, whereas the unique proteins offer an opportunity to design "magic bullets", which would specifically target a narrow group of bacteria or even one particular pathogen [266].

Table 7.1. Cellular targets of most commonly used antibiotics

Antibiotic groups, examples	Bacterial target	Resistance mechanisms
β-Lactams: Penicillins, cephalosporins, carbapenems, monobactams	Peptidoglycan transpeptidase, other proteins	Hydrolysis by β-lactamase, alterations in penicillin-binding proteins
Glycopeptides: Bacitracin, colistin, dactinomycin, teichoplanin, vancomycin, virginiamycin	Peptidoglycan transpeptidase, transglycosylase	Modification of the UDP-muramyl pentapeptide
Aminoglycosides: Amikacin, kanamycin, gentamycin, hygromycin, neomycin, puromycin, streptomycin, tobramycin	Ribosomal 30S subunit	Acetylation, adenylation, or phosphorylation of the antibitic by specific modifying enzymes
Tetracyclins: Doxycycline, methacycline, minocycline, tetracycline	Ribosomal 30S subunit	Export by efflux pumps, mutations
Macrolides: Azithromycin, dirithromycin, clarithromycin, spiramycin, erythromycin, oleandomycin	Ribosomal 50S subunit	rRNA methylation; *rrn*, *rplD*, and *rplV* mutations; hydrolysis by esterases, export by efflux pumps
Quinolones: Nalidixic acid, ciprofloxacin	DNA gyrase β-subunit	*gyrB* mutations
Lincosamides: Clindamycin, lincomycin	23S rRNA	rRNA methylation, *rrn* mutations, drug adenylation
Chloramphenicol	Peptidyl-transferase center on the ribosomal 50S subunit	Inactivation by acetylation, export by efflux pumps
Sulfonamides: Sulfamethoxazole	Dihydropteroate synthase	*folP* mutations
Trimethoprim	Dihydrofolate reductase	*folA* mutaions
Nitroimidazoles Metronidazole	Chromosomal DNA	Nitroreductase mutations, preventing drug activation
Rifampin	RNA polymerase β-subunit	*rpoB* mutations

In addition to the lists of probable essential genes, search for potential drug targets in microbial genomes heavily relies on the understanding of bacterial metabolism, which is briefly discussed above.

7.6.1. Potential targets for broad-spectrum drugs

The list of probable essential genes that potentially could be used as targets for broad-range antibiotics can be derived using more or less the same approach as employed for the delineation of the "minimal genome" ([452], ♦2.2.5). Inclusion of certain genes in this list is, of course, affected by non-orthologous gene displacement and enzyme recruitment.

It should be noted that compiling the list of the likely essential genes for each particular group of bacteria (all bacteria, all gram-positive bacteria, all mycobacteria, and so on) by computational means is only one of several ways to accomplish this task, although, arguably, it is the easiest and fastest one. Any predictions of essentiality for a given gene still have to be verified experimentally by checking the lethality of knockout mutants [9,520]. As mentioned above (♦3.5.1), lists of essential *E. coli* genes are available at http://magpie.genome.wisc.edu/~chris/essential.html and http://www.shigen.nig.ac.jp/ecoli/pec/Analyses.jsp?key=0.

In addition to the genes that encode well-characterized essential proteins, the availability of complete genomes allows one to tap into the pool of uncharacterized genes whose wide distribution in microbial genomes marks them as being most likely essential [57,451]. Searches for such genes can be easily performed using the "phyletic patterns search" tool of the COG database. In addition, the COG database contains lists of poorly characterized and uncharacterized protein families, which are listed as functional groups R and S, respectively. A collection of uncharacterized conserved proteins, including those from partially sequenced genomes, is available in PROSITE database (http://www.expasy.org/cgi-bin/lists?upflist.txt).

The diversity of microbial metabolic pathways described above, offers numerous possibilities to look for potential drug targets among the metabolic enzymes. One straightforward approach is to select the pathways that are essential for certain pathogens but are absent in humans. Such pathways include murein biosynthesis, the shikimate pathway of aromatic amino acid biosynthesis (Figure 7.6), and the deoxyxylulose (non-mevalonate) pathway of terpenoid biosynthesis. It is remarkable that certain inhibitors of the latter pathway (fosmidomycin, fluoropyruvate, FR-900098) have been studied as potential antibiotics long before the characterization of their cellular targets [400,518].

7.6.2. Potential targets for pathogen-specific drugs

Although current approaches favor "one-shot" antibacterials that can eliminate bacterial infection irrespective of the nature of the pathogen, it gradually becomes clear that we will soon need a variety of drugs that would be effective against selected groups of organisms or even a single pathogen. There is nothing particularly new in this concept: people have been using antituberculine and anti-syphilis drugs for almost a century without requiring them to also cure common cold or gastrointestinal problems.

The novelty stemming from the availability of complete genome sequences is that now it has become possible to analyze the genome of a pathogen in detail, looking for weak spots or unusual enzymes that are likely to be essential for this particular organism. In addition to the traditional drug targets, such as the cell envelope and the systems for DNA replication, transcription, and translation, this brings into play for consideration as potential drug targets such proteins as host interaction factors, transporters for essential nutrients, enzymes of intermediary metabolism, and many others.

Host interaction factors can be searched for by using the so-called "differential genome display", first proposed by Peer Bork and his colleagues [366,371]. This approach looks for the genes that are present in the genome of a pathogen but not in the genome of a closely related free-living bacterium. Because genomes of parasitic bacteria typically code for fewer proteins than the genomes of their free-living cousins, genes detected by this approach are likely to be important for pathogenicity. Bork and colleagues applied this approach to the identification of potential pathogenicity factors in *H. influenzae* and *H. pylori* through comparison of their genomes against *E. coli* [366,371].

Because many pathogens have reduced biosynthetic capabilities and rely on the host for the supply of certain essential nutrients (♦3.2), the respective membrane transport systems can be valid targets for drug intervention. For example, the still uncharacterized biotin transport system appears to be the only means of biotin acquisition for several pathogens, such as *S. pyogenes*, *R. prowazekii*, *C. trachomatis*, and *T. pallidum* (♦7.5.4). Actually, one could start probing this hypothetical system right away by using various biotin analogs. As an added benefit, such a study would eventually lead to the identification of the transport system components.

Using surface proteins of bacteria as drug targets has an obvious advantage because drugs interacting with these proteins do not have to cross the cytoplasmic membrane, which largely removes the problem of drug efflux-mediated resistance [523,524]. On the other hand, humans also import biotin, therefore, at this stage, it cannot be ruled out that an inhibitor of biotin uptake might be toxic for humans. This emphasizes the need for identification of the genes coding for the bacterial uptake system: once these are known, we will be in a better position to assess the likelihoods of toxic side effects of any drugs targeting this function.

Another approach to searching for pathogen-specific drug targets would rely on the enzymes that are subject to non-orthologous gene displacement and are found in certain pathogens in a different form that is present in humans. The rationale for using enzyme inhibitors as antimicrobial drugs comes from the successful use of sulfamethoxazole and trimethoprim, inhibitors of two different steps of the folate biosynthetic pathway. Indeed, while each of these drugs is only moderately effective against most bacterial pathogens, their combination proved to be effective and reasonably safe. In several instances, detailed analysis of non-orthologous displacement cases has led to suggestions that alternative forms of essential enzymes could be used as drug targets ([266,271], see refs in Table 7.2).

Table 7.2. Examples of pathogen-specific drug targets

Enzymes with limited phyletic distribution	Human pathogens that depend on these enzymes	Ref.
ATP/ADP translocase, bacterial/plant type	*R. prowazekii, C. trachomatis, C. pneumoniae*	[895]
3-Dehydroquinate dehydratase, class II	*C. jejuni, H. influenzae, H. pylori, P. aeruginosa, V. cholerae*	[305]
DhnA-type fructose-1,6-bisphosphate aldolase	*C. trachomatis, C. pneumoniae*	[257]
Lysyl-tRNA synthetase, class I	*B. burgdorferi, R. prowazekii, T. pallidum,*	[375]
Na^+-translocating NADH: ubiquinone oxidoreductase	*C. trachomatis, C. pneumoniae, Cl. perfringens, T. denticola*	[334]
Na^+-translocating oxalo-acetate decarboxylase	*S. pyogenes, T. pallidum,*	[334]
Orotidine 5'-phosphate decarboxylase	*M. leprae, M. tuberculosis*	[12]
Pyridoxine biosynthesis enzymes PDX1, PDX2	*Bacillus anthracis, H. influenzae, L. monocytogenes, M. leprae, M. tuberculosis, S. pneumoniae*	[263,635]
Cofactor-independent phosphoglycerate mutase	*C. jejuni, H. pylori, M. genitalium, P. aeruginosa, V. cholerae*	[258,261]

7.7. Conclusions and Outlook

This chapter shows that central metabolism is the ultimate playground of non-orthologous gene displacement, where the logic of phyletic patterns works best. Metabolic pathways are so amenable to this type of analysis because, if an organism encodes a significant fraction of the enzymes for a particular pathway, it is extremely likely that, in reality, is also has the enzymes for the rest of the steps. Therefore, candidate enzymes for the missing steps may be sought for and, at least in some instances, found among uncharacterized orthologous sets (identified through COGs or otherwise) with phyletic patterns that are, at least in part, complementary to those for known enzymes for the given step. So far, only very few of the computational predictions made by this approach have been tested experimentally, but in those studies that have been conducted, the success rate has been quite high. Conversely, there are enigmatic cases where most of the enzymes of a given pathway are missing in an organism but one or two still stay around (by using this language, we imply loss of a pathway, which is indeed largely the case in parasites and heterotrophs, including ourselves). Most likely, these are cases of exaptation, where an enzyme that is no longer needed in its original metabolic capacity has found another job, thus saving itself from extinction. Elucidation of these exapted functions seems to be an interesting avenue of research.

The finding that metabolic pathways are so prone to non-orthologous gene displacement seems to indirectly convey a message of general biological significance. We know for a fact that enzymes in the same metabolic pathway are connected through reaction intermediates, but, on almost all occasions, precious little is known about the actual macromolecular organization of these enzymes in the cell. Analysis of phyletic patterns shows that many, if not most, metabolic enzymes with different structures but the same reaction chemistry are interchangeable in evolution. This suggests that, most of the time, the chemistry is, after all, the principal aspect of the metabolic functions, whereas the role of co-adaptation of subunits of macromolecular complexes is likely to be limited.

The major contribution of lineage-specific gene loss to the evolution of metabolic pathways is beyond doubt. Horizontal gene transfer is harder to demonstrate but, realistically, it appears certain that this phenomenon also had a substantial role. Indeed, it defies credibility to postulate that LUCA had each one of the alternative forms of metabolic enzymes (and the corresponding reaction intermediates), the existence of which became apparent through the comparative-genomic studies (as well as those, perhaps numerous ones that remain to be discovered). The relative contributions of gene loss and horizontal transfer hopefully will be better understood through the application of algorithmic methods briefly outlined in Chapter 6.

Identification of potential targets for antibacterial drugs using phyletic patterns, the differential genome display technique and other similar approaches is a natural task for comparative genomics and will likely remain one of its most important practical applications for years to come.

7.8. Further Reading

1. Romano AH, Conway T. 1996. Evolution of carbohydrate metabolic pathways. *Research in Microbiology* 147: 448-455.
2. Galperin MY, Walker DR, Koonin EV. 1998. Analogous enzymes: independent inventions in enzyme evolution. *Genome Research* 8: 779-790
3. Dandekar T, Schuster S, Snel B, Huynen M, Bork P. 1999. Pathway alignment: application to the comparative analysis of glycolytic enzymes. *Biochemical Journal* 343: 115-124.
4. Huynen MA, Dandekar T, Bork P. 1999. Variation and evolution of the citric-acid cycle: a genomic perspective. *Trends in Microbiology* 7: 281-291.
5. Cordwell SJ. 1999. Microbial genomes and "missing" enzymes: redefining biochemical pathways. *Archives of Microbiology* 172: 269-279.
6. Galperin MY, Koonin EV. 2001. Comparative genome analysis. In: *Bioinformatics: a practical guide to the analysis of genes and proteins* (Baxevanis AD and Ouellette BFF, eds) pp. 359-392. John Wiley & Sons, New York.
7. Canback B, Andersson SG, Kurland CG. 2002. The global phylogeny of glycolytic enzymes. *Proceedings of the National Academy of Sciences of the United States of America* 99: 6097-6102.
8. Galperin MY, Koonin EV. 1999. Searching for drug targets in microbial genomes. *Current Opinion in Biotechnology* 10, 571-578.

CHAPTER 8
GENOMES AND THE PROTEIN UNIVERSE

We have now surveyed some of the principal methodological approaches of comparative genomics and the major evolutionary conclusions that can be inferred from genome comparisons. In this short chapter, we take a view of genomes from a different vantage point. We briefly describe the current understanding of the organization of the protein Universe and project it on genomes to reveal common and unique patterns.

8.1. The Protein Universe is Highly Structured and there are Few Common Folds

The theoretical size of the sequence space, i.e. the total number of possible protein sequence is, for all practical purposes, infinite. Assuming than average protein length is 200 amino acids, there can be 20^{200} different protein sequences, a number that is much greater than, for example, the number of protons in our Universe (not the protein universe often mentioned in this chapter, but the physical Universe around us). Our current theoretical understanding of protein folding is insufficient to estimate the total possible number of protein structures, but one suspects it is also vast. Obviously, only a miniscule fraction of the practically infinite sequence space is populated by real protein sequences. Still the number of unique sequences encoded in real genes is likely to be substantial. For example, assuming there is 10^7-10^8 species on Earth and the genome of each species consists of 10^3-10^5 genes, there are 10^{10}-10^{13} unique protein sequences, a speck compared to the vast sequence space, but still several orders of magnitude more than contained in today's databases. A question of major fundamental and practical interest is how these sequences are distributed in the sequence and structure spaces. The discussion of numerous homologous relationships between proteins in the preceding chapters should make it obvious that the distribution cannot be random: there certainly are numerous, distinct clusters of homologs separated from other clusters. However, more precise, quantitative answers are necessary and these are usually sought within the framework of hierarchical classification of proteins. Throughout this book, we repeatedly referred to protein folds, superfamilies and families, which are categories within this classification. However, before we proceed further with the discussion of the structure of the protein universe, it is useful to identify the entire hierarchy and to introduce the levels more precisely (Table 8.1). This classification had been introduced largely through analysis of protein structures in the context of the SCOP database construction [590]. Similar categories are adopted in the CATH database [633]. The phylogenetic clusters, such as COGs, which we encountered throughout this book, lie directly below the lowest category in the structural classification, the family.

Perhaps the most important categories in this classification are Fold (near the top of the hierarchy, disregarding the less informative notion of a structural class) and COG, near the bottom, which represent two fundamental levels of evolutionary relationships. In particular, folds are, typically, the largest monophyletic classes of proteins and the number of distinct folds may be considered the central aspect of the structure of the protein universe.

Table 8.1. Hierarchical classification of proteins

Category	Example	Definition, criteria, main features
Structural class	α/β	Overall composition of structural elements. No evolutionary relationship.
Fold	P-loop	Topology of the folded protein backbone. Monophyletic origin?
Superfamily	P-loop containing nucleotide triphosphate hydrolases	Recognizable sequence similarity (at least a conserved motif); conservation of basic biochemical properties. Monophyletic origin;
Family	Nucleotide and nucleoside kinases	Significant sequence similarity; conservation of biochemical function.
Group of orthologs (COG)	Adenylate kinase	Orthologous relationships within the given set of species; conservation of the biological function.
Lineage-specific expansion	DR0202, DR0494 and DR2273 in *D. radiodurans*	Paralogs originating from a lineage-specific duplication

Of course, experimentally determined structures of proteins representing all existing folds are not yet available, so the number of folds needs to be estimated through extrapolation. A number of researchers have attempted to come up with such estimates. In the first such analyses, Zuckerkandl [945] and Barker and Dayhoff [75] examined "independent" sequence superfamilies (i.e. those for which similarity could not be detected with the methods available at the time) and converged at ~1,000 such families. The first estimate based on independent structures rather than sequences was produced in 1992 by Cyrus Chothia [147] who used a very straightforward extrapolation approach. He found that ~1/4 of the sequences encoded in each (then partially sequenced) genome showed significant similarity to sequences in the SWISS-PROT database, and ~1/3 of the sequences from SWISS-PROT were related to one of the 83 folds available at the time. From these data, elementary extrapolation: 83x3x4≈1,000 gives the expected total number of folds, which remarkably agreed with the early estimates despite the different approach. Subsequent, more sophisticated estimates based on theoretical analysis of the sampling of sequence families from the structure database or from genome-specific sequence sets produced estimates of the total number of folds between 700 and 4,000 [306,912,938].

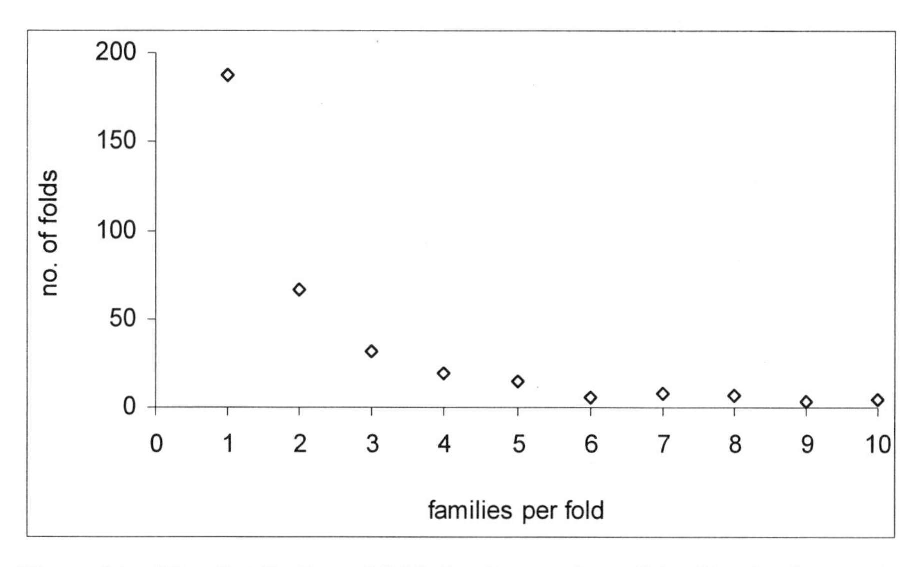

Figure 8.1. The distribution of folds by the number of families in the protein structural database (PDB). The families were obtained by clustering the sequences with the cut-off 0.3 bit/position, which was shown to give the best fit to the data (see [912]).

A recent study by Coulson and Moult, which explicitly incorporated the division of protein folds into three categories: superfolds that consist of numerous families, mesofolds that include a limited number of families, and unifolds that consist of only one, compact family, produced an even higher estimate, at least 10,000 unique folds [161]. The discrepancy between these estimates becomes less dramatic and easier to explain when one examines the distribution of folds by the number of protein families (Figure 8.1). This distribution shows that there is a small number of folds with a large number of families (mostly these are the well-known superfolds, such as P-loop NTPases, the Rossmann fold or TIM-barrels) and an increasing number of folds that consist of a small number of families. By far the largest class are the "unifolds". Thus, it seems certain that the great majority (>95%) of the protein families belong to ~1000 common folds. What is still in dispute is the number of unifolds that encompass the rest of the proteins. Approximately, one half of the more common folds are already represented by at least one experimentally determined structure, which means that at least rough mapping of the protein universe is already at an advanced stage.

The curve in Figure 8.1 shows that the protein universe is extremely well-structured: not only are most of the sequences clustered in a small number of densely populated areas (folds), but the distribution of sequences among the folds is highly non-random. Why are a few folds (superfolds) so common whereas the majority are rare? As on many occasions in biology, there are two types of explanation, a physical/functional and an evolutionary one.

Table 8.2. Predicted number of protein folds in complete genomes

Organism	No. of proteins in the genome	No. of detected folds	Predicted no. of folds	
			Low	High
M. genitalium	480	101	239	352
R. prowazekii	834	138	335	495
A. aeolicus	1522	173	368	542
M. jannaschii	1715	135	276	412
A. pernix	1760[a]	141	298	438
Synechocystis sp.	3169	199	447	658
M. tuberculosis	3918	208	457	673
B. subtilis	4100	221	462	680
E. coli	4289	231	548	808
S. cerevisiae	6530	215	488	720
The Protein Universe	N/A	364	908	1357

The first view maintains that the common folds are good for protein functions because they are particularly stable and/or because they are well suited to accommodate catalytic and binding site (e. g. in the loops flanking the β-sheet in the P-loop or Rossmann fold). The evolutionary approach interprets the distribution from the point of view of the simple "the rich get richer" principle: an already common domain is more likely to be adapted for a new function via duplication with subsequent diversification simply because the chance of a duplication such a domain is greater (in network analysis, this is called "preferential attachment"). Probably more realistically, the two views may be combined in the "the fit get fitter" principle [74]: domains that are functionally versatile and stable are favored by selection, but once they become common, purely stochastic processes help them proliferate farther.

Once we have obtained an approximate but apparently reasonable count of the folds in the entire protein universe, it is of interest to see how does this set of folds project on genomes from different walks of life. An extrapolation from the number of folds actually detected in the protein sequences encoded in each genome shows that unicellular organisms encode from ~30% to ~60% of the folds (Table 8.2), which shows that even these simple life forms extensively sample the protein universe.

8.2. Counting the Beans: Structural Genomics, Distributions of Protein Folds and Superfamilies in Genomes and Some Models of Genome Evolution

Accessible structural classifications of proteins, such as SCOP and CATH, became available almost simultaneously with multiple genome sequences. So it was a natural idea to take a structural census of the genomes, i.e. determine and compare the distributions of protein folds and superfamilies in them [280,281,367,833,911]. These "surveys of finite part lists", to use the lucky phrase of Mark Gerstein [281], are vital to ***structural genomics***, the rapidly growing research direction, which we cannot cover in this book in any adequate depth, but must mention at least in passing. In principle, the goal of structural genomics is, no more no less, to determine all protein structures existing in nature. However, since this goal is unattainable for all practical purposes, structural genomics aims at determining a representative set of structures, which would allow the rest to be modeled on the basis of homology [145,733,872].

The clustered organization of the protein universe discussed in the previous section makes this feasible provided that a strategy of target prioritization is well defined. This strategy is to ensure that newly determined structures are non-redundant, that is, each of them represents a family or at least a COG (see Table 8.1), which so far has been missing in the structural database. For this purpose, it is essential that the structural census of genomes is as complete and accurate as possible. A recent conservative estimate suggests that "it would take approximately 16,000 carefully selected structure determinations to construct useful atomic models for the vast majority of all proteins" [872], and there is little doubt that this research program will be carried out well within the first quarter of the 21st century. So far, few structures have been actually determined within the structural-genomic paradigm, but these indeed turned out to be novel and led to functional and evolutionary insights (e.g. [572,937]).

Beyond the indisputably important practical goals of structural genomics, the genome-specific catalogues of protein folds and superfamilies are of fundamental interest: their qualitative examination may highlight important differences in the life styles of different organisms, whereas mathematical analysis of the distributions has the potential of revealing hidden regularities in genome evolution. Table 8.3 shows the list of the top 10 protein folds for a number of bacterial, archaeal and eukaryotic genomes. The counts of the folds were obtained by running domain-specific PSSMs (based on the SCOP classification) against the predicted protein set from each genome [911]. What immediately strikes one when perusing this list is, first, how similar are the rankings for organisms with very different genome sizes and lifestyles and the cumulative rankings for the three domains of life. The other prominent feature of these distributions is the overwhelming domination of the P-loops, particularly, in prokaryotes. The smaller the proteome the greater the fraction of P-loops, which emphasizes the involvement of P-loop NTPases in vital, housekeeping functions (e.g. translation and replication). At least for the top 30 folds, the distribution of the fraction in genomes depending on the rank gives an excellent fit to the exponent, which is what should be expected if the probability of duplication is the same for all folds (Figure 8.2; [911]). However, the most abundant fold, the P-loops (rank 1) is ***much more*** abundant than predicted on the basis of this straightforward assumption (Figure 8.2). To paraphrase the famous quip of J. B. S. Haldane about beetles, "God seems to have an inordinate fondness for P-loops". P-loops NTPases are the motors associated with so many diverse functions that, when a new function emerges, it is extremely likely that a duplication of a P-loop domain will provide the necessary engine. It is only logical to surmise that this domain was among the first, if not ***the first*** one, which took shape at the dawn of life; this is, of course, compatible with all reconstructions of ancestral gene repertories (Chapter 6).

Table 8.3. Top 10 protein folds in complete genomes from the three domains of life

Fold	Bacteria									Archaea				Eukaryotes				All
	Mg	Bb	Aa	Hp	Hi	Bs	Ec	x, %	Rank	Mj	Af	x, %	Rank	Sc	Ce	x, %	Rank	x, %
P-loop NTPases	53	83	102	92	108	177	186	22.5	1	125	125	22.5	1	239	216	11.4	1	18.8%
TIM-barrel	5	13	35	26	36	89	105	6.0	2	37	47	7.7	3	93	98	4.7	4	6.1%
Ferredoxin-like	3	6	22	18	20	20	55	3.0	9	50	67	11.1	2	68	85	3.7	5	5.9%
SAM-dependent methyltransferases	7	11	27	39	32	37	39	5.1	3	40	30	6.0	4	45	50	2.3	12	4.5%
Protein kinases, catalytic core	1	0	2	1	1	3	3	0.4	27	3	3	0.5	26	128	305	9.5	2	3.5%
Rossmann-fold domains	2	6	24	16	19	68	53	3.6	6	15	22	3.5	8	51	89	3.2	9	3.4%
ATP pyrophosphatases	13	13	18	17	15	22	23	3.9	4	27	22	4.4	5	25	16	1.1	19	3.1%
Rossmann-like fold domain	6	5	21	9	16	43	63	3.6	7	10	42	3.8	7	37	32	1.7	14	3.0%
Flavodoxin-like	3	7	16	15	16	57	66	3.8	5	11	35	4.3	6	22	11	0.9	20	3.0%
7- and 8-bladed β-propeller	0	1	1	0	2	5	6	0.3	28	2	3	0.4	27	132	131	6.5	3	2.4%

x,% indicates the average share of the given fold among the complete genomes of the corresponding superkingdom; All shows its fraction across all the complete genomes (as of 2001).

In 1998, Martijn Huynen and Erik van Nimwegen [374] made the seminal observation that the frequency distributions of paralogous protein families in genomes seemed to follow a negative power law, i.e. the dependence of the general form:

$$F(i)=ci^{-\gamma} \tag{8.1}$$

where $F(i)$ is the frequency of a family with i members and c and r are coefficients. This observation is extremely interesting and provocative because power laws are found in an enormous variety of biological, physical and other contexts, which seem to have little if anything in common. Examples of quantities that show power distributions include the number of acquaintances or sexual contacts people have, the number of links between documents in the Internet or the number of species that become extinct within a year. The classic Pareto law in economics describing the distribution of people by their income and the even more famous Zipf law in linguistics describing the frequency distribution of words in texts belong in the same category. What all these phenomena do have in common is that they are based on ensembles or networks with preferential attachment, which evolve according to "the rich get richer" or perhaps "the fit get fitter" principles already mentioned above [74]. More specifically, power laws can be potentially explained in terms of self-criticality phenomena, but they also allow simple explanations through gene (protein, domain) birth, death and "invention" models (which we designate BDIMs for short) [685,726].

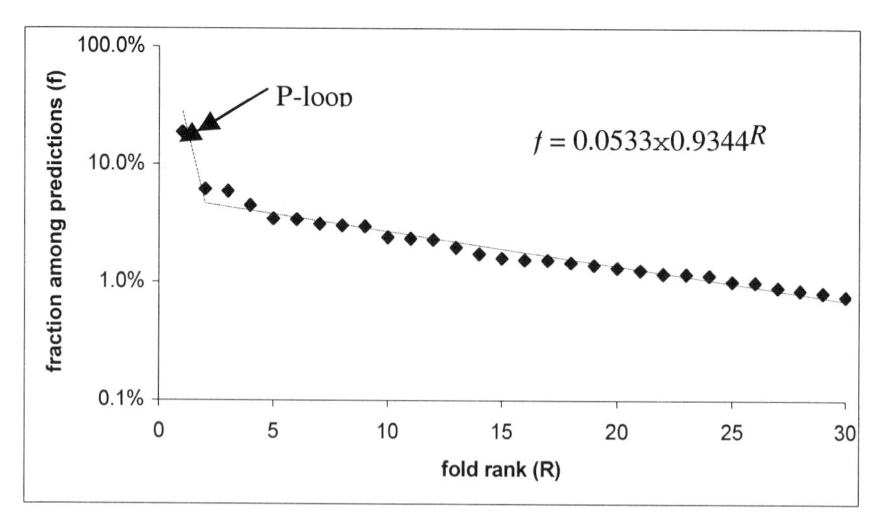

Figure 8.2. Distribution of the top 30 protein folds in combined proteomes. The vertical axis gives the average fraction of a fold in all analyzed proteomes.

Gene birth occurs via duplication; gene death is, of course, the loss process discussed in detail in Chapter 6, and innovation may involve gene acquisition via horizontal gene transfer, emergence of genes from non-coding sequences or emergence of globular domains from non-globular sequences.

A more detailed analysis indicates that domain family distributions are actually best described by so-called generalized Pareto distributions, which include power laws as asymptotics (i.e. the distribution fits a power law for large families) [419]. Figure 8.3 shows the domain family size distribution for the plant *A. thaliana* with the corresponding fits. It has been shown that such distributions can be generated by a specific class of models (linear BDIMs), in which domain birth rate is equal to the death rate and each of these rates depends on the family size in such a way that members of small families are somewhat more likely to either duplicate or die than members of large families. Furthermore, it can be proved that linear BDIMs rapidly reach equilibrium from any initial conditions [419]. Translating this into the language of genome evolution, although genomes are "in flux" (this is also supported by the BDIM analysis, which reveals fairly high innovation rates), the number of families of a given size remains constant over extended periods of evolution. This may be considered yet another manifestation of the genomic clock discussed in Chapter 6.

For sure, the above is "bean bag genomics", which ignores the biological identity of protein families. It is notable, however, that linear BDIMs describe with considerable accuracy even the number of large, expanded families present in each genome (Figure 8.3 and [419]), although the proliferation of such families is usually regarded as adaptation.

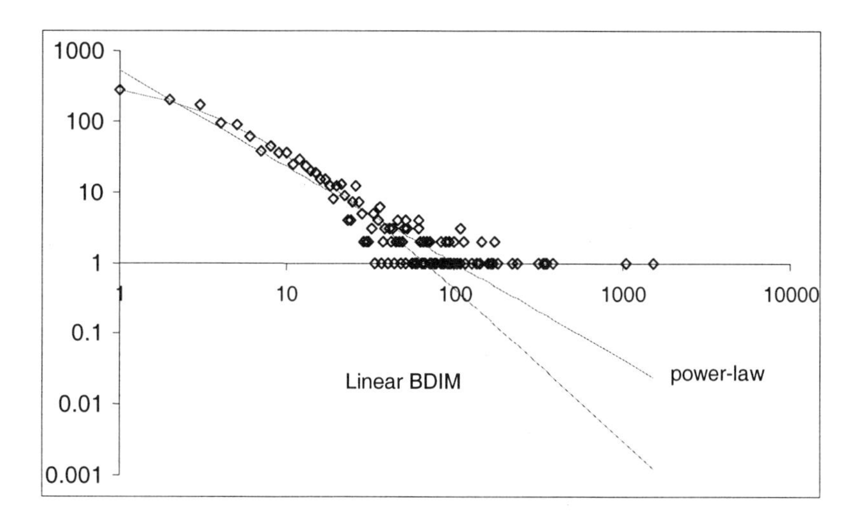

Figure 8.3. Domain family size distribution for *Arabidopsis thaliana*. Horizontal axis: number of family members, vertical axis: number of families.

The observations described here suggest a more nuanced view: natural selection "chooses" families that are allowed to proliferate, but the general dynamics of protein family evolution can be described as a purely stochastic process. It seems that even over-simplified models that disregard selection are starting to reveal fundamental aspects of genome evolution.

8.3. Evolutionary Dynamics of Multidomain Proteins and Domain Accretion.

Protein domains do not exist in isolation. In the course of evolution, they often combine to form multidomain architectures. As we have seen, analysis of such architectures may be helpful for predicting functions of uncharacterized domains (the guilt by association approach, see Chapter 5). Multidomain proteins play critical roles in the cell by providing effective links between different functional systems. Because of this ability, complex multidomain architectures are particularly common in all kinds of signaling systems. In many orthologous sets of proteins, a distinct trend can be traced toward increased complexity of domain architectures in more complex organisms. This tendency, which was dubbed "domain accretion" [459], is illustrated in Figure 8.4 for a set of orthologous eukaryotic transcription factors. Various domains added to the conserved core, typically at the ends, provide tethering to other chromatin-associated proteins and, in the case of *A. thaliana*, apparently to the ubiquitin-dependent protein degradation machinery.

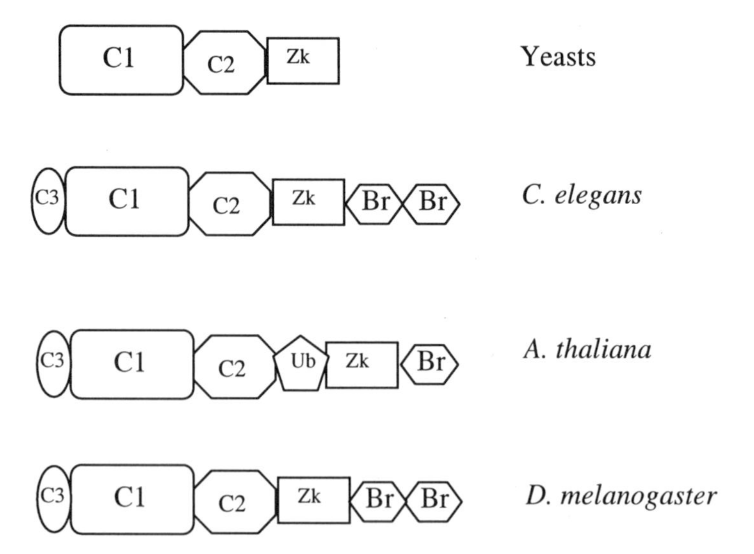

Figure 8.4. Domain accretion in transcription factor TAFii250.
C1,C2,C3, uncharacterized conserved domains; Zk, Zn knuckle; Br, Bromo domain; Ub, ubiquitin.

Since proteins form complex networks (see below), even a modest increase in the number of domains in interacting partners may translate into numerous new interactions, which probably contributes to the solution of the ostensible paradox of "too few" genes in complex organisms [488].
Given the utility of multidomain proteins for a variety of cellular functions, one could think that natural selection would favor their formation to the extent that they would be over-represented with respect to the single-domain proteins. Quantitative analysis does not seem to support this conclusion. The distribution of proteins by the number of different (repetitions of the same domain excluded from the analysis) domains shows an excellent fit to an exponent ([911]; Y.I. Wolf and EVK, unpublished observations; Figure 8.5), which is compatible with a random recombination (joining and break) model of evolution of multidomain proteins. We should note, however, that the slopes of the curves in Figure 8.5 are markedly different for archaea, bacteria and eukaryotes, indicating that the fraction of multidomain proteins or, in terms of the random model, the relative probability of domain joining increases in the order:

Archaea < Bacteria < Eukaryotes.

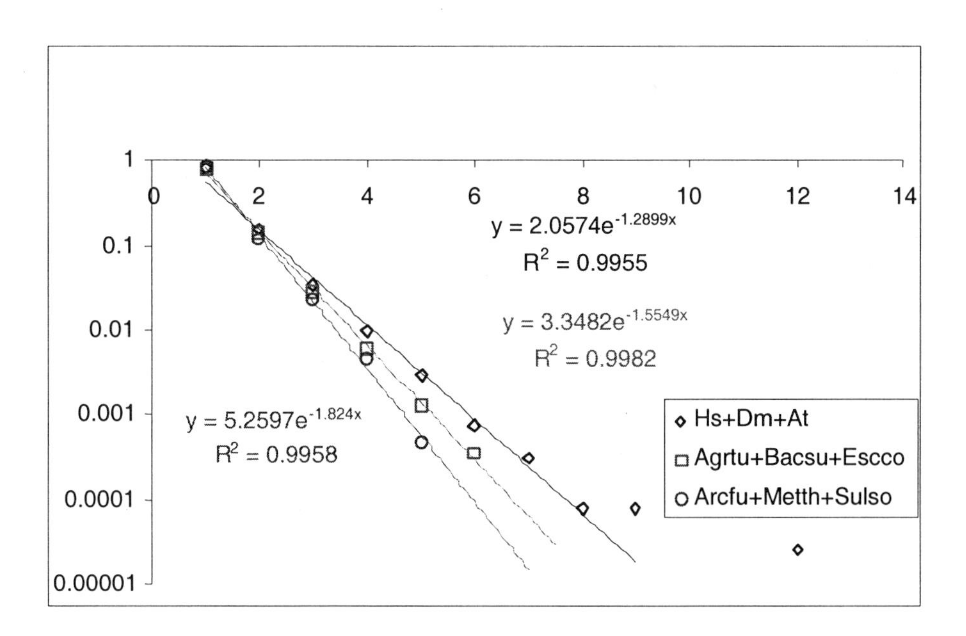

Figure 8.5. Distributions of the number of proteins with different number of domains in bacteria, archaea and eukaryotes. The plot is in double logarithmic scale.

The under-representation of multidomain proteins in archaea compared to the other two domains might be related to the low stability of large proteins in the hyperthermophilic habitats of these organisms. The excess of multidomain proteins in eukaryotes is not unexpected given the complexity considerations above, and we also should note the deviation from the exponent in the right tail of the distribution caused by the presence of proteins with a large number of domains (Figure 8.6).

The above analysis tells us nothing about the propensity of individual domains to form multidomain architectures and these propensities differ widely. Perhaps by now we should not be surprised to learn that the distribution of the number of connections a domain has with other domains in multidomain proteins can be roughly approximated by a power law ([924]; Figure 8.6). This means that a small number of domains are hubs of multidomain connections that hold together cellular interaction networks. We already referred to these domains as "promiscuous" and mentioned some examples when discussing the "guilt by association" approach in Chapter 5. As in the case of the evolutionary dynamics of domain families discussed in section 8.2, although evolution of multidomain proteins seems to occur via random processes of joining and breaking (Figure 8.4), the fit (to form usable multidomain architectures) still "gets fitter and fitter".

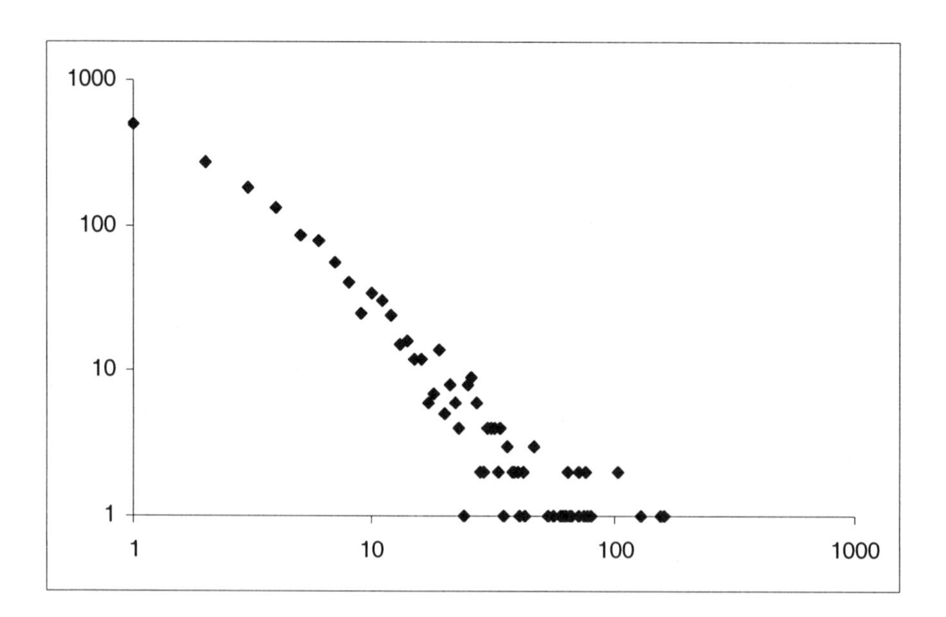

Figure 8.6. Distribution of protein domains by the number of links in multidomain proteins. The number of links is the number of different domains with which the given domain combines in multidomain proteins. The data from 13 analyzed bacterial and archaeal genomes were combined. The plot is in double logarithmic scale.

8.4. Conclusions and Outlook

In this brief chapter, we only could cast a very superficial and formal glance at the protein universe. Even so, we could notice several major features of this world. Although the theoretical sequence and probably structure spaces are virtually infinite, the populated parts are definitely finite and, more importantly, extremely non-homogeneous. Because of this concentration of proteins in a relatively small number of hubs – folds and superfamilies – a properly equipped party of explorers, such as the structural genomics programme, can visit at least all the major ones in a reasonable time.

We have further seen that superimposing the structure of the protein universe upon genomes and quantitative analysis of the results may give us unexpected insights into some general principles (may we tentatively say "laws"?) of genome evolution.

8.5. Further reading

1. Barabasi AL. 2002. *Linked: The New Science of Networks.* Perseus Publishing, Cambridge, MA.

2. Coulson AF, Moult J. 2002. A unifold, mesofold, and superfold model of protein fold use. *Proteins* 46, 61-71.

3. Huynen MA, van Nimwegen E. 1998. The frequency distribution of gene family sizes in complete genomes. *Molecular Biology and Evolution* 15, 583-589.

4. Koonin EV, Aravind L, Kondrashov AS. 2000. The impact of comparative genomics on our understanding of evolution. *Cell* 101, 573-576.

5. Qian J, Luscombe NM, Gerstein M. 2001. Protein family and fold occurrence in genomes: power-law behaviour and evolutionary model. *Journal of Molecular Biology* 313, 673-681.

6. Vitkup D, Melamud E, Moult J, Sander C. 2001. Completeness in structural genomics. *Nature Structural Biology* 8, 559-566.

IX. EPILOGUE: PEERING THROUGH THE CRYSTAL BALL

It's hard to make predictions, especially about the future.
Attributed to Yogi Berra, Niels Bohr, Mark Twain, and many others
(http://larry.denenberg.com/predictions.html)

If you come to a fork in the road, take it.
Yogi Berra

Throughout this book, we discussed various aspects of computational genomics, all of which, more or less, fit under the broad categories of ***functional prediction*** or ***evolutionary reconstruction***. Before we put the pen down, it is tempting to dream of the future. We hope that, when presenting on these pages several research directions in computational genomics, be it exploitation of genomic context for prediction of gene functions or reconstruction of early evolutionary events, we managed to convey to the reader the openness of the most important problems in each of these areas. It does not seem to make sense to recapitulate these. However, it is attractive to think of some real new avenues of research, those that are in their infancy now, but may be expected to flourish, say, in the year 2012. The images in the crystal ball are vague and we may easily take them for something that they are not, but since we are trying to think big, it will not be too embarrassing to miss the target.

9.1. Functional Genomics: a Programme of Prediction-driven Research?

Improving our understanding of how the cell works through prediction of gene functions and interpretation of experimental results using genomic information is usually considered the main goal of genomics, computational genomics in particular. But what is functional genomics? Currently, the phrase is used in the broadest possible sense to describe any experimental work that involves a contribution from genome analysis. This is probably well justified, but we believe that functional genomics also could be defined in a much more specific and focused way.

Namely, functional genomics could be a coherent research programme build around predictions made by genome comparison, conceptually very much like structural genomics (see Chapter 8). Structural genomics strives to prioritize targets for protein structure determination on the basis of two principal criteria: (i) probable structural novelty and (ii) importance of the biological function. Functional genomics could select targets for experimental analysis in exactly the same way except that "functional" should be substituted for "structural" under (i). Clearly, the targets of the highest priority for detailed experimental study should be those uncharacterized genes whose functions are essential and, on top of this, are likely to reveal novel biochemical mechanisms. How do we find essential genes among those with unknown functions?

The reader who went through the previous chapters (actually Chapter 2 alone would suffice, but Chapters 5, 6 and 7 are also helpful in providing numerous cases in point) should already have the answer. We select the genes with phyletic patterns that show wide spread among many diverse lineages of organisms; these are most likely to be essential. For one last time in this book, we will employ the COG database to select a set of high priority targets for functional genomics. We have seen already that there are only ~65 truly universal COGs and the majority of these have well-characterized functions (however, see below). Therefore let us select those COGs that are represented in all archaea and eukaryotes (and perhaps in some of the bacteria); this will give us an extra shot at finding new important eukaryotic genes, which is what often interests a biologist the most. The search selects 186 COGs, of which ~20 have not been characterized in terms of their specific cellular function(s). Typically, the biochemical activity of these proteins has been predicted at different levels of specificity, although but four remain completely mysterious.

In Table 9.1, we show a sample of these conserved archaeo-eukaryotic genes, which we ranked, more or less subjectively (that is, without any further computation, on the basis of "biological considerations" alone), in the decreasing order of "excitement", in terms of identifying new, important functions. For most of the genes on this list, we could make additional predictions on the basis of genomic context (♦5.3) and the phyletic pattern itself. Not unexpectedly, most of the proteins encoded by the genes on our list are predicted to perform various roles in translation. There are several GTPases (some shown in Table 9.1), which are predicted to function as translation factors, unknown translation regulators, and RNA-binding proteins.

We seem to have an indisputable leader in terms of general importance and novelty, COG0533. This is one of only two universal COGs, along with COG0012 (already discussed briefly in Chapter 5), with virtually unknown functions. Since COG0012 is firmly predicted to consist of RNA-binding GTPases, i.e. ubiquitous and essential translation factors, the palm goes to COG0533. The proteins comprising this COG have a clearly predicted fold, the HSP70-like ATPase, with a metal-dependent protease active site inserted into the ATPase domain. On the basis of these features, it may be predicted that this is a previously undetected ATP-dependent protease with probable chaperone functions, perhaps involved in co-translational degradation of misfolded and/or prematurely terminated proteins. Moreover, genomic context suggests that it might be a subunit of a chaperone complex, which also includes an inactivated paralog of the predicted protease and several other proteins ([916] and EVK, unpublished observations). It seems that the data presented in Table 9.1 and other similar observations tell us something truly fundamental: we have so many essential pieces missing from our current picture of archaeal and eukaryotic translation that we cannot claim a proper understanding of this crucial process. A concerted, prediction-based experimental programme could go a long way toward an adequate description of translation in its real complexity. To our knowledge, such a programme does not exist.

Given the phyletic pattern we choose, the search outlined above was clearly geared toward detecting unknown, essential translation components. Similar straightforward computational approaches can be readily applied for identifying other kinds of functional systems. For example, a recent examination of the COGs has shown that, strikingly, there is only one protein that is unique for hyperthermophiles (both bacteria and archaea), the reverse gyrase [238], which leads to the inevitable conclusion that this protein is essential for life under hyperthermophilic conditions. In all likelihood, however, it is not sufficient. Given the wide spread of HGT in the prokaryotic world, other genes that are important for thermophiles have probably "leaked" into mesophiles. Therefore a slightly more sophisticated approach can be employed to identify those additional determinants of the thermophilic phenotype. In Chapter 6, we briefly discussed a predicted repair system that is characteristic of thermophiles, although its individual components are found in some mesophiles also [541]. One could use the genes comprising this system as a training set to define the threshold phyletic pattern characteristic of functional systems important for thermophily (e.g. a pattern, in which 2/3 of the represented species are thermophiles, could be a reasonable cut-off). This approach reveals several proteins that are likely to perform thermophile-specific functions, e.g. distinct molecular chaperones (K.S. Makarova, Y.I. Wolf and EVK, unpublished observations). Clearly, the inquiries into phyletic patterns can be formulated in many different ways (e.g. directed toward detection of potential drug targets, see ♦7.6) and, accordingly, will reveal gene sets that can be reasonably implicated in various types of biological functions.

Table 9.1. Conserved protein families with predicted functions in search of experimental verification

COG No., predicted function	Phyletic pattern	Domain architecture	Additional functional inference from context analysis	Ref.
COG0533 Predicted ATP-dependent metalloprotease, possible chaperone activity	Universal	HSP70-like ATPase domain, inserted metalloprotease active site	Might belong to a chaperone complex involved in translation regulation. The only characterized member is sialoglycoprotease from Pasteurella	[1,43].
COG0589 Nucleotide-binding proteins (domains) of the UspA superfamily	Nearly universal, missing in several bacteria	USPA superfamily nucleotide-binding domains, several fusions, e.g. with protein kinases, CBS domains	Implicated in various forms of signal transduction and stress response	[540,937]
COG0061: Predicted (sugar) kinase	Nearly universal, missing in some bacterial parasites with small genomes	Predicted kinase domain distantly related to diacylglycerol kinases and phosphofructokinase	Genomic context points to multiple functions, i.e. in sugar metabolism, but possibly also in translation and repair.	L. Aravind and EVK, unpublished observations
COG0012 Predicted GTPase, probably essential, universal translation factor	Universal	GTPase + RNA-binding TGS domain	Involvement in translation supported by presence in a predicted operon with peptidyl-tRNA hydrolase	[135,505]

COG1163: predicted GTPase, probably essential, archaeo-eukaryote-specific translation factor	All archaea and eukaryotes, not found in bacteria	GTPase + RNA-binding TGS domain	No hints from genomic context, but involvement in translation supported by the phyletic pattern and the presence of TGS domain	[505]
COG2102: predicted thiouridine synthase	All archaea and eukaryotes, not found in bacteria	PP-loop ATPase domain	Involvement in translation supported by fusion with translation initiation factor in eukaryotes	[27]
COG1245 Predicted ATPase, probably translation regulator	All archaea and eukaryotes, not found in bacteria	ABC-class ATPase	Involvement in translation supported by fusion with queuine/archaeosine tRNA-ribosyltransferase	
COG2016: predicted RNA-binding protein	All archaea and eukaryotes	PUA domain (predicted RNA-binding)		

We believe that this straightforward approach could serve as the foundation of a major direction in functional genomics. This direction certainly will not capture the entire complexity of life and is not supposed to supplant the predominant current approach based on purely biological considerations and experimental contingency. However, it could be a strong complement to the traditional methodology.

Is this vision going to materialize? The crystal ball gets dim and we just do not know. The precedent of structural genomics is really heartening: this is where a similar rational approach clearly did work (even if the actual realization is only taking its early steps). Granted, this happened, in large part, because the methods for protein structure determination became much faster and the whole enterprise is turning into a real "technology". Perhaps prediction-directed functional genomics should await similar breakthroughs (if these ever are to come) in experimental study of protein functions. Nevertheless, it seems important for experimentalists to realize that this systematic approach to functional genomics is feasible at least in principle.

9.2. Digging up Genomic Junkyards

We have already apologized in the Preface for dealing almost exclusively with proteins in this book. It is only fair if we devote one or two paragraphs in this epilogue to the other kind of genetic material, the non-coding sequences. Even the largest eukaryotic genomes currently known to us contain only a few times more genes than complex prokaryotes have. Moreover, and almost incredibly, a complex actinomycete like *Streptomyces* has almost as many genes as a fruit fly, and we can by no means rule out that bacteria exist with even larger genomes, so that the ranges of gene numbers overlap. Surely, some of the organismic complexity of eukaryotes comes from domain accretion (Chapter 8) and widespread alternative splicing [118,576]. However, an obvious and truly dramatic difference between prokaryotic and eukaryotic genomes is in the amount and fraction of non-coding DNA (Figure 9.1). It seems that the powerful pressure of selection for genome compactness, that keeps intergenic distances as short as possible in prokaryotes [709] and apparently also in unicellular eukaryotes [425], had been somehow removed in multicellular eukaryotes. Furthermore, as encapsulated in the famous C-value paradox, the complexity of a eukaryotic organism does not at all seem proportionate, even roughly, to the genome size: indeed some bony fish appear to have larger genomes than humans by a factor of hundreds.

What is all this non-coding DNA for? An astonishing (and probably humiliating for those with a feeling of eukaryotic supremacy) explanation appeared in the classic 1980 article of W. Ford Doolittle and Carmen Sapienza [197] and was reaffirmed in the eloquent accompanying paper of Leslie Orgel and Francis Crick [634]. These researchers hypothesized that the bulk of non-coding DNA in multicellular eukaryotes is selfish, has no function useful for the organism and is there simply because the organism does not know how to get rid of it (or finds it too expensive) and has learned to live with all that junk. This is certainly a powerful idea and, to a degree, it is definitely correct: the basic selfishness of mobile elements such as ALUs and LINEs, which comprise a significant fraction of the human genome, is beyond doubt. However, even with these selfish elements, the matter is not quite that simple.

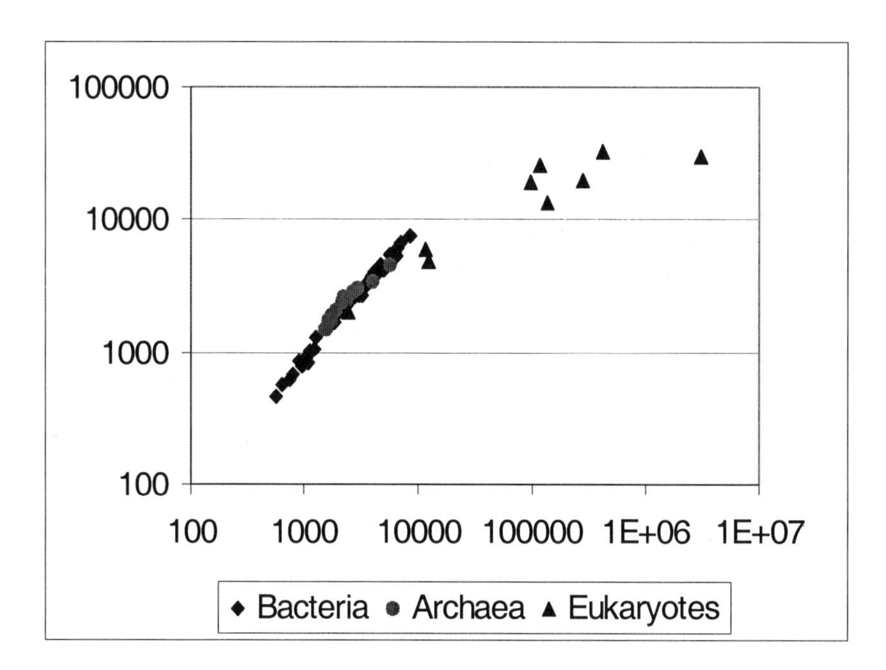

Figure 9.1. Correlation between the genome size (in kb) and number of genes. The number of genes in bacteria and archaea is proportional to the genome size. In eukaryotes, the gene number grows much slower than the genome size, resulting in a large fraction of non-coding DNA. The data are from Table 1.4.

Recent genome-wide analyses showed that a small fraction of transposable elements become exapted to function as protein-coding sequences [539,609] or, to an even greater extent, for regulatory elements of mammalian genes [409]. Obviously, these exapted sequences contribute to innovation, which may be critically important in evolution. So why are complex eukaryotes so tolerant to mobile elements: simply because they cannot get rid of them or because, despite their selfishness, these elements are repeatedly put to a good use? We do not have the answer. Perhaps the most balanced viewpoint is that these explanations are not alternative but rather give us complementary aspects of the real story. The mobile elements might have started off as pure genomic parasites, but later have been exploited by the host and now should be most properly regarded as symbionts.

More generally, recent comparisons of the non-coding DNA sequences in two nematode species and in humans and mice have shown that 20-30% of the "junk" non-coding DNA evolves under purifying selection, i.e. is not junk at all [762,763]. The impetus of these findings, if supported by further analysis, is hard to overestimate: they indicate that ~93% of functionally important DNA in our genomes does not code for proteins. What are the functions of this "dark matter"? As of today, we are not even close to an answer (note that, if the above estimates are correct, we have to account for ~600 megabase of DNA in the human genome, an equivalent of ~50 yeast genomes!). Most likely, there is no single solution to the puzzle because the functions of non-coding DNA are likely to be versatile. Some recent hints are most tantalizing: it seems that a huge fraction of the non-coding DNA is transcribed at some level and some of the transcripts are previously unidentified microRNAs, which are likely to perform a plenitude of regulatory functions (see [180] and the references therein). Whatever are the solutions to the mystery of "junk" DNA, it seems likely that a book on functional genomics of eukaryotes to be written ten years from now is not going to be predominantly about proteins anymore.

9.3. "Dreams of a final theory"[1]

Lord Rutherford famously (and, of course, arrogantly) quipped that "there are two kinds of science: physics and stamp collection". We freely and humbly admit that 99% of the comparative-genomic work discussed in this book is of the second kind. For sure, what genomicists collect are not more or less useless (even if beautiful) stamps, but important empirical findings on genomes, gene functions and homologous relationships. Subsequent analyses of these observations lead to crucial, even if, again, empirical generalizations on the evolution of life forms, such as genome fluidity caused by gene loss and horizontal gene transfer. Nevertheless, it must be admitted that this is still observational science. We think, however, that perhaps some of the analysis presented in Chapter 8 and, possibly, the genomic clock concept discussed in Chapter 6 belong to the other 1%, the part of comparative and evolutionary genomic that, in some ways, starts resembling physics. What we mean is that analysis of certain features of genomes, e.g. the size distributions of protein families, seems to reveal footprints of extremely general evolutionary processes whose mathematical form, such as power law asymptotics, recurs across an astonishing range of phenomena. These are just little steps toward general theory in biology, and there is not even a guarantee that they lead us in the right direction. In principle, however, we believe that a new paradigm of theoretical biology might not be unsustainable: ***through comparative genome analysis, develop a theory(ies) of the major evolutionary processes and apply them to the reconstruction of the history of life***.

Biology is an inherently historical science. In that respect, it is analogous to cosmology, which, in the last half century, has absorbed many advances of theoretical physics and, from a stamp-collection-like activity, has become a legitimate physical discipline. Indeed, the analogy seems to run deep. No theory will ever explain why our galaxy or our solar system look exactly the way they do because this is the result of unique fluctuations, which occurred during the evolution of a tiny corner of the universe. However, theoretical studies combined with increasingly detailed observation have led to a rigorous description of many important aspects of the evolution of the universe as a whole. The famous Penrose-Hawking theorem, which proves that, under a very broad class of conditions, there

---

[1] This is the title of the wonderful popular book of the Nobel Prize winner Steven Weinberg on the physicists' quest for GUTs (Grand Unified Theories) and GTEs (General Theories of Everything) [886].

was indeed in the beginning of the universe, is a good example of this. Similarly, if the evolution of life on earth could be run again or if evolution from similar initial conditions has actually run on another planet (something we would dearly love to know and one day probably will), the outcome would be quite different (and we would not be here to analyze it). The discovery of the major role of HGT in evolution emphasizes this unpredictability of life's evolution (a history with only vertical inheritance would be much easier to reconstruct but, again, we would not be around to do it). It seems almost certain that, in a rerun of life's evolution, the future eukaryotic cell would not form a symbiosis with a particular alpha-proteobacterium and eukaryotic life, as we know, it would not exist. However, it might not be unreasonable to hope that, eventually, the new theory of evolution might be able to answer questions such as: How likely is it that ***some*** archaeon would enter a symbiotic relationship with ***some*** complex bacterium, resulting in the emergence of new life forms of unprecedented complexity? One may further speculate that theoretical analysis could help develop reasonable models of the earliest phase of protein evolution, preceding the "Schmalhausen threshold" postulated in Chapter 6. From there, it might even be possible to glimpse the solution to the mystery that we view as the Holy Grail of evolutionary biology: the origin of genetic coding. If that point is reached, it will be time to say that we are starting to understand life.

APPENDICES

1. Glossary

This glossary does not aim at completeness and defines only those terms of genomics and computational biology that are extensively used in this book and a few additional terms deemed by the authors to be specifically important in comparative genomics. Many addititional definitions can be found at the following WWW sites:
http://www.ncbi.nlm.nih.gov/Education/BLASTinfo/glossary2.html,
http://genomics.phrma.org/lexicon/index.html.

Alignment

The principal form of representation of the similarity between nucleotide or amino acid sequences whereby nucleotides or amino acids inferred to have been derived from the same ancestral residue are superimposed. Identical or similar (in the case of amino acid sequences) residues are typically marked in an alignment. A ***pairwise alignment*** consists of two aligned sequences; pairwise alignments are typically produced as the output of a database search. A ***multiple alignment*** consists of three or more aligned sequences. In a ***global alignment***, sequences are aligned in their entirety, whereas a ***local alignment*** consists of aligned subsequences. *The* ***optimal (global or local) alignment*** of two sequences is the alignment with the highest score under a given scoring system, which is least likely to be found by chance; optimal alignment algorithms for multiple sequences are currently not feasible. ***Structural alignment*** is a sequence alignment produced by superposition of the three-dimensional structures of the respective proteins; structural alignments are generally thought to be the most accurate form of alignment.

Annotation (genome annotation)

Primary analysis of genomes deposited in sequence databases; typically involves prediction of genes and their functions.

Archaea (formerly archaebacteria)

One of the three primary kingdoms ("domains" or "superkingdoms") of life, along with bacteria and eukaryotes. Like bacteria, archaea are prokaryotes, i.e. simple, unicellular organisms with small cells containing no distinct organelles. Archaeal systems of replication, transcription and translation are phylogenetically and mechanistically affiliated with those of eukaryotes, whereas the metabolic and signaling systems largely resemble those of bacteria.

Bioinformatics

Narrowly and most appropriately defined, methods of computer science and informatics as applied to storage and retrieval of biological (particularly sequence and structural) information. More often used to designate almost any application of computers in biology, including the entire field of computational genomics.

BLAST (= Basic Local Alignment Search Tool)

The most commonly used suite of programs for fast similarity searches in nucleotide or protein sequence databases.

BLOSUM (= **BLOcks SUbstitution Matrix**)

A series of amino acid substitution matrices, in which the scores for each pairs of amino acid residues are derived from the frequencies of substitutions in local alignments of homologous proteins from the BLOCKS database. Each matrix is tailored to a particular evolutionary distance. For example, in the BLOSUM62 matrix, which is used by default by BLAST, the scores were derived from the alignments of sequences with no more than 62% identity.

CATH

A hierarchical classification of protein structures that clusters proteins at the levels of Class, Architecture, Topology and Homologous superfamily (http://www.biochem.ucl.ac.uk/bsm/cath_new/index.html).

Composition-based statistics

Modification of the Karlin-Altschul statistics of sequence comparison that involves calculation of the scaling parameters separately for each database sequence. This typically solves problems associated with **low-complexity regions** and obviates the necessity for **filtering**.

Consensus sequence

One of the forms of representation of the sequence conservation in an alignment of DNA or protein sequences. A consensus includes residues that are most frequent in each position of the alignment.

Convergence

Independent origin of similar features in sequences, structures and biological processes due to shared functional constraints.

Domain (protein domain)

A compact, stable, independently folding unit of protein structure. Often used also to designate a portion of a protein with distinct evolutionary history, which may exist in the form of a stand-alone protein or as part of different ***domain architectures*** (linear sequence of domains in a ***multidomain protein***). An evolutionary domain is typically identical to a structural domain, but in some cases, may include more than one structural domain.

Domain accretion

A notable phenomenon in the evolution of complex life forms, primarily eukaryotes, whereby increasingly complex domain architectures are observed within a set of **orthologs**.

E-value (= Expectation value of a given score).

The number of different alignments with scores equal to or greater than a given value that are expected to be found by chance in a search of a given database with a particular method and substitution matrix (e.g. BLAST with BLOSUM62). The lower the E-value the more statistically significant is the score.

Exaptation

Recruitment of a protein for a new function, which may be mechanistically, but not biologically related to the original one

Family (of homologous proteins)

One of the main levels in the hierarchical classification of proteins. There is no generally accepted, formal definition. Typically, proteins with readily detectable, statistically significant sequence similarity to each other are classified as a family.

FASTA

The first fast and sensitive method for sequence database search.

Filtering (sequence filtering)

Masking repeats, regions of **low sequence complexity** or regions with other features in nucleotide or amino acid sequences, typically in the query sequence prior to a database search to avoid spurious hits. Also see **SEG**.

Fold (protein fold, protein structural fold)

The central category in the hierarchical classification of proteins. Proteins with the same basic topology of structural elements are classified as having the same fold.

Functional genomics

Study of biological functions of individual proteins, complexes, pathways etc. based on or at least facilitated by analysis of genome sequences.

Gap

A space introduced into an alignment to compensate for an insertion or deletion in one aligned sequence relative to the other(s); typically designated by a dash.

Gap penalty

A penalty to the overall alignment score (a negative score) for gap introduction and/or extension, intended to prevent accumulation of too many gaps in an alignment (reflecting the fact that insertions and deletions are less common during evolution than substitutions).

Gene

A piece of genomic DNA that encodes the synthesis of a (at least one) mRNA or structural RNA molecule.

GenBank

Archival database of nucleotide sequences; typically, searched for nucleotide sequence similarity.

Genome

The complete DNA sequence of a life form; consists of genes and intergenic regions.

Genomics

The study of genomes.

GenPept

Archival database of protein sequences translated from ORFs annotated in GenBank; the basis for the **NR** database.

Hidden Markov Models (HMM)

Mathematical formalism used to describe sequence alignments or individual sequences in terms of the probability of a given symbol in a given position.

Homology

Origin from a common ancestor. Applies primarily to genes and their products (homologs), but entire complexes, pathways, operons etc. also may be called homologous when they are inferred to have a common origin.

Horizontal (lateral) gene transfer (HGT or LGT)

Transfer of genes from one phylogenetic lineage to another, as opposed to vertical descent along a lineage.

HSP (High-scoring Segment Pair)

A segment of an alignment of two sequences such that its similarity score cannot be improved by adding or trimming any letters; applies to similar sequence segments detected in database searches, such as BLAST.

Karlin-Atlschul statistics

Statistical theory based on the extreme value distribution of the best scores for pairs of sequences; used for E-value calculation in BLAST.

Lineage-specific expansion (of paralogous gene families)

A series of duplications of genes during the evolution of a phylogenetic lineage leading to the emergence of a lineage-specific cluster of paralogs. Typically, lineage-specific expansions are associated with adaptations linked to the lifestyle of the respective organisms.

Lineage-specific gene loss

Elimination of a distinct set of genes in a particular phylogenetic lineage.

Low-complexity regions

Sequence regions of biased nucleotide or amino acid composition, e.g. homopolymeric runs, short repeats, and more subtle overrepresentations of one or a few residues. Low-complexity regions in amino acid sequences typically assume non-globular structure in proteins. These regions cause spurious hits in database searches unless they are masked (e.g. with the **SEG** program) or **composition-based statistics** is applied.

Molecular clock

The evolutionary concept, introduced by Emile Zuckerkandl and Linus Pauling, according to which the rate of a gene's evolution remains constant throughout its history.

Motif (protein sequence motif)

A pattern of evolutionarily conserved amino acid residues in a protein that are important for its function and are located within a certain (typically, short) distance from each other in the protein sequence. A single domain may contain several conserved sequence motifs.

Non-orthologous gene displacement (NOGD)

Displacement, in the course of evolution, of a gene coding for a protein responsible for a particular function with a non-orthologous (unrelated or distantly related), but functionally analogous gene.

NR (non-redundant database)

A non-redundant subset of nucleotide sequences from GenBank or protein sequences from GenPept, derived by combining all identical sequence entries from the same organism into a single entry. Used as the default database for BLAST searches.

ORF (Open Reading Frame)

A piece of DNA from a (potential) start codon to a stop codon that may code for a protein.

Orthologs

Homologous genes from different species related via speciation (vertical descent), i.e. originating from the same ancestral gene in the last common ancestor of the compared species.

P-value

The probability of an alignment with a score at least as great as the observed one being produced by chance in a search of a given database with a particular method and substitution matrix (e.g. BLAST with BLOSUM62). The lower the P-value the more likely are the genes in question to be homologous. See also **E-value**.

PAM (Accepted Point Mutation per 100 links)

A measure of evolutionary divergence of protein sequences, introduced by Margaret Dayhoff. One PAM corresponds to one substitution in 100 amino acids in a protein sequence. The series of PAM matrices reflects evolutionary distances between protein sequences measured in PAMs (e.g. PAM80 corresponds to 80 expected changes in 100 residues).

Paralogs

Homologous genes related via duplication.

Parsimony

A general principle of evolutionary reconstruction aimed at constructing scenarios with the minimal number of event required to account for the available data. Specifically, embedded in maximum parsimony methods for phylogenetic tree construction.

Proteome

The entire set of proteins expressed by a given organism. Sometimes also used to designate the set of potential proteins inferred from a genome sequence. For this usage, *conceptual proteome* or *predicted proteome* probably are more appropriate terms.

Percent identity

The simplest measure of similarity between two nucleotide or protein sequences. A poor and potentially misleading criterion for assessing distant evolutionary relationships.

Phylogenomics

An approach in evolutionary studies combining genome comparison with phylogenetic tree analysis.

PSI-BLAST (Position-Specific Iterating BLAST)

An iterative method for sequence similarity search based on the BLAST algorithm; a powerful approach for studying distant evolutionary relationships between proteins.

Position-Specific Scoring Matrix (PSSM)

A matrix of scores for each amino acid in each position of a multiple alignment. The typical size of a PSSM is 20x(alignment length).

Query

The input sequence that is fed into a program to search a database for homologs, predict structural features etc.

SCOP (=Structural Classification of Proteins)

Hierarchical classification of known protein structures that is based on the notions of structural class, fold, superfamily and family and employs both structural and sequence similarity as classification criteria (http://scop.mrc-lmb.cam.ac.uk/scop/).

SEG

A program for finding and filtering **low-complexity regions** in protein sequences. In its most common implementation, replaces masked residues with "X".

Similarity score

Quantitative measure of the similarity between two sequences.

Structural genomics

The research program aiming at determination of a representative set of protein structures, with target selection and prioritization based on genome comparison.

Substitution matrix

A matrix of the substitution scores for each of the 20 amino acids.

Superfamily

The level in the hierarchical classification of proteins that is intermediate between **fold** and **family**.

SWISS-PROT

A popular curated protein sequence database.

Threading

A family of methods aiming at predicting protein structure by determining the compatibility of a given sequence with available structural templates.

Transcriptome

The complete set of RNA transcripts expressed in a given organism.

2. Useful Web Sites

2.1. Databases

This is an abbreviated list of the databases discussed in Chapter 3.

General Purpose Sequence Databases.

Nucleotide sequence databases

GenBank	http://www.ncbi.nlm.nih.gov/Entrez
EMBL	http://www.ebi.ac.uk
DDBJ	http://www.ddbj.nig.ac.jp

Protein sequence databases

GenPept	http://www.ncbi.nlm.nih.gov/entrez
SWISS-PROT	http://www.expasy.org/sprot
ENZYME	http://www.expasy.org/enzyme
TrEMBL	http://www.expasy.org/sprot
PIR	http://pir.georgetown.edu
MIPS	http://mips.gsf.de/proj/protseqdb
PRF	http://www.prf.or.jp/en

Motif and domain databases

PROSITE	http://www.expasy.org/prosite
BLOCKS	http://www.blocks.fhcrc.org
Pfam	http://www.sanger.ac.uk/Software/Pfam
	http://pfam.wustl.edu
	http://www.cgr.ki.se/Pfam
	http://pfam.jouy.inra.fr
SMART	http://smart.embl-heidelberg.de
ProDom	http://www.toulouse.inra.fr/prodom.html
COGs	http://www.ncbi.nlm.nih.gov/COG
InterPro	http://www.ebi.ac.uk/interpro
CDD	http://www.ncbi.nlm.nih.gov/Structure/cdd/cdd.shtml

Structure databases

PDB	http://www.rcsb.org/pdb
	http://rutgers.rcsb.org/pdb
	http://nist.rcsb.org/pdb
MMDB	http://www.ncbi.nlm.nih.gov/Structure
FSSP	http://www.ebi.ac.uk/dali/fssp
SCOP	http://scop.mrc-lmb.cam.ac.uk/scop
	http://scop.berkeley.edu
CATH	http://www.biochem.ucl.ac.uk/bsm/cath_new

General genomics databases

NCBI Genomes	http://www.ncbi.nlm.nih.gov/PMGifs/Genomes/bact.html
NCBI FTP site	ftp://ncbi.nlm.nih.gov/Entrez/Genomes
DDBJ GIB	http://gib.genes.nig.ac.jp
EBI Genomes	http://www.ebi.ac.uk/genomes
TIGR	http://www.tigr.org/tdb/mdb/mdb.html
GOLD	http://wit.integratedgenomics.com/GOLD

Specialized genomics databases

COGs	http://www.ncbi.nlm.nih.gov/COG
KEGG	http://www.genome.ad.jp/kegg
TIGR	http://www.tigr.org/tdb
WIT	http://wit.mcs.anl.gov
ERGO	http://ergo.integratedgenomics.com/ERGO
MBGD	http://mbgd.genome.ad.jp
PEDANT	http://pedant.gsf.de

Organism-specific databases

Escherichia coli

E. coli genome	http://www.genome.wisc.edu
Essential genes	http://magpie.genome.wisc.edu/~chris/essential.html
GenoBase	http://ecoli.aist-nara.ac.jp/docs/genobase/index.html
PEC	http://shigen.lab.nig.ac.jp/ecoli/pec
EcoGene	http://bmb.med.miami.edu/ecogene
E. coli index	http://web.bham.ac.uk/bcm4ght6
CGSC	http://cgsc.biology.yale.edu
EcoCyc	http://ecocyc.doubletwist.com
RegulonDB	http://www.cifn.unam.mx/Computational_Genomics/regulondb/
Colibri	http://bioweb.pasteur.fr/GenoList/Colibri

Bacillus subtilis

Subtilist	http://bioweb.pasteur.fr/GenoList/SubtiList
Micado	http://locus.jouy.inra.fr
Sporulation	http://www.rhul.ac.uk/Biological-Sciences/cutting/

Saccharomyces cerevisiae

SGD	http://genome-www.stanford.edu/Saccharomyces
MIPS	http://mips.gsf.de/proj/yeast
YPD	http://www.proteome.com/databases .
TRIPLES	http://ygac.med.yale.edu
MitoPD	http://bmerc-www.bu.edu/mito

Genome Deletion Project	http://genomics.stanford.edu
Saccharomyces Promoter Database	http://cgsigma.cshl.org/jian

Unicellular eukaryotes

Candida albicans	http://sequence-www.stanford.edu/group/candida
Dictyostelium discoideum	http://www.uni-koeln.de/dictyostelium
Entamoeba histolytica	http://www.lshtm.ac.uk/mp/bcu/enta/homef.htm
Giardia lamblia	http://www.mbl.edu/Giardia
Leishmania major	http://www.ebi.ac.uk/parasites/leish.html,
Neurospora crassa	http://mips.gsf.de/proj/Neurospora
Plasmodium falciparum	http://www.plasmodb.org
	http://www.ncbi.nlm.nih.gov/Malaria
	http://www.sanger.ac.uk/Projects/P_falciparum
Pneumocystis carinii	http://biology.uky.edu/Pc

Arabidopsis thaliana

TAIR	http://www.arabidopsis.org
TIGR	http://www.tigr.org/tdb/e2k1/ath1
MIPS	http://mips.gsf.de/proj/thal/db/index.html
DAtA	http://sequence-www.stanford.edu/ara/SPP.html
KAOS	http://www.kazusa.or.jp/kaos
CSHL	http://nucleus.cshl.org/protarab

Caenorhabditis elegans

WormBase	http://www.wormbase.org
	http://wormbase.sanger.ac.uk
WormPD	http://www.proteome.com/databases/index.html
Sanger Centre	http://www.sanger.ac.uk/Projects/C_elegans
WashU	http://genome.wustl.edu/gsc/Projects/C.elegans

Drosophila melanogaster

FlyBase	http://flybase.bio.indiana.edu/
GadFly	http://www.fruitfly.org
FlyBrain	http://flybrain.neurobio.arizona.edu
InterActive Fly	http://sdb.bio.purdue.edu/fly/aimain/1aahome.htm
Drosophila gene expression	http://quantgen.med.yale.edu
Drosophila Community Portal	http://www.cybergenome.com/drosophila

Human

OMIM™	http://www.ncbi.nlm.nih.gov/Omim
	http://www.ncbi.nlm.nih.gov/entrez/query.fcgi?db=OMIM
LocusLink	http://www.ncbi.nlm.nih.gov/LocusLink
HomoloGene	http://www.ncbi.nlm.nih.gov/HomoloGene
euGenes	http://iubio.bio.indiana.edu/eugenes

Genes and Disease http://www.ncbi.nlm.nih.gov/disease

Various databases

NCBI Taxonomy http://www.ncbi.nlm.nih.gov/Taxonomy
Ribosomal Database Project http://rdp.cme.msu.edu
http://wdcm.nig.ac.jp/RDP

TRANSFAC http://www.gene-regulation.de
http://transfac.gbf.de
BRITE http://www.genome.ad.jp/brite_old
http://www.genome.ad.jp/brite
DIP http://dip.doe-mbi.ucla.edu
BIND http://www.bind.ca
BioCarta http://www.biocarta.com/genes/allPathways.asp
EPD http://www.epd.isb-sib.ch
ENZYME http://www.expasy.org/enzyme
IUBMB http://www.chem.qmw.ac.uk/iubmb/enzyme
Klotho http://www.ibc.wustl.edu/klotho
BRENDA http://www.brenda.uni-koeln.de
LIGAND http://www.genome.ad.jp/dbget/ligand.html
AAindex http://www.genome.ad.jp/dbget/aaindex.html
GtRDB http://rna.wustl.edu/GtRDB
Protein Mutant Database http://pmd.ddbj.nig.ac.jp

2.2. Major genome sequencing centers

Publicly-funded genome centers in the USA

Baylor College of Medicine Human Genome Sequencing Center (human, mouse, *Drosophila*, *Dictyostelium*):
http://www.hgsc.bcm.tmc.edu

DOE Joint Genome Insitute (human, mouse, pufferfish, sea squirt, white rot fungus, many bacteria):
http://www.jgi.doe.gov

Lawrence Berkeley Laboratory Life Sciences Division:
http://www.lbl.gov/lifesciences

Lawrence Livermore National Laboratory (human, *Drosophila*):
http://bbrp.llnl.gov

Los Alamos National Laboratory Center for Human Genome Studies
http://jgi-lanl-public.lanl.gov

Stanford University Human Genome Center (human, mouse):
http://www-shgc.stanford.edu

Stanford Genome Technology Center (yeast, *Arabidopsis*, malaria, *Candida*, maize, human, *Cryptococcus neoformans*):
http://sequence-www.stanford.edu

The Institute for Genomic Research (human, mouse, rat, *Drosophila*, rice, *Arabidopsis*, zebrafish, *Plasmodium*, *Trypanosoma*, microbial genomes): http://www.tigr.org

University of California at Berkeley Drosophila Genome Center http://www.fruitfly.org

University of Illinois at Urbana-Champaign Biotechnology Center (*Arabidopsis*, *Salmonella*, soybean EST, cattle EST, honey bees EST): http://www.biotec.uiuc.edu

University of Minnesota Computational Biology Center (*Arabidopsis*, loblolly pine, rice, maize, *Brassica napus*, soybean, *Pasteurella*): http://www.cbc.umn.edu/ResearchProjects/index.html

University of Oklahoma's Advanced Center for Genome Technology (human, mouse, *Actinobacillus actinomycetemcomitans*, *Neisseria gonorrhoeae*, *Streptococcus mutans*, *Strep. pyogenes*, *Staphylococcus aureus*, *Aspergillus nidulans*, *Cryptococcus neoformans*, *Neurospora crassa*, *Fusarium sporotrichioides*): http://www.genome.ou.edu

University of Utah Genome Center (human, mouse): http://www.genome.utah.edu

University of Washington Genome Center: http://www.genome.washington.edu

University of Wisconsin *E. coli* Genome Center (*E. coli*, *Shigella*, *Salmonella*, *Yersinia*): http://www.genome.wisc.edu

Washington University Genome Sequencing Center (human, *C. elegans*, *C. briggsae*, *Arabidopsis*, yeast, *Salmonella*): http://genome.wustl.edu/gsc

Whitehead Institute for Biomedical Research, MIT (human, mouse): http://www-genome.wi.mit.edu

USA private companies

AstraZeneca Pharmaceuticals (*Helicobacter pylori*): http://scriabin.astrazeneca-boston.com/hpylori

Celera Genomics (human, mouse, human SNPs): http://www.celera.com

Cereon Genomics (*Myxococcus xanthus*, *Aspergillus nidulans*): http://microbial.cereon.com

Genome Therapeutics (human, *Methanobacterium thermoautotrophicum*, *Clostridium acetobutylicum*): http://www.cric.com

Diversa (*Aquifex aeolicus*, *Streptomyces diversa*, *Pyrolobus fumarii*) http://www.diversa.com/

Intergrated Genomics (*Bacillus cereus*, *Thiobacillus ferrooxidans*): http://www.integratedgenomics.com

United Kingdom:

The Sanger Centre (human, mouse, chicken, pufferfish, zebrafish, *Drosophila*, *C. elegans*, *Dictyostellium*, *Leishmania*, *Plasmodium*, *Trypanosma*, yeast, *Candida*, *S. pombe*, many bacterial genomes):
http://www.sanger.ac.uk/Projects

Canada

University of Britich Columbia Genome Sequence Centre (human, mouse, rat, cow, *Cryptococcus neoformans*):
http://www.bcgsc.bc.ca/

France

Genoscope (human, mouse, pig, frog, rice, wheat, *Arabidopsis*, many bacterial genomes):
http://www.genoscope.fr/externe/English/Projets/projets.html

Germany

Göttingen Genomics Laboratory (*Thermus thermophilus*, *Methanosarcina mazei*, *Clostridium tetani*):
http://www.g2l.bio.uni-goettingen.de

Japan

Japan Marine Science and Technology Center (JAMSTEC):
http://www.jamstec.go.jp/jamstec-e/bio/DEEPSTAR/FResearch.html
Kazusa DNA Research Institute (human, *Arabidopsis*, cyanobacteria)
http://www.kazusa.or.jp/cyano
National Institute of Agrobiological Sciences (rice):
http://rgp.dna.affrc.go.jp
National Institute of Technology and Evaluation, Biotechnology Center (*Pyrococcus horikoshii*, *Aeropyrum pernix*, *Staphylococcus aureus*)
http://www.bio.nite.go.jp/E-home/biomenu-e2.html
RIKEN Genomic Sciences Center (*Buchnera* sp.):
http://www.gsc.riken.go.jp

3. Problems

We included only 20 problems but most of them are large, include multiple questions and, to a considerable extent, are open-ended. We suggest checking with different chapters of this book after solving the problem, but of course the reader should also consult them in case of difficulties in addressing these questions.

1. Rank the following criteria of similarity between protein sequences in the order of increasing utility for protein function prediction:

- Percent similarity
- Percent identity
- Annotation in SWISS-PROT
- Bit score reported by BLAST
- E-value reported by BLAST
- Annotation in GenBank
- E-value for conserved domains reported by CDD search
- Presence of PROSITE patterns

To assess you choice, go back to Chapter 4 and, if necessary, further reading listed at the end of this chapter.

2. Are the following statements on protein sequence similarity true or false?

- the greater E-value the more similar two protein sequences are
- the greater the BLAST score the more similar two sequence are
- E-value reported by PSI-BLAST at convergence is the best measure of similarity between two protein sequences
- Sequences with $E > 10^{-5}$ in BLAST searches are not homologous
- Sequences with $E < 10^{-10}$ in BLAST searches are homologous
- Sequences with $E > 10$ in BLAST searches are not homologous.

Check your answers using Chapter 4 and, if necessary, further reading listed at the end of this chapter.

3. Take the texts of the first and fourth stanza of "The Raven" (available, e.g. at http://www.nps.gov/edal/raven.htm) and align them using LALIGN (a variant of FASTA at http://fasta.bioch.virginia.edu/fasta/lalign2.htm) and BLAST2seq (at http://www.ncbi.nlm.nih.gov/blast/bl2seq/bl2.html; both LALIGN and BLAST2seq are also available at the INFOBIOGEN web site, http://www.infobiogen.fr/services/menuserv.html). Compare the alignments and similarity scores obtained by these two tools. Scroll down the LALIGN output and explore alternative alignments. Submit these texts to the PLALIGN tool, http://fasta.bioch.virginia.edu/fasta/plalign2.htm and inspect the graphical output of the program.

Edit the texts by removing the o's and u's and align them using the Needleman-Wunsh and Smith-Waterman algorithms at the EBI's EMBOSS server http://www.ebi.ac.uk/emboss/align. Compare the results with BLAST2seq and LALIGN outputs.

4. Use SCOP and CATH databases to find 5-10 examples of distant homologs that have less than 10% identity over the entire protein length. Can you identify functional motifs in these proteins given that they are homologous?

5. Use all databases in your disposal to collect the information about the following proteins that are mutated in the human diseases (you will be able to utilize this information in sequence analysis of these proteins in problem 20):

Bcl10, frataxin, fukutin, gigaxonin, harmonin, nyctalopin, paracaspase, parkin, pendrin, VEGF, VEGFR.

6. For each protein listed in the left column, obtain the sequence from the NCBI protein database and apply analysis tools available on the web until you arrive at annotation you feel comfortable with (***do not look*** at SWISS-PROT or COG annotations!)

Once you have finished, compare your conclusions with SWISS-PROT and COG annotations and with the comments in the papers listed in the right column. If you come across other useful examples like these, the authors will be happy to learn about them.

Protein gi	Assigned function, ref.	Comment, ref.
740170	Regulator of mitotic spindle assembly [933]	[548,852,934]
3844713	Arginine deiminase [639]	[264,466]
1592164	Mannose-sensitive hemagglutinin E [31]	[264]
1591024	Dihydropteroate synthase [928]	[387]
1591468	Thymidylate synthase [928]	[387]
1592116	Cysteinyl-tRNA synthetase [218]	[387]
10580454	Inosine-5'-monophosphate dehydrogenase	[616]
Protein gi	**Putative homolog, ref.**	**Comment, ref.**
5803005	Cellular homolog of the hepatitis delta antigen (gi|112963) [115]	[525,705]
6968351	Homolog of type III secreted proteins SipB, IpaB and YopB [449]	[648]
1592287	IMP dehydrogenase homolog [31]	[264]

7. List all the conserved domains you can find in the following proteins. What can you say about these proteins based on their domain architectures? Compare your conclusions to those in the original publications and try to explain any discrepancies. After that, look at the comments in the papers listed in the right column.

Protein name, gi	**Conserved domain, ref.**	**Comment, ref.**
ClpX, 2506299, 2506300	C-terminal PDZ-like domain [222,508],	[613,648].
FLASH, 4754905	CED-4-like P-loop ATPase domain, DED domain [378,560]	[458]
BRCA1, 728984	Granin [395,802,809,817]	[454]

8. Using COGs, compile a list of proteins that form

- "Archaeal genomic signature"
- "Archaeo-eukaryotic core"
- "Genomic determinants of hyperthermophily "
- "Pathogen-specific gene set"
- "Uncharacterized bacteria-specific proteins"

How would the results change if you allow one or two exceptions? Compare your conclusions to those in refs. [307,540,541]

9. Which protein families are most conserved in evolution? Why? Give specific examples. Are enzymes typically more or less conserved than transcription regulators? Why?

10. Make an alignment and build a phylogenetic tree of class I lysyl-tRNA synthetases (COG1384). How reliable is this tree? What can you say about the evolution of this enzyme? Also, compare the phyletic patterns of the COGs for Class I and II lysyl-tRNA synthetases (COG1384 and COG1190, respectively). What conclusions can be drawn from the relationship between these patterns? Compare the results of your analysis to those described in refs. [25,907,909].

11. Identify all TCA cycle enzymes in:

a) *Archaeoglobus fulgidus*
b) *Methanococcus jannaschii*
c) *Helicobacter pylori*

How would you explain the presence of these particular enzymes and absence of the others? Compare your conclusions to those in ref. [370].

12. Resolution of Holliday junction is an essential step in DNA repair and recombination in all organisms. Compare the phyletic patterns of three (predicted) Holliday junction resolvases, which comprise COG0816, COG0817 and COG1591. How would you describe the relationship between the proteins in these COGs? Which of them are:

- homologs?
- orthologs?
- paralogs?

Do you believe that the following phenomena are involved in the evolution of this function:

- non-orthologous gene displacement?
- partial functional redundancy?
- Lineage-specific gene loss?
- Horizontal gene transfer?

Compare your conclusions to those in ref. [50].

13. COG2965 includes the PriB proteins, which are present only in some proteobacterial species and are involved in the aseembly of the primosome, a large protein complex involved in replication initiation [11]. Examine the genomic context of the *priB* gene. Do you find it surprising? Investigate the relationship between PriB and single-stranded DNA-binding protein Ssb (COG0629) using several sequence analysis tool (note that the crystal structure of Ssb is also available). Are PriB and Ssb:

- Homologs?
- Orthologs?
- Paralogs?
- All of the above?
- None of the above?

Should you identify any meaningful relationship between PriB and Ssb, examine the genomic context of Ssb. By combining all the information obtained in the above analyses, can you come up with an evolutionary scenario(s) explaining the origin of PriB and the genomic location of both genes? There is no publication describing these relationships. One of the authors (EVK) is only preparing a manuscript on this subject, to be published in 2003, so the reader has ample opportunity to address the problem before this prepared publication appears in print.

14. Identify prokaryotic homologs of the yeast DNA repair protein RAD50. What is the easiest way to do this using the built-in functionalities of BLAST? Examine first the results of the CDD search. Does this search suggest that RAD50 is an ATPase? If so, do the hits to the following ATPase families reflect homology to RAD50:

- V-ATPase subunit A
- ATPase subunits of ABC-transporters
- SMC-like ATPases

What is the nature of the similarity between myosin and RAD50? Is this similarity spurious? Is it evidence of homology? Of the top 10 prokaryotic hits, which proteins are real homologs of RAD50? Answer this question for the search run with

- composition-based statistics turned on
- composition-based statistics turned off
- composition-based statistics off and low-complexity filtering on

If the results are not fully satisfactory, can you improve them by

- applying both composition-based statistics and low-complexity filtering
- running BLAST with a fragment(s) of the RAD50 sequence
- running FASTA

Identify the conserved motifs that are most important for the function of RAD50. On the basis of the identified motif, can you predict the domain architecture and the three-dimensional structure of this protein? What is unusual about this structure?

Once you have the answers, check with Chapter 4, SCOP and PDB databases and ref. [296].

15. Archaea, bacteria and eukaryotes synthesize a variety of isoprenoid compounds. Reconstruct and compare the pathways leading to the formation of a common intermediate, isopentenyl pyrophosphate, in

- *Escherichia coli*
- *Borrelia burgdorferi*
- *Methanococcus jannaschii*
- *Saccharomyces cerevisiae*

What evolutionary events could account for such patterns? Which of these enzymes are present in humans? What are the biochemical mechanisms of mevalonic aciduria (OMIM: 251170) and hyperimmunoglobulinaemia D (OMIM: 2609200)? Compare your conclusions to those in refs. [111,780].

16. In the genome(s) of *Streptococcus pneumoniae*, identify

- potential targets for antibiotics against *S. pneumoniae*
- surface-exposed proteins that could serve as vaccine candidates.

How many of these proteins are shared between

- *S. pneumoniae* and *S. pyogenes*?
- *S. pneumoniae* and *Staphylococcus aureus*?
- *S. pneumoniae* and other gram-positive bacteria?

Which ones appear to be reasonable candidates for broad-range antibiotics? Compare your conclusions to those in [836,839]

17. In Chapter 5, prediction of the archaeal exosome using genome context approaches is mentioned. Try to reproduce this prediction starting with the sequences of eukaryotic exosome subunits available in GenBank. Compare the protein compositions and the gene organizations for the predicted exosomes of *Methanothermobacter thermoautotrophicus* and *Thermoplasma volcanii*. How do you explain the differences? Compare the results to those in [469].

18. Of the 30 COGs that include GTPases, 13 are marked as "Predicted", which means that, at least at the time of the latest COG release, the GTPase activity has not been determined experimentally. What is the basis for this prediction in each case? Do you agree with these predictions? What are the unusual features of proteins in COG1162? Are these proteins indeed likely to be GTPases? For several of the "Predicted GTPase" COGs, specific biological activity, at least in general terms, is also predicted on the COG web site. What is the support for these predictions? Are there always at least two independent lines of evidence? Can you make additional predictions? Once you have answers, check with Chapters 4 and 5, and with refs. [135,505].

19. DNA primases in bacteria and in archaea-eukaryotes are unrelated (see ♦6.4.1.1.). Can you identify archaeal and/or eukaryotic homologs of the bacterial primase and, conversely, bacterial homologs of at least one subunit of the archaeal-eukaryotic primase? If yes, what are the domain architectures and probable functions of these proteins? Do you believe that the following phenomena have been involved in the evolution of primases?

- non-orthologous gene displacement
- lineage-specific gene loss
- horizontal gene transfer
- exaptation

Compare your answers to ♦6.4.1.1 and refs. [46,504].

20. For each of the following inherited human diseases, read the detailed description in OMIM and perform complete sequence analysis of the corresponding protein. Try explaining how mutations in these proteins could contribute to the observed phenotypes. Compare your conclusions to those in refs. [593,597,796] (see also http://www.ncbi.nlm.nih.gov/Disease_Genes and http://www.ncbi.nlm.nih.gov/XREFdb/pubtools.html).

Disease name	OMIM entry	Protein gi
Aarskog-Scott syndrome (faciogenital dysplasia)	305400	595425
Achondroplasia	100800	4885233
Addison disease (X- linked adrenoleukodystrophy)	300100	38591
Agammaglobulinemia, X- linked	300300	1684915
Alzheimer disease	104300	1709856
Amyotrophic lateral sclerosis	602433	338276
Aniridia type II	106210	189354
Ataxia telangiectasia	208900	870786
Autosomal chronic granulomatosis	233700	189051
Campomelic dysplasia	114290	758103
Brachydactyly type B1	113000	4514620
Barth Syndrome	302060	1263110
Bloom Syndrome	210900	1072122
Ceroid lipofuscinosis	256730	1314355
Chondrodysplasia punctata	302950	4502241
Choroideremia	303100	4261520|
Pitutary hormone deficiency	262600	6572501
Congenital adrenal hyperplasia:	201910	180964
Congenital chloride diarrhea	214700	4557535
Congenital stationary night blindness	310500	12007646
Craniosynostosis	604757	1321638
Diabetes mellitus, type II (noninsulin-dependent)	125853 604641	4885433
Diastrophic dysplasia	222600	4557539
Dominant optic atrophy	165500	17380163

Duchenne muscular dystrophy	310200	181857
Emery-Dreifuss muscular dystrophy	181350	600619
Endotoxin hyporesponsiveness	603030	9622357
Familial adenomatous polyposis	175100	4557319
Familial British dementia	176500	11527402
Familial cylindromatosis	132700	8250236
Focal segmental glomerulosclerosis	603278	12025678
Fragile X mental retardation	600819	1730139
Friedreich ataxia	229300	4503785
Fukuyama-type muscular distrophy	253800	3370993
Giant axonal neuropathy	256850	11545731
Hailey-Hailey disease	169600	7656910
Lafora's disease	254780	6005986
Macular corneal dystrophy	217800	11023146
MALT lymphoma	604860	5706378
MALT lymphoma	603517	4502379
May-Hegglin anomaly	155100	6166599
Mulibrey nanism	253250	15147333
McKusick-Kaufman syndrome	236700	10946954
Netherton syndrome	256500	4585699
Niemann-Pick C1 disease	257220	8134594
Parietal foramina	168500	1321638
Parkinsonism	168600	4758884
Pendred syndrome	274600	4505697
Primary congenital lymphedema	153100	1718189
Retinitis pigmentosa	604705	14133726
Stargardt disease	248200	6707663
Trichorhinophalangeal syndrome	190350	7657659
Spondylocostal dysostosis	277300	7417347
Usher syndrome type 1c	276904	12963769
Visceral heterotaxy	605376	14211837
Wolcott-Rallison syndrome	226980	9652337
X-linked mental retardation	300267	11036529

References

1. Abdullah KM, Lo RY, Mellors A. 1991. Cloning, nucleotide sequence, and expression of the *Pasteurella haemolytica* A1 glycoprotease gene. J. Bacteriol. 173: 5597-5603.

2. Abell CW, Kwan SW. 2001. Molecular characterization of monoamine oxidases A and B. Prog. Nucleic Acid. Res. Mol. Biol. 65: 129-156.

3. Adams MD, Celniker SE, Holt RA, Evans CA, Gocayne JD, Amanatides PG, Scherer SE, Li PW, Hoskins RA, Galle RF, et al. 2000. The genome sequence of *Drosophila melanogaster*. Science 287: 2185-2195.

4. Adams MD, Kelley JM, Gocayne JD, Dubnick M, Polymeropoulos MH, Xiao H, Merril CR, Wu A, Olde B, Moreno RF, et al. 1991. Complementary DNA sequencing: expressed sequence tags and human genome project. Science 252: 1651-1656.

5. Adrain C, Martin SJ. 2001. The mitochondrial apoptosome: a killer unleashed by the cytochrome seas. Trends Biochem. Sci. 26: 390-397.

6. Ahlquist P. 2002. RNA-dependent RNA polymerases, viruses, and RNA silencing. Science 296: 1270-1273.

7. Ahlquist P, Strauss EG, Rice CM, Strauss JH, Haseloff J, Zimmern D. 1985. Sindbis virus proteins nsP1 and nsP2 contain homology to nonstructural proteins from several RNA plant viruses. J. Virol. 53: 536-542.

8. Aitken DM, Brown AD. 1969. Citrate and glyoxylate cycles in the halophile, *Halobacterium salinarium*. Biochim. Biophys. Acta 177: 351-354.

9. Akerley BJ, Rubin EJ, Camilli A, Lampe DJ, Robertson HM, Mekalanos JJ. 1998. Systematic identification of essential genes by in vitro mariner mutagenesis. Proc. Natl. Acad. Sci. USA 95: 8927-8932.

10. Akhmanova A, Voncken F, van Alen T, van Hoek A, Boxma B, Vogels G, Veenhuis M, Hackstein JH. 1998. A hydrogenosome with a genome. Nature 396: 527-528.

11. Alberts B, Johnson A, Lewis J, Raff M, Roberts K, Walter P. 2002. Molecular Biology of the Cell. Garland Science, New York.

12. Aldovini A, Husson RN, Young RA. 1993. The *uraA* locus and homologous recombination in *Mycobacterium bovis* BCG. J. Bacteriol. 175: 7282-7289.

13. Alexandrov NN, Luethy R. 1998. Alignment algorithm for homology modeling and threading. Protein Sci. 7: 254-258.

14. Alexandrov NN, Nussinov R, Zimmer RM. 1996. Fast protein fold recognition via sequence to structure alignment and contact capacity potentials. Pac. Symp. Biocomput. 53-72.

15. Alifano P, Fani R, Lio P, Lazcano A, Bazzicalupo M, Carlomagno MS, Bruni CB. 1996. Histidine biosynthetic pathway and genes: structure, regulation, and evolution. Microbiol. Rev. 60: 44-69.

16. Allam AM, Hassan MM, Elzainy TA. 1975. Formation and cleavage of 2-keto-3-deoxygluconate by 2-keto-3- deoxygluconate aldolase of *Aspergillus niger*. J. Bacteriol. 124: 1128-1131.

17. Allard J, Grochulski P, Sygusch J. 2001. Covalent intermediate trapped in 2-keto-3-deoxy-6- phosphogluconate (KDPG) aldolase structure at 1.95-A resolution. Proc. Natl. Acad. Sci. USA 98: 3679-3684.

18. Altschul SF, Boguski MS, Gish W, Wootton JC. 1994. Issues in searching molecular sequence databases. Nat. Genet. 6: 119-129.

19. Altschul SF, Gish W. 1996. Local alignment statistics. Methods Enzymol. 266: 460-480.

20. Altschul SF, Gish W, Miller W, Myers EW, Lipman DJ. 1990. Basic local alignment search tool. J. Mol. Biol. 215: 403-410.

21. Altschul SF, Koonin EV. 1998. Iterated profile searches with PSI-BLAST--a tool for discovery in protein databases. Trends Biochem. Sci. 23: 444-447.

22. Altschul SF, Madden TL, Schaffer AA, Zhang J, Zheng Z, Miller W, Lipman DJ. 1997. Gapped BLAST and PSI-BLAST - A new generation of protein database search programs. Nucl. Acids Res. 25: 3389-3402.

23. Altshtein AD. 1987. Origin of the genetic system - the progene hypothesis. Mol. Biol. (Moscow) 21: 257-268.

24. Altshtein AD, Efimov AV. 1988. Physicochemical basis of origin of the genetic code - stereochemical analysis of interaction of amino acids and nucleotides on the basis of the progene hypothesis. Mol. Biol. (Moscow) 22: 1133-1149.

25. Ambrogelly A, Korencic D, Ibba M. 2002. Functional annotation of class I lysyl-tRNA synthetase phylogeny indicates a limited role for gene transfer. J. Bacteriol. 184: 4594-4600.

26. Anantharaman V, Koonin EV, Aravind L. 2001. Regulatory potential, phyletic distribution and evolution of ancient, intracellular small-molecule-binding domains. J. Mol. Biol. 307: 1271-1292.

27. Anantharaman V, Koonin EV, Aravind L. 2002. Comparative genomics and evolution of proteins involved in RNA metabolism. Nucleic Acids Res. 30: 1427-1464.

28. Anderson S, Bankier AT, Barrell BG, de Bruijn MH, Coulson AR, Drouin J, Eperon IC, Nierlich DP, Roe BA, Sanger F, et al. 1981. Sequence and organization of the human mitochondrial genome. Nature 290: 457-465.

29. Andersson JO, Doolittle WF, Nesbo CL. 2001. Genomics. Are there bugs in our genome? Science 292: 1848-1850.

30. Andersson SG, Zomorodipour A, Andersson JO, Sicheritz-Ponten T, Alsmark UC, Podowski RM, Naslund AK, Eriksson AS, Winkler HH, Kurland CG. 1998. The genome sequence of *Rickettsia prowazekii* and the origin of mitochondria. Nature 396: 133-140.

31. Andrade M, Casari G, de Daruvar A, Sander C, Schneider R, Tamames J, Valencia A, Ouzounis C. 1997. Sequence analysis of the *Methanococcus jannaschii* genome and the prediction of protein function. Comput. Appl. Biosci. 13: 481-483.

32. Angell S, Schwarz E, Bibb MJ. 1992. The glucose kinase gene of *Streptomyces coelicolor* A3(2): its nucleotide sequence, transcriptional analysis and role in glucose repression. Mol. Microbiol. 6: 2833-2844.

33. Apweiler R. 2000. Protein sequence databases. Adv. Protein. Chem. 54: 31-71.

34. Apweiler R, Attwood TK, Bairoch A, Bateman A, Birney E, Biswas M, Bucher P, Cerutti L, Corpet F, Croning MD, et al. 2001. The InterPro database, an integrated documentation resource for protein families, domains and functional sites. Nucleic Acids Res. 29: 37-40.

35. Arabidopsis Genome Initiative. 2000. Analysis of the genome sequence of the flowering plant *Arabidopsis thaliana*. Nature 408: 796-815.

36. Aravind L. 2000. Guilt by association: contextual information in genome analysis. Genome Res 10: 1074-1077.

37. Aravind L, Anantharaman V, Koonin EV. 2002. Monophyly of class I aminoacyl tRNA synthetase, USPA, ETFP, photolyase, and PP-ATPase nucleotide-binding domains: implications for protein evolution in the RNA. Proteins 48: 1-14.

38. Aravind L, Dixit VM, Koonin EV. 1999. The domains of death: evolution of the apoptosis machinery. Trends Biochem. Sci. 24: 47-53.

39. Aravind L, Dixit VM, Koonin EV. 2001. Apoptotic molecular machinery: vastly increased complexity in vertebrates revealed by genome comparisons. Science 291: 1279-1284.

40. Aravind L, Koonin EV. 1998. The HD domain defines a new superfamily of metal-dependent phosphohydrolases. Trends Biochem. Sci. 23: 469-472.

41. Aravind L, Koonin EV. 1998. Phosphoesterase domains associated with DNA polymerases of diverse origins. Nucleic Acids Res. 26: 3746-3752.

42. Aravind L, Koonin EV. 1999. DNA-binding proteins and evolution of transcription regulation in the archaea. Nucleic Acids Res. 27: 4658-4670.

43. Aravind L, Koonin EV. 1999. Gleaning non-trivial structural, functional and evolutionary information about proteins by iterative database searches. J Mol Biol 287: 1023-1040.

44. Aravind L, Koonin EV. 2000. The alpha/beta fold uracil DNA glycosylases: a common origin with diverse fates. Genome Biol 1: RESEARCH0007.

45. Aravind L, Koonin EV. 2000. Eukaryote-specific domains in translation initiation factors: implications for translation regulation and evolution of the translation system. Genome Res 10: 1172-1184.

46. Aravind L, Koonin EV. 2001. Prokaryotic homologs of the eukaryotic DNA-end-binding protein Ku, novel domains in the Ku protein and prediction of a prokaryotic double-strand break repair system. Genome Res. 11: 1365-1374.

47. Aravind L, Koonin EV. 2002. Classification of the caspase-hemoglobinase fold: detection of new families and implications for the origin of the eukaryotic separins. Proteins 46: 355-367.

48. Aravind L, Landsman D. 1998. AT-hook motifs identified in a wide variety of DNA-binding proteins. Nucleic Acids Res. 26: 4413-4421.

49. Aravind L, Leipe DD, Koonin EV. 1998. Toprim--a conserved catalytic domain in type IA and II topoisomerases, DnaG-type primases, OLD family nucleases and RecR proteins. Nucleic Acids Res. 26: 4205-4213.

50. Aravind L, Makarova KS, Koonin EV. 2000. Holliday junction resolvases and related nucleases: identification of new families, phyletic distribution and evolutionary trajectories. Nucleic Acids Res. 28: 3417-3432.

51. Aravind L, Mazumder R, Vasudevan S, Koonin EV. 2002. Trends in protein evolution inferred from sequence and structure analysis. Curr. Opin. Struct. Biol. 12: 392-399.

52. Aravind L, Tatusov RL, Wolf YI, Walker DR, Koonin EV. 1998. Evidence for massive gene exchange between archaeal and bacterial hyperthermophiles. Trends Genet. 14: 442-444.

53. Aravind L, Tatusov RL, Wolf YI, Walker DR, Koonin EV. 1999. Reply. Trends Genet 15: 299-300.

54. Aravind L, Walker DR, Koonin EV. 1999. Conserved domains in DNA repair proteins and evolution of repair systems. Nucleic Acids Res. 27: 1223-1242.

55. Aravind L, Watanabe H, Lipman DJ, Koonin EV. 2000. Lineage-specific loss and divergence of functionally linked genes in eukaryotes. Proc. Natl. Acad. Sci. USA 97: 11319-11324.

56. Argos P, Kamer G, Nicklin MJ, Wimmer E. 1984. Similarity in gene organization and homology between proteins of animal picornaviruses and a plant comovirus suggest common ancestry of these virus families. Nucleic Acids Res. 12: 7251-7267.

57. Arigoni F, Talabot F, Peitsch M, Edgerton MD, Meldrum E, Allet E, Fish R, Jamotte T, Curchod ML, Loferer H. 1998. A genome-based approach for the identification of essential bacterial genes. Nat. Biotechnol. 16: 851-856.

58. Arnoult D, Tatischeff I, Estaquier J, Girard M, Sureau F, Tissier JP, Grodet A, Dellinger M, Traincard F, Kahn A, et al. 2001. On the evolutionary conservation of the cell death pathway: mitochondrial release of an apoptosis-inducing factor during *Dictyostelium discoideum* cell death. Mol. Biol. Cell 12: 3016-3030.

59. Artymiuk PJ, Poirrette AR, Rice DW, Willett P. 1996. Biotin carboxylase comes into the fold. Nat Struct Biol 3: 128-132.

60. Ashburner M, Ball CA, Blake JA, Botstein D, Butler H, Cherry JM, Davis AP, Dolinski K, Dwight SS, Eppig JT, et al. 2000. Gene ontology: tool for the unification of biology. The Gene Ontology Consortium. Nat Genet 25: 25-29.

61. Attwood TK, Blythe MJ, Flower DR, Gaulton A, Mabey JE, Maudling N, McGregor L, Mitchell AL, Moulton G, Paine K, et al. 2002. PRINTS and PRINTS-S shed light on protein ancestry. Nucleic Acids Res. 30: 239-241.

62. Attwood TK, Croning MD, Flower DR, Lewis AP, Mabey JE, Scordis P, Selley JN, Wright W. 2000. PRINTS-S: the database formerly known as PRINTS. Nucleic Acids Res. 28: 225-227.

63. Augustin MA, Huber R, Kaiser JT. 2001. Crystal structure of a DNA-dependent RNA polymerase (DNA primase). Nat. Struct. Biol. 8: 57-61.

64. Babbitt PC, Hasson MS, Wedekind JE, Palmer DR, Barrett WC, Reed GH, Rayment I, Ringe D, Kenyon GL, Gerlt JA. 1996. The enolase superfamily: a general strategy for enzyme-catalyzed abstraction of the alpha-protons of carboxylic acids. Biochemistry 35: 16489-16501.

65. Backstrom D, Sjoberg RM, Lundberg LG. 1986. Nucleotide sequence of the structural gene for dihydroorotase of *Escherichia coli* K12. Eur. J. Biochem. 160: 77-82.

66. Bader GD, Donaldson I, Wolting C, Ouellette BF, Pawson T, Hogue CW. 2001. BIND--The Biomolecular Interaction Network Database. Nucleic Acids Res. 29: 242-245.

67. Badger JH, Olsen GJ. 1999. CRITICA: coding region identification tool invoking comparative analysis. Mol. Biol. Evol. 16: 512-524.

68. Baer R, Bankier AT, Biggin MD, Deininger PL, Farrell PJ, Gibson TJ, Hatfull G, Hudson GS, Satchwell SC, Seguin C, et al. 1984. DNA sequence and expression of the B95-8 Epstein-Barr virus genome. Nature 310: 207-211.

69. Bairoch A, Apweiler R. 2000. The SWISS-PROT protein sequence database and its supplement TrEMBL in 2000. Nucleic Acids Res 28: 45-48.

70. Baldi P, Brunak S, Frasconi P, Soda G, Pollastri G. 1999. Exploiting the past and the future in protein secondary structure prediction. Bioinformatics 15: 937-946.

71. Baldi P, Chauvin Y. 1994. Hidden Markov Models of the G-protein-coupled receptor family. J. Comput. Biol. 1: 311-336.

72. Baldi P, Chauvin Y, Hunkapiller T, McClure MA. 1994. Hidden Markov models of biological primary sequence information. Proc. Natl. Acad. Sci. USA 91: 1059-1063.

73. Bao Q, Tian Y, Li W, Xu Z, Xuan Z, Hu S, Dong W, Yang J, Chen Y, Xue Y, et al. 2002. A complete sequence of the *T. tengcongensis* genome. Genome Res. 12: 689-700.

74. Barabasi AL. 2002. Linked: The New Science of Networks. Perseus Pr, New York.

75. Barker WC, Dayhoff MO. 1979. Role of gene duplication in the evolution of complex physiological mechanisms: an assessment based on protein sequence data. In: Stadler Symposium, pp. 125-144. University of Missouri, Columbia.

76. Barker WC, Garavelli JS, Hou Z, Huang H, Ledley RS, McGarvey PB, Mewes HW, Orcutt BC, Pfeiffer F, Tsugita A, et al. 2001. Protein Information Resource: a community resource for expert annotation of protein data. Nucleic Acids Res. 29: 29-32.

77. Barnes MR, Russell RB, Copley RR, Ponting CP, Bork P, Cumberledge S, Reichsman F, Moore HM. 1999. A lipid-binding domain in Wnt: a case of mistaken identity? Curr. Biol. 9: R717-R719.

78. Bartolucci S, Rella R, Guagliardi A, Raia CA, Gambacorta A, De Rosa M, Rossi M. 1987. Malic enzyme from archaebacterium *Sulfolobus solfataricus*. Purification, structure, and kinetic properties. J. Biol. Chem. 262: 7725-7731.

79. Bateman A. 1997. The structure of a domain common to archaebacteria and the homocystinuria disease protein. Trends Biochem. Sci. 22: 12-13.

80. Bateman A, Birney E, Durbin R, Eddy SR, Howe KL, Sonnhammer EL. 2000. The Pfam protein families database. Nucleic Acids Res. 28: 263-266.

81. Begley TP, Downs DM, Ealick SE, McLafferty FW, Van Loon AP, Taylor S, Campobasso N, Chiu HJ, Kinsland C, Reddick JJ, et al. 1999. Thiamin biosynthesis in prokaryotes. Arch. Microbiol. 171: 293-300.

82. Begley TP, Kinsland C, Mehl RA, Osterman A, Dorrestein P. 2001. The biosynthesis of nicotinamide adenine dinucleotides in bacteria. Vitam. Horm. 61: 103-119.

83. Begley TP, Xi J, Kinsland C, Taylor S, McLafferty F. 1999. The enzymology of sulfur activation during thiamin and biotin biosynthesis. Curr. Opin. Chem. Biol. 3: 623-629.

84. Benner SA, Cohen MA, Gonnet GH. 1993. Empirical and structural models for insertions and deletions in the divergent evolution of proteins. J. Mol. Biol. 229: 1065-1082.

85. Bentley SD, Chater KF, Cerdeno-Tarraga AM, Challis GL, Thomson NR, James KD, Harris DE, Quail MA, Kieser H, Harper D, et al. 2002. Complete genome sequence of the model actinomycete *Streptomyces coelicolor* A3(2). Nature 417: 141-147.

86. Bergh S, Uhlen M. 1992. Analysis of a polyketide synthesis-encoding gene cluster of *Streptomyces curacoi*. Gene 117: 131-136.

87. Bertolaet BL, Clarke DJ, Wolff M, Watson MH, Henze M, Divita G, Reed SI. 2001. UBA domains of DNA damage-inducible proteins interact with ubiquitin. Nat. Struct. Biol. 8: 417-422.

88. Besemer J, Lomsadze A, Borodovsky M. 2001. GeneMarkS: a self-training method for prediction of gene starts in microbial genomes. Implications for finding sequence motifs in regulatory regions. Nucleic Acids Res. 29: 2607-2618.

89. Bhatia U, Robison K, Gilbert W. 1997. Dealing with database explosion: a cautionary note. Science 276: 1724-1725.

90. Biagini GA, Finlay BJ, Lloyd D. 1997. Evolution of the hydrogenosome. FEMS Microbiol. Lett. 155: 133-140.

91. Binieda A, Fuhrmann M, Lehner B, Rey-Berthod C, Frutiger-Hughes S, Hughes G, Shaw NM. 1999. Purification, characterization, DNA sequence and cloning of a pimeloyl-CoA synthetase from *Pseudomonas mendocina* 35. Biochem. J. 340: 793-801.

92. Binstock A, Rex J. 1995. Practical algorithms for programmers. Addiison-Wesley Publishing Co., Reading, MA.

93. Birney E, Durbin R. 2000. Using GeneWise in the *Drosophila* annotation experiment. Genome Res. 10: 547-548.

94. Blattner FR, Plunkett G, 3rd, Bloch CA, Perna NT, Burland V, Riley M, Collado-Vides J, Glasner JD, Rode CK, Mayhew GF, et al. 1997. The complete genome sequence of *Escherichia coli* K-12. Science 277: 1453-1474.

95. Blom N, Sygusch J. 1997. Product binding and role of the C-terminal region in class I D-fructose 1,6-bisphosphate aldolase. Nat. Struct. Biol. 4: 36-39.

96. Bolotin A, Khazak V, Stoynova N, Ratmanova K, Yomantas Y, Kozlov Y. 1995. Identical amino acid sequence of the *aroA(G)* gene products of *Bacillus subtilis* 168 and *B. subtilis* Marburg strain. Microbiology 141: 2219-2222.

97. Bolotin A, Wincker P, Mauger S, Jaillon O, Malarme K, Weissenbach J, Ehrlich SD, Sorokin A. 2001. The complete genome sequence of the lactic acid bacterium *Lactococcus lactis* ssp. *lactis* IL1403. Genome Res. 11: 731-753.

98. Bond CS, Clements PR, Ashby SJ, Collyer CA, Harrop SJ, Hopwood JJ, Guss JM. 1997. Structure of a human lysosomal sulfatase. Structure 5: 277-289.

99. Bork P, Bairoch A. 1996. Go hunting in sequence databases but watch out for the traps. Trends Genet. 12: 425-427.

100. Bork P, Hofmann K, Bucher P, Neuwald AF, Altschul SF, Koonin EV. 1997. A superfamily of conserved domains in DNA damage-responsive cell cycle checkpoint proteins. FASEB J. 11: 68-76.

101. Bork P, Holm L, Koonin EV, Sander C. 1995. The cytidylyltransferase superfamily: identification of the nucleotide-binding site and fold prediction. Proteins 22: 259-266.

102. Bork P, Koonin EV. 1994. A P-loop-like motif in a widespread ATP pyrophosphatase domain: implications for the evolution of sequence motifs and enzyme activity. Proteins 20: 347-355.

103. Bork P, Koonin EV. 1996. Protein sequence motifs. Curr. Opin. Struct. Biol. 6: 366-376.

104. Bork P, Koonin EV. 1998. Predicting functions from protein sequences--where are the bottlenecks? Nat. Genet. 18: 313-318.

105. Bork P, Ouzounis C, Sander C, Scharf M, Schneider R, Sonnhammer E. 1992. What's in a genome? Nature 358: 287.

106. Bork P, Sander C, Valencia A. 1992. An ATPase domain common to prokaryotic cell cycle proteins, sugar kinases, actin, and hsp70 heat shock proteins. Proc. Natl. Acad. Sci. USA 89: 7290-7294.

107. Borodovskii MI, Sprizhitskii IA, Golovanov EI, Aleksandrov AA. 1986. Statistical characteristics in primary structures of functional regions of *Escherichia coli* genome. II. Non-stationary Markov chains. Mol. Biol. (Moscow) 20: 1024-1033.

108. Borodovsky M, McIninch J. 1993. GeneMark: parallel gene recognition for both DNA strands. Comp. Chem. 17: 123-133.

109. Borodovsky M, McIninch JD, Koonin EV, Rudd KE, Medigue C, Danchin A. 1995. Detection of new genes in a bacterial genome using Markov models for three gene classes. Nucleic Acids Res. 23: 3554-3562.

110. Borodovsky M, Rudd KE, Koonin EV. 1994. Intrinsic and extrinsic approaches for detecting genes in a bacterial genome. Nucleic Acids Res. 22: 4756-4767.

111. Boucher Y, Doolittle WF. 2000. The role of lateral gene transfer in the evolution of isoprenoid biosynthesis pathways. Mol. Microbiol. 37: 703-716.

112. Boysen RI, Hearn MT. 2001. The metal binding properties of the CCCH motif of the 50S ribosomal protein L36 from *Thermus thermophilus*. J. Pept.Res. 57: 19-28.

113. Bratton SB, Cohen GM. 2001. Apoptotic death sensor: an organelle's alter ego? Trends Pharmacol. Sci. 22: 306-315.

114. Braun EL, Halpern AL, Nelson MA, Natvig DO. 2000. Large-scale comparison of fungal sequence information: mechanisms of innovation in *Neurospora crassa* and gene loss in *Saccharomyces cerevisiae*. Genome Res. 10: 416-430.

115. Brazas R, Ganem D. 1996. A cellular homolog of hepatitis delta antigen: implications for viral replication and evolution. Science 274: 90-94.

116. Brenner SE. 1999. Errors in genome annotation. Trends Genet. 15: 132-133.

117. Brenner SE, Chothia C, Hubbard TJ. 1998. Assessing sequence comparison methods with reliable structurally identified distant evolutionary relationships. Proc. Natl. Acad. Sci. USA 95: 6073-6078.

118. Brett D, Pospisil H, Valcarcel J, Reich J, Bork P. 2002. Alternative splicing and genome complexity. Nat. Genet. 30: 29-30.

119. Brochier C, Bapteste E, Moreira D, Philippe H. 2002. Eubacterial phylogeny based on translational apparatus proteins. Trends Genet. 18: 1-5.

120. Brochier C, Philippe H, Moreira D. 2000. The evolutionary history of ribosomal protein RpS14: horizontal gene transfer at the heart of the ribosome. Trends Genet. 16: 529-533.

121. Brody T. 1999. The Interactive Fly: gene networks, development and the Internet. Trends Genet. 15: 333-334.

122. Brown JR, Doolittle WF. 1997. Archaea and the prokaryote-to-eukaryote transition. Microbiol. Mol. Biol. Rev. 61: 456-502.

123. Brunak S, Engelbrecht J, Knudsen S. 1991. Prediction of human mRNA donor and acceptor sites from the DNA sequence. J. Mol. Biol. 220: 49-65.

124. Brunner NA, Brinkmann H, Siebers B, Hensel R. 1998. NAD+-dependent glyceraldehyde-3-phosphate dehydrogenase from *Thermoproteus tenax*. The first identified archaeal member of the aldehyde dehydrogenase superfamily is a glycolytic enzyme with unusual regulatory properties. J. Biol. Chem. 273: 6149-6156.

125. Bryant SH, Altschul SF. 1995. Statistics of sequence-structure threading. Curr. Opin. Struct. Biol. 5: 236-244.

126. Buchanan CL, Connaris H, Danson MJ, Reeve CD, Hough DW. 1999. An extremely thermostable aldolase from *Sulfolobus solfataricus* with specificity for non-phosphorylated substrates. Biochem. J. 343 Pt 3: 563-570.

127. Bui ET, Bradley PJ, Johnson PJ. 1996. A common evolutionary origin for mitochondria and hydrogenosomes. Proc. Natl. Acad. Sci. USA 93: 9651-9656.

128. Bujnicki JM, Elofsson A, Fischer D, Rychlewski L. 2001. Structure prediction meta server. Bioinformatics 17: 750-751.

129. Bujnicki JM, Radlinska M, Rychlewski L. 2000. Atomic model of the 5-methylcytosine-specific restriction enzyme McrA reveals an atypical zinc finger and structural similarity to betabetaalphaMe endonucleases. Mol. Microbiol. 37: 1280-1281.

130. Bult CJ, White O, Olsen GJ, Zhou L, Fleischmann RD, Sutton GG, Blake JA, FitzGerald LM, Clayton RA, Gocayne JD, et al. 1996. Complete genome sequence of the methanogenic archaeon, *Methanococcus jannaschii.* Science 273: 1058-1073.

131. Burge C, Karlin S. 1997. Prediction of complete gene structures in human genomic DNA. J. Mol. Biol. 268: 78-94.

132. Burge CB, Karlin S. 1998. Finding the genes in genomic DNA. Curr. Opin. Struct. Biol. 8: 346-354.

133. Burset M, Guigo R. 1996. Evaluation of gene structure prediction programs. Genomics 34: 353-367.

134. Bushman F. 2001. Lateral DNA Transfer: Mechanisms and Consequences. Cold Spring Harbor Laboratory, Cold Spring Harbor, NY.

135. Caldon CE, Yoong P, March PE. 2001. Evolution of a molecular switch: universal bacterial GTPases regulate ribosome function. Mol. Microbiol. 41: 289-297.

136. Callebaut I, Mornon JP. 1997. From BRCA1 to RAP1: a widespread BRCT module closely associated with DNA repair. FEBS Lett. 400: 25-30.

137. Capela D, Barloy-Hubler F, Gouzy J, Bothe G, Ampe F, Batut J, Boistard P, Becker A, Boutry M, Cadieu E, et al. 2001. Analysis of the chromosome sequence of the legume symbiont *Sinorhizobium meliloti* strain 1021. Proc. Natl. Acad. Sci. USA 98: 9877-9882.

138. Carreras J, Mezquita J, Bosch J, Bartrons R, Pons G. 1982. Phylogeny and ontogeny of the phosphoglycerate mutases--IV. Distribution of glycerate-2,3-P2 dependent and independent phosphoglycerate mutases in algae, fungi, plants and animals. Comp. Biochem. Physiol. [B] 71: 591-597.

139. Carter CA. 2002. Tsg101: HIV-1's ticket to ride. Trends Microbiol 10: 203-205.

140. Cavalier-Smith T. 1989. Molecular phylogeny. Archaebacteria and Archezoa. Nature 339: 100-101.

141. Cavener DR. 1992. GMC oxidoreductases. A newly defined family of homologous proteins with diverse catalytic activities. J. Mol. Biol. 223: 811-814.

142. Cecconi F. 1999. Apaf1 and the apoptotic machinery. Cell Death. Differ. 6: 1087-1098.

143. Cech TR. 2000. Structural biology. The ribosome is a ribozyme. Science 289: 878-879.

144. Chambaud I, Heilig R, Ferris S, Barbe V, Samson D, Galisson F, Moszer I, Dybvig K, Wroblewski H, Viari A, et al. 2001. The complete genome sequence of the murine respiratory pathogen *Mycoplasma pulmonis*. Nucleic Acids Res. 29: 2145-2153.

145. Chance MR, Bresnick AR, Burley SK, Jiang JS, Lima CD, Sali A, Almo SC, Bonanno JB, Buglino JA, Boulton S, et al. 2002. Structural genomics: a pipeline for providing structures for the biologist. Protein Sci. 11: 723-738.

146. Chinnaiyan AM. 1999. The apoptosome: heart and soul of the cell death machine. Neoplasia 1: 5-15.

147. Chothia C. 1992. Proteins. One thousand families for the molecular biologist. Nature 357: 543-544.

148. Clarke GD, Beiko RG, Ragan MA, Charlebois RL. 2002. Inferring genome trees by using a filter to eliminate phylogenetically discordant sequences and a distance matrix based on mean normalized BLASTP scores. J. Bacteriol. 184: 2072-2080.

149. Claros MG, Brunak S, von Heijne G. 1997. Prediction of N-terminal protein sorting signals. Curr. Opin. Struct. Biol. 7: 394-398.

150. Claros MG, Vincens P. 1996. Computational method to predict mitochondrially imported proteins and their targeting sequences. Eur. J. Biochem. 241: 779-786.

151. Claros MG, von Heijne G. 1994. TopPred II: an improved software for membrane protein structure predictions. Comput Appl. Biosci. 10: 685-686.

152. Cole ST, Brosch R, Parkhill J, Garnier T, Churcher C, Harris D, Gordon SV, Eiglmeier K, Gas S, Barry CE, 3rd, et al. 1998. Deciphering the biology of *Mycobacterium tuberculosis* from the complete genome sequence. Nature 393: 537-544.

153. Cole ST, Eiglmeier K, Parkhill J, James KD, Thomson NR, Wheeler PR, Honore N, Garnier T, Churcher C, Harris D, et al. 2001. Massive gene decay in the leprosy bacillus. Nature 409: 1007-1011.

154. Collard F, Collet J, Gerin I, Veiga-da-Cunha M, Van Schaftingen E. 1999. Identification of the cDNA encoding human 6-phosphogluconolactonase, the enzyme catalyzing the second step of the pentose phosphate pathway. FEBS Lett. 459: 223-226.

155. Combet C, Blanchet C, Geourjon C, Deleage G. 2000. NPS@: network protein sequence analysis. Trends Biochem. Sci. 25: 147-150.

156. Cook WJ, Jeffrey LC, Sullivan ML, Vierstra RD. 1992. Three-dimensional structure of a ubiquitin-conjugating enzyme (E2). J. Biol. Chem. 267: 15116-15121.

157. Copley RR, Bork P. 2000. Homology among $(\beta\alpha)_8$ barrels: implications for the evolution of metabolic pathways. J. Mol. Biol. 303: 627-641.

158. Corpet F, Gouzy J, Kahn D. 1999. Recent improvements of the ProDom database of protein domain families. Nucleic Acids Res. 27: 263-267.

159. Corpet F, Servant F, Gouzy J, Kahn D. 2000. ProDom and ProDom-CG: tools for protein domain analysis and whole genome comparisons. Nucleic Acids Res. 28: 267-269.

160. Costanzo MC, Crawford ME, Hirschman JE, Kranz JE, Olsen P, Robertson LS, Skrzypek MS, Braun BR, Hopkins KL, Kondu P, et al. 2001. YPD, PombePD and WormPD: model

organism volumes of the BioKnowledge library, an integrated resource for protein information. Nucleic Acids Res. 29: 75-79.

161. Coulson AF, Moult J. 2002. A unifold, mesofold, and superfold model of protein fold use. Proteins 46: 61-71.

162. Crick FH. 1968. The origin of the genetic code. J Mol Biol 38: 367-379.

163. Crick FHC. 1958. On protein synthesis. Symp. Soc. Exp. Biol. XII: 139-163.

164. Cserzo M, Wallin E, Simon I, von Heijne G, Elofsson A. 1997. Prediction of transmembrane alpha-helices in prokaryotic membrane proteins: the dense alignment surface method. Protein Eng. 10: 673-676.

165. Cuff JA, Barton GJ. 2000. Application of multiple sequence alignment profiles to improve protein secondary structure prediction. Proteins 40: 502-511.

166. Cuff JA, Clamp ME, Siddiqui AS, Finlay M, Barton GJ. 1998. JPred: a consensus secondary structure prediction server. Bioinformatics 14: 892-893.

167. Da Silva AC, Ferro JA, Reinach FC, Farah CS, Furlan LR, Quaggio RB, Monteiro-Vitorello CB, Sluys MA, Almeida NF, Alves LM, et al. 2002. Comparison of the genomes of two *Xanthomonas* pathogens with differing host specificities. Nature 417: 459-463.

168. Dandekar T, Huynen M, Regula JT, Ueberle B, Zimmermann CU, Andrade MA, Doerks T, Sanchez-Pulido L, Snel B, Suyama M, et al. 2000. Re-annotating the *Mycoplasma pneumoniae* genome sequence: adding value, function and reading frames. Nucleic Acids Res. 28: 3278-3288.

169. Dandekar T, Snel B, Huynen M, Bork P. 1998. Conservation of gene order: a fingerprint of proteins that physically interact. Trends Biochem. Sci. 23: 324-328.

170. Darwin C. 1859. The Origin of Species. Murray, London.

171. Daugherty M, Vonstein V, Overbeek R, Osterman A. 2001. Archaeal shikimate kinase, a new member of the GHMP-kinase family. J. Bacteriol. 183: 292-300.

172. Dayhoff MO, Eck RV. 1968. Atlas of Protein Sequence and Structure. Vol. 3 National Biomedical Research Foundation, Silver Spring, MD.

173. Dayhoff MO, Eck RV, Chang MA, Sochard MR. 1965. Atlas of Protein Sequence and Structure. Vol. 1 National Biomedical Research Foundation, Silver Spring, MD.

174. Dayhoff MO, Schwartz RM, Orcutt BC. 1978. A model of evolutionary change in proteins. In: Atlas of Protein Sequence and Structure. M.O. Dayhoff, ed., pp. 345-352. National Biomedical Research Foundation, Washington, DC.

175. Deckert G, Warren PV, Gaasterland T, Young WG, Lenox AL, Graham DE, Overbeek R, Snead MA, Keller M, Aujay M, et al. 1998. The complete genome of the hyperthermophilic bacterium *Aquifex aeolicus*. Nature 392: 353-358.

176. Delarue M, Poch O, Tordo N, Moras D, Argos P. 1990. An attempt to unify the structure of polymerases. Protein Eng 3: 461-467.

177. Delcher AL, Harmon D, Kasif S, White O, Salzberg SL. 1999. Improved microbial gene identification with GLIMMER. Nucleic Acids Res. 27: 4636-4641.

178. Delcher AL, Kasif S, Fleischmann RD, Peterson J, White O, Salzberg SL. 1999. Alignment of whole genomes. Nucleic Acids Res. 27: 2369-2376.

179. DelVecchio VG, Kapatral V, Redkar RJ, Patra G, Mujer C, Los T, Ivanova N, Anderson I, Bhattacharyya A, Lykidis A, et al. 2002. The genome sequence of the facultative intracellular pathogen *Brucella melitensis*. Proc. Natl. Acad. Sci. USA 99: 443-448.

180. Dennis C. 2002. Gene regulation: The brave new world of RNA. Nature 418: 122-124.

181. Deppenmeier U, Johann A, Hartsch T, Merkl R, Schmitz RA, Martinez-Arias R, Henne A, Wiezer A, Baumer S, Jacobi C, et al. 2002. The genome of *Methanosarcina mazei*: evidence for lateral gene transfer between bacteria and archaea. J. Mol. Microbiol. Biotechnol. 4: 453-461.

182. Deshpande KL, Seubert PH, Tillman DM, Farkas WR, Katze JR. 1996. Cloning and characterization of cDNA encoding the rabbit tRNA-guanine transglycosylase 60-kilodalton subunit. Arch. Biochem. Biophys. 326: 1-7.

183. Dessaux Y, Petit A, Tempe J, Demarez M, Legrain C, Wiame JM. 1986. Arginine catabolism in Agrobacterium strains: role of the Ti plasmid. J. Bacteriol. 166: 44-50.

184. DiRuggiero J, Dunn D, Maeder DL, Holley-Shanks R, Chatard J, Horlacher R, Robb FT, Boos W, Weiss RB. 2000. Evidence of recent lateral gene transfer among hyperthermophilic archaea. Mol. Microbiol. 38: 684-693.

185. Doolittle RF. 1981. Similar amino acid sequences: chance or common ancestry? Science 214: 149-159.

186. Doolittle RF. 1986. Of Urfs and Orfs: A Primer on How to Analyze Derived Amino Acid Sequences. University Science Books, Mill Valley, CA.

187. Doolittle RF. 1994. Convergent evolution: the need to be explicit. Trends Biochem. Sci. 19: 15-18.

188. Doolittle RF. 1999. Do you dig my groove? Nat. Genet. 23: 6-8.

189. Doolittle RF, Feng DF, Tsang S, Cho G, Little E. 1996. Determining divergence times of the major kingdoms of living organisms with a protein clock. Science 271: 470-477.

190. Doolittle RF, Handy J. 1998. Evolutionary anomalies among the aminoacyl-tRNA synthetases. Curr. Opin. Genet. Dev. 8: 630-636.

191. Doolittle RF, York AL. 2002. Bacterial actins? An evolutionary perspective. Bioessays 24: 293-296.

192. Doolittle WF. 1998. You are what you eat: a gene transfer ratchet could account for bacterial genes in eukaryotic nuclear genomes. Trends Genet. 14: 307-311.

193. Doolittle WF. 1999. Lateral genomics. Trends Cell. Biol. 9: M5-M8.

194. Doolittle WF. 1999. Phylogenetic classification and the universal tree. Science 284: 2124-2129.

195. Doolittle WF. 2000. Uprooting the tree of life. Scientific American 282: 90-95.

196. Doolittle WF, Brown JR. 1994. Tempo, mode, the progenote, and the universal root. Proc. Natl. Acad. Sci. USA 91: 6721-6728.

197. Doolittle WF, Sapienza C. 1980. Selfish genes, the phenotype paradigm and genome evolution. Nature 284: 601-603.

198. Drewke C, Leistner E. 2001. Biosynthesis of vitamin B6 and structurally related derivatives. Vitam. Horm. 61: 121-155.

199. Dujon B. 1998. European Functional Analysis Network (EUROFAN) and the functional analysis of the *Saccharomyces cerevisiae* genome. Electrophoresis 19: 617-624.

200. Dunwell JM, Khuri S, Gane PJ. 2000. Microbial relatives of the seed storage proteins of higher plants: conservation of structure and diversification of function during evolution of the cupin superfamily. Microbiol. Mol. Biol. Rev. 64: 153-179.

201. Durbecq V, Legrain C, Roovers M, Pierard A, Glansdorff N. 1997. The carbamate kinase-like carbamoyl phosphate synthetase of the hyperthermophilic archaeon *Pyrococcus furiosus*, a missing link in the evolution of carbamoyl phosphate biosynthesis. Proc. Natl. Acad. Sci. USA 94: 12803-12808.

202. Durbin R, Eddy S, Krogh A, Mitchison A. 1998. Biological Sequence Analysis: Probabilistic Models of Proteins and Nucleic Acids. Cambridge University Press, Cambridge.

203. Dyall SD, Johnson PJ. 2000. Origins of hydrogenosomes and mitochondria: evolution and organelle biogenesis. Curr. Opin. Microbiol. 3: 404-411.

204. Dyall SD, Koehler CM, Delgadillo-Correa MG, Bradley PJ, Plumper E, Leuenberger D, Turck CW, Johnson PJ. 2000. Presence of a member of the mitochondrial carrier family in hydrogenosomes: conservation of membrane-targeting pathways between hydrogenosomes and mitochondria. Mol. Cell. Biol. 20: 2488-2497.

205. Dyer WE, Weaver LM, Zhao JM, Kuhn DN, Weller SC, Herrmann KM. 1990. A cDNA encoding 3-deoxy-D-arabino-heptulosonate 7-phosphate synthase from *Solanum tuberosum* L. J. Biol. Chem. 265: 1608-1614.

206. Dynes JL, Firtel RA. 1989. Molecular complementation of a genetic marker in *Dictyostelium* using a genomic DNA library. Proc. Natl. Acad. Sci. USA 86: 7966-7970.

207. Eberhardt S, Korn S, Lottspeich F, Bacher A. 1997. Biosynthesis of riboflavin: an unusual riboflavin synthase of *Methanobacterium thermoautotrophicum*. J. Bacteriol. 179: 2938-2943.

208. Eddy SR. 1996. Hidden Markov models. Curr. Opin. Struct. Biol. 6: 361-365.

209. Edgell DR, Doolittle WF. 1997. Archaea and the origin(s) of DNA replication proteins. Cell 89: 995-998.

210. Egelman EH. 2001. Molecular evolution: actin's long lost relative found. Curr. Biol. 11: R1022-R1024.

211. Ehrenshaft M, Bilski P, Li MY, Chignell CF, Daub ME. 1999. A highly conserved sequence is a novel gene involved in de novo vitamin B6 biosynthesis. Proc. Natl. Acad. Sci. USA 96: 9374-9378.

212. Emanuelsson O, Nielsen H, Brunak S, von Heijne G. 2000. Predicting subcellular localization of proteins based on their N-terminal amino acid sequence. J. Mol. Biol. 300: 1005-1016.

213. Enright AJ, Illopoulos I, Kyrpides NC, Ouzounis CA. 1999. Protein interaction maps for complete genomes based on gene fusion events. Nature 402: 86-90.

214. Erickson HP. 1995. FtsZ, a prokaryotic homolog of tubulin? Cell 80: 367-370.

215. Erickson HP. 1998. Atomic structures of tubulin and FtsZ. Trends Cell Biol. 8: 133-137.

216. Esser L, Wang CR, Hosaka M, Smagula CS, Sudhof TC, Deisenhofer J. 1998. Synapsin I is structurally similar to ATP-utilizing enzymes. EMBO J. 17: 977-984.

217. Evrard C, Fastrez J, Declercq JP. 1998. Crystal structure of the lysozyme from bacteriophage lambda and its relationship with V and C-type lysozymes. J. Mol. Biol. 276: 151-164.

218. Fabrega C, Farrow MA, Mukhopadhyay B, de Crecy-Lagard V, Ortiz AR, Schimmel P. 2001. An aminoacyl tRNA synthetase whose sequence fits into neither of the two known classes. Nature 411: 110-114.

219. Faisst S, Meyer S. 1992. Compilation of vertebrate-encoded transcription factors. Nucleic Acids Res. 20: 3-26.

220. Falquet L, Pagni M, Bucher P, Hulo N, Sigrist CJ, Hofmann K, Bairoch A. 2002. The PROSITE database, its status in 2002. Nucleic Acids Res. 30: 235-238.

221. Feng DF, Doolittle RF. 1987. Progressive sequence alignment as a prerequisite to correct phylogenetic trees. J. Mol. Evol. 25: 351-360.

222. Feng HP, Gierasch LM. 1998. Molecular chaperones: clamps for the Clps? Curr. Biol. 8: R464-R467.

223. Ferretti JJ, McShan WM, Ajdic D, Savic DJ, Savic G, Lyon K, Primeaux C, Sezate S, Suvorov AN, Kenton S, et al. 2001. Complete genome sequence of an M1 strain of *Streptococcus pyogenes*. Proc. Natl. Acad. Sci. USA 98: 4658-4663.

224. Fichant GA, Quentin Y. 1995. A frameshift error detection algorithm for DNA sequencing projects. Nucleic Acids Res. 23: 2900-2908.

225. Fiers W, Contreras R, Duerinck F, Haegeman G, Iserentant D, Merregaert J, Min Jou W, Molemans F, Raeymaekers A, Van den Berghe A, et al. 1976. Complete nucleotide sequence of bacteriophage MS2 RNA: primary and secondary structure of the replicase gene. Nature 260: 500-507.

226. Filee J, Forterre P, Sen-Lin T, Laurent J. 2002. Evolution of DNA polymerase families: evidences for multiple gene exchange between cellular and viral proteins. J. Mol. Evol. 54: 763-773.

227. Fischer D, Eisenberg D. 1996. Protein fold recognition using sequence-derived predictions. Protein Sci. 5: 947-955.

228. Fitch WM. 1970. Distinguishing homologous from analogous proteins. Syst. Zool. 19: 99-113.

229. Fitch WM. 2000. Homology a personal view on some of the problems. Trends Genet. 16: 227-231.

230. Fitz-Gibbon ST, House CH. 1999. Whole genome-based phylogenetic analysis of free-living microorganisms. Nucleic Acids Res. 27: 4218-4222.

231. Fitz-Gibbon ST, Ladner H, Kim UJ, Stetter KO, Simon MI, Miller JH. 2002. Genome sequence of the hyperthermophilic crenarchaeon *Pyrobaculum aerophilum*. Proc. Natl. Acad. Sci. USA 99: 984-989.

232. Fleischmann RD, Adams MD, White O, Clayton RA, Kirkness EF, Kerlavage AR, Bult CJ, Tomb J-F, Dougherty BA, Merrick JM, et al. 1995. Whole-genome random sequencing and assembly of *Haemophilus influenzae* Rd. Science 269: 496-512.

233. Floriano B, Bibb M. 1996. afsR is a pleiotropic but conditionally required regulatory gene for antibiotic production in Streptomyces coelicolor A3(2). Mol. Microbiol. 21: 385-396.

234. Floyd EE, Jones ME. 1985. Isolation and characterization of the orotidine 5'-monophosphate decarboxylase domain of the multifunctional protein uridine 5'-monophosphate synthase. J. Biol. Chem. 260: 9443-9451.

235. Forsdyke DR. 1995. Sense in antisense? J. Mol. Evol. 41: 582-586.

236. Forterre P. 1999. Displacement of cellular proteins by functional analogues from plasmids or viruses could explain puzzling phylogenies of many DNA informational proteins. Mol. Microbiol. 33: 457-465.

237. Forterre P. 2001. Genomics and early cellular evolution. The origin of the DNA world. C R Acad Sci III 324: 1067-1076.

238. Forterre P. 2002. A hot story from comparative genomics: reverse gyrase is the only hyperthermophile-specific protein. Trends Genet 18: 236-237.

239. Fothergill-Gilmore LA, Michels PA. 1993. Evolution of glycolysis. Prog. Biophys. Mol. Biol. 59: 105-235.

240. Frade JM, Michaelidis TM. 1997. Origin of eukaryotic programmed cell death: a consequence of aerobic metabolism? Bioessays 19: 827-832.

241. Fraser CM, Casjens S, Huang WM, Sutton GG, Clayton R, Lathigra R, White O, Ketchum KA, Dodson R, Hickey EK, et al. 1997. Genomic sequence of a Lyme disease spirochaete, *Borrelia burgdorferi.* Nature 390: 580-586.

242. Fraser CM, Gocayne JD, White O, Adams MD, Clayton RA, Fleischmann RD, Bult CJ, Kerlavage AR, Sutton G, Kelley JM, et al. 1995. The minimal gene complement of *Mycoplasma genitalium.* Science 270: 397-403.

243. Fraser CM, Norris SJ, Weinstock GM, White O, Sutton GG, Dodson R, Gwinn M, Hickey EK, Clayton R, Ketchum KA, et al. 1998. Complete genome sequence of *Treponema pallidum*, the syphilis spirochete. Science 281: 375-388.

244. Fraser HI, Kvaratskhelia M, White MF. 1999. The two analogous phosphoglycerate mutases of *Escherichia coli.* FEBS Lett. 455: 344-348.

245. Frishman D, Albermann K, Hani J, Heumann K, Metanomski A, Zollner A, Mewes HW. 2001. Functional and structural genomics using PEDANT. Bioinformatics 17: 44-57.

246. Frishman D, Argos P. 1997. The future of protein secondary structure prediction accuracy. Fold Des 2: 159-162.

247. Frishman D, Argos P. 1997. Seventy-five percent accuracy in protein secondary structure prediction. Proteins 27: 329-335.

248. Frishman D, Mewes HW. 1997. PEDANTic genome analysis. Trends Genet. 13: 415-416.

249. Frishman D, Mironov A, Mewes HW, Gelfand M. 1998. Combining diverse evidence for gene recognition in completely sequenced bacterial genomes. Nucleic Acids Res. 26: 2941-2947.

250. Fujibuchi W, Ogata H, Matsuda H, Kanehisa M. 2000. Automatic detection of conserved gene clusters in multiple genomes by graph comparison and P-quasi grouping. Nucleic Acids Res. 28: 4029-4036.

251. Fujita Y, Yoshida K, Miwa Y, Yanai N, Nagakawa E, Kasahara Y. 1998. Identification and expression of the *Bacillus subtilis* fructose-1, 6-bisphosphatase gene (*fbp*). J. Bacteriol. 180: 4309-4313.

252. Gaasterland T, Sczyrba A, Thomas E, Aytekin-Kurban G, Gordon P, Sensen CW. 2000. MAGPIE/EGRET annotation of the 2.9-Mb *Drosophila melanogaster* Adh region. Genome Res. 10: 502-510.

253. Gaasterland T, Sensen CW. 1996. MAGPIE: automated genome interpretation. Trends Genet. 12: 76-78.

254. Galagan JE, Nusbaum C, Roy A, Endrizzi MG, Macdonald P, FitzHugh W, Calvo S, Engels R, Smirnov S, Atnoor D, et al. 2002. The genome of *M. acetivorans* reveals extensive metabolic and physiological diversity. Genome Res. 12: 532-542.

255. Galibert F, Finan TM, Long SR, Puhler A, Abola P, Ampe F, Barloy-Hubler F, Barnett MJ, Becker A, Boistard P, et al. 2001. The composite genome of the legume symbiont *Sinorhizobium meliloti*. Science 293: 668-672.

256. Galperin MY. 2001. Conserved "hypothetical" proteins: new hints and new puzzles. Comp. Funct. Genomics 2: 14-18.

257. Galperin MY, Aravind L, Koonin EV. 2000. Aldolases of the DhnA family: a possible solution to the problem of pentose and hexose biosynthesis in archaea. FEMS Microbiol. Lett. 183: 259-264.

258. Galperin MY, Bairoch A, Koonin EV. 1998. A superfamily of metalloenzymes unifies phosphopentomutase and cofactor- independent phosphoglycerate mutase with alkaline phosphatases and sulfatases. Protein Sci. 7: 1829-1835.

259. Galperin MY, Brenner SE. 1998. Using metabolic pathway databases for functional annotation. Trends Genet. 14: 332-333.

260. Galperin MY, Grishin NV. 2000. The synthetase domains of cobalamin biosynthesis amidotransferases cobB and cobQ belong to a new family of ATP-dependent amidoligases, related to dethiobiotin synthetase. Proteins 41: 238-247.

261. Galperin MY, Jedrzejas MJ. 2001. Conserved core structure and active site residues in alkaline phosphatase superfamily enzymes. Proteins 45: 318-324.

262. Galperin MY, Koonin EV. 1997. A diverse superfamily of enzymes with ATP-dependent carboxylate-amine/thiol ligase activity. Protein Sci. 6: 2639-2643.

263. Galperin MY, Koonin EV. 1997. Sequence analysis of an exceptionally conserved operon suggests enzymes for a new link between histidine and purine biosynthesis. Mol. Microbiol. 24: 443-445.

264. Galperin MY, Koonin EV. 1998. Sources of systematic error in functional annotation of genomes: domain rearrangement, non-orthologous gene displacement, and operon disruption. In Silico Biol. 1: 55-67.

265. Galperin MY, Koonin EV. 1999. Functional genomics and enzyme evolution. Homologous and analogous enzymes encoded in microbial genomes. Genetica 106: 159-170.

266. Galperin MY, Koonin EV. 1999. Searching for drug targets in microbial genomes. Curr Opin Biotechnol 10: 571-578.

267. Galperin MY, Koonin EV. 2000. Who's your neighbor? new computational approaches for functional genomics. Nat. Biotechnol. 18: 609-613.

268. Galperin MY, Koonin EV. 2001. Comparative genome analysis. In: Bioinformatics: a practical guide to the analysis of genes and proteins. A.D. Baxevanis and B.F.F. Ouellette, ed., pp. 359-392. John Wiley & Sons, New York.

269. Galperin MY, Nikolskaya AN, Koonin EV. 2001. Novel domains of the prokaryotic two-component signal transduction systems. FEMS Microbiol Lett 203: 11-21.

270. Galperin MY, Tatusov RL, Koonin EV. 1999. Comparing microbial genomes: how the gene set determines the lifestyle. In: Organization of the Prokaryotic Genome. R.L. Charlebois, ed., pp. 91-108. ASM Press, Washington, D.C.

271. Galperin MY, Walker DR, Koonin EV. 1998. Analogous enzymes: independent inventions in enzyme evolution. Genome Res 8: 779-790.

272. Gardner MJ, Tettelin H, Carucci DJ, Cummings LM, Aravind L, Koonin EV, Shallom S, Mason T, Yu K, Fujii C, et al. 1998. Chromosome 2 sequence of the human malaria parasite *Plasmodium falciparum*. Science 282: 1126-1132.

273. Garnier J, Gibrat JF, Robson B. 1996. GOR method for predicting protein secondary structure from amino acid sequence. Methods Enzymol. 266: 540-553.

274. Garnier J, Osguthorpe DJ, Robson B. 1978. Analysis of the accuracy and implications of simple methods for predicting the secondary structure of globular proteins. J. Mol. Biol. 120: 97-120.

275. Garrus JE, von Schwedler UK, Pornillos OW, Morham SG, Zavitz KH, Wang HE, Wettstein DA, Stray KM, Cote M, Rich RL, et al. 2001. TSG101 and the vacuolar protein sorting pathway are essential for HIV-1 budding. Cell 107: 55-65.

276. Gelfand MS, Mironov AA, Pevzner PA. 1996. Gene recognition via spliced sequence alignment. Proc. Natl. Acad. Sci. USA 93: 9061-9066.

277. Geourjon C, Deleage G. 1994. SOPM: a self-optimized method for protein secondary structure prediction. Protein Eng. 7: 157-164.

278. Geourjon C, Deleage G. 1995. SOPMA: significant improvements in protein secondary structure prediction by consensus prediction from multiple alignments. Comput. Appl. Biosci. 11: 681-684.

279. Gerdes SY, Scholle MD, D'Souza M, Bernal A, Baev MV, Farrell M, Kurnasov OV, Daugherty MD, Mseeh F, Polanuyer BM, et al. 2002. From genetic footprinting to antimicrobial drug targets: examples in cofactor biosynthetic pathways. J. Bacteriol. 184: 4555-4572.

280. Gerstein M. 1997. A structural census of genomes: comparing bacterial, eukaryotic, and archaeal genomes in terms of protein structure. J. Mol. Biol. 274: 562-576.

281. Gerstein M, Hegyi H. 1998. Comparing genomes in terms of protein structure: surveys of a finite parts list. FEMS Microbiol. Rev. 22: 277-304.

282. Gibrat JF, Garnier J, Robson B. 1987. Further developments of protein secondary structure prediction using information theory. New parameters and consideration of residue pairs. J. Mol. Biol. 198: 425-443.

283. Gibrat JF, Madej T, Bryant SH. 1996. Surprising similarities in structure comparison. Curr. Opin. Struct. Biol. 6: 377-385.

284. Gilbert DG. 2002. euGenes: a eukaryote genome information system. Nucleic Acids Res. 30: 145-148.

285. Giraud MF, Leonard GA, Field RA, Berlind C, Naismith JH. 2000. RmlC, the third enzyme of dTDP-L-rhamnose pathway, is a new class of epimerase. Nat. Struct. Biol. 7: 398-402.

286. Glaser P, Frangeul L, Buchrieser C, Rusniok C, Amend A, Baquero F, Berche P, Bloecker H, Brandt P, Chakraborty T, et al. 2001. Comparative genomics of *Listeria* species. Science 294: 849-852.

287. Glass JI, Lefkowitz EJ, Glass JS, Heiner CR, Chen EY, Cassell GH. 2000. The complete sequence of the mucosal pathogen *Ureaplasma urealyticum*. Nature 407: 757-762.

288. Glockner G, Eichinger L, Szafranski K, Pachebat JA, Bankier AT, Dear PH, Lehmann R, Baumgart C, Parra G, Abril JF, et al. 2002. Sequence and analysis of chromosome 2 of *Dictyostelium discoideum*. Nature 418: 79-85.

289. Goff SA, Ricke D, Lan TH, Presting G, Wang R, Dunn M, Glazebrook J, Sessions A, Oeller P, Varma H, et al. 2002. A draft sequence of the rice genome (*Oryza sativa* L. ssp. *japonica*). Science 296: 92-100.

290. Goffeau A, Barrell BG, Bussey H, Davis RW, Dujon B, Feldmann H, Galibert F, Hoheisel JD, Jacq C, Johnston M, et al. 1996. Life with 6000 genes. Science 274: 546-567.

291. Gogarten JP, Kibak H, Dittrich P, Taiz L, Bowman EJ, Bowman BJ, Manolson MF, Poole RJ, Date T, Oshima T, et al. 1989. Evolution of the vacuolar H^+-ATPase: implications for the origin of eukaryotes. Proc. Natl. Acad. Sci. USA 86: 6661-6665.

292. Golding GB, Gupta RS. 1995. Protein-based phylogenies support a chimeric origin for the eukaryotic genome. Mol. Biol. Evol. 12: 1-6.

293. Goodner B, Hinkle G, Gattung S, Miller N, Blanchard M, Qurollo B, Goldman BS, Cao Y, Askenazi M, Halling C, et al. 2001. Genome sequence of the plant pathogen and biotechnology agent *Agrobacterium tumefaciens* C58. Science 294: 2323-2328.

294. Gopalan V, Vioque A, Altman S. 2002. RNase P: variations and uses. J. Biol. Chem. 277: 6759-6762.

295. Gorbalenya AE, Donchenko AP, Blinov VM, Koonin EV. 1989. Cysteine proteases of positive strand RNA viruses and chymotrypsin-like serine proteases. A distinct protein superfamily with a common structural fold. FEBS Lett. 243: 103-114.

296. Gorbalenya AE, Koonin EV. 1990. Superfamily of UvrA-related NTP-binding proteins. Implications for rational classification of recombination/repair systems. J. Mol. Biol. 213: 583-591.

297. Gorina S, Pavletich NP. 1996. Structure of the p53 tumor suppressor bound to the ankyrin and SH3 domains of 53BP2. Science 274: 1001-1005.

298. Gosset G, Bonner CA, Jensen RA. 2001. Microbial origin of plant-type 2-keto-3-deoxy-D-arabino-heptulosonate 7-phosphate synthases, exemplified by the chorismate- and tryptophan-regulated enzyme from *Xanthomonas campestris*. J. Bacteriol. 183: 4061-4070.

299. Gottschalk G, Bender R. 1982. D-Gluconate dehydratase from *Clostridium pasteurianum*. Methods Enzymol. 90: 283-287.

300. Gough J, Chothia C. 2002. SUPERFAMILY: HMMs representing all proteins of known structure. SCOP sequence searches, alignments and genome assignments. Nucleic Acids Res. 30: 268-272.

301. Gough J, Karplus K, Hughey R, Chothia C. 2001. Assignment of homology to genome sequences using a library of hidden Markov models that represent all proteins of known structure. J. Mol. Biol. 313: 903-919.

302. Gould SJ. 1997. The exaptive excellence of spandrels as a term and prototype. Proc. Natl. Acad. Sci. USA 94: 10750-10755.

303. Gould SJ. 1997. Full House. Random House, New York.

304. Gould SJ. 2002. The Structure of Evolutionary Theory. Harvard Univ. Press, Cambrdige, MA.

305. Gourley DG, Shrive AK, Polikarpov I, Krell T, Coggins JR, Hawkins AR, Isaacs NW, Sawyer L. 1999. The two types of 3-dehydroquinase have distinct structures but catalyze the same overall reaction. Nat. Struct. Biol. 6: 521-525.

306. Govindarajan S, Recabarren R, Goldstein RA. 1999. Estimating the total number of protein folds. Proteins 35: 408-414.

307. Graham DE, Overbeek R, Olsen GJ, Woese CR. 2000. An archaeal genomic signature. Proc. Natl. Acad. Sci. USA 97: 3304-3308.

308. Graham DE, Xu H, White RH. 2002. A divergent archaeal member of the alkaline phosphatase binuclear metalloenzyme superfamily has phosphoglycerate mutase activity. FEBS Lett. 517: 190-194.

309. Graupner M, White RH. 2001. *Methanococcus jannaschii* generates L-proline by cyclization of L-ornithine. J. Bacteriol. 183: 5203-5205.

310. Graur D, Li W-H. 2000. Fundamentals of Molecular Evolution. Sinauer Associates, Sunderland, MA.

311. Gray MW. 1992. The endosymbiont hypothesis revisited. Int. Rev. Cytol. 141: 233-357.

312. Gray MW. 1999. Evolution of organellar genomes. Curr. Opin. Genet. Dev. 9: 678-687.

313. Gray MW, Burger G, Lang BF. 2001. The origin and early evolution of mitochondria. Genome Biol 2: REVIEWS1018.

314. Gribskov M. 1992. Translational initiation factors IF-1 and eIF-2 alpha share an RNA-binding motif with prokaryotic ribosomal protein S1 and polynucleotide phosphorylase. Gene 119: 107-111.

315. Gribskov M, McLachlan AD, Eisenberg D. 1987. Profile analysis: detection of distantly related proteins. Proc. Natl. Acad. Sci. USA 84: 4355-4358.

316. Grimaud R, Ezraty B, Mitchell JK, Lafitte D, Briand C, Derrick PJ, Barras F. 2001. Repair of oxidized proteins. Identification of a new methionine sulfoxide reductase. J. Biol. Chem. 276: 48915-48920.

317. Grishin NV. 2001. Fold change in evolution of protein structures. J. Struct. Biol. 134: 167-185.

318. Grishin NV, Wolf YI, Koonin EV. 2000. From complete genomes to measures of substitution rate variability within and between proteins. Genome Res. 10: 991-1000.

319. Gronewold TM, Kaiser D. 2001. The act operon controls the level and time of C-signal production for Myxococcus xanthus development. Mol. Microbiol. 40: 744-756.

320. Grossebuter W, Hartl T, Gorisch H, Stezowski JJ. 1986. Purification and properties of malate dehydrogenase from the thermoacidophilic archaebacterium *Thermoplasma acidophilum*. Biol. Chem. Hoppe Seyler 367: 457-463.

321. Guermeur Y. 1999. Combinaison de classifieurs statistiques. Application a la prediction de structure secondaire des proteines. Vol. 15.

322. Guermeur Y, Geourjon C, Gallinari P, Deleage G. 1999. Improved performance in protein secondary structure prediction by inhomogeneous score combination. Bioinformatics 15: 413-421.

323. Gulick AM, Hubbard BK, Gerlt JA, Rayment I. 2001. Evolution of enzymatic activities in the enolase superfamily: identification of the general acid catalyst in the active site of D-glucarate dehydratase from *Escherichia coli*. Biochemistry 40: 10054-10062.

324. Gupta RS, Golding GB. 1996. The origin of the eukaryotic cell. Trends Biochem. Sci 21: 166-171.

325. Guyonvarch A, Nguyen-Juilleret M, Hubert JC, Lacroute F. 1988. Structure of the *Saccharomyces cerevisiae* URA4 gene encoding dihydroorotase. Mol Gen Genet 212: 134-141.

326. Habenicht A, Hellman U, Cerff R. 1994. Non-phosphorylating GAPDH of higher plants is a member of the aldehyde dehydrogenase superfamily with no sequence homology to phosphorylating GAPDH. J. Mol. Biol. 237: 165-171.

327. Haeckel E. 1997. The Wonders of Life: A Popular Study of Biological Philosophy. De Young Press, New York.

328. Hager PW, Calfee MW, Phibbs PV. 2000. The *Pseudomonas aeruginosa devB*/SOL homolog, *pgl*, is a member of the hex regulon and encodes 6-phosphogluconolactonase. J. Bacteriol. 182: 3934-3941.

329. Hall SL, Padgett RA. 1994. Conserved sequences in a class of rare eukaryotic nuclear introns with non-consensus splice sites. J. Mol. Biol. 239: 357-365.

330. Handy J, Doolittle RF. 1999. An attempt to pinpoint the phylogenetic introduction of glutaminyl-tRNA synthetase among bacteria. J. Mol. Evol. 49: 709-715.

331. Hansen T, Oehlmann M, Schonheit P. 2001. Novel type of glucose-6-phosphate isomerase in the hyperthermophilic archaeon *Pyrococcus furiosus*. J. Bacteriol. 183: 3428-3435.

332. Hansmann S, Martin W. 2000. Phylogeny of 33 ribosomal and six other proteins encoded in an ancient gene cluster that is conserved across prokaryotic genomes: influence of excluding poorly alignable sites from analysis. Int. J. Syst. Evol. Microbiol. 50: 1655-1663.

333. Hard T, Rak A, Allard P, Kloo L, Garber M. 2000. The solution structure of ribosomal protein L36 from T*hermus thermophilus* reveals a zinc-ribbon-like fold. J. Mol. Biol. 296: 169-180.

334. Hase CC, Fedorova ND, Galperin MY, Dibrov PA. 2001. Sodium ion cycle in bacterial pathogens: evidence from cross-genome comparisons. Microbiol. Mol. Biol. Rev. 65: 353-370.

335. Haseloff J, Goelet P, Zimmern D, Ahlquist P, Dasgupta R, Kaesberg P. 1984. Striking similarities in amino acid sequence among nonstructural proteins encoded by RNA viruses that have dissimilar genomic organization. Proc. Natl. Acad. Sci. USA 81: 4358-4362.

336. Hegde R, Srinivasula SM, Zhang Z, Wassell R, Mukattash R, Cilenti L, DuBois G, Lazebnik Y, Zervos AS, Fernandes-Alnemri T, et al. 2002. Identification of Omi/HtrA2 as a mitochondrial apoptotic serine protease that disrupts inhibitor of apoptosis protein-caspase interaction. J. Biol. Chem. 277: 432-438.

337. Heidelberg JF, Eisen JA, Nelson WC, Clayton RA, Gwinn ML, Dodson RJ, Haft DH, Hickey EK, Peterson JD, Umayam LA, et al. 2000. DNA sequence of both chromosomes of the cholera pathogen *Vibrio cholerae*. Nature 406: 477-483.

338. Heikinheimo P, Goldman A, Jeffries C, Ollis DL. 1999. Of barn owls and bankers: a lush variety of alpha/beta hydrolases. Structure Fold. Des. 7: R141-146.

339. Heine A, DeSantis G, Luz JG, Mitchell M, Wong CH, Wilson IA. 2001. Observation of covalent intermediates in an enzyme mechanism at atomic resolution. Science 294: 369-374.

340. Hengartner MO. 2000. The biochemistry of apoptosis. Nature 407: 770-776.

341. Henikoff S. 1991. Playing with blocks: some pitfalls of forcing multiple alignments. New Biol. 3: 1148-1154.

342. Henikoff S, Henikoff JG. 1992. Amino acid substitution matrices from protein blocks. Proc. Natl. Acad. Sci. USA 89: 10915-10919.

343. Henikoff S, Henikoff JG. 1993. Performance evaluation of amino acid substitution matrices. Proteins 17: 49-61.

344. Henikoff S, Henikoff JG. 1994. Position-based sequence weights. J. Mol. Biol. 243: 574-578.

345. Henikoff S, Henikoff JG, Pietrokovski S. 1999. Blocks+: a non-redundant database of protein alignment blocks derived from multiple compilations. Bioinformatics 15: 471-479.

346. Hillis DM, Moritz C, Mable BK. 1996. Molecular Systematics. Sinauer Assoc., Sunderland, MA.

347. Himmelreich R, Hilbert H, Plagens H, Pirkl E, Li BC, Herrmann R. 1996. Complete sequence analysis of the genome of the bacterium *Mycoplasma pneumoniae*. Nucleic Acids Res. 24: 4420-4449.

348. Himmelreich R, Plagens H, Hilbert H, Reiner B, Herrmann R. 1997. Comparative analysis of the genomes of the bacteria *Mycoplasma pneumoniae* and *Mycoplasma genitalium*. Nucleic Acids Res. 25: 701-712.

349. Hirokawa T, Boon-Chieng S, Mitaku S. 1998. SOSUI: classification and secondary structure prediction system for membrane proteins. Bioinformatics 14: 378-379.

350. Hoersch S, Leroy C, Brown NP, Andrade MA, Sander C. 2000. The GeneQuiz web server: protein functional analysis through the Web. Trends Biochem. Sci. 25: 33-35.

351. Hofmann K. 2000. Sensitive protein comparisons with profiles and hidden Markov models. Brief. Bioinform. 1: 167-178.

352. Hofmann K, Stoffel W. 1993. TMbase - A database of membrane spanning proteins segments. Biol. Chem. Hoppe-Seyler 374: 166.

353. Holm L, Sander C. 1995. Dali: a network tool for protein structure comparison. Trends Biochem. Sci. 20: 478-480.

354. Holm L, Sander C. 1996. Mapping the protein universe. Science 273: 595-603.

355. Holm L, Sander C. 1997. An evolutionary treasure: unification of a broad set of amidohydrolases related to urease. Proteins 28: 72-82.

356. Horner DS, Hirt RP, Kilvington S, Lloyd D, Embley TM. 1996. Molecular data suggest an early acquisition of the mitochondrion endosymbiont. Proc. R. Soc. Lond. B Biol. Sci. 263: 1053-1059.

357. Hoskins J, Alborn WE, Jr., Arnold J, Blaszczak LC, Burgett S, DeHoff BS, Estrem ST, Fritz L, Fu DJ, Fuller W, et al. 2001. Genome of the bacterium *Streptococcus pneumoniae* strain R6. J. Bacteriol. 183: 5709-5717.

358. House CH, Fitz-Gibbon ST. 2002. Using homolog groups to create a whole-genomic tree of free-living organisms: an update. J. Mol. Evol. 54: 539-547.

359. Hoyle F, Wickramasinghe NC. 1999. Astronomical Origins of Life. Kluwer Academic Publishers, New York.

360. Huala E, Dickerman AW, Garcia-Hernandez M, Weems D, Reiser L, LaFond F, Hanley D, Kiphart D, Zhuang M, Huang W, et al. 2001. The Arabidopsis Information Resource (TAIR): a comprehensive database and web-based information retrieval, analysis, and visualization system for a model plant. Nucleic Acids Res. 29: 102-105.

361. Huang X, Adams MD, Zhou H, Kerlavage AR. 1997. A tool for analyzing and annotating genomic sequences. Genomics 46: 37-45.

362. Huber H, Hohn MJ, Rachel R, Fuchs T, Wimmer VC, Stetter KO. 2002. A new phylum of Archaea represented by a nanosized hyperthermophilic symbiont. Nature 417: 63-67.

363. Huerta AM, Salgado H, Thieffry D, Collado-Vides J. 1998. RegulonDB: a database on transcriptional regulation in Escherichia coli. Nucleic Acids Res 26: 55-59.

364. Hutchison CA, Peterson SN, Gill SR, Cline RT, White O, Fraser CM, Smith HO, Venter JC. 1999. Global transposon mutagenesis and a minimal *Mycoplasma* genome. Science 286: 2165-2169.

365. Hutvagner G, McLachlan J, Pasquinelli AE, Balint E, Tuschl T, Zamore PD. 2001. A cellular function for the RNA-interference enzyme Dicer in the maturation of the let-7 small temporal RNA. Science 293: 834-838.

366. Huynen M, Dandekar T, Bork P. 1998. Differential genome analysis applied to the species-specific features of *Helicobacter pylori*. FEBS Lett. 426: 1-5.

367. Huynen M, Doerks T, Eisenhaber F, Orengo C, Sunyaev S, Yuan Y, Bork P. 1998. Homology-based fold predictions for *Mycoplasma genitalium* proteins. J. Mol. Biol. 280: 323-326.

368. Huynen M, Snel B, Lathe W, 3rd, Bork P. 2000. Predicting protein function by genomic context: quantitative evaluation and qualitative inferences. Genome Res. 10: 1204-1210.

369. Huynen M, Snel B, Lathe W, Bork P. 2000. Exploitation of gene context. Curr. Opin. Struct. Biol. 10: 366-370.

370. Huynen MA, Dandekar T, Bork P. 1999. Variation and evolution of the citric-acid cycle: a genomic perspective. Trends Microbiol. 7: 281-291.

371. Huynen MA, Diaz-Lazcoz Y, Bork P. 1997. Differential genome display. Trends Genet. 13: 389-390.

372. Huynen MA, Snel B. 2000. Gene and context: integrative approaches to genome analysis. Adv. Protein. Chem. 54: 345-379.

373. Huynen MA, Snel B, Bork P. 1999. Lateral gene transfer, genome surveys, and the phylogeny of prokaryotes. Science 286: 1443a.

374. Huynen MA, van Nimwegen E. 1998. The frequency distribution of gene family sizes in complete genomes. Mol. Biol. Evol. 15: 583-589.

375. Ibba M, Bono JL, Rosa PA, Soll D. 1997. Archaeal-type lysyl-tRNA synthetase in the Lyme disease spirochete *Borrelia burgdorferi*. Proc. Natl. Acad. Sci. USA 94: 14383-14388.

376. Ibba M, Soll D. 2000. Aminoacyl-tRNA synthesis. Annu. Rev. Biochem. 69: 617-650.

377. Illangasekare M, Yarus M. 1999. Specific, rapid synthesis of Phe-RNA by RNA. Proc. Natl. Acad. Sci. USA 96: 5470-5475.

378. Imai Y, Kimura T, Murakami A, Yajima N, Sakamaki K, Yonehara S. 1999. The CED-4-homologous protein FLASH is involved in Fas-mediated activation of caspase-8 during apoptosis. Nature 398: 777-785.

379. Iserentant D, Verachtert H. 1995. Cloning and sequencing of the LEU2 homologue gene of *Schwanniomyces occidentalis*. Yeast 11: 467-473.

380. Ishikawa J, Hotta K. 1999. FramePlot: a new implementation of the frame analysis for predicting protein-coding regions in bacterial DNA with a high G+C content. FEMS Microbiol. Lett. 174: 251-253.

381. Ishikawa K, Mihara Y, Gondoh K, Suzuki E, Asano Y. 2000. X-ray structures of a novel acid phosphatase from Escherichia blattae and its complex with the transition-state analog molybdate. EMBO J. 19: 2412-2423.

382. Ishimoto LK, Ishimoto KS, Cascino A, Cipollaro M, Eiserling FA. 1988. The structure of three bacteriophage T4 genes required for tail-tube assembly. Virology 164: 81-90.

383. Ito S, Fushinobu S, Yoshioka I, Koga S, Matsuzawa H, Wakagi T. 2001. Structural basis for the ADP-specificity of a novel glucokinase from a hyperthermophilic archaeon. Structure 9: 205-214.

384. Ito T, Chiba T, Ozawa R, Yoshida M, Hattori M, Sakaki Y. 2001. A comprehensive two-hybrid analysis to explore the yeast protein interactome. Proc. Natl. Acad. Sci. USA 98: 4569-4574.

385. Itoh T, Takemoto K, Mori H, Gojobori T. 1999. Evolutionary instability of operon structures disclosed by sequence comparisons of complete microbial genomes. Mol. Biol. Evol. 16: 332-346.

386. Iwabe N, Kuma K, Hasegawa M, Osawa S, Miyata T. 1989. Evolutionary relationship of archaebacteria, eubacteria, and eukaryotes inferred from phylogenetic trees of duplicated genes. Proc. Natl. Acad. Sci. USA 86: 9355-9359.

387. Iyer LM, Aravind L, Bork P, Hofmann K, Mushegian AR, Zhulin IB, Koonin EV. 2001. Quod erat demonstrandum? The mystery of experimental validation of erroneous computational analyses of protein sequences. Genome Biology 2: RESEARCH0051.

388. Jacob F. 1977. Evolution and tinkering. Science 196: 1161-1166.

389. Jacobs MD, Harrison SC. 1998. Structure of an IkappaBalpha/NF-kappaB complex. Cell 95: 749-758.

390. Jain R, Rivera MC, Lake JA. 1999. Horizontal gene transfer among genomes: the complexity hypothesis. Proc. Natl. Acad. Sci. USA 96: 3801-3806.

391. Javid-Majd F, Stapleton MA, Harmon MF, Hanks BA, Mullins LS, Raushel FM. 1996. Comparison of the functional differences for the homologous residues within the carboxy

phosphate and carbamate domains of carbamoyl phosphate synthetase. Biochemistry 35: 14362-14369.

392. Jeanmougin F, Thompson JD, Gouy M, Higgins DG, Gibson TJ. 1998. Multiple sequence alignment with Clustal X. Trends Biochem. Sci. 23: 403-405.

393. Jedrzejas MJ. 2000. Structure, function, and evolution of phosphoglycerate mutases: comparison with fructose-2,6-bisphosphatase, acid phosphatase, and alkaline phosphatase. Prog. Biophys. Mol. Biol. 73: 263-287.

394. Jedrzejas MJ, Chander M, Setlow P, Krishnasamy G. 2000. Structure and mechanism of action of a novel phosphoglycerate mutase from *Bacillus stearothermophilus*. EMBO J. 19: 1419-1431.

395. Jensen RA, Thompson ME, Jetton TL, Szabo CI, van der Meer R, Helou B, Tronick SR, Page DL, King MC, Holt JT. 1996. BRCA1 is secreted and exhibits properties of a granin. Nat. Genet. 12: 303-308.

396. Jia J, Huang W, Schorken U, Sahm H, Sprenger GA, Lindqvist Y, Schneider G. 1996. Crystal structure of transaldolase B from *Escherichia coli* suggests a circular permutation of the alpha/beta barrel within the class I aldolase family. Structure 4: 715-724.

397. John J, Crennell SJ, Hough DW, Danson MJ, Taylor GL. 1994. The crystal structure of glucose dehydrogenase from *Thermoplasma acidophilum*. Structure 2: 385-393.

398. Johnsen U, Selig M, Xavier KB, Santos H, Schonheit P. 2001. Different glycolytic pathways for glucose and fructose in the halophilic archaeon *Halococcus saccharolyticus*. Arch. Microbiol. 175: 52-61.

399. Johnson KA, Chen L, Yang H, Roberts MF, Stec B. 2001. Crystal structure and catalytic mechanism of the MJ0109 gene product: a bifunctional enzyme with inositol monophosphatase and fructose 1,6-bisphosphatase activities. Biochemistry 40: 618-630.

400. Jomaa H, Wiesner J, Sanderbrand S, Altincicek B, Weidemeyer C, Hintz M, Turbachova I, Eberl M, Zeidler J, Lichtenthaler HK, et al. 1999. Inhibitors of the nonmevalonate pathway of isoprenoid biosynthesis as antimalarial drugs. Science 285: 1573-1576.

401. Jones DT. 1999. GenTHREADER: an efficient and reliable protein fold recognition method for genomic sequences. J. Mol. Biol. 287: 797-815.

402. Jones DT. 1999. Protein secondary structure prediction based on position-specific scoring matrices. J. Mol. Biol. 292: 195-202.

403. Jones DT, Taylor WR, Thornton JM. 1992. The rapid generation of mutation data matrices from protein sequences. Comput. Appl. Biosci. 8: 275-282.

404. Jones DT, Taylor WR, Thornton JM. 1994. A mutation data matrix for transmembrane proteins. FEBS Lett. 339: 269-275.

405. Jones DT, Thornton JM. 1996. Potential energy functions for threading. Curr. Opin. Struct. Biol. 6: 210-216.

406. Jones DT, Tress M, Bryson K, Hadley C. 1999. Successful recognition of protein folds using threading methods biased by sequence similarity and predicted secondary structure. Proteins 37: 104-111.

407. Joo WS, Jeffrey PD, Cantor SB, Finnin MS, Livingston DM, Pavletich NP. 2002. Structure of the 53BP1 BRCT region bound to p53 and its comparison to the Brca1 BRCT structure. Genes Dev. 16: 583-593.

408. Jordan IK, Makarova KS, Spouge JL, Wolf YI, Koonin EV. 2001. Lineage-specific gene expansions in bacterial and archaeal genomes. Genome Res. 11: 555-565.

409. Jordan IK, Rogozin IB, Glazko GV, Koonin EV. 2002. Origin of human regulatory sequences from transposable elements. Trends Genet. in press.

410. Joza N, Susin SA, Daugas E, Stanford WL, Cho SK, Li CY, Sasaki T, Elia AJ, Cheng HY, Ravagnan L, et al. 2001. Essential role of the mitochondrial apoptosis-inducing factor in programmed cell death. Nature 410: 549-554.

411. Kaeberlein T, Lewis K, Epstein SS. 2002. Isolating "uncultivable" microorganisms in pure culture in a simulated natural environment. Science 296: 1127-1129.

412. Kalman S, Mitchell W, Marathe R, Lammel C, Fan J, Hyman RW, Olinger L, Grimwood J, Davis RW, Stephens RS. 1999. Comparative genomes of *Chlamydia pneumoniae* and *C. trachomatis*. Nat. Genet. 21: 385-389.

413. Kanagasundaram V, Scopes R. 1992. Isolation and characterization of the gene encoding gluconolactonase from *Zymomonas mobilis*. Biochim Biophys Acta 1171: 198-200.

414. Kanehisa M, Goto S. 2000. KEGG: Kyoto Encyclopedia of Genes and Genomes. Nucleic Acids Res. 28: 27-30.

415. Kaneko T, Nakamura Y, Sato S, Asamizu E, Kato T, Sasamoto S, Watanabe A, Idesawa K, Ishikawa A, Kawashima K, et al. 2000. Complete genome structure of the nitrogen-fixing symbiotic bacterium *Mesorhizobium loti*. DNA Res 7: 331-338.

416. Kaneko T, Nakamura Y, Wolk CP, Kuritz T, Sasamoto S, Watanabe A, Iriguchi M, Ishikawa A, Kawashima K, Kimura T, et al. 2001. Complete genomic sequence of the filamentous nitrogen-fixing cyanobacterium *Anabaena* sp. strain PCC 7120. DNA Res. 8: 205-253.

417. Kaneko T, Sato S, Kotani H, Tanaka A, Asamizu E, Nakamura Y, Miyajima N, Hirosawa M, Sugiura M, Sasamoto S, et al. 1996. Sequence analysis of the genome of the unicellular cyanobacterium *Synechocystis* sp. strain PCC6803. II. Sequence determination of the entire genome and assignment of potential protein-coding regions. DNA Res. 3: 109-136.

418. Kapatral V, Anderson I, Ivanova N, Reznik G, Los T, Lykidis A, Bhattacharyya A, Bartman A, Gardner W, Grechkin G, et al. 2002. Genome sequence and analysis of the oral bacterium *Fusobacterium nucleatum* strain ATCC 25586. J. Bacteriol. 184: 2005-2018.

419. Karev GP, Wolf YI, Koonin EV. 2002. Mathematical modeling of the evolution of domain composition of proteomes: A birth-and-death process with innovation. In: Computational genomics: from sequence to function. M.Y. Galperin and E.V. Koonin, ed., Caister Academic Press, Wymondham, UK.

420. Karlberg O, Canback B, Kurland CG, Andersson SG. 2000. The dual origin of the yeast mitochondrial proteome. Yeast 17: 170-187.

421. Karlin S, Altschul SF. 1990. Methods for assessing the statistical significance of molecular sequence features by using general scoring schemes. Proc. Natl. Acad. Sci. USA 87: 2264-2268.

422. Karplus K. 1995. Evaluating regularizers for estimating distributions of amino acids. Proc. Int. Conf. Intell. Syst. Mol. Biol. 3: 188-196.

423. Karplus K, Barrett C, Hughey R. 1998. Hidden Markov models for detecting remote protein homologies. Bioinformatics 14: 846-856.

424. Karplus K, Hu B. 2001. Evaluation of protein multiple alignments by SAM-T99 using the BAliBASE multiple alignment test set. Bioinformatics 17: 713-720.

425. Katinka MD, Duprat S, Cornillot E, Metenier G, Thomarat F, Prensier G, Barbe V, Peyretaillade E, Brottier P, Wincker P, et al. 2001. Genome sequence and gene compaction of the eukaryote parasite *Encephalitozoon cuniculi*. Nature 414: 450-453.

426. Kawarabayasi Y, Hino Y, Horikawa H, Jin-no K, Takahashi M, Sekine M, Baba S, Ankai A, Kosugi H, Hosoyama A, et al. 2001. Complete genome sequence of an aerobic thermoacidophilic crenarchaeon, *Sulfolobus tokodaii* strain7. DNA Res. 8: 123-140.

427. Kawarabayasi Y, Hino Y, Horikawa H, Yamazaki S, Haikawa Y, Jin-no K, Takahashi M, Sekine M, Baba S, Ankai A, et al. 1999. Complete genome sequence of an aerobic hyper-thermophilic crenarchaeon, *Aeropyrum pernix* K1. DNA Res. 6: 83-101.

428. Kawarabayasi Y, Sawada M, Horikawa H, Haikawa Y, Hino Y, Yamamoto S, Sekine M, Baba S, Kosugi H, Hosoyama A, et al. 1998. Complete sequence and gene organization of the genome of a hyper-thermophilic archaebacterium, *Pyrococcus horikoshii* OT3. DNA Res. 5: 147-155.

429. Kawashima S, Kanehisa M. 2000. AAindex: amino acid index database. Nucleic Acids Res. 28: 374.

430. Kawashima T, Yamamoto Y, Aramaki H, Nunoshiba T, Kawamoto T, Watanabe K, Yamazaki M, Kanehori K, Amano N, Ohya K, et al. 1999. Determination of the complete genomic DNA sequence of *Thermoplasma volcanium* GSS1. Proc. Jpn. Acad. 75: 213-218.

431. Keck JL, Roche DD, Lynch AS, Berger JM. 2000. Structure of the RNA polymerase domain of *E. coli* primase. Science 287: 2482-2486.

432. Keeling PJ, Doolittle WF. 1997. Evidence that eukaryotic triosephosphate isomerase is of alpha-proteobacterial origin. Proc. Natl. Acad. Sci. USA 94: 1270-1275.

433. Keith B, Dong XN, Ausubel FM, Fink GR. 1991. Differential induction of 3-deoxy-D-arabino-heptulosonate 7-phosphate synthase genes in *Arabidopsis thaliana* by wounding and pathogenic attack. Proc. Natl. Acad. Sci. USA 88: 8821-8825.

434. Kelley LA, MacCallum RM, Sternberg MJ. 2000. Enhanced genome annotation using structural profiles in the program 3D-PSSM. J. Mol. Biol. 299: 499-520.

435. Kengen SW, Tuininga JE, de Bok FA, Stams AJ, de Vos WM. 1995. Purification and characterization of a novel ADP-dependent glucokinase from the hyperthermophilic archaeon *Pyrococcus furiosus*. J. Biol. Chem. 270: 30453-30457.

436. Ketting RF, Fischer SE, Bernstein E, Sijen T, Hannon GJ, Plasterk RH. 2001. Dicer functions in RNA interference and in synthesis of small RNA involved in developmental timing in *C. elegans*. Genes Dev. 15: 2654-2659.

437. Khuri S, Bakker FT, Dunwell JM. 2001. Phylogeny, function, and evolution of the cupins, a structurally conserved, functionally diverse superfamily of proteins. Mol. Biol. Evol. 18: 593-605.

438. Kim RY, Gasser R, Wistow GJ. 1992. mu-crystallin is a mammalian homologue of *Agrobacterium* ornithine cyclodeaminase and is expressed in human retina. Proc. Natl. Acad. Sci. USA 89: 9292-9296.

439. Kimsey HH, Kaiser D. 1992. The orotidine-5'-monophosphate decarboxylase gene of *Myxococcus xanthus*. Comparison to the OMP decarboxylase gene family. J. Biol. Chem. 267: 819-824.

440. Kimura M. 1983. The Neutral Theory of Molecular Evolution. Cambridge University Press, Cambridge, UK.

441. King RD, Saqi M, Sayle R, Sternberg MJ. 1997. DSC: public domain protein secondary structure predication. Comput. Appl. Biosci. 13: 473-474.

442. Kitagawa M, Oyama T, Kawashima T, Yedvobnick B, Kumar A, Matsuno K, Harigaya K. 2001. A human protein with sequence similarity to *Drosophila* mastermind coordinates the nuclear form of notch and a CSL protein to build a transcriptional activator complex on target promoters. Mol. Cell. Biol. 21: 4337-4346.

443. Klena JD, Pradel E, Schnaitman CA. 1992. Comparison of lipopolysaccharide biosynthesis genes rfaK, rfaL, rfaY, and rfaZ of *Escherichia coli* K-12 and *Salmonella typhimurium*. J. Bacteriol. 174: 4746-4752.

444. Klenk HP, Clayton RA, Tomb J-F, White O, Nelson KE, Ketchum KA, Dodson RJ, Gwinn M, Hickey EK, Peterson JD, et al. 1997. The complete genome sequence of the hyperthermophilic, sulphate- reducing archaeon *Archaeoglobus fulgidus*. Nature 390: 364-370.

445. Kletzin A, Adams MW. 1996. Molecular and phylogenetic characterization of pyruvate and 2-ketoisovalerate ferredoxin oxidoreductases from *Pyrococcus furiosus* and pyruvate ferredoxin oxidoreductase from *Thermotoga maritima*. J. Bacteriol. 178: 248-257.

446. Kneller DG, Cohen FE, Langridge R. 1990. Improvements in protein secondary structure prediction by an enhanced neural network. J. Mol. Biol. 214: 171-182.

447. Kolesov G, Mewes H-W, Frishman D. 2001. SNAPping up functionally related genes based on context information: A colinearity-free approach. J. Mol. Biol. 311: 639-656.

448. Kondrashov FA, Rogozin IB, Wolf YI, Koonin EV. 2002. Selection in the evolution of gene duplications. Genome Biol. 3: RESEARCH0008.

449. Konkel ME, Kim BJ, Rivera-Amill V, Garvis SG. 1999. Bacterial secreted proteins are required for the internaliztion of *Campylobacter jejuni* into cultured mammalian cells. Mol. Microbiol. 32: 691-701.

450. Koonin EV. 1994. Conserved sequence pattern in a wide variety of phosphoesterases. Protein Sci. 3: 356-358.

451. Koonin EV. 1998. Genomic microbiology: right on target? Nat. Biotechnol. 16: 821-822.

452. Koonin EV. 2000. How many genes can make a cell: the minimal-gene-set concept. Annu. Rev. Genomics Hum. Genet. 1: 99-116.

453. Koonin EV, Abagyan RA. 1997. TSG101 may be the prototype of a class of dominant negative ubiquitin regulators. Nat. Genet. 16: 330-331.

454. Koonin EV, Altschul SF, Bork P. 1996. BRCA1 protein products ... Functional motifs. Nat. Genet. 13: 266-268.

455. Koonin EV, Aravind L. 2000. Dynein light chains of the Roadblock/LC7 group belong to an ancient protein superfamily implicated in NTPase regulation. Curr. Biol. 10: R774-776.

456. Koonin EV, Aravind L. 2000. The NACHT family - a new group of predicted NTPases implicated in apoptosis and MHC transcription activation. Trends Biochem. Sci. 25: 223-224.

457. Koonin EV, Aravind L. 2002. Origin and evolution of eukaryotic apoptosis: the bacterial connection. Cell Death Differ. 9: 394-404.

458. Koonin EV, Aravind L, Hofmann K, Tschopp J, Dixit VM. 1999. Apoptosis. Searching for FLASH domains. Nature 401: 662-663.

459. Koonin EV, Aravind L, Kondrashov AS. 2000. The impact of comparative genomics on our understanding of evolution. Cell 101: 573-576.

460. Koonin EV, Dolja VV. 1993. Evolution and taxonomy of positive-strand RNA viruses: implications of comparative analysis of amino acid sequences. Crit. Rev. Biochem. Mol. Biol. 28: 375-430.

461. Koonin EV, Galperin MY. 1997. Prokaryotic genomes: the emerging paradigm of genome-based microbiology. Curr. Opin. Genet. Dev. 7: 757-763.

462. Koonin EV, Makarova KS, Aravind L. 2001. Horizontal gene transfer in prokaryotes: quantification and classification. Annu. Rev. Microbiol 55: 709-742.

463. Koonin EV, Makarova KS, Wolf YI, Aravind L. 2002. Horizontal gene transfer and its role in the evolution of prokaryotes. In: Horizontal Gene Transfer. M. Syvanen and C.I. Kado, ed., pp. 277-304. Academic Press, London & San Diego.

464. Koonin EV, Mushegian AR. 1996. Complete genome sequences of cellular life forms: glimpses of theoretical evolutionary genomics. Curr. Opin. Genet. Dev. 6: 757-762.

465. Koonin EV, Mushegian AR, Bork P. 1996. Non-orthologous gene displacement. Trends Genet. 12: 334-336.

466. Koonin EV, Mushegian AR, Galperin MY, Walker DR. 1997. Comparison of archaeal and bacterial genomes: computer analysis of protein sequences predicts novel functions and suggests a chimeric origin for the archaea. Mol. Microbiol. 25: 619-637.

467. Koonin EV, Mushegian AR, Rudd KE. 1996. Sequencing and analysis of bacterial genomes. Curr. Biol. 6: 404-416.

468. Koonin EV, Tatusov RL, Galperin MY. 1998. Beyond the complete genomes: from sequence to structure and function. Curr. Opin. Struct. Biol. 8: 355-363.

469. Koonin EV, Wolf YI, Aravind L. 2001. Prediction of the archaeal exosome and its connections with the proteasome and the translation and transcription machineries by a comparative-genomic approach. Genome Res. 11: 240-252.

470. Korbel JO, Snel B, Huynen MA, Bork P. 2002. SHOT: a web server for the construction of genome phylogenies. Trends Genet. 18: 158-162.

471. Kornegay JR, Schilling JW, Wilson AC. 1994. Molecular adaptation of a leaf-eating bird: stomach lysozyme of the hoatzin. Mol. Biol. Evol. 11: 921-928.

472. Krogh A. 2000. Using database matches with for HMMGene for automated gene detection in *Drosophila*. Genome Res. 10: 523-528.

473. Krogh A, Brown M, Mian IS, Sjolander K, Haussler D. 1994. Hidden Markov models in computational biology. Applications to protein modeling. J. Mol. Biol. 235: 1501-1531.

474. Krogh A, Larsson B, von Heijne G, Sonnhammer EL. 2001. Predicting transmembrane protein topology with a hidden Markov model: application to complete genomes. J. Mol. Biol. 305: 567-580.

475. Krylov DM, Koonin EV. 2001. A novel family of predicted retroviral-like aspartyl proteases with a possible key role in eukaryotic cell cycle control. Curr. Biol. 11: R584-587.

476. Kryukov GV, Kumar RA, Koc A, Sun Z, Gladyshev VN. 2002. Selenoprotein R is a zinc-containing stereo-specific methionine sulfoxide reductase. Proc. Natl. Acad. Sci. USA 99: 4245-4250.

477. Kunst F, Ogasawara N, Moszer I, Albertini AM, Alloni G, Azevedo V, Bertero MG, Bessieres P, Bolotin A, Borchert S, et al. 1997. The complete genome sequence of the gram-positive bacterium *Bacillus subtilis*. Nature 390: 249-256.

478. Kuriyan J, O'Donnell M. 1993. Sliding clamps of DNA polymerases. J. Mol. Biol. 234: 915-925.

479. Kurland CG. 2000. Something for everyone. Horizontal gene transfer in evolution. EMBO Rep 1: 92-95.

480. Kurland CG, Andersson SG. 2000. Origin and evolution of the mitochondrial proteome. Microbiol. Mol. Biol. Rev. 64: 786-820.

481. Kuroda M, Ohta T, Uchiyama I, Baba T, Yuzawa H, Kobayashi I, Cui L, Oguchi A, Aoki K, Nagai Y, et al. 2001. Whole genome sequencing of meticillin-resistant *Staphylococcus aureus*. Lancet 357: 1225-1240.

482. Kuznetsov VA. 2002. Statistics of the numbers of transcripts and protein sequences encoded in the genome. In: Computational and Statistical Approaches to Genomics. W.Zhang and I. Shmulevich, ed. Kluwer, Boston.

483. Kyrpides NC, Olsen GJ. 1999. Archaeal and bacterial hyperthermophiles: horizontal gene exchange or common ancestry? Trends Genet. 15: 298-299.

484. Labedan B, Riley M. 1995. Gene products of *Escherichia coli*: sequence comparisons and common ancestries. Mol. Biol. Evol. 12: 980-987.

485. Laber B, Maurer W, Scharf S, Stepusin K, Schmidt FS. 1999. Vitamin B6 biosynthesis: formation of pyridoxine 5'-phosphate from 4-(phosphohydroxy)-L-threonine and 1-deoxy-D-xylulose-5-phosphate by PdxA and PdxJ protein. FEBS Lett. 449: 45-48.

486. Lake MW, Wuebbens MM, Rajagopalan KV, Schindelin H. 2001. Mechanism of ubiquitin activation revealed by the structure of a bacterial MoeB-MoaD complex. Nature 414: 325-329.

487. Lam E, Kato N, Lawton M. 2001. Programmed cell death, mitochondria and the plant hypersensitive response. Nature 411: 848-853.

488. Lander ES, Linton LM, Birren B, Nusbaum C, Zody MC, Baldwin J, Devon K, Dewar K, Doyle M, FitzHugh W, et al. 2001. Initial sequencing and analysis of the human genome. Nature 409: 860-921.

489. Lang BF, Gray MW, Burger G. 1999. Mitochondrial genome evolution and the origin of eukaryotes. Annu. Rev. Genet. 33: 351-397.

490. Larsen TM, Wedekind JE, Rayment I, Reed GH. 1996. A carboxylate oxygen of the substrate bridges the magnesium ions at the active site of enolase: structure of the yeast enzyme complexed with the equilibrium mixture of 2-phosphoglycerate and phosphoenolpyruvate at 1.8 A resolution. Biochemistry 35: 4349-4358.

491. Lathe WC, 3rd, Snel B, Bork P. 2000. Gene context conservation of a higher order than operons. Trends Biochem. Sci. 25: 474-479.

492. Laub MT, Smith DW. 1998. Finding intron/exon splice junctions using INFO, INterruption Finder and Organizer. J. Comput. Biol. 5: 307-321.

493. Lawrence CE, Altschul SF, Boguski MS, Liu JS, Neuwald AF, Wootton JC. 1993. Detecting subtle sequence signals: a Gibbs sampling strategy for multiple alignment. Science 262: 208-214.

494. Lawrence JG. 1997. Selfish operons and speciation by gene transfer. Trends Microbiol. 5: 355-359.

495. Lawrence JG. 1999. Gene transfer, speciation, and the evolution of bacterial genomes. Curr. Opin. Microbiol. 2: 519-523.

496. Lawrence JG, Ochman H. 1997. Amelioration of bacterial genomes: rates of change and exchange. J. Mol. Evol. 44: 383-397.

497. Lawrence JG, Ochman H. 1998. Molecular archaeology of the *Escherichia coli* genome. Proc. Natl. Acad. Sci. USA 95: 9413-9417.

498. Lawrence JG, Ochman H. 2002. Reconciling the many faces of lateral gene transfer. Trends Microbiol. 10: 1-4.

499. Lawrence MC, Barbosa JA, Smith BJ, Hall NE, Pilling PA, Ooi HC, Marcuccio SM. 1997. Structure and mechanism of a sub-family of enzymes related to N-acetylneuraminate lyase. J. Mol. Biol. 266: 381-399.

500. le Coq D, Fillinger S, Aymerich S. 1999. Histidinol phosphate phosphatase, catalyzing the penultimate step of the histidine biosynthesis pathway, is encoded by *ytvP* (*hisJ*) in *Bacillus subtilis*. J. Bacteriol. 181: 3277-3280.

501. Lee SH, Hidaka T, Nakashita H, Seto H. 1995. The carboxyphosphonoenolpyruvate synthase-encoding gene from the bialaphos-producing organism *Streptomyces hygroscopicus*. Gene 153: 143-144.

502. Lehnherr H, Maguin E, Jafri S, Yarmolinsky MB. 1993. Plasmid addiction genes of bacteriophage P1: *doc*, which causes cell death on curing of prophage, and *phd*, which prevents host death when prophage is retained. J. Mol. Biol. 233: 414-428.

503. Leipe DD, Aravind L, Grishin NV, Koonin EV. 2000. The bacterial replicative helicase DnaB evolved from a RecA duplication. Genome Res. 10: 5-16.

504. Leipe DD, Aravind L, Koonin EV. 1999. Did DNA replication evolve twice independently? Nucleic Acids Res. 27: 3389-3401.

505. Leipe DD, Wolf YI, Koonin EV, Aravind L. 2002. Classification and evolution of P-loop GTPases and related ATPases. J. Mol. Biol. 317: 41-72.

506. Lespinet O, Koonin EV, Wolf YI, Aravind L. 2002. The role of lineage-specific gene family expansion in the evolution of eukaryotes. Genome Res. 12: 1048-1059.

507. Letunic I, Goodstadt L, Dickens NJ, Doerks T, Schultz J, Mott R, Ciccarelli F, Copley RR, Ponting CP, Bork P. 2002. Recent improvements to the SMART domain-based sequence annotation resource. Nucleic Acids Res. 30: 242-244.

508. Levchenko I, Smith CK, Walsh NP, Sauer RT, Baker TA. 1997. PDZ-like domains mediate binding specificity in the Clp/Hsp100 family of chaperones and protease regulatory subunits. Cell 91: 939-947.

509. Levin JM. 1997. Exploring the limits of nearest neighbour secondary structure prediction. Protein Eng. 10: 771-776.

510. Levin JM, Robson B, Garnier J. 1986. An algorithm for secondary structure determination in proteins based on sequence similarity. FEBS Lett. 205: 303-308.

511. Levy CW, Buckley PA, Sedelnikova S, Kato Y, Asano Y, Rice DW, Baker PJ. 2002. Insights into enzyme evolution revealed by the structure of methylaspartate ammonia lyase. Structure 10: 105-113.

512. Lewis K. 2000. Programmed death in bacteria. Microbiol. Mol. Biol. Rev. 64: 503-514.

513. Lewis S, Ashburner M, Reese MG. 2000. Annotating eukaryote genomes. Curr. Opin. Struct. Biol. 10: 349-354.

514. Li L, Cohen SN. 1996. Tsg101: a novel tumor susceptibility gene isolated by controlled homozygous functional knockout of allelic loci in mammalian cells. Cell 85: 319-329.

515. Li L, Li X, Francke U, Cohen SN. 1997. The TSG101 tumor susceptibility gene is located in chromosome 11 band p15 and is mutated in human breast cancer. Cell 88: 143-154.

516. Li L, Liao J, Ruland J, Mak TW, Cohen SN. 2001. A TSG101/MDM2 regulatory loop modulates MDM2 degradation and MDM2/p53 feedback control. Proc. Natl. Acad. Sci. USA 98: 1619-1624.

517. Li WH, Gu Z, Wang H, Nekrutenko A. 2001. Evolutionary analyses of the human genome. Nature 409: 847-849.

518. Lichtenthaler HK. 2000. Non-mevalonate isoprenoid biosynthesis: enzymes, genes and inhibitors. Biochem. Soc. Trans. 28: 785-789.

519. Lin J, Gerstein M. 2000. Whole-genome trees based on the occurrence of folds and orthologs: implications for comparing genomes on different levels. Genome Res. 10: 808-818.

520. Link AJ, Phillips D, Church GM. 1997. Methods for generating precise deletions and insertions in the genome of wild-type *Escherichia coli*: application to open reading frame characterization. J. Bacteriol. 179: 6228-6237.

521. Lipman DJ, Pearson WR. 1985. Rapid and sensitive protein similarity searches. Science 227: 1435-1441.

522. Lo Conte L, Ailey B, Hubbard TJ, Brenner SE, Murzin AG, Chothia C. 2000. SCOP: a structural classification of proteins database. Nucleic Acids Res. 28: 257-259.

523. Lomovskaya O, Warren MS, Lee A, Galazzo J, Fronko R, Lee M, Blais J, Cho D, Chamberland S, Renau T, et al. 2001. Identification and characterization of inhibitors of multidrug resistance efflux pumps in *Pseudomonas aeruginosa*: novel agents for combination therapy. Antimicrob. Agents Chemother. 45: 105-116.

524. Lomovskaya O, Watkins W. 2001. Inhibition of efflux pumps as a novel approach to combat drug resistance in bacteria. J. Mol. Microbiol. Biotechnol. 3: 225-236.

525. Long M, de Souza SJ, Gilbert W. 1997. Delta-interacting protein A and the origin of hepatitis delta antigen. Science 276: 824-825.

526. Lowther WT, Weissbach H, Etienne F, Brot N, Matthews BW. 2002. The mirrored methionine sulfoxide reductases of *Neisseria gonorrhoeae pilB*. Nat. Struct. Biol. 9: 348-352.

527. Luban J. 2001. HIV-1 and Ebola virus: the getaway driver nabbed. Nat. Med. 7: 1278-1280.

528. Lukashin AV, Borodovsky M. 1998. GeneMark.hmm: new solutions for gene finding. Nucleic Acids Res. 26: 1107-1115.

529. Lukatela G, Krauss N, Theis K, Selmer T, Gieselmann V, von Figura K, Saenger W. 1998. Crystal structure of human arylsulfatase A: the aldehyde function and the metal ion at the active site suggest a novel mechanism for sulfate ester hydrolysis. Biochemistry 37: 3654-3664.

530. Lundstrom J, Rychlewski L, Bujnicki J, Elofsson A. 2001. Pcons: a neural-network-based consensus predictor that improves fold recognition. Protein Sci. 10: 2354-2362.

531. Lupas A. 1996. Coiled coils: new structures and new functions. Trends Biochem. Sci. 21: 375-382.

532. Lupas A. 1996. Prediction and analysis of coiled-coil structures. Methods Enzymol. 266: 513-525.

533. Lupas A. 1997. Predicting coiled-coil regions in proteins. Curr. Opin. Struct. Biol. 7: 388-393.

534. Lynch M, Conery JS. 2000. The evolutionary fate and consequences of duplicate genes. Science 290: 1151-1155.

535. Madej T, Gibrat JF, Bryant SH. 1995. Threading a database of protein cores. Proteins 23: 356-369.

536. Maden BE. 1995. No soup for starters? Autotrophy and the origins of metabolism. Trends Biochem. Sci. 20: 337-341.

537. Madeo F, Herker E, Maldener C, Wissing S, Lachelt S, Herlan M, Fehr M, Lauber K, Sigrist SJ, Wesselborg S, et al. 2002. A caspase-related protease regulates apoptosis in yeast. Mol Cell. 9: 911-917.

538. Mai X, Adams MW. 1996. Characterization of a fourth type of 2-keto acid-oxidizing enzyme from a hyperthermophilic archaeon: 2-ketoglutarate ferredoxin oxidoreductase from *Thermococcus litoralis*. J. Bacteriol. 178: 5890-5896.

539. Makalowski W. 2000. Genomic scrap yard: how genomes utilize all that junk. Gene 259: 61-67.

540. Makarova KS, Aravind L, Galperin MY, Grishin NV, Tatusov RL, Wolf YI, Koonin EV. 1999. Comparative genomics of the archaea (Euryarchaeota): Evolution of conserved protein families, the stable core, and the variable shell. Genome Res. 9: 608-628.

541. Makarova KS, Aravind L, Grishin NV, Rogozin IB, Koonin EV. 2002. A DNA repair system specific for thermophilic Archaea and bacteria predicted by genomic context analysis. Nucleic Acids Res 30: 482-496.

542. Makarova KS, Ponomarev VA, Koonin EV. 2001. Two C or not two C: recurrent disruption of Zn-ribbons, gene duplication, lineage-specific gene loss, and horizontal gene transfer in evolution of bacterial ribosomal proteins. Genome Biol. 2: RESEARCH 0033.

543. Mallick P, Goodwill KE, Fitz-Gibbon S, Miller JH, Eisenberg D. 2000. Selecting protein targets for structural genomics of *Pyrobaculum aerophilum*: validating automated fold assignment methods by using binary hypothesis testing. Proc. Natl. Acad. Sci. USA 97: 2450-2455.

544. Marchal S, Cobessi D, Rahuel-Clermont S, Tete-Favier F, Aubry A, Branlant G. 2001. Chemical mechanism and substrate binding sites of NADP-dependent aldehyde dehydrogenase from *Streptococcus mutans*. Chem. Biol. Interact. 130-132: 15-28.

545. Marchler-Bauer A, Panchenko AR, Shoemaker BA, Thiessen PA, Geer LY, Bryant SH. 2002. CDD: a database of conserved domain alignments with links to domain three-dimensional structure. Nucleic Acids Res. 30: 281-283.

546. Marcotte EM, Pellegrini M, Ng HL, Rice DW, Yeates TO, Eisenberg D. 1999. Detecting protein function and protein-protein interactions from genome sequences. Science 285: 751-753.

547. Marcotte EM, Xenarios I, van Der Bliek AM, Eisenberg D. 2000. Localizing proteins in the cell from their phylogenetic profiles. Proc. Natl. Acad. Sci. USA 97: 12115-12120.

548. Margalit H, Nadir E, Ben-Sasson SA. 1994. A complete Alu element within the coding sequence of a central gene. Cell 78: 173-174.

549. Marsh JJ, Lebherz HG. 1992. Fructose-bisphosphate aldolases: an evolutionary history. Trends Biochem. Sci. 17: 110-113.

550. Martins LM, Iaccarino I, Tenev T, Gschmeissner S, Totty NF, Lemoine NR, Savopoulos J, Gray CW, Creasy CL, Dingwall C, et al. 2002. The serine protease Omi/HtrA2 regulates apoptosis by binding XIAP through a reaper-like motif. J. Biol. Chem. 277: 439-444.

551. Mason PJ, Stevens D, Diez A, Knight SW, Scopes DA, Vulliamy TJ. 1999. Human hexose-6-phosphate dehydrogenase (glucose 1-dehydrogenase) encoded at 1p36: coding sequence and expression. Blood Cells Mol. Dis. 25: 30-37.

552. Matte-Tailliez O, Brochier C, Forterre P, Philippe H. 2002. Archaeal phylogeny based on ribosomal proteins. Mol Biol Evol 19: 631-639.

553. Maxam AM, Gilbert W. 1977. A new method for sequencing DNA. Proc. Natl. Acad. Sci. USA 74: 560-564.

554. May BJ, Zhang Q, Li LL, Paustian ML, Whittam TS, Kapur V. 2001. Complete genomic sequence of *Pasteurella multocida*, Pm70. Proc. Natl. Acad. Sci. USA 98: 3460-3465.

555. Maynard Smith J, Szathmary E. 1998. The Major Transitions in Evolution. Oxford University Press, Oxford.

556. Mazauric MH, Roy H, Kern D. 1999. tRNA glycylation system from *Thermus thermophilus*. $tRNA^{Gly}$ identity and functional interrelation with the glycylation systems from other phylae. Biochemistry 38: 13094-13105.

557. McClelland M, Sanderson KE, Spieth J, Clifton SW, Latreille P, Courtney L, Porwollik S, Ali J, Dante M, Du F, et al. 2001. Complete genome sequence of *Salmonella enterica* serovar *Typhimurium* LT2. Nature 413: 852-856.

558. McGeoch DJ, Davison AJ. 1986. DNA sequence of the herpes simplex virus type 1 gene encoding glycoprotein gH, and identification of homologues in the genomes of varicella-zoster virus and Epstein-Barr virus. Nucleic Acids Res. 14: 4281-4292.

559. McGuffin LJ, Jones DT. 2002. Targeting novel folds for structural genomics. Proteins 48: 44-52.

560. Medema JP. 1999. Apoptosis. Life and death in a FLASH. Nature 398: 756-757.

561. Medigue C, Rechenmann F, Danchin A, Viari A. 1999. Imagene: an integrated computer environment for sequence annotation and analysis. Bioinformatics 15: 2-15.

562. Medigue C, Rouxel T, Vigier P, Henaut A, Danchin A. 1991. Evidence for horizontal gene transfer in *Escherichia coli* speciation. J. Mol. Biol. 222: 851-856.

563. Mehl RA, Kinsland C, Begley TP. 2000. Identification of the *Escherichia coli* nicotinic acid mononucleotide adenylyltransferase gene. J. Bacteriol. 182: 4372-4374.

564. Meier P, Finch A, Evan G. 2000. Apoptosis in development. Nature 407: 796-801.

565. Meller J, Elber R. 2001. Linear programming optimization and a double statistical filter for protein threading protocols. Proteins 45: 241-261.

566. Messerschmidt A, Prade L, Wever R. 1997. Implications for the catalytic mechanism of the vanadium-containing enzyme chloroperoxidase from the fungus Curvularia inaequalis by X-ray structures of the native and peroxide form. Biol. Chem. 378: 309-315.

567. Meyer E, Kappock TJ, Osuji C, Stubbe J. 1999. Evidence for the direct transfer of the carboxylate of N5-carboxyaminoimidazole ribonucleotide (N5-CAIR) to generate 4-carboxy-5-aminoimidazole ribonucleotide catalyzed by *Escherichia coli* PurE, an N5-CAIR mutase. Biochemistry 38: 3012-3018.

568. Michels S, Scagliarini S, Della Seta F, Carles C, Riva M, Trost P, Branlant G. 1994. Arguments against a close relationship between non-phosphorylating and phosphorylating glyceraldehyde-3-phosphate dehydrogenases. FEBS Lett 339: 97-100.

569. Milanesi L, D'Angelo D, Rogozin IB. 1999. GeneBuilder: interactive in silico prediction of gene structure. Bioinformatics 15: 612-621.

570. Miller BG, Hassell AM, Wolfenden R, Milburn MV, Short SA. 2000. Anatomy of a proficient enzyme: the structure of orotidine 5'-monophosphate decarboxylase in the presence and absence of a potential transition state analog. Proc. Natl. Acad. Sci. USA 97: 2011-2016.

571. Miller W. 2001. Comparison of genomic DNA sequences: solved and unsolved problems. Bioinformatics 17: 391-397.

572. Minasov G, Teplova M, Stewart GC, Koonin EV, Anderson WF, Egli M. 2000. Functional implications from crystal structures of the conserved *Bacillus subtilis* protein Maf with and without dUTP. Proc. Natl. Acad. Sci. USA 97: 6328-6333.

573. Miramar MD, Costantini P, Ravagnan L, Saraiva LM, Haouzi D, Brothers G, Penninger JM, Peleato ML, Kroemer G, Susin SA. 2001. NADH oxidase activity of mitochondrial apoptosis-inducing factor. J. Biol. Chem. 276: 16391-16398.

574. Mirny L, Shakhnovich E. 2001. Protein folding theory: from lattice to all-atom models. Annu. Rev. Biophys. Biomol. Struct. 30: 361-396.

575. Mirny LA, Finkelstein AV, Shakhnovich EI. 2000. Statistical significance of protein structure prediction by threading. Proc. Natl. Acad. Sci. USA 97: 9978-9983.

576. Mironov AA, Fickett JW, Gelfand MS. 1999. Frequent alternative splicing of human genes. Genome Res. 9: 1288-1293.

577. Mitaku S, Ono M, Hirokawa T, Boon-Chieng S, Sonoyama M. 1999. Proportion of membrane proteins in proteomes of 15 single-cell organisms analyzed by the SOSUI prediction system. Biophys. Chem. 82: 165-171.

578. Mittenhuber G. 2001. Phylogenetic analyses and comparative genomics of vitamin B6 (pyridoxine) and pyridoxal phosphate biosynthesis pathways. J. Mol. Microbiol. Biotechnol. 3: 1-20.

579. Miyazaki K, Eguchi H, Yamagishi A, Wakagi T, Oshima T. 1992. Molecular cloning of the isocitrate dehydrogenase gene of an extreme thermophile, *Thermus thermophilus* HB8. Appl Environ Microbiol 58: 93-98.

580. Miyazawa S, Jernigan RL. 1999. An empirical energy potential with a reference state for protein fold and sequence recognition. Proteins 36: 357-369.

581. Moran NA, Mira A. 2001. The process of genome shrinkage in the obligate symbiont *Buchnera aphidicola*. Genome Biol 2: RESEARCH0054.

582. Mount SM. 2000. Genomic sequence, splicing, and gene annotation. Am. J. Hum. Genet. 67: 788-792.

583. Mueller EJ, Meyer E, Rudolph J, Davisson VJ, Stubbe J. 1994. N5-carboxyamino-imidazole ribonucleotide: evidence for a new intermediate and two new enzymatic activities in the de novo purine biosynthetic pathway of *Escherichia coli*. Biochemistry 33: 2269-2278.

584. Mukund S, Adams MW. 1995. Glyceraldehyde-3-phosphate ferredoxin oxidoreductase, a novel tungsten-containing enzyme with a potential glycolytic role in the hyper-thermophilic archaeon *Pyrococcus furiosus*. J. Biol. Chem. 270: 8389-8392.

585. Muller A, MacCallum RM, Sternberg MJ. 1999. Benchmarking PSI-BLAST in genome annotation. J. Mol. Biol. 293: 1257-1271.

586. Murakami KS, Masuda S, Campbell EA, Muzzin O, Darst SA. 2002. Structural basis of transcription initiation: an RNA polymerase holoenzyme-DNA complex. Science 296: 1285-1290.

587. Murakami KS, Masuda S, Darst SA. 2002. Structural basis of transcription initiation: RNA polymerase holoenzyme at 4 A resolution. Science 296: 1280-1284.

588. Murzin AG. 1996. Structural classification of proteins: new superfamilies. Curr. Opin. Struct. Biol. 6: 386-394.

589. Murzin AG. 2002. DNA building block reinvented. Science 297: 61-62.

590. Murzin AG, Brenner SE, Hubbard T, Chothia C. 1995. SCOP: a structural classification of proteins database for the investigation of sequences and structures. J. Mol. Biol. 247: 536-540.

591. Mushegian A. 1999. The minimal genome concept. Curr. Opin. Genet. Dev. 9: 709-714.

592. Mushegian A. 1999. The Purloined Letter: bacterial orthologs of archaeal NMN adenylyltransferase are domains within multifunctional transcription regulator NadR. J. Mol. Microbiol. Biotechnol. 1: 127-128.

593. Mushegian AR, Bassett DE, Jr., Boguski MS, Bork P, Koonin EV. 1997. Positionally cloned human disease genes: patterns of evolutionary conservation and functional motifs. Proc. Natl. Acad. Sci. USA 94: 5831-5836.

594. Mushegian AR, Fullner KJ, Koonin EV, Nester EW. 1996. A family of lysozyme-like virulence factors in bacterial pathogens of plants and animals. Proc. Natl. Acad. Sci. USA 93: 7321-7326.

595. Mushegian AR, Koonin EV. 1996. Gene order is not conserved in bacterial evolution. Trends Genet. 12: 289-290.

596. Mushegian AR, Koonin EV. 1996. A minimal gene set for cellular life derived by comparison of complete bacterial genomes. Proc. Natl. Acad. Sci. U S A 93: 10268-10273.

597. Mushegian AR, Koonin EV. 1996. Sequence analysis of eukaryotic developmental proteins: ancient and novel domains. Genetics 144: 817-828.

598. Myllykallio H, Lipowski G, Leduc D, Filee J, Forterre P, Liebl U. 2002. An alternative flavin-dependent mechanism for thymidylate synthesis. Science 297: 105-107.

599. Myllykallio H, Lopez P, Lopez-Garcia P, Heilig R, Saurin W, Zivanovic Y, Philippe H, Forterre P. 2000. Bacterial mode of replication with eukaryotic-like machinery in a hyperthermophilic archaeon. Science 288: 2212-2215.

600. Nakai K, Horton P. 1999. PSORT: a program for detecting sorting signals in proteins and predicting their subcellular localization. Trends Biochem Sci. 24: 34-36.

601. Nakai K, Kanehisa M. 1991. Expert system for predicting protein localization sites in gram-negative bacteria. Proteins 11: 95-110.

602. Nakai K, Kanehisa M. 1992. A knowledge base for predicting protein localization sites in eukaryotic cells. Genomics 14: 897-911.

603. Nakashita H, Kozuka K, Hidaka T, Hara O, Seto H. 2000. Identification and expression of the gene encoding phosphonopyruvate decarboxylase of *Streptomyces hygroscopicus*. Biochim. Biophys. Acta 1490: 159-162.

604. Nakashita H, Watanabe K, Hara O, Hidaka T, Seto H. 1997. Studies on the biosynthesis of bialaphos. Biochemical mechanism of C-P bond formation: discovery of phosphonopyruvate decarboxylase which catalyzes the formation of phosphonoacetaldehyde from phosphonopyruvate. J. Antibiot. (Tokyo) 50: 212-219.

605. Natale DA, Shankavaram UT, Galperin MY, Wolf YI, Aravind L, Koonin EV. 2000. Towards understanding the first genome sequence of a crenarchaeon by genome annotation using clusters of orthologous groups of proteins (COGs). Genome Biol 1: RESEARCH0009.

606. Needleman SB, Wunsch CD. 1970. A general method applicable to the search for similarities in the amino acid sequence of two proteins. J. Mol. Biol. 48: 443-453.

607. Nei M, Kumar S. 2000. Molecular Evolution and Phylogenetics. Oxford University Press, Oxford.

608. Neidhart DJ, Kenyon GL, Gerlt JA, Petsko GA. 1990. Mandelate racemase and muconate lactonizing enzyme are mechanistically distinct and structurally homologous. Nature 347: 692-694.

609. Nekrutenko A, Li WH. 2001. Transposable elements are found in a large number of human protein-coding genes. Trends Genet. 17: 619-621.

610. Nelson KE, Clayton RA, Gill SR, Gwinn ML, Dodson RJ, Haft DH, Hickey EK, Peterson JD, Nelson WC, Ketchum KA, et al. 1999. Evidence for lateral gene transfer between Archaea and bacteria from genome sequence of *Thermotoga maritima*. Nature 399: 323-329.

611. Nesbo CL, Boucher Y, Doolittle WF. 2001. Defining the core of nontransferable prokaryotic genes: the euryarchaeal core. J Mol Evol 53: 340-350.

612. Neuwald AF. 1997. An unexpected structural relationship between integral membrane phosphatases and soluble haloperoxidases. Protein Sci. 6: 1764-1767.

613. Neuwald AF, Aravind L, Spouge JL, Koonin EV. 1999. AAA+: A class of chaperone-like ATPases associated with the assembly, operation, and disassembly of protein complexes. Genome Res. 9: 27-43.

614. Neuwald AF, Liu JS, Lawrence CE. 1995. Gibbs motif sampling: detection of bacterial outer membrane protein repeats. Protein Sci. 4: 1618-1632.

615. Neuwald AF, Liu JS, Lipman DJ, Lawrence CE. 1997. Extracting protein alignment models from the sequence database. Nucleic Acids Res. 25: 1665-1677.

616. Ng WV, Kennedy SP, Mahairas GG, Berquist B, Pan M, Shukla HD, Lasky SR, Baliga NS, Thorsson V, Sbrogna J, et al. 2000. Genome sequence of *Halobacterium* species NRC-1. Proc. Natl. Acad. Sci. USA 97: 12176-12181.

617. Nielsen H, Engelbrecht J, Brunak S, von Heijne G. 1997. Identification of prokaryotic and eukaryotic signal peptides and prediction of their cleavage sites. Protein Eng. 10: 1-6.

618. Nierman WC, Feldblyum TV, Laub MT, Paulsen IT, Nelson KE, Eisen J, Heidelberg JF, Alley MR, Ohta N, Maddock JR, et al. 2001. Complete genome sequence of *Caulobacter crescentus*. Proc. Natl. Acad. Sci. USA 98: 4136-4141.

619. Nishimura K, Nakayashiki T, Inokuchi H. 1995. Cloning and identification of the *hemG* gene encoding protoporphyrinogen oxidase (PPO) of *Escherichia coli* K-12. DNA Res. 2: 1-8.

620. Nissen P, Hansen J, Ban N, Moore PB, Steitz TA. 2000. The structural basis of ribosome activity in peptide bond synthesis. Science 289: 920-930.

621. Nogales E, Downing KH, Amos LA, Lowe J. 1998. Tubulin and FtsZ form a distinct family of GTPases. Nat. Struct. Biol. 5: 451-458.

622. Nolling J, Breton G, Omelchenko MV, Makarova KS, Zeng Q, Gibson R, Lee HM, Dubois J, Qiu D, Hitti J, et al. 2001. Genome sequence and comparative analysis of the solvent-producing bacterium *Clostridium acetobutylicum*. J. Bacteriol. 183: 4823-4838.

623. Notredame C, Higgins DG, Heringa J. 2000. T-Coffee: A novel method for fast and accurate multiple sequence alignment. J. Mol. Biol. 302: 205-217.

624. Nowitzki U, Flechner A, Kellermann J, Hasegawa M, Schnarrenberger C, Martin W. 1998. Eubacterial origin of nuclear genes for chloroplast and cytosolic glucose-6-phosphate isomerase from spinach: sampling eubacterial gene diversity in eukaryotic chromosomes through symbiosis. Gene 214: 205-213.

625. Ochman H, Lawrence JG, Groisman EA. 2000. Lateral gene transfer and the nature of bacterial innovation. Nature 405: 299-304.

626. Ogata H, Audic S, Renesto-Audiffren P, Fournier PE, Barbe V, Samson D, Roux V, Cossart P, Weissenbach J, Claverie JM, et al. 2001. Mechanisms of evolution in *Rickettsia conorii* and *R. prowazekii*. Science 293: 2093-2098.

627. Ohno S. 1970. Evolution by gene duplication. Springer, New York.

628. Okajima T, Goto S, Tanizawa K, Tagaya M, Fukui T, Shimofuruya H, Suzuki J. 1995. Cloning, sequencing, and expression in *Escherichia coli* of cDNA encoding porcine brain UMP-CMP kinase. J. Biochem. (Tokyo) 117: 980-986.

629. Oliver SG, van der Aart QJ, Agostoni-Carbone ML, Aigle M, Alberghina L, Alexandraki D, Antoine G, Anwar R, Ballesta JP, Benit P, et al. 1992. The complete DNA sequence of yeast chromosome III. Nature 357: 38-46.

630. Olsen GJ, Woese CR, Overbeek R. 1994. The winds of (evolutionary) change: breathing new life into microbiology. J. Bacteriol. 176: 1-6.

631. Olsen TC, Eiken HG, Knappskog PM, Kase BF, Mansson JE, Boman H, Apold J. 1996. Mutations in the iduronate-2-sulfatase gene in five Norwegians with Hunter syndrome. Hum. Genet. 97: 198-203.

632. Oparin AI. 1953. The Origin of Life. Dover Pubns, New York.

633. Orengo CA, Michie AD, Jones S, Jones DT, Swindells MB, Thornton JM. 1997. CATH--a hierarchic classification of protein domain structures. Structure 5: 1093-1108.

634. Orgel LE, Crick FH. 1980. Selfish DNA: the ultimate parasite. Nature 284: 604-607.

635. Osmani AH, May GS, Osmani SA. 1999. The extremely conserved pyroA gene of *Aspergillus nidulans* is required for pyridoxine synthesis and is required indirectly for resistance to photosensitizers. J. Biol. Chem. 274: 23565-23569.

636. Ouali M, King RD. 2000. Cascaded multiple classifiers for secondary structure prediction. Protein Sci. 9: 1162-1176.

637. Oubrie A, Rozeboom HJ, Kalk KH, Olsthoorn AJ, Duine JA, Dijkstra BW. 1999. Structure and mechanism of soluble quinoprotein glucose dehydrogenase. EMBO J. 18: 5187-5194.

638. Ouzounis C, Casari G, Sander C, Tamames J, Valencia A. 1996. Computational comparisons of model genomes. Trends Biotechnol. 14: 280-285.

639. Ouzounis C, Casari G, Valencia A, Sander C. 1996. Novelties from the complete genome of *Mycoplasma genitalium*. Mol. Microbiol. 20: 898-900.

640. Overbeek R, Fonstein M, D'Souza M, Pusch GD, Maltsev N. 1998. The use of contiguity on the chromosome to predict functional coupling. In Silico Biol. 1: 93-108, http://www.bioinfo.de/isb/1998/1901/0009/.

641. Overbeek R, Fonstein M, D'Souza M, Pusch GD, Maltsev N. 1999. The use of gene clusters to infer functional coupling. Proc. Natl. Acad. Sci. USA 96: 2896-2901.

642. Overbeek R, Larsen N, Pusch GD, D'Souza M, Selkov E, Jr., Kyrpides N, Fonstein M, Maltsev N, Selkov E. 2000. WIT: integrated system for high-throughput genome sequence analysis and metabolic reconstruction. Nucleic Acids Res. 28: 123-125.

643. Overbeek R, Larsen N, Smith W, Maltsev N, Selkov E. 1997. Representation of function: the next step. Gene 191: GC1-GC9.

644. Pace NR. 1997. A molecular view of microbial diversity and the biosphere. Science 276: 734-740.

645. Page AP. 1999. A highly conserved nematode protein folding operon in *Caenorhabditis elegans* and *Caenorhabditis briggsae*. Gene 230: 267-275.

646. Page RD. 1996. TreeView: an application to display phylogenetic trees on personal computers. Comput. Appl. Biosci. 12: 357-358.

647. Page RD, Holmes E. 1998. Molecular Evolution : A Phylogenetic Approach. Blackwell Science, London.

648. Pallen M, Wren B, Parkhill J. 1999. 'Going wrong with confidence': misleading sequence analyses of CiaB and ClpX. Mol. Microbiol. 34: 195.

649. Panchenko A, Marchler-Bauer A, Bryant SH. 1999. Threading with explicit models for evolutionary conservation of structure and sequence. Proteins 37: 133-140.

650. Panchenko AR, Bryant SH. 2002. A comparison of position-specific score matrices based on sequence and structure alignments. Protein Sci. 11: 361-370.

651. Panchenko AR, Marchler-Bauer A, Bryant SH. 2000. Combination of threading potentials and sequence profiles improves fold recognition. J. Mol. Biol. 296: 1319-1331.

652. Park J, Karplus K, Barrett C, Hughey R, Haussler D, Hubbard T, Chothia C. 1998. Sequence comparisons using multiple sequences detect three times as many remote homologues as pairwise methods. J. Mol. Biol. 284: 1201-1210.

653. Parkhill J, Achtman M, James KD, Bentley SD, Churcher C, Klee SR, Morelli G, Basham D, Brown D, Chillingworth T, et al. 2000. Complete DNA sequence of a serogroup A strain of *Neisseria meningitidis* Z2491. Nature 404: 502-506.

654. Parkhill J, Dougan G, James KD, Thomson NR, Pickard D, Wain J, Churcher C, Mungall KL, Bentley SD, Holden MT, et al. 2001. Complete genome sequence of a multiple drug resistant *Salmonella enterica* serovar *Typhi* CT18. Nature 413: 848-852.

655. Parkhill J, Wren BW, Mungall K, Ketley JM, Churcher C, Basham D, Chillingworth T, Davies RM, Feltwell T, Holroyd S, et al. 2000. The genome sequence of the food-borne pathogen *Campylobacter jejuni* reveals hypervariable sequences. Nature 403: 665-668.

656. Parkhill J, Wren BW, Thomson NR, Titball RW, Holden MT, Prentice MB, Sebaihia M, James KD, Churcher C, Mungall KL, et al. 2001. Genome sequence of *Yersinia pestis*, the causative agent of plague. Nature 413: 523-527.

657. Parra G, Blanco E, Guigo R. 2000. GeneID in *Drosophila*. Genome Res. 10: 511-515.

658. Parrish J, Li L, Klotz K, Ledwich D, Wang X, Xue D. 2001. Mitochondrial endonuclease G is important for apoptosis in C. elegans. Nature 412: 90-94.

659. Patterson DJ. 1999. The Diversity of Eukaryotes. Am. Nat. 154: S96-S124.

660. Patthy L. 1999. Genome evolution and the evolution of exon-shuffling--a review. Gene 238: 103-114.

661. Pearl FM, Lee D, Bray JE, Sillitoe I, Todd AE, Harrison AP, Thornton JM, Orengo CA. 2000. Assigning genomic sequences to CATH. Nucleic Acids Res 28: 277-282.

662. Pearson WR. 1996. Effective protein sequence comparison. Methods Enzymol. 266: 227-258.

663. Pearson WR. 2000. Flexible sequence similarity searching with the FASTA3 program package. Methods Mol. Biol. 132: 185-219.

664. Pearson WR, Lipman DJ. 1988. Improved tools for biological sequence comparison. Proc. Natl. Acad. Sci. USA 85: 2444-2448.

665. Pellegrini M, Marcotte EM, Thompson MJ, Eisenberg D, Yeates TO. 1999. Assigning protein functions by comparative genome analysis: protein phylogenetic profiles. Proc. Natl. Acad. Sci. USA 96: 4285-4288.

666. Pennisi E. 1998. Genome data shake tree of life. Science 280: 672-674.

667. Pennisi E. 1999. Is it time to uproot the tree of life? Science 284: 1305-1307.

668. Perier RC, Praz V, Junier T, Bonnard C, Bucher P. 2000. The eukaryotic promoter database (EPD). Nucleic Acids Res. 28: 302-303.

669. Perna NT, Plunkett G, 3rd, Burland V, Mau B, Glasner JD, Rose DJ, Mayhew GF, Evans PS, Gregor J, Kirkpatrick HA, et al. 2001. Genome sequence of enterohaemorrhagic *Escherichia coli* O157:H7. Nature 409: 529-533.

670. Persson B, Argos P. 1994. Prediction of transmembrane segments in proteins utilising multiple sequence alignments. J. Mol. Biol. 237: 182-192.

671. Persson B, Argos P. 1997. Prediction of membrane protein topology utilizing multiple sequence alignments. J. Protein Chem 16: 453-457.

672. Pertea M, Lin X, Salzberg SL. 2001. GeneSplicer: a new computational method for splice site prediction. Nucleic Acids Res 29: 1185-1190.

673. Peterson JD, Umayam LA, Dickinson T, Hickey EK, White O. 2001. The Comprehensive Microbial Resource. Nucleic Acids Res. 29: 123-125.

674. Peterson SN, Fraser CM. 2001. The complexity of simplicity. Genome Biol. 2: COMMENT2002.

675. Ploux O, Soularue P, Marquet A, Gloeckler R, Lemoine Y. 1992. Investigation of the first step of biotin biosynthesis in *Bacillus sphaericus*. Purification and characterization of the pimeloyl-CoA synthase, and uptake of pimelate. Biochem. J. 287: 685-690.

676. Podobnik M, McInerney P, O'Donnell M, Kuriyan J. 2000. A TOPRIM domain in the crystal structure of the catalytic core of *Escherichia coli* primase confirms a structural link to DNA topoisomerases. J. Mol. Biol. 300: 353-362.

677. Pollastri G, Przybylski D, Rost B, Baldi P. 2002. Improving the prediction of protein secondary structure in three and eight classes using recurrent neural networks and profiles. Proteins 47: 228-235.

678. Ponting CP. 2001. Plagiarized bacterial genes in the human book of life. Trends Genet 17: 235-237.

679. Ponting CP, Aravind L, Schultz J, Bork P, Koonin EV. 1999. Eukaryotic signalling domain homologues in archaea and bacteria. Ancient ancestry and horizontal gene transfer. J. Mol. Biol. 289: 729-745.

680. Ponting CP, Cai YD, Bork P. 1997. The breast cancer gene product TSG101: a regulator of ubiquitination? J Mol Med 75: 467-469.

681. Poon KK, Chen CL, Wong SL. 2001. Roles of glucitol in the GutR-mediated transcription activation process in Bacillus subtilis: tight binding of GutR to tis binding site. J. Biol. Chem. 276: 9620-9625.

682. Pornillos O, Alam SL, Rich RL, Myszka DG, Davis DR, Sundquist WI. 2002. Structure and functional interactions of the Tsg101 UEV domain. EMBO J. 21: 2397-2406.

683. Purcarea C, Herve G, Cunin R, Evans DR. 2001. Cloning, expression, and structure analysis of carbamate kinase-like carbamoyl phosphate synthetase from Pyrococcus abyssi. Extremophiles 5: 229-239.

684. Purcarea C, Simon V, Prieur D, Herve G. 1996. Purification and characterization of carbamoylphosphate synthetase from the deep-sea hyperthermophilic archaebacterium *Pyrococcus abyssi*. Eur. J. Biochem. 236: 189-199.

685. Qian J, Luscombe NM, Gerstein M. 2001. Protein family and fold occurrence in genomes: power-law behaviour and evolutionary model. J. Mol. Biol. 313: 673-681.

686. Quinn CL, Stephenson BT, Switzer RL. 1991. Functional organization and nucleotide sequence of the *Bacillus subtilis* pyrimidine biosynthetic operon. J. Biol. Chem. 266: 9113-9127.

687. Ragan MA. 2001. Detection of lateral gene transfer among microbial genomes. Curr. Opin. Genet. Dev. 11: 620-626.

688. Ragan MA. 2001. On surrogate methods for detecting lateral gene transfer. FEMS Microbiol. Lett. 201: 187-191.

689. Ragan MA, Gaasterland T. 1998. Microbial genescapes: a prokaryotic view of the yeast genome. Microb Comp. Genomics 3: 219-235.

690. Raghava GP. 2000. Protein secondary structure prediction using nearest neighbor and neural network approach. In: CASP4. p. 75.

691. Rajasekaran S, Jin X, Spouge JL. 2002. The efficient computation of position-specific match scores with the fast fourier transform. J. Comput. Biol. 9: 23-33.

692. Ramon-Maiques S, Marina A, Uriarte M, Fita I, Rubio V. 2000. The 1.5 A resolution crystal structure of the carbamate kinase-like carbamoyl phosphate synthetase from the hyperthermophilic archaeon *Pyrococcus furiosus*, bound to ADP, confirms that this thermostable enzyme is a carbamate kinase, and provides insight into substrate binding and stability in carbamate kinases. J. Mol. Biol. 299: 463-476.

693. Rayl EA, Moroson BA, Beardsley GP. 1996. The human *purH* gene product, 5-aminoimidazole-4-carboxamide ribonucleotide formyltransferase/IMP cyclohydrolase. Cloning, sequencing, expression, purification, kinetic analysis, and domain mapping. J. Biol. Chem. 271: 2225-2233.

694. Read TD, Brunham RC, Shen C, Gill SR, Heidelberg JF, White O, Hickey EK, Peterson J, Utterback T, Berry K, et al. 2000. Genome sequences of *Chlamydia trachomatis* MoPn and *Chlamydia pneumoniae* AR39. Nucleic Acids Res. 28: 1397-1406.

695. Reeck GR, de Haen C, Teller DC, Doolittle RF, Fitch WM, Dickerson RE, Chambon P, McLachlan AD, Margoliash E, Jukes TH. 1987. "Homology" in proteins and nucleic acids: a terminology muddle and a way out of it. Cell 50: 667.

696. Reese MG, Eeckman FH, Kulp D, Haussler D. 1997. Improved splice site detection in Genie. J. Comput. Biol. 4: 311-323.

697. Reese MG, Hartzell G, Harris NL, Ohler U, Abril JF, Lewis SE. 2000. Genome annotation assessment in *Drosophila melanogaster*. Genome Res. 10: 483-501.

698. Reese MG, Kulp D, Tammana H, Haussler D. 2000. Genie--gene finding in *Drosophila melanogaster*. Genome Res. 10: 529-538.

699. Reichsman F, Moore HM, Cumberledge S. 1999. Sequence homology between Wingless/Wnt-1 and a lipid-binding domain in secreted phospholipase A2. Curr. Biol. 9: R353-R355.

700. Remm M, Storm CE, Sonnhammer EL. 2001. Automatic clustering of orthologs and in-paralogs from pairwise species comparisons. J. Mol. Biol. 314: 1041-1052.

701. Rich T, Allen RL, Wyllie AH. 2000. Defying death after DNA damage. Nature 407: 777-783.

702. Rigden DJ, Bagyan I, Lamani E, Setlow P, Jedrzejas MJ. 2001. A cofactor-dependent phosphoglycerate mutase homolog from *Bacillus* species is actually a broad specificity phosphatase. Protein Sci. 10: 1835-1846.

703. Rivera MC, Jain R, Moore JE, Lake JA. 1998. Genomic evidence for two functionally distinct gene classes. Proc. Natl. Acad. Sci. USA 95: 6239-6244.

704. Robb FT, Maeder DL, Brown JR, DiRuggiero J, Stump MD, Yeh RK, Weiss RB, Dunn DM. 2001. Genomic sequence of hyperthermophile, *Pyrococcus furiosus*: implications for physiology and enzymology. Methods Enzymol. 330: 134-157.

705. Robertson HD. 1996. How did replicating and coding RNAs first get together? Science 274: 66-67.

706. Roelofs J, Van Haastert PJ. 2001. Genes lost during evolution. Nature 411: 1013-1014.

707. Roger AJ. 1999. Reconstructing early events in eukaryotic evolution. Amer. Natur. 154: S146-S163.

708. Rogozin IB, D'Angelo D, Milanesi L. 1999. Protein-coding regions prediction combining similarity searches and conservative evolutionary properties of protein-coding sequences. Gene 226: 129-137.

709. Rogozin IB, Makarova KS, Murvai J, Czabarka E, Wolf YI, Tatusov RL, Szekely LA, Koonin EV. 2002. Connected gene neighborhoods in prokaryotic genomes. Nucleic Acids Res. 30: 2212-2223.

710. Rogozin IB, Milanesi L. 1997. Analysis of donor splice sites in different eukaryotic organisms. J. Mol. Evol. 45: 50-59.

711. Romano AH, Conway T. 1996. Evolution of carbohydrate metabolic pathways. Res. Microbiol. 147: 448-455.

712. Ronimus RS, de Heus E, Morgan HW. 2001. Sequencing, expression, characterisation and phylogeny of the ADP-dependent phosphofructokinase from the hyperthermophilic, euryarchaeal *Thermococcus zilligii*. Biochim. Biophys. Acta. 1517: 384-391.

713. Rosenthal B, Mai Z, Caplivski D, Ghosh S, de la Vega H, Graf T, Samuelson J. 1997. Evidence for the bacterial origin of genes encoding fermentation enzymes of the amitochondriate protozoan parasite *Entamoeba histolytica*. J. Bacteriol. 179: 3736-3745.

714. Ross-Macdonald P, Coelho PS, Roemer T, Agarwal S, Kumar A, Jansen R, Cheung KH, Sheehan A, Symoniatis D, Umansky L, et al. 1999. Large-scale analysis of the yeast genome by transposon tagging and gene disruption. Nature 402: 413-418.

715. Rost B, Casadio R, Fariselli P, Sander C. 1995. Transmembrane helices predicted at 95% accuracy. Protein Sci. 4: 521-533.

716. Rost B, Fariselli P, Casadio R. 1996. Topology prediction for helical transmembrane proteins at 86% accuracy. Protein Sci. 5: 1704-1718.

717. Rost B, Sander C. 1993. Improved prediction of protein secondary structure by use of sequence profiles and neural networks. Proc. Natl. Acad. Sci. USA 90: 7558-7562.

718. Rost B, Sander C. 1993. Prediction of protein secondary structure at better than 70% accuracy. J. Mol. Biol. 232: 584-599.

719. Ruepp A, Graml W, Santos-Martinez ML, Koretke KK, Volker C, Mewes HW, Frishman D, Stocker S, Lupas AN, Baumeister W. 2000. The genome sequence of the thermoacidophilic scavenger *Thermoplasma acidophilum*. Nature 407: 508-513.

720. Rujan T, Martin W. 2001. How many genes in *Arabidopsis* come from cyanobacteria? An estimate from 386 protein phylogenies. Trends Genet. 17: 113-120.

721. Ruland J, Sirard C, Elia A, MacPherson D, Wakeham A, Li L, de la Pompa JL, Cohen SN, Mak TW. 2001. p53 accumulation, defective cell proliferation, and early embryonic lethality in mice lacking TSG101. Proc. Natl. Acad. Sci. USA 98: 1859-1864.

722. Russell RB. 1998. Detection of protein three-dimensional side-chain patterns: new examples of convergent evolution. J. Mol. Biol. 279: 1211-1227.

723. Russell RB, Marquez JA, Hengstenberg W, Scheffzek K. 2002. Evolutionary relationship between the bacterial HPr kinase and the ubiquitous PEP-carboxykinase: expanding the P-loop nucleotidyl transferase superfamily. FEBS Lett. 517: 1-6.

724. Russell RB, Sasieni PD, Sternberg MJ. 1998. Supersites within superfolds. Binding site similarity in the absence of homology. J. Mol. Biol. 282: 903-918.

725. Rychlewski L, Jaroszewski L, Li W, Godzik A. 2000. Comparison of sequence profiles. Strategies for structural predictions using sequence information. Protein Sci. 9: 232-241.

726. Rzhetsky A, Gomez SM. 2001. Birth of scale-free molecular networks and the number of distinct DNA and protein domains per genome. Bioinformatics 17: 988-996.

727. Sagan L. 1967. On the origin of mitosing cells. J. Theor. Biol. 14: 255-274.

728. Salamov AA, Solovyev VV. 1995. Prediction of protein secondary structure by combining nearest-neighbor algorithms and multiple sequence alignments. J. Mol. Biol. 247: 11-15.

729. Salamov AA, Solovyev VV. 1997. Protein secondary structure prediction using local alignments. J. Mol. Biol. 268: 31-36.

730. Salamov AA, Solovyev VV. 2000. *Ab initio* gene finding in *Drosophila* genomic DNA. Genome Res. 10: 516-522.

731. Salanoubat M, Genin S, Artiguenave F, Gouzy J, Mangenot S, Arlat M, Billault A, Brottier P, Camus JC, Cattolico L, et al. 2002. Genome sequence of the plant pathogen *Ralstonia solanacearum*. Nature 415: 497-502.

732. Salgado H, Moreno-Hagelsieb G, Smith TF, Collado-Vides J. 2000. Operons in *Escherichia coli*: genomic analyses and predictions. Proc. Natl. Acad. Sci. USA 97: 6652-6657.

733. Sali A. 1998. 100,000 protein structures for the biologist. Nat. Struct. Biol. 5: 1029-1032.

734. Salwinski L, Eisenberg D. 2001. Motif-based fold assignment. Protein Sci. 10: 2460-2469.

735. Salzberg SL, Delcher AL, Kasif S, White O. 1998. Microbial gene identification using interpolated Markov models. Nucleic Acids Res. 26: 544-548.

736. Salzberg SL, Pertea M, Delcher AL, Gardner MJ, Tettelin H. 1999. Interpolated Markov models for eukaryotic gene finding. Genomics 59: 24-31.

737. Salzberg SL, White O, Peterson J, Eisen JA. 2001. Microbial genes in the human genome: lateral transfer or gene loss? Science 292: 1903-1906.

738. Sanchez LB, Galperin MY, Muller M. 2000. Acetyl-CoA synthetase from the amitochondriate eukaryote *Giardia lamblia* belongs to the newly recognized superfamily of acyl-CoA synthetases (Nucleoside diphosphate-forming). J. Biol. Chem. 275: 5794-5803.

739. Sanchez LB, Muller M. 1998. Cloning and heterologous expression of *Entamoeba histolytica* adenylate kinase and uridylate/cytidylate kinase. Gene 209: 219-228.

740. Sandigursky M, Franklin WA. 1999. Thermostable uracil-DNA glycosylase from Thermotoga maritima a member of a novel class of DNA repair enzymes. Curr. Biol. 9: 531-534.

741. Sandigursky M, Franklin WA. 2000. Uracil-DNA glycosylase in the extreme thermophile *Archaeoglobus fulgidus*. J. Biol. Chem. 275: 19146-19149.

742. Sanger F, Coulson AR, Hong GF, Hill DF, Petersen GB. 1982. Nucleotide sequence of bacteriophage lambda DNA. J. Mol. Biol. 162: 729-773.

743. Sanger F, Nicklen S, Coulson AR. 1977. DNA sequencing with chain-terminating inhibitors. Proc. Natl. Acad. Sci. USA 74: 5463-5467.

744. Saraste M, Sibbald PR, Wittinghofer A. 1990. The P-loop -- a common motif in ATP- and GTP-binding proteins. Trends Biochem. Sci. 15: 430-434.

745. Sasarman A, Echelard Y, Letowski J, Tardif D, Drolet M. 1988. Nucleotide sequence of the *hemX* gene, the third member of the *uro* operon of *Escherichia coli* K12. Nucleic Acids Res. 16: 11835.

746. Sasarman A, Letowski J, Czaika G, Ramirez V, Nead MA, Jacobs JM, Morais R. 1993. Nucleotide sequence of the *hemG* gene involved in the protoporphyrinogen oxidase activity of *Escherichia coli* K12. Can. J. Microbiol. 39: 1155-1161.

747. Saxild HH, Nygaard P. 2000. The *yexA* gene product is required for phosphoribosyl-formylglycinamidine synthetase activity in *Bacillus subtilis*. Microbiology 146: 807-814.

748. Schaffer AA, Aravind L, Madden TL, Shavirin S, Spouge JL, Wolf YI, Koonin EV, Altschul SF. 2001. Improving the accuracy of PSI-BLAST protein database searches with composition-based statistics and other refinements. Nucleic Acids Res. 29: 2994-3005.

749. Scharf M, Schneider R, Casari G, Bork P, Valencia A, Ouzounis C, Sander C. 1994. GeneQuiz: a workbench for sequence analysis. Intel. Syst. Mol. Biol. 2: 348-353.

750. Schmalhausen II, Dordick I, Dobzhansky T. 1987. Factors in Evolution. University of Chicago Press, Chicago.

751. Schmidt DM, Hubbard BK, Gerlt JA. 2001. Evolution of enzymatic activities in the enolase superfamily: functional assignment of unknown proteins in *Bacillus subtilis* and *Escherichia coli* as L-Ala-D/L-Glu epimerases. Biochemistry 40: 15707-15715.

752. Schneider TD, Stephens RM. 1990. Sequence logos: a new way to display consensus sequences. Nucleic Acids Res. 18: 6097-6100.

753. Schopf JW. 1993. Microfossils of the early archean apex chert: new evidence of the antiquity of life. Science 260: 640-646.

754. Schopf JW, Packer BM. 1987. Early Archean (3.3-billion to 3.5-billion-year-old) microfossils from Warrawoona Group, Australia. Science 237: 70-73.

755. Schuldt AJ, Brand AH. 1999. Mastermind acts downstream of notch to specify neuronal cell fates in the *Drosophila* central nervous system. Dev. Biol. 205: 287-295.

756. Schuler GD, Altschul SF, Lipman DJ. 1991. A workbench for multiple alignment construction and analysis. Proteins 9: 180-190.

757. Schultz J, Milpetz F, Bork P, Ponting CP. 1998. SMART, a simple modular architecture research tool: identification of signaling domains. Proc. Natl. Acad. Sci. USA 95: 5857-5864.

758. Schurmann M, Sprenger GA. 2001. Fructose-6-phosphate aldolase is a novel class I aldolase from *Escherichia coli* and is related to a novel group of bacterial transaldolases. J. Biol. Chem. 276: 11055-11061.

759. Schwartz S, Zhang Z, Frazer KA, Smit A, Riemer C, Bouck J, Gibbs R, Hardison R, Miller W. 2000. PipMaker -- a web server for aligning two genomic DNA sequences. Genome Res. 10: 577-586.

760. Scott-Ram NR. 1990. Transformed Cladistics, Taxonomy and Evolution. Cambridge University Press, Cambridge, MA.

761. Selkov E, Maltsev N, Olsen GJ, Overbeek R, Whitman WB. 1997. A reconstruction of the metabolism of *Methanococcus jannaschii* from sequence data. Gene 197: GC11-GC26.

762. Shabalina SA, Kondrashov AS. 1999. Pattern of selective constraint in *C. elegans* and *C. briggsae* genomes. Genet. Res. 74: 23-30.

763. Shabalina SA, Ogurtsov AY, Kondrashov VA, Kondrashov AS. 2001. Selective constraint in intergenic regions of human and mouse genomes. Trends Genet. 17: 373-376.

764. She Q, Singh RK, Confalonieri F, Zivanovic Y, Allard G, Awayez MJ, Chan-Weiher CC, Clausen IG, Curtis BA, De Moors A, et al. 2001. The complete genome of the crenarchaeon *Sulfolobus solfataricus* P2. Proc. Natl. Acad. Sci. USA 98: 7835-7840.

765. Shi J, Blundell TL, Mizuguchi K. 2001. FUGUE: sequence-structure homology recognition using environment-specific substitution tables and structure-dependent gap penalties. J. Mol. Biol. 310: 243-257.

766. Shigenobu S, Watanabe H, Hattori M, Sakaki Y, Ishikawa H. 2000. Genome sequence of the endocellular bacterial symbiont of aphids *Buchnera* sp. APS. Nature 407: 81-86.

767. Shimizu T, Ohtani K, Hirakawa H, Ohshima K, Yamashita A, Shiba T, Ogasawara N, Hattori M, Kuhara S, Hayashi H. 2002. Complete genome sequence of *Clostridium perfringens*, an anaerobic flesh-eater. Proc. Natl. Acad. Sci. USA 99: 996-1001.

768. Shmatkov AM, Melikyan AA, Chernousko FL, Borodovsky M. 1999. Finding prokaryotic genes by the 'frame-by-frame' algorithm: targeting gene starts and overlapping genes. Bioinformatics 15: 874-886.

769. Sicheritz-Ponten T, Andersson SG. 2001. A phylogenomic approach to microbial evolution. Nucleic Acids Res. 29: 545-552.

770. Siebers B, Brinkmann H, Dorr C, Tjaden B, Lilie H, van Der Oost J, Verhees CH. 2001. Archaeal fructose-1,6-bisphosphate aldolases constitute a new family of archaeal type class I aldolase. J. Biol. Chem. 276: 28710-28718.

771. Simmer JP, Kelly RE, Rinker AG, Jr., Zimmermann BH, Scully JL, Kim H, Evans DR. 1990. Mammalian dihydroorotase: nucleotide sequence, peptide sequences, and evolution of the dihydroorotase domain of the multifunctional protein CAD. Proc. Natl. Acad. Sci. USA 87: 174-178.

772. Simpson AJ, Reinach FC, Arruda P, Abreu FA, Acencio M, Alvarenga R, Alves LM, Araya JE, Baia GS, Baptista CS, et al. 2000. The genome sequence of the plant pathogen *Xylella fastidiosa*. Nature 406: 151-157.

773. Singer TP, Johnson MK. 1985. The prosthetic groups of succinate dehydrogenase: 30 years from discovery to identification. FEBS Lett. 190: 189-198.

774. Singleton MR, Hakansson K, Timson DJ, Wigley DB. 1999. Structure of the adenylation domain of an NAD+-dependent DNA ligase. Structure Fold. Des. 7: 35-42.

775. Sippl MJ, Flockner H. 1996. Threading thrills and threats. Structure 4: 15-19.

776. Skaar EP, Tobiason DM, Quick J, Judd RC, Weissbach H, Etienne F, Brot N, Seifert HS. 2002. The outer membrane localization of the *Neisseria gonorrhoeae* MsrA/B is involved in survival against reactive oxygen species. Proc. Natl. Acad. Sci. USA.

777. Skovgaard M, Jensen LJ, Brunak S, Ussery D, Krogh A. 2001. On the total number of genes and their length distribution in complete microbial genomes. Trends Genet 17: 425-428.

778. Skulachev VP. 2001. The programmed death phenomena, aging, and the Samurai law of biology. Exp. Gerontol. 36: 995-1024.

779. Slesarev AI, Mezhevaya KV, Makarova KS, Polushin NN, Shcherbinina OV, Shakhova VV, Belova GI, Aravind L, Natale DA, Rogozin IB, et al. 2002. The complete genome of hyperthermophile *Methanopyrus kandleri* AV19 and monophyly of archaeal methanogens. Proc. Natl. Acad. Sci. USA 99: 4644-4649.

780. Smit A, Mushegian A. 2000. Biosynthesis of isoprenoids via mevalonate in Archaea: the lost pathway. Genome Res. 10: 1468-1484.

781. Smith DR, Doucette-Stamm LA, Deloughery C, Lee H, Dubois J, Aldredge T, Bashirzadeh R, Blakely D, Cook R, Gilbert K, et al. 1997. Complete genome sequence of *Methanobacterium thermoautotrophicum* deltaH: functional analysis and comparative genomics. J. Bacteriol. 179: 7135-7155.

782. Smith MW, Feng DF, Doolittle RF. 1992. Evolution by acquisition: the case for horizontal gene transfers. Trends Biochem. Sci. 17: 489-493.

783. Smith TF, Lo Conte L, Bienkowska J, Gaitatzes C, Rogers RG, Jr., Lathrop R. 1997. Current limitations to protein threading approaches. J. Comput. Biol. 4: 217-225.

784. Smith TF, Waterman MS. 1981. Identification of common molecular subsequences. J. Mol. Biol. 147: 195-197.

785. Smoller D, Friedel C, Schmid A, Bettler D, Lam L, Yedvobnick B. 1990. The *Drosophila* neurogenic locus mastermind encodes a nuclear protein unusually rich in amino acid homopolymers. Genes Dev. 4: 1688-1700.

786. Snel B, Bork P, Huynen MA. 1999. Genome phylogeny based on gene content. Nat. Genet. 21: 108-110.

787. Snel B, Bork P, Huynen MA. 2002. Genomes in flux: the evolution of archaeal and proteobacterial gene content. Genome Res. 12: 17-25.

788. Snel B, Lehmann G, Bork P, Huynen MA. 2000. STRING: a web-server to retrieve and display the repeatedly occurring neighbourhood of a gene. Nucleic Acids Res. 28: 3442-3444.

789. Snyder EE, Stormo GD. 1995. Identification of protein coding regions in genomic DNA. J. Mol. Biol. 248: 1-18.

790. Sogin ML. 1991. Early evolution and the origin of eukaryotes. Curr. Opin. Genet. Dev. 1: 457-463.

791. Solovyev VV, Salamov AA. 1994. Predicting alpha-helix and beta-strand segments of globular proteins. Comput. Appl. Biosci. 10: 661-669.

792. Solovyev VV, Salamov AA, Lawrence CB. 1994. Predicting internal exons by oligonucleotide composition and discriminant analysis of spliceable open reading frames. Nucleic Acids Res. 22: 5156-5163.

793. Sorensen KI, Hove-Jensen B. 1996. Ribose catabolism of *Escherichia coli*: characterization of the *rpiB* gene encoding ribose phosphate isomerase B and of the *rpiR* gene, which is involved in regulation of *rpiB* expression. J. Bacteriol. 178: 1003-1011.

794. Souciet JL, Nagy M, Le Gouar M, Lacroute F, Potier S. 1989. Organization of the yeast URA2 gene: identification of a defective dihydroorotase-like domain in the multifunctional carbamoylphosphate synthetase-aspartate transcarbamylase complex. Gene 79: 59-70.

795. Spath C, Kraus A, Hillen W. 1997. Contribution of glucose kinase to glucose repression of xylose utilization in *Bacillus megaterium.* J. Bacteriol. 179: 7603-7605.

796. Sreekumar KR, Aravind L, Koonin EV. 2001. Computational analysis of human disease-associated genes and their protein products. Curr. Opin. Genet. Dev. 11: 247-257.

797. Staden R. 1977. Sequence data handling by computer. Nucleic Acids Res. 4: 4037-4051.

798. Stalon V, Vander Wauven C, Momin P, Legrain C. 1987. Catabolism of arginine, citrulline and ornithine by *Pseudomonas* and related bacteria. J. Gen. Microbiol. 133: 2487-2495.

799. Stanhope MJ, Lupas A, Italia MJ, Koretke KK, Volker C, Brown JR. 2001. Phylogenetic analyses do not support horizontal gene transfers from bacteria to vertebrates. Nature 411: 940-944.

800. Stapleton MA, Javid-Majd F, Harmon MF, Hanks BA, Grahmann JL, Mullins LS, Raushel FM. 1996. Role of conserved residues within the carboxy phosphate domain of carbamoyl phosphate synthetase. Biochemistry 35: 14352-14361.

801. Stec B, Yang H, Johnson KA, Chen L, Roberts MF. 2000. MJ0109 is an enzyme that is both an inositol monophosphatase and the 'missing' archaeal fructose-1,6-bisphosphatase. Nat. Struct. Biol. 7: 1046-1050.

802. Steeg PS. 1996. Granin expectations in breast cancer? Nat. Genet. 12: 223-225.

803. Steen IH, Madsen MS, Birkeland NK, Lien T. 1998. Purification and characterization of a monomeric isocitrate dehydrogenase from the sulfate-reducing bacterium *Desulfobacter vibrioformis* and demonstration of the presence of a monomeric enzyme in other bacteria. FEMS Microbiol Lett 160: 75-79.

804. Stein L, Sternberg P, Durbin R, Thierry-Mieg J, Spieth J. 2001. WormBase: network access to the genome and biology of *Caenorhabditis elegans*. Nucleic Acids Res. 29: 82-86.

805. Stephens RS, Kalman S, Lammel C, Fan J, Marathe R, Aravind L, Mitchell W, Olinger L, Tatusov RL, Zhao Q, et al. 1998. Genome sequence of an obligate intracellular pathogen of humans: *Chlamydia trachomatis*. Science 282: 754-759.

806. Stewart CB, Schilling JW, Wilson AC. 1987. Adaptive evolution in the stomach lysozymes of foregut fermenters. Nature 330: 401-404.

807. Stover CK, Pham X-QT, Erwin AL, Mizoguchi SD, Warrener P, Hickey MJ, Brinkman FSL, Hufnagle WO, Kowalik DJ, Lagrou M, et al. 2000. Complete genome sequence of *Pseudomonas aeruginosa* PA01, an opportunistic pathogen. Nature 406: 959-964.

808. Strasser A, O'Connor L, Dixit VM. 2000. Apoptosis signaling. Annu. Rev. Biochem. 69: 217-245.

809. Stratton MR. 1996. Recent advances in understanding of genetic susceptibility to breast cancer. Hum. Mol. Genet. 5: 1515-1519.

810. Stukey J, Carman GM. 1997. Identification of a novel phosphatase sequence motif. Protein Sci. 6: 469-472.

811. Stultz CM, White JV, Smith TF. 1993. Structural analysis based on state-space modeling. Protein Sci. 2: 305-314.

812. Sunyaev S, Kuznetsov E, Rodchenkov I, Tumanyan V. 1997. Protein sequence-structure compatibility criteria in terms of statistical hypothesis testing. Protein Eng. 10: 635-646.

813. Suyama M, Bork P. 2001. Evolution of prokaryotic gene order: genome rearrangements in closely related species. Trends Genet. 17: 10-13.

814. Suzuki M, Sahara T, Tsuruha J, Takada Y, Fukunaga N. 1995. Differential expression in *Escherichia coli* of the *Vibrio* sp. strain ABE-1 *icdI* and *icdII* genes encoding structurally different isocitrate dehydrogenase isozymes. J. Bacteriol. 177: 2138-2142.

815. Suzuki Y, Imai Y, Nakayama H, Takahashi K, Takio K, Takahashi R. 2001. A serine protease, HtrA2, is released from the mitochondria and interacts with XIAP, inducing cell death. Mol Cell 8: 613-621.

816. Swanson KW, Irwin DM, Wilson AC. 1991. Stomach lysozyme gene of the langur monkey: tests for convergence and positive selection. J. Mol. Evol. 33: 418-425.

817. Szabo CI, Wagner LA, Francisco LV, Roach JC, Argonza R, King MC, Ostrander EA. 1996. Human, canine and murine BRCA1 genes: sequence comparison among species. Hum. Mol. Genet. 5: 1289-1298.

818. Szallies A, Kubata BK, Duszenko M. 2002. A metacaspase of *Trypanosoma brucei* causes loss of respiration competence and clonal death in the yeast *Saccharomyces cerevisiae*. FEBS Lett 517: 144-150.

819. Takahashi N, Yamada T. 1999. Glucose and lactate metabolism by *Actinomyces naeslundii*. Crit. Rev. Oral. Biol. Med. 10: 487-503.

820. Takami H, Nakasone K, Takaki Y, Maeno G, Sasaki R, Masui N, Fuji F, Hirama C, Nakamura Y, Ogasawara N, et al. 2000. Complete genome sequence of the alkaliphilic bacterium *Bacillus halodurans* and genomic sequence comparison with *Bacillus subtilis*. Nucleic Acids Res. 28: 4317-4331.

821. Takayama S, McGarvey GJ, Wong CH. 1997. Microbial aldolases and transketolases: new biocatalytic approaches to simple and complex sugars. Annu. Rev. Microbiol. 51: 285-310.

822. Tamames J, Casari G, Ouzounis C, Valencia A. 1997. Conserved clusters of functionally related genes in two bacterial genomes. J. Mol. Evol. 44: 66-73.

823. Tamoi M, Ishikawa T, Takeda T, Shigeoka S. 1996. Molecular characterization and resistance to hydrogen peroxide of two fructose-1,6-bisphosphatases from *Synechococcus* PCC 7942. Arch. Biochem. Biophys. 334: 27-36.

824. Tamoi M, Murakami A, Takeda T, Shigeoka S. 1998. Acquisition of a new type of fructose-1,6-bisphosphatase with resistance to hydrogen peroxide in cyanobacteria: molecular characterization of the enzyme from *Synechocystis* PCC 6803. Biochim. Biophys. Acta 1383: 232-244.

825. Tanaka K, Tazuya K, Yamada K, Kumaoka H. 2000. Biosynthesis of pyridoxine: origin of the nitrogen atom of pyridoxine in microorganisms. J. Nutr Sci. Vitaminol. (Tokyo) 46: 55-57.

826. Tatusov RL, Altschul SF, Koonin EV. 1994. Detection of conserved segments in proteins: iterative scanning of sequence databases with alignment blocks. Proc. Natl. Acad. Sci. USA 91: 12091-12095.

827. Tatusov RL, Galperin MY, Natale DA, Koonin EV. 2000. The COG database: a tool for genome-scale analysis of protein functions and evolution. Nucleic Acids Res. 28: 33-36.

828. Tatusov RL, Koonin EV, Lipman DJ. 1997. A genomic perspective on protein families. Science 278: 631-637.

829. Tatusov RL, Mushegian AR, Bork P, Brown NP, Hayes WS, Borodovsky M, Rudd KE, Koonin EV. 1996. Metabolism and evolution of *Haemophilus influenzae* deduced from a whole-genome comparison with *Escherichia coli*. Curr. Biol. 6: 279-291.

830. Tatusova TA, Madden TL. 1999. BLAST 2 Sequences, a new tool for comparing protein and nucleotide sequences. FEMS Microbiol. Lett. 174: 247-250.

831. Tauch A, Homann I, Mormann S, Ruberg S, Billault A, Bathe B, Brand S, Brockmann-Gretza O, Ruckert C, Schischka N, et al. 2002. Strategy to sequence the genome of *Corynebacterium glutamicum* ATCC 13032: use of a cosmid and a bacterial artificial chromosome library. J. Biotechnol. 95: 25-38.

832. Tazoe M, Ichikawa K, Hoshino T. 2002. Biosynthesis of vitamin B6 in *Rhizobium*: in vitro synthesis of pyridoxine from 1-deoxy-D-xylulose and 4-hydroxy-L-threonine. Biosci. Biotechnol. Biochem. 66: 934-936.

833. Teichmann SA, Chothia C, Gerstein M. 1999. Advances in structural genomics. Curr. Opin. Struct. Biol. 9: 390-399.

834. Teichmann SA, Mitchison G. 1999. Is there a phylogenetic signal in prokaryote proteins? J. Mol. Evol. 49: 98-107.

835. Tekaia F, Lazcano A, Dujon B. 1999. The genomic tree as revealed from whole proteome comparisons. Genome Res. 9: 550-557.

836. Tettelin H, Nelson KE, Paulsen IT, Eisen JA, Read TD, Peterson S, Heidelberg J, DeBoy RT, Haft DH, Dodson RJ, et al. 2001. Complete genome sequence of a virulent isolate of *Streptococcus pneumoniae*. Science 293: 498-506.

837. Tettelin H, Saunders NJ, Heidelberg J, Jeffries AC, Nelson KE, Eisen JA, Ketchum KA, Hood DW, Peden JF, Dodson RJ, et al. 2000. Complete genome sequence of *Neisseria meningitidis* serogroup B strain MC58. Science 287: 1809-1815.

838. Thanaraj TA, Robinson AJ. 2000. Prediction of exact boundaries of exons. Brief. Bioinform. 1: 343-356.

839. Thanassi JA, Hartman-Neumann SL, Dougherty TJ, Dougherty BA, Pucci MJ. 2002. Identification of 113 conserved essential genes using a high-throughput gene disruption system in *Streptococcus pneumoniae*. Nucleic Acids Res. 30: 3152-3162.

840. The *C. elegans* Sequencing Consortium. 1998. Genome sequence of the nematode *C. elegans*: a platform for investigating biology. Science 282: 2012-2018.

841. Thoden JB, Holden HM, Wesenberg G, Raushel FM, Rayment I. 1997. Structure of carbamoyl phosphate synthetase: a journey of 96 A from substrate to product. Biochemistry 36: 6305-6316.

842. Thompson H, Tersteegen A, Thauer RK, Hedderich R. 1998. Two malate dehydrogenases in *Methanobacterium thermoautotrophicum*. Arch Microbiol 170: 38-42.

843. Thompson JD, Higgins DG, Gibson TJ. 1994. CLUSTAL W: improving the sensitivity of progressive multiple sequence alignment through sequence weighting, position-specific gap penalties and weight matrix choice. Nucleic Acids Res. 22: 4673-4680.

844. Thomson GJ, Howlett GJ, Ashcroft AE, Berry A. 1998. The *dhnA* gene of *Escherichia coli* encodes a class I fructose bisphosphate aldolase. Biochem. J. 331: 437-445.

845. Thornberry NA, Lazebnik Y. 1998. Caspases: enemies within. Science 281: 1312-1316.

846. Todd AE, Orengo CA, Thornton JM. 2001. Evolution of function in protein superfamilies, from a structural perspective. J. Mol. Biol. 307: 1113-1143.

847. Toh H, Hayashida H, Miyata T. 1983. Sequence homology between retroviral reverse transcriptase and putative polymerases of hepatitis B virus and cauliflower mosaic virus. Nature 305: 827-829.

848. Tomb J-F, White O, Kerlavage AR, Clayton RA, Sutton GG, Fleishmann RF, Ketchum KA, Klenk HP, Gill S, Dougherty BA, et al. 1997. The complete genome sequence of the gastric pathogen *Helicobacter pylori*. Nature 388: 539-547.

849. Trach K, Hoch JA. 1989. The *Bacillus subtilis* spo0B stage 0 sporulation operon encodes an essential GTP-binding protein. J. Bacteriol. 171: 1362-1371.

850. Trovato M, Maras B, Linhares F, Costantino P. 2001. The plant oncogene *rolD* encodes a functional ornithine cyclodeaminase. Proc. Natl. Acad. Sci. USA 98: 13449-13453.

851. Trzcinska-Danielewicz J, Fronk J. 2000. Exon-intron organization of genes in the slime mold *Physarum polycephalum*. Nucleic Acids Res. 28: 3411-3416.

852. Tugendreich S, Feng Q, Kroll J, Sears DD, Boeke JD, Hieter P. 1994. Alu sequences in RMSA-1 protein? Nature 370: 106.

853. Tuininga JE, Verhees CH, van der Oost J, Kengen SW, Stams AJ, de Vos WM. 1999. Molecular and biochemical characterization of the ADP-dependent phosphofructokinase from the hyperthermophilic archaeon *Pyrococcus furiosus*. J. Biol. Chem. 274: 21023-21028.

854. Tumbula D, Vothknecht UC, Kim HS, Ibba M, Min B, Li T, Pelaschier J, Stathopoulos C, Becker H, Soll D. 1999. Archaeal aminoacyl-tRNA synthesis: diversity replaces dogma. Genetics 152: 1269-1276.

855. Tumbula DL, Becker HD, Chang WZ, Soll D. 2000. Domain-specific recruitment of amide amino acids for protein synthesis. Nature 407: 106-110.

856. Turnbough CL, Jr., Kerr KH, Funderburg WR, Donahue JP, Powell FE. 1987. Nucleotide sequence and characterization of the *pyrF* operon of *Escherichia coli* K12. J. Biol. Chem. 262: 10239-10245.

857. Tusnady GE, Simon I. 2001. The HMMTOP transmembrane topology prediction server. Bioinformatics 17: 849-850.

858. Uberbacher EC, Xu Y, Mural RJ. 1996. Discovering and understanding genes in human DNA sequence using GRAIL. Methods Enzymol. 266: 259-281.

859. Uetz P, Giot L, Cagney G, Mansfield TA, Judson RS, Knight JR, Lockshon D, Narayan V, Srinivasan M, Pochart P, et al. 2000. A comprehensive analysis of protein-protein interactions in *Saccharomyces cerevisiae*. Nature 403: 623-627.

860. Uhlmann F, Wernic D, Poupart MA, Koonin EV, Nasmyth K. 2000. Cleavage of cohesin by the CD clan protease separin triggers anaphase in yeast. Cell 103: 375-386.

861. Uren AG, O'Rourke K, Aravind LA, Pisabarro MT, Seshagiri S, Koonin EV, Dixit VM. 2000. Identification of paracaspases and metacaspases: two ancient families of caspase-like proteins, one of which plays a key role in MALT lymphoma. Mol. Cell 6: 961-967.

862. Valenzuela JG, Charlab R, Galperin MY, Ribeiro JM. 1998. Purification, cloning, and expression of an apyrase from the bed bug *Cimex lectularius*. A new type of nucleotide-binding enzyme. J. Biol. Chem. 273: 30583-30590.

863. van Asselt EJ, Dijkstra AJ, Kalk KH, Takacs B, Keck W, Dijkstra BW. 1999. Crystal structure of *Escherichia coli* lytic transglycosylase Slt35 reveals a lysozyme-like catalytic domain with an EF-hand. Structure Fold. Des. 7: 1167-1180.

864. van den Ent F, Amos L, Lowe J. 2001. Bacterial ancestry of actin and tubulin. Curr. Opin. Microbiol. 4: 634-638.

865. van den Ent F, Amos LA, Lowe J. 2001. Prokaryotic origin of the actin cytoskeleton. Nature 413: 39-44.

866. van der Oost J, Huynen MA, Verhees CH. 2002. Molecular characterization of phosphoglycerate mutase in archaea. FEMS Microbiol. Lett. 212: 111-120.

867. van der Oost J, Schut G, Kengen SW, Hagen WR, Thomm M, de Vos WM. 1998. The ferredoxin-dependent conversion of glyceraldehyde-3-phosphate in the hyperthermophilic archaeon *Pyrococcus furiosus* represents a novel site of glycolytic regulation. J. Biol. Chem. 273: 28149-28154.

868. Vander Horn PB, Backstrom AD, Stewart V, Begley TP. 1993. Structural genes for thiamine biosynthetic enzymes (*thiCEFGH*) in *Escherichia coli* K-12. J. Bacteriol. 175: 982-992.

869. Venclovas, Zemla A, Fidelis K, Moult J. 2001. Comparison of performance in successive CASP experiments. Proteins 45: 163-170.

870. Venter JC, Adams MD, Myers EW, Li PW, Mural RJ, Sutton GG, Smith HO, Yandell M, Evans CA, Holt RA, et al. 2001. The sequence of the human genome. Science 291: 1304-1351.

871. Venter JC, Smith HO, Hood L. 1996. A new strategy for genome sequencing. Nature 381: 364-366.

872. Vitkup D, Melamud E, Moult J, Sander C. 2001. Completeness in structural genomics. Nat. Struct. Biol. 8: 559-566.

873. von Heijne G. 1984. How signal sequences maintain cleavage specificity. J. Mol. Biol. 173: 243-251.

874. von Heijne G. 1985. Signal sequences. The limits of variation. J. Mol. Biol. 184: 99-105.

875. von Heijne G. 1992. Membrane protein structure prediction. Hydrophobicity analysis and the positive-inside rule. J. Mol. Biol. 225: 487-494.

876. von Mering C, Krause R, Snel B, Cornell M, Oliver SG, Fields S, Bork P. 2002. Comparative assessment of large-scale data sets of protein-protein interactions. Nature 417: 399-403.

877. Wagner A. 2002. Selection and gene duplication: a view from the genome. Genome Biol. 3: REVIEWS1012.

878. Walker DR, Koonin EV. 1997. SEALS: a system for easy analysis of lots of sequences. ISMB 5: 333-339.

879. Walker GE, Dunbar B, Hunter IS, Nimmo HG, Coggins JR. 1996. Evidence for a novel class of microbial 3-deoxy-D-arabino-heptulosonate-7-phosphate synthase in

Streptomyces coelicolor A3(2), *Streptomyces rimosus* and *Neurospora crassa.* Microbiology 142: 1973-1982.

880. Walker JE, Saraste M, Runswick MJ, Gay NJ. 1982. Distantly related sequences in the alpha- and beta-subunits of ATP synthase, myosin, kinases and other ATP-requiring enzymes and a common nucleotide binding fold. EMBO J. 1: 945-951.

881. Warburg O, Christian W. 1943. Isolierung und kristallization des garungsferments zymohexase. Biochem. Z. 314: 149-176.

882. Ward CD, Flanegan JB. 1992. Determination of the poliovirus RNA polymerase error frequency at eight sites in the viral genome. J. Virol. 66: 3784-3793.

883. Ward CD, Stokes MA, Flanegan JB. 1988. Direct measurement of the poliovirus RNA polymerase error frequency in vitro. J. Virol. 62: 558-562.

884. Watanabe H, Mori H, Itoh T, Gojobori T. 1997. Genome plasticity as a paradigm of eubacteria evolution. J. Mol. Evol. 44: S57-S64.

885. Webb BJ, Liu JS, Lawrence CE. 2002. BALSA: Bayesian algorithm for local sequence alignment. Nucleic Acids Res. 30: 1268-1277.

886. Weinberg S. 1993. Dreams of a Final Theory. Pantheon Books.

887. Weinstock GM, Hardham JM, McLeod MP, Sodergren EJ, Norris SJ. 1998. The genome of *Treponema pallidum*: new light on the agent of syphilis. FEMS Microbiol. Rev. 22: 323-332.

888. Wells VR, Plotch SJ, DeStefano JJ. 2001. Determination of the mutation rate of poliovirus RNA-dependent RNA polymerase. Virus Res. 74: 119-132.

889. Whelan S, Goldman N. 2001. A general empirical model of protein evolution derived from multiple protein families using a maximum-likelihood approach. Mol. Biol. Evol. 18: 691-699.

890. White JV, Stultz CM, Smith TF. 1994. Protein classification by stochastic modeling and optimal filtering of amino-acid sequences. Math. Biosci. 119: 35-75.

891. White O, Eisen JA, Heidelberg JF, Hickey EK, Peterson JD, Dodson RJ, Haft DH, Gwinn ML, Nelson WC, Richardson DL, et al. 1999. Genome sequence of the radioresistant bacterium *Deinococcus radiodurans* R1. Science 286: 1571-1517.

892. Wilbur WJ, Lipman DJ. 1983. Rapid similarity searches of nucleic acid and protein data banks. Proc. Natl. Acad. Sci. USA 80: 726-730.

893. Wilhite SE, Elden TC, Brzin J, Smigocki AC. 2000. Inhibition of cysteine and aspartyl proteinases in the alfalfa weevil midgut with biochemical and plant-derived proteinase inhibitors. Insect Biochem. Mol. Biol. 30: 1181-1188.

894. Williams C, Xu L, Blumenthal T. 1999. SL1 trans splicing and 3'-end formation in a novel class of *Caenorhabditis elegans* operon. Mol. Cell. Biol. 19: 376-383.

895. Williamson LR, Plano GV, Winkler HH, Krause DC, Wood DO. 1989. Nucleotide sequence of the *Rickettsia prowazekii* ATP/ADP translocase-encoding gene. Gene 80: 269-278.

896. Wilson PJ, Morris CP, Anson DS, Occhiodoro T, Bielicki J, Clements PR, Hopwood JJ. 1990. Hunter syndrome: isolation of an iduronate-2-sulfatase cDNA clone and analysis of patient DNA. Proc. Natl. Acad. Sci. USA 87: 8531-8535.

897. Wingender E, Chen X, Fricke E, Geffers R, Hehl R, Liebich I, Krull M, Matys V, Michael H, Ohnhauser R, et al. 2001. The TRANSFAC system on gene expression regulation. Nucleic Acids Res. 29: 281-283.

898. Wingender E, Chen X, Hehl R, Karas H, Liebich I, Matys V, Meinhardt T, Pruss M, Reuter I, Schacherer F. 2000. TRANSFAC: an integrated system for gene expression regulation. Nucleic Acids Res. 28: 316-319.

899. Winkler HH, Neuhaus HE. 1999. Non-mitochondrial ATP transport. Trends Biochem. Sci. 24: 64-68.

900. Woese C. 1998. The universal ancestor. Proc. Natl. Acad. Sci. USA 95: 6854-6859.

901. Woese CR. 1987. Bacterial evolution. Microbiol. Rev. 51: 221-271.

902. Woese CR. 1990. Evolutionary questions: the "progenote". Science 247: 789.

903. Woese CR. 2000. Interpreting the universal phylogenetic tree. Proc. Natl. Acad. Sci. USA 97: 8392-8396.

904. Woese CR. 2002. On the evolution of cells. Proc. Natl. Acad. Sci. USA 99: 8742-8747.

905. Woese CR, Fox GE. 1977. Phylogenetic structure of the prokaryotic domain: the primary kingdoms. Proc. Natl. Acad. Sci. USA 74: 5088-5090.

906. Woese CR, Kandler O, Wheelis ML. 1990. Towards a natural system of organisms: proposal for the domains Archaea, Bacteria, and Eucarya. Proc. Natl. Acad. Sci. USA 87: 4576-4579.

907. Woese CR, Olsen GJ, Ibba M, Soll D. 2000. Aminoacyl-tRNA synthetases, the genetic code, and the evolutionary process. Microbiol. Mol. Biol. Rev. 64: 202-236.

908. Wolf E, Kim PS, Berger B. 1997. MultiCoil: a program for predicting two- and three-stranded coiled coils. Protein Sci. 6: 1179-1189.

909. Wolf YI, Aravind L, Grishin NV, Koonin EV. 1999. Evolution of aminoacyl-tRNA synthetases-analysis of unique domain architectures and phylogenetic trees reveals a complex history of horizontal gene transfer events. Genome Res. 9: 689-710.

910. Wolf YI, Aravind L, Koonin EV. 1999. Rickettsiae and Chlamydiae: evidence of horizontal gene transfer and gene exchange. Trends Genet 15: 173-175.

911. Wolf YI, Brenner SE, Bash PA, Koonin EV. 1999. Distribution of protein folds in the three superkingdoms of life. Genome Res. 9: 17-26.

912. Wolf YI, Grishin NV, Koonin EV. 2000. Estimating the number of protein folds and families from complete genome data. J. Mol. Biol. 299: 897-905.

913. Wolf YI, Kondrashov FA, Koonin EV. 2000. No footprints of primordial introns in a eukaryotic genome. Trends Genet. 16: 333-334.

914. Wolf YI, Rogozin IB, Grishin NV, Koonin EV. 2002. Genome-trees and the Tree of Life. Trends Genet. 18: 472.

915. Wolf YI, Rogozin IB, Grishin NV, Tatusov RL, Koonin EV. 2001. Genome trees constructed using five different approaches suggest new major bacterial clades. BMC Evol Biol 1: 8.

916. Wolf YI, Rogozin IB, Kondrashov AS, Koonin EV. 2001. Genome alignment, evolution of prokaryotic genome organization and prediction of gene function using genomic context. Genome Res. 11: 356-372.

917. Wood DW, Setubal JC, Kaul R, Monks DE, Kitajima JP, Okura VK, Zhou Y, Chen L, Wood GE, Almeida NF, Jr., et al. 2001. The genome of the natural genetic engineer *Agrobacterium tumefaciens* C58. Science 294: 2317-2323.

918. Wood V, Gwilliam R, Rajandream MA, Lyne M, Lyne R, Stewart A, Sgouros J, Peat N, Hayles J, Baker S, et al. 2002. The genome sequence of *Schizosaccharomyces pombe*. Nature 415: 871-880.

919. Wootton JC. 1994. Non-globular domains in protein sequences: automated segmentation using complexity measures. Comput. Chem. 18: 269-285.

920. Wootton JC, Drummond MH. 1989. The Q-linker: a class of interdomain sequences found in bacterial multidomain regulatory proteins. Protein Eng. 2: 535-543.

921. Wootton JC, Federhen S. 1996. Analysis of compositionally biased regions in sequence databases. Methods Enzymol. 266: 554-571.

922. Wu CH, Xiao C, Hou Z, Huang H, Barker WC. 2001. iProClass: an integrated, comprehensive and annotated protein classification database. Nucleic Acids Res. 29: 52-54.

923. Wu L, Aster JC, Blacklow SC, Lake R, Artavanis-Tsakonas S, Griffin JD. 2000. MAML1, a human homologue of *Drosophila mastermind*, is a transcriptional co-activator for NOTCH receptors. Nat. Genet. 26: 484-489.

924. Wuchty S. 2001. Scale-free behavior in protein domain networks. Mol. Biol. Evol. 18: 1694-1702.

925. Xenarios I, Fernandez E, Salwinski L, Duan XJ, Thompson MJ, Marcotte EM, Eisenberg D. 2001. DIP: The Database of Interacting Proteins: 2001 update. Nucleic Acids Res. 29: 239-241.

926. Xenarios I, Rice DW, Salwinski L, Baron MK, Marcotte EM, Eisenberg D. 2000. DIP: the database of interacting proteins. Nucleic Acids Res. 28: 289-291.

927. Xi J, Ge Y, Kinsland C, McLafferty FW, Begley TP. 2001. Biosynthesis of the thiazole moiety of thiamin in *Escherichia coli*: identification of an acyldisulfide-linked protein--protein conjugate that is functionally analogous to the ubiquitin/E1 complex. Proc. Natl. Acad. Sci. USA 98: 8513-8518.

928. Xu H, Aurora R, Rose GD, White RH. 1999. Identifying two ancient enzymes in Archaea using predicted secondary structure alignment. Nat. Struct. Biol. 6: 750-754.

929. Yamamoto K, Kurisu G, Kusunoki M, Tabata S, Urabe I, Osaki S. 2001. Crystal structure of glucose dehydrogenase from *Bacillus megaterium* IWG3 at 1.7 A resolution. J. Biochem. (Tokyo) 129: 303-312.

930. Yanai I, Derti A, DeLisi C. 2001. Genes linked by fusion events are generally of the same functional category: a systematic analysis of 30 microbial genomes. Proc. Natl. Acad. Sci. USA 98: 7940-7945.

931. Yanai I, Wolf YI, Koonin EV. 2002. Evolution of gene fusions: horizontal transfer versus independent events. Genome Biol 3: RESEARCH0024.

932. Yarmolinsky MB. 1995. Programmed cell death in bacterial populations. Science 267: 836-837.

933. Yeo JP, Alderuccio F, Toh BH. 1994. A new chromosomal protein essential for mitotic spindle assembly. Nature 367: 288-291.

934. Yeo JP, Alderuccio F, Toh BH. 1997. Erratum: A new chromosomal protein essential for mitotic spindle assembly. Nature 388: 697.

935. Yuan J, Yankner BA. 2000. Apoptosis in the nervous system. Nature 407: 802-809.

936. Zalkin H, Smith JL. 1998. Enzymes utilizing glutamine as an amide donor. Adv. Enzymol. Relat. Areas Mol. Biol. 72: 87-144.

937. Zarembinski TI, Hung LW, Mueller-Dieckmann HJ, Kim KK, Yokota H, Kim R, Kim SH. 1998. Structure-based assignment of the biochemical function of a hypothetical protein: a test case of structural genomics. Proc. Natl. Acad. Sci. USA 95: 15189-15193.

938. Zhang C, DeLisi C. 1998. Estimating the number of protein folds. J. Mol. Biol. 284: 1301-1305.

939. Zhang C, Kim SH. 2000. Environment-dependent residue contact energies for proteins. Proc. Natl. Acad. Sci. USA 97: 2550-2555.

940. Zhang MQ. 1997. Identification of protein coding regions in the human genome by quadratic discriminant analysis. Proc. Natl. Acad. Sci. USA 94: 565-568.

941. Zhang R, Evans G, Rotella FJ, Westbrook EM, Beno D, Huberman E, Joachimiak A, Collart FR. 1999. Characteristics and crystal structure of bacterial inosine-5'-monophosphate dehydrogenase. Biochemistry 38: 4691-4700.

942. Zhang Z, Schaffer AA, Miller W, Madden TL, Lipman DJ, Koonin EV, Altschul SF. 1998. Protein sequence similarity searches using patterns as seeds. Nucleic Acids Res. 26: 3986-3990.

943. Zimmermann BH, Evans DR. 1993. Cloning, overexpression, and characterization of the functional dihydroorotase domain of the mammalian multifunctional protein CAD. Biochemistry 32: 1519-1527.

944. Zorio DA, Blumenthal T. 1999. U2AF35 is encoded by an essential gene clustered in an operon with RRM/cyclophilin in *Caenorhabditis elegans*. Rna 5: 487-494.

945. Zuckerkandl E. 1975. The appearance of new structures and functions in proteins during evolution. J Mol Evol 7: 1-57.

946. Zuckerkandl E, Pauling L. 1965. Molecules as documents of evolutionary history. J. Theor. Biol. 8: 357-366.

947. Zvelebil MJ, Barton GJ, Taylor WR, Sternberg MJ. 1987. Prediction of protein secondary structure and active sites using the alignment of homologous sequences. J. Mol. Biol. 195: 957-961.

948. Rigden DJ, Jedrzejas MJ, Galperin MY. 2003. Amidase domains from bacterial and phage autolysins define a family of gamma-D,L-glutamate-specific amidohydrolases. Trends Biochem. Sci. 28: 230-234.

949. Crooks GE, Hon G, Chandonia JM, Brenner SE 2004. WebLogo: A sequence logo generator. Genome Res. *in press.*

950. Keeling PJ, Charlebois RL, Doolittle WF. 1994. Archaebacterial genomes: eubacterial form and eukaryotic content. Curr. Opin. Genet. Dev. 1994. 4: 816-822

951. Mirkin BG, Fenner TI, Galperin MY, Koonin EV. 2003. Algorithms for computing parsimo-nious evolutionary scenarios for genome evolution, the last universal common ancestor and dominance of horizontal gene transfer in the evolution of prokaryotes. BMC Evol Biol. 3: 2.

952. Rodionov DA, Mironov AA, Gelfand MS. 2002. Conservation of the biotin regulon and the BirA regulatory signal in Eubacteria and Archaea. Genome Res. 12: 1507-1516.

INDEX

D

E

F

G

N

O

P

R

S

T

U

V

W

X

Y

Z